Katrin Kleemann
A Mist Connection

Historical Catastrophe Studies / Historische Katastrophenforschung

Edited by/Herausgegeben von
Dominik Collet, Christopher Gerrard, Christian Rohr

Katrin Kleemann

A Mist Connection

An Environmental History of the Laki Eruption
of 1783 and Its Legacy

DE GRUYTER

First published in 2023.
Accepted as a dissertation at Ludwig-Maximilians-Universität München 2020.

The Andrea von Braun Foundation generously supported the research for this doctoral thesis.
Printed with the kind assistance of the Academic Society of Freiburg im Breisgau.
Printed with the kind assistance of Siblings Boehringer Ingelheim Foundation for the Humanities.

ISBN 978-3-11-162043-5
e-ISBN (PDF) 978-3-11-073192-7
e-ISBN (EPUB) 978-3-11-073202-3
ISSN 2699-7223
DOI https://doi.org/10.1515/9783110731927

Library of Congress Control Number: 2023932784

Bibliographic information published by the Deutsche Nationalbibliothek
The Deutsche Nationalbibliothek lists this publication in the Deutsche Nationalbibliografie;
detailed bibliographic data are available on the internet at http://dnb.dnb.de.

Cover image: Dash_med/iStock/Getty Images Plus
Typesetting: Integra Software Services Pvt. Ltd.

www.degruyter.com

For my parents
Helga and Wolfgang Kleemann

Acknowledgments

In 2010, the Icelandic volcano Eyjafjallajökull erupted and grounded transatlantic air traffic. The whole of Europe came to a standstill. I followed the news closely, as I had planned a short trip to France in a few weeks' time. I was fascinated by the images coming out of Iceland; ash turned day into night and the streets hid under a thick layer of soot. Some mentions were made of past volcanic eruptions on the island, including the disastrous flood basalt event of 1783. This event, and this year, held my interest for quite some time. So it was, three years later, when deciding on a topic to tackle for my master's thesis, I landed on Laki. I delved further into the topic for my doctoral thesis. While I scoured the past, new fissure eruptions occurred in the Holuhraun lava field and on the Reykjanes Peninsula. As I read of past eruptions, I witnessed, thankfully from a safe distance, these sublime natural occurrences in real life. I feel privileged to have "experienced" these fissure eruptions from such a perspective.

I also feel privileged to have had such tremendous support over the past few years, without which I would have been unable to complete this book. I sincerely thank Claudia Jarzebowksi and Claudia Ulbrich, who supervised my master's thesis at the Freie Universität Berlin. During my time there, I was lucky enough to land a position as a student assistant at the Max Planck Institute for the History of Science. For this invaluable experience, I would like to thank Veronika Lipphardt. These years formed the foundation for all that came after; I am incredibly grateful for them.

I am extremely fortunate to have been a member of the "Environment and Society" doctoral program at the Rachel Carson Center at LMU Munich. It was the perfect environment to pursue a research project in environmental history. My deepest gratitude goes to my doctoral supervisors: Christof Mauch of the Rachel Carson Center, Christian Rohr from the University of Bern, and Anke Friedrich from LMU Munich. To say these three helped me immensely is a massive understatement. It was during my first few weeks there that Christof introduced me to Anke, who was kind enough to invite me on a geological field trip to Spain. Shortly after, I took on geology as my minor, and a whole new way of researching the history of Icelandic volcanism suddenly unfolded before my eyes. I would like to thank all those who commented on drafts of my chapters at the Oberseminar in Schönwag, and everyone at the workshops, conferences, and colloquia that I was lucky enough to attend. A big thank you to the other members of the "Environment and Society" doctoral program, RCC staff members, and Carson fellows. Particular thanks go to Kimberly Coulter, Carmen Dines, Rob Emmett, Lena Engel, Arielle Helmick, Katie Ritson, and Helmuth Trischler. At the department of Earth and Environmental Sciences, I thank Sara Carena, Simon Kübler, Andrea Mazon Carro, and Stefanie Rieger.

It was, and is, a great pleasure to be part of the Climate History Network. I thank Dagomar Degroot, Bathsheba Demuth, and Sam White for their invaluable advice over the years. Connecting with fellow researchers in the "Volcanic Impacts on

Climate and Society" working group, part of Past Global Changes, also proved particularly fruitful. A big thanks to all those I collaborated with from there. In particular, I thank Kevin Anchukaitis, Martin Bauch, Christophe Corona, Stephan Ebert, Heli Huhtamaa, Francis Ludlow, Timothy Newfield, Gill Plunkett, Alan Robock, Michael Sigl, Markus Stoffel, Matthew Toohey, Rob Wilson, and Brian Zambri for sharing their expertise, data, and graphics. I thank Anja Schmidt, in particular, for sharing her sources with me.

I thoroughly appreciate being invited to present my research in various colloquia: here, many thanks are owed to Melanie Arndt, Arndt Brendecke, John Haldon, Wolfgang Jürries, Clive Oppenheimer, Christian Rohr, Milena Viceconte, and Céline Vidal. Throughout my project, I received helpful input from Rudolf Brázdil, Gaston Demarée, Karl-Erik Frandsen, Ólöf Garðarsdóttir, Jürg Luterbacher, Alan MacEachern, Franz Mauelshagen, Ruth Morgan, Astrid Ogilvie, Sandra Swart, and Elena Xoplaki.

I am much obliged to all the research and reading room staff at the many libraries and archives I visited whilst undertaking research for this book; in particular, the reading room staff of the newspaper department at the Berlin State Library, Ingrid Rack at the Stadtarchiv Würzburg, Mark Beswick at the Met Office National Meteorological Archive in Exeter, Rupert Baker and Virginia Mills at the Royal Society of London, Zoe Stansell at the British Library, Samantha Smart at the National Records of Scotland, Ashley Cataldo and Kimberly Toney at the American Antiquarian Society, Tracey deJong and Earle Spamer at the American Philosophical Society, and Anna Clutterbuck-Cook and Brendan Kieran at the Massachusetts Historical Society.

Thanks to Dorothea Hutterer, Óðinn Melsted, David Patterson, Marie Prott, and Georgina Walsh, who read the final draft of my dissertation. For additional support, I thank Aurélie Beauchet, Morgan Costigan, Ian Fenwick, Alice Gording, Danny Hall, Chris Handisides, Thérèse Kielty, Claire Lagier, Conor McGee, Patrick McLoone, Eveline de Smalen, Patrick Walsh, Rosa Walsh, Sarah Walsh, and Bernice Wildhaber.

I am deeply indebted to the Christoph von Braun and the Andrea von Braun Foundation. They very generously supported my research project. A big thank you to the Rachel Carson Center, the Amerikainstitut at LMU, and the European Society for Environmental History for their financial support. Without this, the numerous archival research trips to Germany, the UK, and the United States would not have been possible. Furthermore, I thank the Wissenschaftliche Gesellschaft Freiburg im Breisgau and the Geschwister Boehringer Ingelheim Stiftung für Geisteswissenschaften for their generous help with the printing costs of this book. At the German Maritime Museum / Leibniz Institute for Maritime History, I thank Ruth Schilling for her support and patience during the final months of preparation for this book's publication.

Thanks to Dominik Collet, Christopher Gerrard, and Christian Rohr for accepting my book manuscript as part of their Historical Catastrophe Studies series with De Gruyter. I also must express how happy I am to have been given the opportunity to publish my book open access. To all those involved in this decision, many, many thanks. I also thank the anonymous reviewer for their comments. I thoroughly appreciate the hard work of Elisabeth Kempf, Eva Locher, Julie Miess, and Gabriela Rus; their assistance and endless patience made the process much easier.

To my grandmother, Ursula Kleemann, who sadly is not here anymore to finally see this book in print, I send my deepest thanks. She always supported my academic endeavors with kind words and cold, hard cash. Without her, I would not have made it this far. A big thanks to my sister and brother-in-law, Svenja and Lennart Heinemann, for their support. For their many kind words, I thank Stephanie and Terry Walsh. I appreciate the patient support of my husband, Jack Walsh, and thank him for the time he took preparing many of the graphics and maps. I dedicate this book to my parents, Helga and Wolfgang Kleemann. They have generously supported me throughout my studies and always ensured I had a healthy supply of books and materials to satisfy my curiosity throughout my childhood. On a shelf in my family home, my first set of nature encyclopedias still occupies space. One of these, a volume covering volcanoes, features the Laki eruption. I imagine I browsed this for a moment or two, then turned the page.

Bremerhaven, January 2023
Katrin Kleemann

Contents

Acknowledgments —— VII

1   Introduction: 1783 – A Year of Wonders —— 1
    Previous Scholarship —— 4
    Research Focus —— 16
    Methods —— 17
        Environmental History —— 17
        Climate History and the Little Ice Age —— 19
        Disaster History —— 22
        History of Science —— 24
        Deep Geological Time —— 25
        Teleconnections —— 27
    Sources —— 28
    Structure —— 33

2   A Volcano Comes to Life —— 35
    The Geological Formation of Iceland —— 35
    Icelandic Volcanism in the Holocene —— 41
    Icelandic Volcanoes in Historical Times —— 46
    Before the Laki Eruption —— 54
    The Laki Eruption —— 60
        Kirkjubæjarklaustur and the Fire Districts —— 60
        The Fire Priest —— 63
        Pollution and Environmental Impact of the Laki Eruption in Iceland —— 73
        Móðuharðindin, the Famine of the Mist —— 77
        The Danish Response to the Crisis —— 81
        The Road to Recovery —— 84

3   Shaking the World —— 88
    History of Geology —— 88
        Volcanoes: A Topic of Meteorology —— 88
        Natural Sciences in the Late Eighteenth Century —— 91
    Historical Context —— 94
        1783: The End of the American Revolution and "Ballomania" in Europe —— 94
        Earthquakes in Lisbon and Calabria —— 100
        Nýey: A Burning "New Island" —— 107
        The Weather of the 1780s —— 109
    The Summer of 1783 —— 112
        The Extraordinary Dry Fog of 1783 —— 112
        The Extreme Heat of Summer —— 132

Thunderclaps and Lightning —— 135
Earthquakes in Europe and Beyond —— 144
Fever and Mortality —— 150
The End of the Fog —— 153
The Harvest —— 156
A Year of Awe —— 159
Reactions to the Unusual Weather —— 160
The Response to the Dry Fog —— 160
The Voice of Reason —— 161
A Search for Precedents —— 164
Blaming the Superstitious —— 167
Fears of the Fog —— 170
The Speculation about the Cause of the Unusual Weather —— 171
A Naturalist's Perspective —— 172
Active Volcanoes, Active Imaginations —— 174
A Time of a Subsurface Revolution —— 184
Elektrizitätstaumel: Are Lightning Rods to Blame? —— 188
The Fireball —— 191
Experiments on the Origin of the Dry Fog —— 196
News from Iceland —— 198
Speculating about a Connection —— 200
The Summer of 1783 outside of Europe —— 205
The Western Hemisphere —— 205
The Eastern Hemisphere —— 211
The Winter of 1783/1784: A Touch of Frost —— 215
The Winter in Europe —— 215
Ice Drift and Flooding in Europe —— 221
The Winter in North America —— 226
Searching for a Connection —— 230
Outlook —— 231
Cold Temperatures Continued —— 231
Annus Arcanus —— 235

4   The Mystery Remains —— 238
Myth and Legend —— 238
A Search in Vain —— 245
Sveinn PÁLSSON —— 248
A Clearer Picture —— 257
Rediscovering Sveinn PÁLSSON —— 264
The Krakatau Eruption of 1883 —— 266
The Dots Connected —— 272

Field Trips to the Laki Fissure ▬ **277**
Lifting the Fog of Ignorance ▬ **281**

5   Conclusions: A Modern Perspective on the Laki Eruption ▬ 283
Continental Drift, Plate Tectonics, and Mantle Plumes ▬ **289**
History and Geology ▬ **291**
Future Research ▬ **294**
The Bigger Picture: Lessons for the Present and Future ▬ **295**
Outlook: The Present and the Future ▬ **298**
    The 200-Year Anniversary of the Laki Eruption ▬ **298**
    Skaftáreldahraun – The Laki Lava Field ▬ **301**
    Iceland and Its Volcanoes in the Future ▬ **305**

Illustrations ▬ **309**

Bibliography ▬ **321**
    Archived Primary Sources ▬ **321**
    Published Primary Sources ▬ **323**
    Published Secondary Sources ▬ **332**
    Online Primary and Secondary Resources ▬ **366**

Index ▬ **373**

1 Introduction: 1783 – A Year of Wonders

On 4 June 1783, the residents of Annonay in southern France gathered to witness the launch of the first hot-air balloon. With their invention, the MONTGOLFIER brothers had kick-started a new epoch; that of air travel. The orbicular silhouettes of hot-air balloons and hydrogen balloons, which took to the sky that same year, would not remain the only reason Europeans looked skyward that summer. A strange dry fog with a sulfuric odor appeared, as if from nowhere, and blanketed Europe, stretching as far as North America, Asia, and North Africa. This ominous mist lasted for several months, throughout the summer and into the autumn of 1783. It turned the sun blood-red and robbed its rays of their strength. An apparent increase in the frequency of thunderstorms, earthquakes, and other unusual phenomena coincided with this enduring fog. In 1784, reflecting on this marvelous period, the French writer Louis-Sébastien MERCIER (1740–1814) dubbed 1783 a year of wonders (*l'année des merveilles*).[1] Indeed, 1783 played host to many anomalies that did make it extraordinary – even for a year situated within the Little Ice Age (1250/1300–1850).[2]

What was causing this upheaval of the elements? Many Europeans thought about this question. These wonders were documented in newspaper reports, weather diaries, publications of learned societies, scholarly monographs, and private communications of the period. The fact that these phenomena were widely discussed and reflected upon is indicative of the great interest contemporaries had in them.[3] Emotions concerning these events ranged from giddy excitement to existential fear. Was all this a sign of an impending disaster? Could the natural sciences shed light on these phenomena? Speculation ran wild. The summer of 1783 was sweltering; was this the root of the problem? Could extraordinary events have an ordinary explanation? Were the earthquakes in Calabria to blame? Was the inclemency an expression of a much larger natural event that was taking place? The then-recent invention and installation of lightning rods prompted many to suggest that they were to blame. Could the emergence of a new, smoking island off the coast of Iceland have anything to do with it all? Was the tumultuous weather in any way related to the apparent eruptions of Gleichberg and Cottaberg, two mountains of volcanic origin in the German Territories?[4] Many even strove to connect a meteor, seen scorching a trail across the upper atmosphere, to these terrestrial events.

---

1 MERCIER 1784: 406.
2 DEMARÉE 2006: 878–879; GLASER 2008: 238; PAYNE 2010: 2–3.
3 A contemporary, in the context of this book, means a person living at the time of the discussed event, unless otherwise stated.
4 The official name of the German Territories is the Holy Roman Empire of the German Nation; in this book, I refer to it as the German Territories for the sake of brevity.

While Europeans on the continent debated these ideas, Icelanders in the so-called Fire Districts had begun to fight for their very survival. Between 8 June 1783 and 7 February 1784, a 27-kilometer-long fissure, which would eventually consist of 140 craters and cones, tore through the highlands of south-central Iceland. This event produced the largest volume of lava of any volcanic eruption on the planet in the last millennium. It triggered the worst disaster in Icelandic history: one-fifth of the population perished in the aftermath, succumbing to disease or starvation. Volcanic ash and gases poisoned the fields and the waters, killing livestock and fish, the very things Icelanders depended upon.

This particular volcanic eruption came to have many names; in Icelandic, it is mainly referred to as *Skaftáreldar*, meaning "Skaftá Fires," after the Skaftá, the riverbed in which its lava flowed.[5] The Laki fissure itself is called *Lakagígar* in Icelandic, meaning "Laki craters."[6] The Icelandic people remember the eruption for its devastating consequence, which they call *móðuharðindin*, the famine of the mist. Internationally, the eruption of the Laki fissure is most often referred to as the "Laki eruption," the tradition I will follow. Mount Laki, the mountain at the center of the Laki fissure, did not erupt in 1783.[7] Scholars now agree that the Laki eruption was the cause of the dry fog that later haunted Europe.[8]

Volcanic ash and gases do not respect political borders. The eruption injected massive amounts of sulfur dioxide and other gases into Earth's atmosphere. These gases were transported further afield by the jet stream, strong winds in the upper atmosphere, and formed the dry fog that caused the frisson of excitement in mainland Europe that year. It is possible that the aerosols of the eruption, or even the dry fog itself, reached the Southern Hemisphere.[9] Despite the enormity of the Laki eruption, at the time, the outside world was oblivious.[10]

---

5 KARLSSON 2000b: 178. *Eldar* is the plural form of *eldur* and means "fire"; BRAGADÓTTIR 2008: 281. Despite the fact that the suffix *-á* in Skaftá indicates that this is the name of a river, I have nonetheless referred to it as the Skaftá River for the reader's convenience.

6 The word *gígar* is the plural of *gígur*, which means "crater"; BRAGADÓTTIR 2008: 456.

7 DE BOER, SANDERS 2002: 118; VASOLD 2004: 603; WITZE, KANIPE 2014: 21, 148.

8 DEMARÉE 1997: 879; THORDARSON, SELF 2003: 1–13.

9 According to a study by TRIGO, VAQUERO, and STOTHERS (2010), a Portuguese astronomer, Bento SANCHES DORTA, who was based in Rio de Janeiro, witnessed an unusually high number of days with a dry fog or haze from 1784 to 1786; he did not observe this, however, in 1783. It is unlikely that the Laki haze descended upon Brazil a full year after the eruption. While Greenlandic ice cores show a signal of the Laki eruption, Antarctic ice cores do not reveal traces of this eruption. This is to be expected; usually, only strong tropical eruptions leave a signal in the ice of both poles. GAO et al. 2007; ZAMBRI et al. 2019a.

10 For essays on the topic of environmental ignorance, see also UEKÖTTER, LÜBKEN 2014. In the last two decades, an emerging body of scholarship has embraced the history of ignorance in conjunction with the history of knowledge; ZWIERLEIN 2016: 40; DASTON 2017; DÜRR 2021; VERBUGT, BURKE 2021: 1.

News of the eruption reached Europe in September 1783, by which time the dry fog had all but disappeared. Another decade would pass until the Icelandic naturalist Sveinn PÁLSSON (1762–1840) discovered the scarred landscape of the fissure, hidden in the remote highlands. Unfortunately, the Danish Natural History Society, the organization that supported his expedition, ran into financial difficulties; this led to an earlier-than-planned termination of his funding. As a result, PÁLSSON's research remained unpublished and – for the most part – unread for almost a century. Thus, the connection between the Laki eruption and the haze in Europe would remain a mystery until the 1880s. In that decade, Norwegian geologist Amund HELLAND (1846–1918) and Icelandic geologist Þorvaldur THORODDSEN (1855–1921) published on the Laki eruption, which – in the context of the colossal eruption of Krakatau in the Dutch East Indies in 1883, and its far-reaching effects – lifted the fog of ignorance.[11] That the unusual weather remained a mystery outside of Iceland for so long is what made this volcanic eruption and its aftermath so fascinating to research.

Contemporaries of the Laki eruption lived in an extraordinary time: a time of invention and uncertainty, trial and sometimes fatal error, a time of ingenuity and superstition. It was, for some, a time of exciting change, but for most, a time of great hardship. In 1783, the American diplomat, inventor, and polymath Benjamin FRANKLIN (1706–1790) wrote a letter to Joseph BANKS (1743–1820), British naturalist, botanist, and the president of the Royal Society of London that offers insight into the state of science in Europe at the time. FRANKLIN was thrilled by the pace of discovery and the resources available for experimentation.

> Furnish'd as all Europe now is with Academies of Science, with nice Instruments and the Spirit of Experiment, the Progress of human Knowledge will be rapid, and Discoveries made of which we have at present no Conception. I begin to be almost sorry I was born so soon since I cannot have the Happiness of knowing what will be known 100 Years hence.[12]

The Enlightenment – an eighteenth-century social and intellectual movement concerning human rationality and autonomy – influenced much of western Europe by 1783.[13] The term means, at its core, to illuminate one's mind. It was a contemporary term used by German philosopher Immanuel KANT (1724–1804) in 1784 when he explained that "the Enlightenment is the human's emancipation from their self-incurred immaturity" through reason.[14] New research into the Enlightenment reveals that it was far from a homogeneous movement; the eighteenth century was a period characterized by complexity and contradiction. The reach of the Enlightenment differed from region to region; in some areas, it began earlier than in others.[15] Many naturalists, amateur

---

11 HELLAND 1886; THORODDSEN 1914.
12 FRANKLIN 2011: 399, Benjamin FRANKLIN in a letter to Sir Joseph BANKS, Passy, 27 July 1783.
13 D'APRILE, SIEBERS 2008: 13–14; STOLLBERG-RILINGER 2011: 14.
14 ALT 2007: 3; STOLLBERG-RILINGER 2011: 9–10.
15 HOCHADEL 2003: 29; STOLLBERG-RILINGER 2011: 15.

weather observers, and philosophers of the time were in a good position and well-equipped to observe, record, and interpret the many unique weather phenomena of the year. Several learned societies existed that regularly published their findings, including numerous reports on the strange fog of the summer of 1783.

Previous Scholarship

Ever since Amund HELLAND and Þorvaldur THORODDSEN's work connected the Laki eruption to the dry fog of 1783, several scholars have tasked themselves with the study of the Laki fissure using the new scientific methods available to them. I will elaborate on the discoveries of PÁLSSON, HELLAND, and THORODDSEN and several other scholarly expeditions to Iceland and the Laki fissure in detail in Chapter Four.

Guðmundur G. BÁRÐARSON (1880 –1933) was an Icelandic scientist who, in 1929, recognized that the volcanoes of the Reykjanes Peninsula belonged to distinct volcanic systems. He refers to these systems as *vulkanbaelter* (volcanic belts). Later studies independently reached the same conclusions regarding the overall volcano-tectonic architecture of Iceland.[16] Icelandic geologist Sigurður ÞÓRARINSSON (1912–1983) mentioned the Laki eruption in his 1952 lecture, *The Thousand Years Struggle against Ice and Fire,* and in a subsequent 1956 publication.[17] ÞÓRARINSSON's work on the Laki eruption intensified from the 1960s to the 1980s; he contributed much to the event's re-emergence from the dustbin of history. ÞÓRARINSSON researched physical descriptions of the eruption and its environmental impact. In addition to studying Laki's effects on Iceland, he also worked on the impact of the eruption on the wider world, particularly Scandinavia. ÞÓRARINSSON also described the wide-ranging consequences of the "bluish gray haze or mist" that spread over much of Europe, Asia, and North Africa.[18] Photographs taken by ÞÓRARINSSON show that he visited the Laki fissure in 1938, 1958, 1962, and 1967.[19] In 1979, Icelandic geologist Sveinn JAKOBSSON first argued that the Laki eruption was part of the Grímsvötn system.[20]

In 1965, historian Vilhjálmur BJARNAR published a paper titled "The Laki Eruption and the Famine of the Mist." In the essay, he cites ÞÓRARINSSON's and THORODDSEN's research. His paper asserts that the "same type of mist [as in Iceland] was seen in the air over a large part of the northern hemisphere, [. . .] from Siberia to North America

---

16 BÁRÐARSON 1929: 187; THORDARSON, LARSEN 2007: 123.
17 ÞÓRARINSSON 1956b.
18 ÞÓRARINSSON 1969. The paper was read at the IAVCEI international symposium on Volcanology in 1968. For his other papers on the Laki eruption, see ÞÓRARINSSON 1953; ÞÓRARINSSON 1979: 150–156; ÞÓRARINSSON 1981: 112–117.
19 ÞÓRARINSSON 1969.
20 JAKOBSSON 1979; GUÐMUNDSSON 1989; THORDARSON, SELF 1993: 236.

and from Europe to North Africa."[21] However, this paper mainly analyzes the impact of the eruption and the fog on Iceland and its population. In 1972, climate historian Christian PFISTER conducted an early study that examined the dry fog outside of Iceland with a focus on Switzerland.[22] Historian Otto MÄUSSNEST briefly studied the dry fog's effects on Germany in a 1983 paper.[23]

Volcanology progressed along with the wider field of geology; in the 1970s, glaciologists began to look for traces of volcanic eruptions, particularly sulfur dioxide and tephra, in ice cores drilled from Antarctica and Greenland.[24] This work, in combination with other proxy data, helped volcanologists to date volcanic eruptions more precisely. Generally, the rule is that traces of strong tropical eruptions can be found in ice cores from both poles, whereas traces from high-latitude eruptions can be found in the ice sheets of their respective poles. A very strong sulfate signal from the Laki eruption has been found in Greenland, and a sulfate layer from it has been found in the northeastern Canadian Arctic and Spitsbergen. Tests of ice cores from western China proved inconclusive.[25]

In 1970, climatologist Hubert Horace LAMB invented the dust veil index (DVI), in which a number represents the volume of dust and aerosols released by a volcanic eruption. The output of Krakatau in 1883 was used as a reference value and had a DVI of 1,000. LAMB's paper discusses various past dust veil events, including the one from 1783. Several maps in the paper show how different volcanoes produce different dust veil spreading patterns. LAMB gives the "Laki-Skaptar Jökull" eruption a DVI value of 2,300.[26] Unbeknownst to Europeans at the time, Mount Asama in Japan also erupted in 1783. LAMB estimated that the combined effect of the "two very great eruptions" in Iceland and Japan generated a cooling in the Northern Hemisphere of 1.3 °C.[27] Although the Asama eruption had devastating local consequences, new research conducted after LAMB developed his DVI shows that the Asama eruption did not significantly influence the weather in the Northern Hemisphere.[28]

---

21 BJARNAR 1965: 415.

22 PFISTER 1972.

23 MÄUSSNEST 1983.

24 CLAUSEN, HAMMER 1988; FIACCO et al. 1994; ZIELINSKI et al. 1994; ZIELINSKI 1995.

25 STOTHERS 1996: 82. Volcanic signals in ice cores: HAMMER 1977; HAMMER, CLAUSEN, DANSGAARD 1981; HAMMER 1984; CLAUSEN, HAMMER 1988; MAYEWSKI et al. 1990; DE ANGELIS, LEGRAND 1994. Volcanic signals in the northeastern Canadian Arctic: KOERNER, FISHER 1982; FISHER, KOERNER 1994. Volcanic signals in Spitsbergen: FUJII et al. 1990. Volcanic signals in western China: THOMPSON 1989; THOMPSON 1990.

26 LAMB 1970: 509 (Laki). This is the combined value of the Laki eruption and the Nýey eruption (here referred to as "Eldeyjar"; for more information on Nýey, see Chapter Three). LAMB (1970: 512) gave the Tambora eruption a DVI of 3,000. See also LAMB 1977; LAMB 1983; LAMB 1985; KELLY, SEAR 1982.

27 LAMB 1995: 297.

28 ZIELINSKI et al. 1994; STOTHERS 1996: 86. The Asama eruption lasted from 9 May to 5 August 1783 and reached a VEI 4; Global Volcanism Program: Asamayama.

In the early 1980s, geologist Tom L. SIMKIN and his colleagues published a catalog of all the known past volcanic eruptions in the world.[29] This catalog uses the volcanic explosivity index (VEI) to categorize these eruptions. The VEI is a logarithmic scale ranging from zero to eight. Apart from VEI 0 to 2, every number represents a tenfold increase in explosivity. Different criteria, such as the volume of erupted ejecta, the height of the eruption cloud, and other observations, influence the classification of a given volcanic eruption. The VEI is not perfect, as all forms of output – for example, ash, lava, and lava bombs – are treated the same and it does not consider sulfur dioxide emissions.[30] A rating of VEI 0 indicates a constant, effusive eruption like Kīlauea on the island of Hawai'i, while a rating of VEI 8 indicates a mega-colossal eruption like the Toba event in 72,000 BC.[31] The most recent large eruption was the VEI 7 super-colossal eruption of Tambora in 1815, a volcano in today's Indonesia, which famously caused a year without a summer in 1816 in North America and Europe.[32] This kind of eruption only occurs once or twice per millennium. Eruptions on the scale of the 1991 Pinatubo eruption in the Philippines (VEI 6, colossal) occur every 50 to 100 years. Despite the fact that it released an astonishingly large volume of lava, the Laki eruption ranks low on the index of volcanic explosivity (VEI 4, cataclysmic). This kind of eruption is known as a flood basalt event.[33] Volcanic eruptions can have an impact far beyond their immediate vicinity, with their gases potentially affecting the climate for several years.[34]

In 1983, the Laki eruption had its 200-year anniversary. Many Icelandic scholars from a variety of disciplines contributed to a book on the eruption and its consequences for Iceland. The volume is titled *Skaftáreldar 1783–84: Ritgerðir og Heimildir* (The 1783–1784 Laki Eruption: Essays and Sources). More than 400 pages long, this book features new research on various aspects of the Laki eruption, ranging from geology to society and health. It also includes a collection of transcriptions of primary materials. The book is in Icelandic, with short English-language summaries of each article.[35] Historians Gaston DEMARÉE and Astrid OGILVIE give credit to the Laki eruption's bicentennial for once again piquing the interest of scholars. After 1984, several studies

---

29 SIMKIN et al. 1981; NEWHALL, SELF 1982.
30 MILES, GRAINGER, HIGHWOOD 2003. Further estimates of the magnitude and stratospheric sulfur injections for eruptions between 500 BCE and 1900 CE can be found in the eVolv2k database; TOOHEY, SIGL 2017.
31 NEWHALL, SELF 1982; AMBROSE 1998; VOGRIPA: Toba; DE BOER, SANDERS 2002: 258.
32 WOOD 2014; BEHRINGER 2019.
33 SIMKIN et al. 1981: 123; NEWHALL, SELF 1982.
34 Global Volcanism Program: Eyjafjallajökull. For an overview of recent scholarship on the effects of volcanic eruptions on climate, see MARSHALL et al. 2022.
35 GUNNLAUGSSON et al. 1984.

were published on Laki, mostly concerning the eruption itself and its aftermath within Iceland.[36]

In the 1990s, scholars were increasingly inclined to broaden the scope of their research. Many of these studies employed historical sources to reconstruct the eruption's impact on continental Europe, mainly Britain and France, and to reveal how the dry fog was perceived at the time. The first major texts analyzing the effects of the Laki eruption on regions outside of Iceland were written by historian Charles WOOD and physical scientist Sigurður STEINÞÓRSSON in 1992.[37] In 1993, historians Roland RABARTIN and Philippe ROCHER studied the possible impact of the Laki eruption on the French weather and harvest prior to the French Revolution.[38] In 1994, environmental and archaeological scientist John GRATTAN began studying the environmental impact of, and social responses to, the eruption with the help of historical British newspaper reports. Numerous papers throughout the 1990s and 2000s, with various co-authors, such as environmental scientist Daniel CHARMAN, scientist F. Brian PYATT, human geographer Mark BRAYSHAY, geoarchaeologist David GILBERTSON, Earth scientist Michael DURAND, and biogeographer John SADLER, analyzed the effects of Laki's volcanic gases on human health and vegetation in Britain and France.[39] Other work on the dry fog, from 1996 onward, was carried out by scientist Richard STOTHERS and historians Astrid OGILVIE and Gaston DEMARÉE, amongst other scholars.[40]

In their 2003 paper, volcanologists Thorvaldur THORDARSON and Stephen SELF also discuss in great detail the phenomenon they call the "Laki haze" using a variety of historical sources. From these sources, they reconstructed when and where the haze occurred, how high it might have reached, the optical effects it produced, and the heat wave with which it coincided.[41] In 2004, historian of science Manfred VASOLD published a short paper on the Laki eruption, which touches on the dry fog, warm temperatures, and health complaints in the German Territories in the summer of 1783, as well as the cold weather, floods, and ice drifts during the winter of 1783/1784.[42] Historian Oliver HOCHADEL briefly looked at the Laki eruption and the dry fog of 1783 as part of his paper on the introduction of lightning rods to the German Territories.[43] Prior to this book, and apart from GRATTAN's paper on the Gleichberg eruption and THORDARSON and

---

36 DEMARÉE, OGILVIE, ZHANG 1998. Studies published after 1984: OGILVIE 1986; STOTHERS et al. 1986; WOODS 1993.
37 STEINÞÓRSSON 1992; WOOD 1992.
38 RABARTIN, ROCHER 1993.
39 GRATTAN, CHARMAN 1994; GRATTAN, PYATT 1994; GRATTAN, BRAYSHAY 1995; GRATTAN, BRAYSHAY, SADLER 1998; BRAYSHAY, GRATTAN 1999; GRATTAN, SADLER 1999; GRATTAN, PYATT 1999; GRATTAN, GILBERTSON, DILL 2000; GRATTAN, SADLER 2001; GRATTAN, BRAYSHAY, SCHÜTTENHELM 2002.
40 STOTHERS 1996; DEMARÉE 1997; DEMARÉE, OGILVIE, ZHANG 1998 (comment on STOTHERS 1996); OGILVIE, JÓNSSON 2000; DEMARÉE, OGILVIE 2001.
41 THORDARSON, SELF 2003: 6–25.
42 VASOLD 2004.
43 HOCHADEL 2009: 55–58.

SELF's research on some historical sources from Germany, the papers mentioned above were the only works conducted on the impact of the Laki eruption on the German Territories.[44]

More work on the eruption's consequences for France was carried out by historians Emmanuel LE ROY LADURIE in 2006 and Emmanuel GARNIER in 2009.[45] LE ROY LADURIE describes Laki's effects as rather unspectacular, particularly when compared to Tambora.[46] In 2014, science journalist Alexandra WITZE and science writer Jeff KANIPE published a popular science book called *The Extraordinary Story of Laki, the Volcano that Turned Eighteenth-Century Europe Dark*.[47] This was the first book exclusively dedicated to the history of the Laki eruption. In 2019, literary scholars such as David HIGGINS and David McCALLAM took a closer look at the Laki eruption.[48] In the 2000s, the aftermath of the Laki eruption was further studied; in particular, scholars evaluated the cold temperatures of the winter of 1783/1784, the flooding of several European rivers, and the ice drifts.[49]

One particular aim of current research is to understand whether the Laki eruption caused a mortality crisis.[50] In 2004, atmospheric scientist Claire WITHAM and volcanologist Clive OPPENHEIMER analyzed mortality rates in England during the eruption; they were drawn to conclude that the eruption probably contributed to extra deaths.[51] In 2011, Earth scientist Sabina MICHNOWICZ further explored this mortality crisis within Great Britain in her thesis. She argues that the data does not point to a surge in mortality that can be linked to the eruption.[52] In 2021, Geoffrey HELLMAN analyzed nearly 1,500 parish registers from England, Wales, the Isle of Man, and Jersey, among others, to ascertain the mortality rate in these places during and after the Laki eruption; his findings suggest that "the Laki eruption was unlikely to have caused a huge surge in the rate of mortality in Britain."[53]

In 2011, volcanologist Anja SCHMIDT modeled how much excess mortality Europe would face in the case of a Laki-style eruption in the present. SCHMIDT assumes that

---

44 GRATTAN, GILBERTSON, DILL 2000; THORDARSON, SELF 2003.

45 LE ROY LADURIE 2006: 111–122; GARNIER 2009: 72–77.

46 LE ROY LADURIE 2006: 119.

47 WITZE, KANIPE 2014.

48 McCALLAM 2013; HIGGINS 2019; McCALLAM 2019.

49 GLASER, HAGEDORN 1990; MUNZAR, ELLEDER, DEUTSCH 2005; DEMARÉE 2006; POLIWODA 2007: 59–84; BRÁZDIL et al. 2010.

50 GRATTAN 1998; DURAND, GRATTAN 1999; DURAND, GRATTAN 2001; GRATTAN, DURAND, SCHÜTTENHELM 2001; GRATTAN, DURAND 2002; GRATTAN, DURAND, TAYLOR 2003; GRATTAN et al. 2003; COURTILLOT 2005; GRATTAN et al. 2005.

51 WITHAM, OPPENHEIMER 2004.

52 MICHNOWICZ 2011.

53 HELLMAN 2021: 239.

with long-term exposure to PM 2.5, particulate matter smaller than 2.5 micrometers, an additional 142,000 people in Europe would perish from cardiopulmonary diseases.[54]

Volcanology changed in the 1990s when satellites began to monitor the atmosphere from space and observe changes in atmospheric chemistry over time. The VEI 6 volcanic eruption of Pinatubo in Philippines was observed by satellite, as was the spread of its volcanic gases into the atmosphere: this provided further proof that ash and gases ejected by volcanic eruptions can have far-reaching global impacts that influence the climate.[55] In 2000, climate modeler Alan ROBOCK published a paper on the mechanisms of volcanic eruptions and how they affect the climate, mainly based on observations of recent eruptions such as Mount St. Helens in 1980, El Chichón in 1982, and Mount Pinatubo in 1991. He asserts that volcanic eruptions that release large amounts of sulfur dioxide are now known to have the potential to perturb the climate significantly.[56] Other relevant scholarly works on the impact of volcanic eruptions on the atmosphere include published papers by Clive OPPENHEIMER, Jelle Zeilinga DE BOER, Donald T. SANDERS, and Haraldur SIGURÐSSON.[57]

Thor THORDARSON and scientists Ármann HÖSKULDSSON and Guðrún LARSEN have contributed to the breadth of knowledge of Icelandic volcanism.[58] New methods and technologies have produced new insights into eruption sequences and effusion rates. THORDARSON and Stephen SELF have come to a different conclusion than Sigurður ÞÓRARINSSON.[59] Initially, flood basalt events did not figure much into the debate regarding the possible effects of volcanic eruptions on climate. It was not until the 1990s that further research on them changed the discourse.[60] It was now accepted that numerous eruptive episodes during one flood basalt event could release large amounts of sulfur from basaltic magma that can result in sulfuric aerosols (H_2SO_4) staying in the atmosphere for months to years.[61] In 1993, THORDARSON and SELF reconstructed the different eruptive episodes of the Laki eruption, estimated the output of gases, tephra, and lava, and even considered the environmental impact beyond Iceland.[62]

At present, several satellites orbit Earth, taking images, observing weather patterns, measuring gas concentrations, and much more. Scholars know that at any

---

54 SCHMIDT et al. 2011; LOUGHLIN et al. 2012; SCHMIDT 2013: 114.
55 MCCORMICK, THOMASON, TREPTE 1995; SELF et al. 1996; PAYNE 2010; GRATTAN, DURAND, TAYLOR 2003: 401–402.
56 ROBOCK 2000; OMAN 2006b.
57 SIGURÐSSON 1999; DE BOER, SANDERS 2002; OPPENHEIMER, PYLE, BARCLAY 2003; ROBOCK, OPPENHEIMER 2003; FRANCIS, OPPENHEIMER 2004; OPPENHEIMER 2011.
58 THORDARSON 1990; THORDARSON 1995; THORDARSON et al. 1996; THORDARSON, SELF 2003; THORDARSON, LARSEN 2007; THORDARSON, HÖSKULDSSON 2008; THORDARSON, HÖSKULDSSON 2014.
59 ÞÓRARINSSON 1969; THORDARSON, SELF 1993: 234.
60 THORDARSON, SELF 1993: 234; SELF et al. 1996; SELF, THORDARSON, KESZTHELYI 1997; THORDARSON 2005: 205.
61 THORDARSON et al. 1996; THORDARSON 2005: 205.
62 THORDARSON, SELF 1993: 258–261.

given time, there are around 40 volcanic eruptions taking place on planet Earth. Currently [January 2023], the Smithsonian Institution's Global Volcanism Program lists 44 eruptions as ongoing.[63] Most of these eruptions do not make the news, as they are small, occur in remote areas, and are detectable only by satellite. Volcanic eruptions have the potential to wreak havoc, especially when they occur near inhabited areas. The direct consequences of these eruptions, such as lava flows, ashfall, pyroclastic flows, and gases that pollute the air, can be devastating. That said, large and explosive volcanic eruptions are relatively rare.

Over the past two decades, climate modelers have studied the Laki eruption in detail. The first models of the atmospheric impact of Laki were conducted in 2003 by atmospheric scientist David S. STEVENSON, atmospheric physicist Ellie J. HIGHWOOD, and Thor THORDARSON.[64] Further modeling followed in 2005 and 2006. Environmental scientist Luke OMAN's work mainly focuses on the impact of high-latitude eruptions on monsoonal rains and the Nile River floods.[65] Anja SCHMIDT and her colleagues also modeled the climate impact of the Laki eruption.[66]

Whether a volcanic eruption has long or short-term effects on the weather depends on several factors. These include the latitude, time of year, and, most importantly, how much sulfur dioxide is released and how high the volcano injects it into the atmosphere (Figure 1).[67] Today we know that volcanic gases in the lower part of the atmosphere (the troposphere) have severe consequences at ground level lasting a few weeks to months. In contrast, volcanic gases in the upper part of the atmosphere (the stratosphere) can alter the climate for up to a few years, a process called climate forcing.[68] If sulfur dioxide reaches the stratosphere, it undergoes oxidization and becomes sulfuric acid aerosol particles. These microscopic particles reflect incoming solar radiation back into space and cause a cooling of the troposphere (known as the albedo effect or volcanic forcing).[69]

---

63 Global Volcanism Program: "Current Eruptions"; Global Volcanism Program: "How Many Volcanoes Are There?"

64 HIGHWOOD, STEVENSON 2003; STEVENSON et al. 2003; THORDARSON et al. 2003a.

65 CHENET, FLUTEAU, COURTILLOT 2005; OMAN et al. 2005; OMAN et al. 2006a; OMAN et al. 2006b.

66 SCHMIDT et al. 2010; SCHMIDT et al. 2012: This study states that 120 Tg of sulfur dioxide were released to the upper troposphere/lower stratosphere, which was followed by three years of below-average temperatures. The climate modeling conducted by this study suggests that the radiative effects produced by the eruption lasted long enough to contribute to the winter cooling. A question that remains is why the summer of 1783 was so warm. On this matter, also see COLE-DAI et al. 2014 (comment on SCHMIDT et al. 2012); SCHMIDT et al. 2014b (response to comment by COLE-DAI et al. 2012); SCHMIDT et al. 2016.

67 KRAVITZ, ROBOCK 2011; WITZE, KANIPE 2014: 131–132.

68 GRATTAN, PYATT 1999; PAYNE 2010: 4.

69 ROBOCK 2000: 191–193; GRATTAN, SADLER 2001: 141; GRATTAN, BRAYSHAY, SCHÜTTENHELM 2002: 88; OPPENHEIMER 2011: 44–46; COOPER et al. 2018: 239.

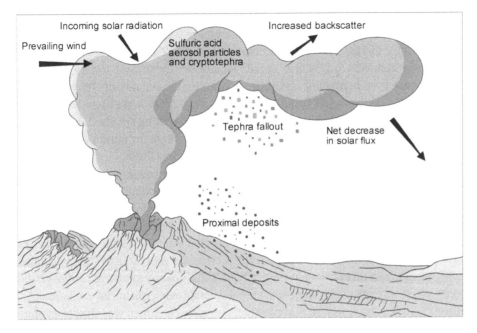

Figure 1: Volcanic outputs injected into the atmosphere.[70]

In Iceland, the stratosphere begins at only nine to 13 kilometers above sea level, as opposed to approximately 18 kilometers in the tropics.[71] To this day, scientists still debate whether the gases released by the Laki eruption reached the stratosphere or only the troposphere.[72] Evidence suggests that they only reached the troposphere: the gases appeared above Europe within a week and were mostly washed out by precipitation within three to four months. Additionally, ice core drillings in Greenland show sulfur particles for 1783 but not for 1784.[73] These points, together with the fact that the gases' effects were noticeable at ground level, all suggest they only reached the troposphere.[74] Although THORDARSON and SELF have done much work reconstructing the volume of sulfur dioxide released by the different eruptive phases, the resulting figures

---

70 For more information and details on the copyright for all of the illustrations, see the list of illustrations at the end of this book.

71 THORDARSON et al. 1996: 207; WITZE, KANIPE 2014: 130.

72 The following scholars argue that the volcanic gases reached the stratosphere: SCARTH 1999: 113; OMAN et al. 2006b: 1; GLASER 2008: 234; OPPENHEIMER 2011: 276; SCHMIDT et al. 2012; WITZE, KANIPE 2014: 134. Brian ZAMBRI et al. argue that most of the aerosols had dissipated by May 1784 (2019a: 6753). These scholars argue that only the troposphere was reached: WOOD 1992: 70–71; GRATTAN, BRAYSHAY 1995: 2; STOTHERS 1996: 79; GRATTAN, SADLER 2001: 138; GRATTAN, BRAYSHAY, SCHÜTTENHELM 2002: 88; GRATTAN, DURAND, TAYLOR 2003: 402.

73 WOOD 1992: 70–71.

74 GRATTAN, BRAYSHAY, SCHÜTTENHELM 2002: 92.

are still uncertain. David STEVENSON and Luke OMAN estimate the uncertainty to be up to 20 percent.[75] Future models might need to adjust these volumes.[76]

The winter of 1783/1784 was 3 °C below the mean (1778–1782), and cooling could be observed from 1784 to 1786; the overall temperature suppression for Europe is estimated to have been between 1° and 2 °C.[77] In contrast, the summer of 1783 was unusually warm: the temperatures in July 1783 in western Europe were almost 3 °C above the mean. Climate models have shown that volcanic cooling should have occurred during that summer. In general, volcanic eruptions increase the albedo effect, as they lead to a decrease in the solar radiation that is absorbed by Earth.[78]

Atmospheric circulation alters heat distribution and causes spatial variation in volcano-related cooling of the planet. Although the Laki eruption likely caused cooling over the year, western European historical sources describe drought and soaring temperatures during the summer.[79] Several temperature reconstructions also confirm this warm weather.[80] For a long time, this puzzled scientists. GRATTAN and SADLER have proposed that the sulfuric gases emitted created a greenhouse effect, which led to higher temperatures.[81] Other studies have subsequently shown that this greenhouse effect was relatively small and probably could not explain the warm temperatures.[82] THORDARSON and SELF argue that the warm summer was due to climate variability.[83]

The latest climate modeling work carried out by Brian ZAMBRI and his colleagues seems to confirm THORDARSON and SELF's hypothesis: a high-pressure system located over northern Europe created atmospheric blocking and caused the heat wave of July 1783. The hot air remained in northern and western Europe, whereas cold polar air traveled to eastern Europe and the Middle East. The Laki eruption caused a phenomenon called hemispherically asymmetric volcanic forcing, which disturbed normal weather patterns

75 STEVENSON et al. 2003; OMAN 2006b.
76 ZAMBRI et al. 2019b: 6787.
77 LAMB 1970; ANGELL, KORSHOVER 1985; BRIFFA et al. 1998; PÍSEK, BRÁZDIL 2006.
78 HANSEN, WAND, LACIS 1978; GRATTAN, SADLER 1999: 162; OMAN et al. 2006b; OPPENHEIMER 2011: 282–283; ZAMBRI et al. 2019b: 6777.
79 THORDARSON, SELF 1993; JACOBY, WORKMAN, D'ARRIGO 1999: 1365.
80 MANLEY 1974; KINGTON 1980; KINGTON 1988; PARKER, LEGG, FOLLAND 1992; LUTERBACHER et al. 2004. Notable exceptions were temperature reconstructions based on maximum latewood density in tree rings from Sämtland in Sweden that showed a cool summer (TINGLEY, HUYBERS 2013; LUTERBACHER et al. 2016; ANCHUKAITIS et al. 2017). EDWARDS et al. (2022) have shown that the acidity of the Laki haze likely created anatomical anomalies in the trees, which means that these tree rings are not suitable for reading the temperature for the summer of 1783; KLEEMANN 2022b.
81 GRATTAN, SADLER 1999: 141, 164; GRATTAN, SADLER 2001.
82 HIGHWOOD, STEVENSON 2003.
83 THORDARSON, SELF 2003.

in the Northern Hemisphere. During the summer of 1783, temperatures were hotter than usual in some areas of the Northern Hemisphere, such as western and northern Europe, and colder than usual in other areas, such as Alaska and northwest Siberia.[84] The close connection between the high-pressure system and the heat seems to be corroborated by the fact that with the dispersion of the high-pressure system, temperatures in western Europe returned to normal.[85] Without the Laki eruption, the heat wave would likely have been even more intense than it was.[86] Thus, in the case of a future Laki-style event, Europe should expect cooling rather than warming.

Another topic of debate in the scientific literature is why the winter of 1783/1784 was so cold. Better multi-proxy records exist for summer temperatures; analyses of tree rings, among the other archives of nature, offer this high-resolution data. Winter temperature records are mainly based on early instrument readings and written documents.[87] The written records are in good agreement that the winter of 1783/1784 was severely cold. Did the Laki eruption cause this freezing winter? A previous study by paleoclimatologist Rosanne D'ARRIGO and colleagues in 2011 suggests that the severely cold winter of 1783/1784 was due to natural climate variability in the shape of a positive El Niño-Southern Oscillation (ENSO) phase and a negative phase of the North Atlantic Oscillation (NAO). They argue that these conditions randomly arose at that time. The same conditions occurred in 2009/2010 and caused an anomalously cold winter in Europe and eastern North America.[88] While ENSO is a variation in wind and sea surface temperatures over the central and eastern Pacific with worldwide teleconnections, the NAO is a fluctuation of sea level pressure between the Icelandic Low and the Azores High. ZAMBRI and his colleagues argue, based on their climate models, that the Laki eruption actually precipitated a positive ENSO phase. This finding is corroborated by results from a study undertaken by climate modeler Francesco PAUSATA and his team: they argue that a southward shift of the Intertropical Convergence Zone (ITCZ) in the aftermath of a high-latitude volcanic eruption leads to a positive ENSO phase, particularly if La Niña or neutral ENSO conditions are present at the time of the eruption.[89] However, there is some debate about how long the ENSO event around the time of the Laki eruption lasted.[90]

Scientists have proposed a connection between Icelandic volcanic eruptions and fluctuations in the Nile River floods. Fluctuations certainly took place in 1783 and

---

84 ZAMBRI et al. 2019b: 6771, 6777–6778.
85 GRATTAN, SADLER 1999: 169.
86 ZAMBRI et al. 2019b: 6771, 6777–6778.
87 FRANKE et al. 2017.
88 D'ARRIGO et al. 2011.
89 PAUSATA et al. 2015; PAUSATA et al. 2016; ZAMBRI et al. 2019b: 6787.
90 D'ARRIGO et al. 2011; DAMODARAN et al. 2018: 521–522; GROVE 2007.

1784.[91] As described above, climate models have shown that the asymmetric cooling in the Northern Hemisphere created a southward shift of the ITCZ. Thus, there was a decrease in precipitation in the tropics; the June eruption, therefore, influenced the monsoonal rains that would usually occur around that time of the year and so triggered droughts in India and eastern Africa. This led to a low Nile River flow in 1783, which resulted in famine in Egypt.[92]

Natural scientists have studied volcanic eruptions worldwide and have established valuable chronologies, which are updated with the availability of data from more precise dating of ice cores.[93] In 2015, glaciologist Michael SIGL and his colleagues re-dated volcanic signals in Greenland and Antarctica's ice cores from the past 2,500 years with the help of multi-proxy records (Figure 2). In particular, spikes of radiocarbon in tree rings caused by extraterrestrial events between 774/775 CE and 993/994 CE serve as precise time markers around the globe.[94]

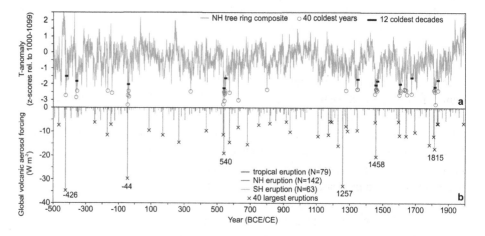

Figure 2: Northern Hemisphere temperature variations (above) and global volcanic aerosol forcing (below), 500 BCE to 2000 CE.

For a long time, in the natural sciences, many volcanologists presumed that tropical eruptions were more likely to have a significant global impact than high-latitude eruptions. High-latitude eruptions were generally believed to be less explosive and, therefore, less likely to eject volcanic gases high enough to reach the stratosphere and have long-lasting climatic impacts (Figure 3). As a consequence, high-latitude eruptions remained

91 OMAN et al. 2006a; MANNING et al. 2017; MIKHAIL 2015; MIKHAIL 2017; ZAMBRI et al. 2019b: 6787.

92 PAUSATA et al. 2016; ZAMBRI et al. 2019b: 6787. On the connections between ENSO and the Nile River, see BELL 1970; ORTLIEB 2004.

93 SIMKIN et al. 1981; Global Volcanism Program.

94 SIGL et al. 2015: 543–544.

comparatively neglected. However, recent studies show that high-latitude eruptions can "have significant impacts on global circulation on seasonal to annual timescales."[95] The Laki eruption is a notable example because of the enormous volumes of lava and sulfur dioxide it produced over several months.[96] The study mentioned above by SIGL and others notes that the Laki eruption was the largest non-tropical eruption in the Northern Hemisphere and the eighth-largest volcanic eruption in the world within the past 2,500 years. It resulted in a summer temperature in Europe and the Arctic that was 0.97 °C below the 1961 to 1990 average.[97] Other volcanoes in the high latitudes that could have potentially significant eruptions can be found in Alaska and Kamchatka in the Northern Hemisphere and Antarctica in the Southern Hemisphere.[98]

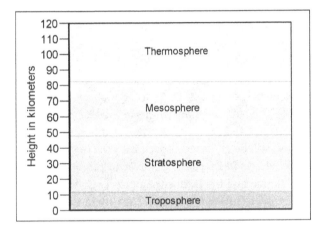

Figure 3: The different layers of the atmosphere.

The Laki eruption is significant because it was the first flood basalt event witnessed and well-documented by humans.[99] In the 2010s, there were other flood basalt events, albeit smaller than the Laki eruption (VEI 4, covering 600 square kilometers), such as the 2014/2015 Holuhraun eruption at nearby Bárðarbunga (VEI 0, covering 82 square kilometers), and the 2018 eruption at Kīlauea in Hawai'i (VEI 3, covering 36 square kilometers).[100] In the 2020s, the first fissure eruptions in Iceland were in 2021 at Geldingadalir and 2022 at

---

95 ZAMBRI et al. 2019a.
96 SCHNEIDER, AMMANN, OTTO-LIESNER 2009; OPPENHEIMER 2011: 269–270.
97 SIGL et al. 2015, Extended Data Table 4: Large volcanic eruptions during the past 2,500 years. Although the volcanoes that caused the 426 BCE and the 1230 CE eruptions are not known, it is known that they were tropical due to their bipolar deposits in the ice cores.
98 OMAN et al. 2005; PAUSATA et al. 2015.
99 THORDARSON 2003: 1.
100 USGS: "Preliminary Summary of Kīlauea Volcano's 2018 Lower East Rift Zone Eruption and Summit Collapse," 2018; Global Volcanism Program: Kīlauea; Global Volcanism Program: Bárðarbunga.

Fagradalsfjall on the Reykjanes Peninsula.[101] Obtaining a better understanding of historical and present-day flood basalt eruptions helps geologists in their study of large igneous
provinces in deep geological time, such as the Deccan Traps in India (500,000 square
kilometers).[102]

Research Focus

While this book starts in deep geological time with the formation of the Iceland mantle plume, possibly as far back as 130 million years ago, and ends with an outlook on
the future of Icelandic volcanism in a warming world, the main focus is the summer
of 1783. During that time, contemporaries in many parts of the Northern Hemisphere
experienced the dry fog by seeing, smelling, and even tasting it. Although the Laki
haze is the principal topic of this book, earthquakes, blood-red sunsets, severe thunderstorms, and meteors will also be discussed.

An environmental history of a volcanic eruption of a certain magnitude – one
that impacted several different countries – ought to be international. Several planetary events play a role in the story of the Laki eruption: from a newly emerging island
in the North Atlantic to earthquakes in Calabria and a volcanic eruption in faraway
Indonesia. The main focus, however, will be on Europe, specifically the German Territories. Previous research on Laki's effects has focused on Great Britain, France, the
Low Countries, the Czech Lands, Alaska, and the Ottoman Empire.[103] Historians Otto
MÄUSSNEST, Manfred VASOLD, and Oliver HOCHADEL have touched upon the German
Territories, but there remains a lot to be uncovered.[104] Thus, in this book, I demonstrate that the German Territories – which I define as a geographic area based on language – was a region very much affected by the Laki eruption. At times, the Laki haze
was present in the German Territories in concentrations strong enough to wither vegetation and cause sore eyes, throats, and breathing difficulties.

Studying a geographic region as vast and diverse as the German Territories in the
eighteenth century reveals that the effects of, and reactions to, the dry fog varied substantially. At different points, naturalists and the media would intervene in an attempt
to circumvent panic in the general public. The discourse between naturalists, editors,
and the public was international. Theories, ideas, and findings were shared and commented on across Europe. For this reason, my book includes not only German primary
sources but also British, Danish, Dutch, French, Italian, and Icelandic sources. I use

---

101 Veðurstofa Íslands: "Fagradalsfjall Eruption," 2021; Global Volcanism Program: Krýsuvík-Trölladyngja.
102 THORDARSON et al. 2003b: 34; PARK 2010: 12; MATHER, SCHMIDT 2021: 104.
103 MICHNOWICZ 2011 (England); HELLMAN 2021 (Wales, Jersey, Isle of Man); COURTILLOT 2005; GRATTAN
et al. 2005 (France); DEMARÉE 1997 (Low Countries); BRÁZDIL et al. 2017 (Czech Lands); JACOBY, WORKMAN,
D'ARRIGO 1999 (Alaska); MIKHAIL 2015 (Egypt).
104 MÄUSSNEST 1983; VASOLD 2004; HOCHADEL 2009.

discourse analysis hermeneutically to examine these contemporary debates about the Laki eruption and the subsequent haze that played out in newspapers and scientific publications, such as monographs and journals of learned societies.

When writing disaster history, the recurrence period of a given event must be established so that reasonable assumptions can be made about when it will happen again. In order to mitigate the potential hazards of a future eruption, knowledge of recurrence periods is crucial. Scientists have determined that a Laki-sized volcanic eruption in Iceland is statistically probable every 200 to 500 years. It has been almost 250 years since the Laki eruption occurred; although the exact timing of the next flood basalt event is unclear, it is safe to say that one will happen again. Can volcanologists, policymakers, public health officials, and the modern global community learn from the events of 1783?

In this book, I explore a number of research questions, including the following. What impact did the Laki eruption have on people in Iceland and the Northern Hemisphere? Moreover, how did the people in 1783 react – physically, emotionally, and intellectually – to the Laki haze? How did naturalists explain the phenomena that they were witnessing? What made the Laki eruption and its effects so extraordinary? What is the Laki eruption's overall legacy, and when did scholars connect the dry fog to the eruption?

Methods

Environmental History

The story of the Laki eruption and its legacy represents one part of this book's contribution to the growing field of environmental history. Historians in this field have the opportunity to work on documented volcanic eruptions from antiquity to modern times.[105] The 1815 Tambora eruption, which has a minor part in the story, was the largest eruption in the last 500 years.[106] The eruption of Krakatau in 1883 plays a significant role in this story due to its occurrence after the invention of telegraphy.[107] Before this point in time, technological limitations meant that, in some parts of the world, an eruption's emissions were visible long before news of the event arrived.

Why should one study a volcanic eruption as a matter of environmental history? This question deserves a nuanced answer. Volcanic eruptions are an ideal subject of study in the field of environmental history because – depending on their explosivity and output of lava and gas – they can have dire local and significant distant effects.

---

105 GUNN 2000; JONES 2000; THÜSEN 2008; WHITE 2011; COCCO 2012; RUDWICK 2014; BAUCH 2015; GUERRA 2015; EBERT 2016; BAUCH 2017; WOZNIAK 2017; MANNING et al. 2017; NEWFIELD 2018; WOZNIAK 2020.
106 WOOD 2014; KRÄMER 2015; PFISTER, WHITE 2018a; BEHRINGER 2019.
107 WINCHESTER 2005.

Their immediate and direct physical consequences, such as lava flow, can potentially change a landscape. Far-reaching gases can adversely affect human health and even cause death; they also affect the vegetation and the well-being of animals. Beyond human health, volcanic ejecta can affect weather and circulation patterns. Obviously, volcanic eruptions affect nature as well as society.[108]

According to historian John MCNEILL, environmental historians "write history as if nature existed. And they recognize that the natural world is not merely the backdrop to human events but evolves in its own right, both of its own accord and in response to human action."[109] Historian Reinhold REITH defines environmental history as the study of the interactions between humans and nature.[110] Nature can be understood as the natural and anthropogenic environment.[111] It is a cultural construct; in reality, it is constantly changing, even without anthropogenic influences. Historian J. Donald HUGHES has identified three themes of environmental history: the first is the influence of environmental factors on human history; the second is the changes in the environment caused by human actions; and the third is the history of what humans think about the environment.[112] Although all three themes play a role in this book, the first and third themes are strongly represented here. In this book, I analyze the influence of a volcanic eruption, an environmental factor, on human history.

For a long time, historians did not consider nature, climate, or weather in their research. Environmental history is changing this paradigm. In the 1970s, the Club of Rome illustrated the limits of growth; consequently, environment, ecology, and conservation have become topics of public debate.[113] In the United States, environmental history emerged in the 1970s and gradually reached other countries and continents. Today, it is perhaps the fastest-growing field of history.[114] Only in the 2010s did climate and weather become leading themes in environmental history: this upward trend was fueled, in part, by the growing concern about anthropogenic climate change and the availability of high-resolution data from paleoclimatological reconstructions.[115] Volcanic eruptions, like the climate, are agents of change; they do not determine people's actions but rather change the number and kind of choices they have.[116]

Volcanic eruptions have far-reaching consequences that pay no heed to political boundaries; in this regard, the study of historical eruptions is quintessentially environmental history. Whereas "traditional" history focuses mainly on the modern nation-state,

---

108 FLANNERY 2007: 15.
109 MCNEILL 2010: 346.
110 REITH 2011: 3.
111 MAUELSHAGEN 2010: 20.
112 HUGHES 2006: 3–8; REITH 2011: 1–4.
113 REITH 2011: 1.
114 MCNEILL 2010: 349–357, 364.
115 PFISTER, WHITE, MAUELSHAGEN 2018: 10.
116 DEGROOT 2018a: 16.

environmental history is transnational and is therefore well-suited for this study.[117] John MCNEILL suggests that there are two potential routes for environmental history in the future: one is imitation (of previous work), and the other one is interdisciplinarity.[118] For this book, I have chosen the latter. Interdisciplinarity is a hallmark and great asset of environmental history because it allows for the examination of a historical topic from different and previously unexplored angles.[119]

Climate History and the Little Ice Age

In the nineteenth century, knowledge of the ice ages and the new concept of deep geological time, rather than biblical timescales, changed the long-held belief that the climates of the past had been stable.[120] Even seemingly small climatic fluctuations of 1 °C can have severe consequences, as is now becoming evident.[121] Up until the 1960s, historians had, for the most part, ignored evidence produced by the physical sciences regarding climate change in historical times. Climate history pioneers such as Emmanuel LE ROY LADURIE, Hubert H. LAMB, and Christian PFISTER deviated from this long-held tendency in the 1960s and 1970s.[122]

Climate historians use approaches of historical climatology to reconstruct the climates of the human past; they treat climate and weather as something that has always influenced the "human experience."[123] They analyze historical documents to study societal, cultural, and economic vulnerability in the face of climatic changes and extreme weather. In addition, they also reconstruct a history of knowledge of the climate.[124] Because of the "shifting-baselines" problem, in many cases, climatic variability can best be understood by reconstructing climates of the past.[125] In the 1980s, climate historians started to contribute to the understanding of climate change by using a historical perspective.[126] Rudolf BRÁZDIL, Christian PFISTER, Heinz WANNER, Hans VON STORCH, and Jürg LUTERBACHER have identified three topics in the field: first, the reconstruction of patterns of climate, weather, and climate-related nature-induced disasters; second, the study of

---

117 MCNEILL 2010: 359.
118 MCNEILL 2010: 365.
119 KLEEMANN 2019b.
120 MAUELSHAGEN 2010: 16–26.
121 DEGROOT 2018a: 8.
122 LAMB 1972; MANLEY 1974; PFISTER 1975; PFISTER 1984; RICHARDS 2003: 64–65; LE ROY LADURIE 2006. Emmanuel LE ROY LADURIE published his book in French in 1967 and it was translated into English in 1971. However, only in the 2000s did LE ROY LADURIE find that seasonal or annual changes in the climate could affect history; PFISTER et al. 2018: 283.
123 PFISTER, WHITE, MAUELSHAGEN 2018: 19–20.
124 MAUELSHAGEN 2010: 19–20.
125 DEGROOT 2018a: 2.
126 PFISTER, WHITE, MAUELSHAGEN 2018: 11.

the vulnerability of societies in the past; and third, the exploration of discourses on weather and climate.[127]

Generally, the weather is what one experiences and can be measured with thermometers and other instruments. The climate, on the other hand, is the average weather calculated statistically over at least 30 years.[128] Climate scientist Michael GLANTZ puts it in a nutshell: "Climate is what you expect. Weather is what you get."[129] Climate change, therefore, alters the average weather and, with it, the frequency and severity of extreme weather events. However, extreme weather events can occur in any climate and are not necessarily caused by climate change.[130]

Climate historians mainly work with the "archives of society," which are historical records that include logbooks, chronicles, or weather diaries with information on harvests, floods, or snowfall. Natural scientists work with the "archives of nature," such as tree rings, ice cores, lake sediments, and stalagmites, to reconstruct past climates.[131] The proxy data retrieved from the archives of nature have advantages and disadvantages and can provide different resolutions, which means they reveal decadal, annual, or seasonal information.[132] In the last two decades, interdisciplinary collaborations between historians and natural scientists have become more common. These collaborations are not without their problems, such as the disparate terminology within each field.[133] Combining the archives of society and the archives of nature is advantageous in that scholars can easily cross-check the reliability of their findings.[134]

Many early climate history studies focused on crisis, disaster, and collapse, but newly emerging scholarship in the field focuses on resilience, adaptation, and complexity.[135] Just as there is no disaster without society, there is also no climate that is bad per se. Even during periods that would seem at first glance to be disadvantageous across the board, there were winners and losers.[136] Many factors influence human thought and behavior; climate alone does not directly result in human action. Instead, interactions between societies and their environments are complex and manifold.[137]

---

127 BRÁZDIL et al. 2005: 366.
128 FLANNERY 2007: 20; MAUELSHAGEN 2010: 7–8.
129 GLANTZ 1996: 1.
130 DEGROOT 2018a: 1, 15.
131 MAUELSHAGEN 2017; BRÖNNIMANN, PFISTER, WHITE 2018; CAMUFFO 2018; PFISTER 2018a; PFISTER 2018b; PFISTER, WHITE 2018b.
132 MAUELSHAGEN 2010: 38–42.
133 PAULING, LUTERBACHER, WANNER 2003; BRÁZDIL et al. 2005; MCNEILL 2010, 364; HALDON et al. 2018; KLEEMANN 2019b: 38–40; WHITE et al. 2022.
134 GLASER 2001: 13.
135 DEGROOT 2018b; SÖRLIN, LANE 2013; WHITE, PFISTER, MAUELSHAGEN 2018; MCCORMICK 2019; STRUNZ, MARSELLE, SCHRÖTER 2019; BAUCH, SCHENK 2020: 1; DEGROOT et al. 2021; DEGROOT et al. 2022.
136 This has recently been illustrated in great detail by Dagomar DEGROOT (2018a) with the example of the Dutch, who thrived during the coldest periods of the Little Ice Age.
137 MAUELSHAGEN 2010: 21–22.

Several factors influence climate; for instance, volcanic eruptions that release large amounts of sulfur dioxide into the atmosphere can warm the stratosphere but cool the surface, resulting in global cooling. Particularly devastating are double eruptions, two or more large volcanic eruptions within the space of a few years.[138] Fluctuations in the sun's activity, in particular so-called sunspot minima, can also produce a cooling effect.[139] Other factors that influence climate include changes in atmospheric and oceanic circulation patterns and orbital deviations, such as the Milanković cycles.[140] Even small temperature fluctuations can create feedback loops: a decrease in temperature creates more snow and sea ice, which increases the Earth's albedo effect.[141]

Throughout the Common Era, there have been several climatic oscillations, including the Late Antique Little Ice Age (410–775/536–660), the Medieval Climatic Anomaly (900–1400), the Little Ice Age (1250/1300–1850), and, with the onset of the industrial revolution in around 1750, anthropogenic climate change.[142] Coincidentally, James WATT introduced his improvements to the steam engine, considered one of the starting points of the Anthropocene, in 1784, the year after the Laki eruption. His engines burned fossil fuels and released carbon dioxide into the atmosphere, traces of which can still be found in air bubbles trapped in the ice of Greenland and Antarctica.[143]

In 1939, geologist François MATTHES coined the term "little ice age" for a period of glacial surges that occurred in the late Holocene. It was *little* compared to the *large* ice ages prior to the Holocene.[144] Glaciologist Jean M. GROVE points out that the term refers explicitly to glacier advances, not the temperature.[145] The appellation is slightly misleading as it suggests a world of ice and snow, but scientists and historians alike use it.[146] Several climate historians have carried out work on the Little Ice Age, a

---

138 ROBOCK 2000; COLE-DAI 2010; SCHMIDT, ROBOCK 2015.
139 EDDY 1976.
140 MAUELSHAGEN 2010: 12–15. The Milanković cycles are the deviations of the Earth's orbital path, its tilt on its axis, and its axial precession, all of which affect the long-term climate.
141 GROVE 2000; ZHONG et al. 2010; MILLER et al. 2012; SIGL et al. 2015; STOFFEL et al. 2015; DEGROOT 2018a: 23.
142 Different scholars offer different definitions of when these anomalies started and ended: MANN 2002; MATTHEWS, BRIFFA 2005; MILLER et al. 2012; WHITE 2014. All of these anomalies were relevant for Europe. However, not all of these anomalies can be found in all regions around the world – with the exception of modern global warming; NEUKOM, STEIGER, GÓMEZ-NAVARRO 2019. Nevertheless, during the Little Ice Age, significant glacial surges occurred in Alaska, central Europe, and Tibet between 1300 and 1850; PFISTER et al. 2018: 268.
143 MENELY 2012: 479.
144 MATTHES 1949; KRÜGER 2008; PFISTER et al. 2018: 268.
145 GROVE 2001.
146 JONES, BRIFFA 2001; REITH 2001: 77; BÜNTGEN, HELLMANN 2014; KELLY, Ó GRÁDA 2014. For a detailed history and discussion of the term, see OGILVIE, JÓNSSON 2000.

period characterized by variable weather with extremes in both directions.[147] Overall, it was a cold climatic regime.[148] Several minima occurred throughout the Little Ice Age, during which temperatures were significantly lower than the average for this period. The most notable minima were the Spörer Minimum (1450–1530), the Grindelwald Fluctuation (1560–1628), the Maunder Minimum (1645–1720), and the Dalton Minimum. The latter lasted from roughly 1760 to 1850 and therefore covered the period of the Laki eruption.[149] Historian Wolfgang BEHRINGER argues that the 1780s saw a density of extreme weather events, the likes of which had not occurred since the Maunder Minimum.[150]

Disaster History

Humankind has always endured setbacks. As we have seen above, historians have only recently begun to study nature-induced disasters in human history. Whereas climate change occurs over a relatively long time, disasters often strike suddenly, unexpectedly, and with brute force. Previously, humans were considered "the only or decisive actor of history."[151] Historians in the field of historical disaster research challenge this view.

As anthropologist Anthony OLIVER-SMITH puts it, "Disasters occur at the intersection of nature and culture and illustrate, often dramatically, the mutuality of each in the constitution of the other."[152] The presence of a society, or people, is required for a force of nature to be considered a disaster. Thus, the idea of a "natural disaster" is a social construct.[153] Historians Dieter GROH, Michael KEMPE, and Franz MAUELSHAGEN suggest that every nature-induced disaster is based upon an extreme natural event; however, not every extreme natural event is considered a disaster.[154] In the last 30

---

147 FAGAN 2000; BEHRINGER, LEHMANN, PFISTER 2005; PFISTER, BRÁZDIL 2006; GLASER 2008; BEHRINGER 2011.
148 FAGAN 2000: 48; DEGROOT 2018a: 22–49; PFISTER et al. 2018: 269.
149 DEGROOT 2018a: 31–41: These minima are named after the researchers who discovered them: astronomer Gustav SPÖRER (1822–1895) discovered a period of low sunspot numbers from historical observations; the Grindelwald Fluctuation is named after a glacier near a Swiss village of the same name; astronomer Edward MAUNDER (1851–1928) discovered a low in sunspot numbers from historical observations for the given period; meteorologist John DALTON (1766–1844) discovered a period of lower-than-average temperatures. The dating of these minima is not set in stone: different scholars give slightly different start and end dates. PFISTER et al. (2018: 269) give 1790 and 1820 as the start and end dates of the Dalton Minimum.
150 BEHRINGER 2011: 214.
151 KEMPE, ROHR 2003: 123 (quote), 123–124.
152 OLIVER-SMITH 2002: 24.
153 KEMPE, ROHR 2003: 124; JUNEJA, MAUELSHAGEN 2007: 2, 14.
154 GROH, KEMPE, MAUELSHAGEN 2003: 15; MAUELSHAGEN 2010: 19, 96, 117–118.

years, historians have begun to understand that disasters have to be regarded as physical *and* socio-cultural events.[155]

The *móðuharðindin* – the famine of the mist – was a disaster for the Icelanders; a fifth of the population perished, and it took decades to recover. The Laki eruption had wide-reaching effects across the Northern Hemisphere and might have caused increased mortality in Europe. Even imagined phenomena that coincided with the eruption, conjured up by people being swept up in the excitement, caused surprise and fear and stimulated the need for explanations. Responses to imagined phenomena were not unlike those shown in the face of real disasters.

Historian Matthias GEORGI studied the English media of 1750. His research shows that in the aftermath of two earthquakes precisely one month apart, news spread of a third earthquake. This third earthquake was imagined. Londoners' reactions to it were real, with many leaving the city to seek refuge. GEORGI shows that although these imagined disasters only affected the lives of the "victims" for a few short days, they still influenced knowledge production. In many ways, 1750 in Britain was comparable to 1783 in Europe: publications, especially newspapers, attempted to spread calm. Despite the media's best efforts, their reports did not always have the intended effect.[156]

Today, the lines between disaster, catastrophe, and calamity are often blurred.[157] The term "natural disaster" is a modern one. The word "disaster" has an astrological origin and means "ill-starred." In 1783, the word of choice was "revolution," which also has an astrological-astronomical origin. Revolution was initially used to describe the rotation of a celestial body on its axis.[158] As early as the sixteenth century, the word "catastrophe," meaning a sudden (down)turn, entered the English language. In the late eighteenth century, "revolution" became synonymous with "catastrophe." Both were used to describe violent geological events, such as volcanic eruptions or earthquakes. A revolution was defined as a "large, important change that is accompanied by unusual events, be it in nature, political relations, or the sciences."[159] Naturalists in 1783 used the word "revolution" to describe what they believed to be an imminent disaster that announced itself with the Laki haze, the blood-red sun, and the many earthquakes. Historian Guido POLIWODA discovered that in 1784, the term "catastrophe" was used for the first time in print by the *Zürcher Zeitung*, a Swiss newspaper, to describe the flooding events throughout Europe earlier that year.[160]

---

155 MAUCH 2009: 5; PFISTER et al. 2010: 283.

156 GEORGI 2009: 15–16, 21, 68–69.

157 IRWIN, SMITH 2020: 98–99.

158 GROH, KEMPE, MAUELSHAGEN 2003: 16–18; WEBER 2015: 11.

159 "Revolution" in KRÜNITZ 1813, vol. 123: 186. "[. . .] eine große, wichtige, von ungewöhnlichen Ereignissen begleitete Veränderung, sey es in der Natur, in den politischen Beziehungen, in den Wissenschaften, etc." See also RIGBY 2015: 16–17.

160 POLIWODA 2007: 30.

History of Science

The Laki eruption struck when the Enlightenment was in full swing; the dry fog and all the other accompanying phenomena proved intriguing subjects of research.[161] Much of the work on the environmental history of the Little Ice Age period has focused on disasters, which has produced a skewed representation of the period.[162] Historian Simon SCHAFFER states that disasters were not more prevalent in the eighteenth century than in other periods; however, the people of this time were "uniquely fitted" to appreciate the meaning of the spectacles of nature.[163]

The Enlightenment was an epoch of change in Europe and elsewhere; it offered the chance to look at the world through a different lens. Thinkers of the time were sometimes at odds with the various local and Church authorities: this clash produced ideas and notions that were far from homogenous. The scientific curiosity of the contemporaries in 1783 – undoubtedly sparked by the Enlightenment – generated an exceptional wealth of primary sources about that fateful summer. Unusual weather has the potential to trigger emotions such as fascination and fear in human beings; these sources provide numerous examples of both.

In 1783, naturalists strove to engage with their environment in a meaningful way. It was apparent that gathering objective data was the principal way, and perhaps the only way, to develop satisfying explanations that addressed the phenomena they were witnessing. With some tried and tested instruments and some exciting recent inventions at their disposal, these naturalists played their part in the unfolding drama of scientific discovery. The tools of their trade included uncontroversial hardware, such as the thermometer, and exciting, high technology, such as the hot-air balloon. Findings were much discussed in learned societies and scholarly journals.

Three topics in the scientific realm stand above all others when investigating the summer of 1783: meteorology, air-travel via hot-air and hydrogen balloons, and electricity. Meteorological networks began to employ standardized equipment; the data produced could be compared and contrasted and served as a foundation for better hypothesizing.[164] Hot-air balloons allowed those who dared to see the world from a different perspective and, while they were at it, measure all sorts of phenomena from places that were previously impossible to reach.[165] And, of course, then there was the lightning rod; hitherto, it would have been inconceivable to even think about capturing

---

161 STOLLBERG-RILINGER 2011: 256.
162 LÜBKEN 2004; REITH 2011: 91.
163 SCHAFFER 1983: 16.
164 GOLINSKI 1999.
165 DE SYON 2002: 7–13; LYNN 2010; THÉBAUD-SORGER 2013.

electricity and rendering it harmless. But in 1783, it became a tantalizing possibility that this could become commonplace; the only resistance to this was a reluctant public.[166]

Deep Geological Time

In his definition of environmental history, John McNEILL states, "More than most varieties of history, environmental history is an interdisciplinary project."[167] Interdisciplinarity is the merging of two or more disciplines, which are not merely combined but draw from one another to influence the research outcome. The idea behind interdisciplinarity is to pave new ground by thinking across traditional boundaries. In this book, I take this idea to heart: while environmental history and climate history are interdisciplinary fields already, this study also combines history with geology. After all, it was Iceland's unique geology that formed the Laki fissure and therefore caused its far-reaching physical and intellectual consequences, including the long-lived missed connection.

The book title, *A Mist Connection*, suggests that the *fog* produced by the eruption and interchangeably called *haze* or *mist* was a *connecting phenomenon*; the entirety of the European continent was burdened, at least to some extent, by its presence. Indeed, those who sought to unravel the mystery of the fog's origin *missed*, or rather overlooked, its connection to the Laki eruption. The potential insights that could be garnered by co-opting knowledge and practices from the field of geology and applying them to environmental history are, as yet, unappreciated; this is another missed connection. *A Mist Connection* also utilizes the approaches and methods of climate history, disaster studies, discourse analysis, and history of science. Environmental history is an ideal home for this topic as it is an interdisciplinary field that accommodates this novel approach.

I will show that at various points of the story, the theories and conclusions of the contemporaries were, at times, far off the mark and, at times, incredibly close to the truth. With an understanding of geological mechanisms, it is possible to separate fact from fiction. If one is to understand conclusions drawn in 1783, one needs an understanding of geology as they understood it then and as we do today.

Geology was a young discipline in 1783. The Swiss naturalist Jean-André DELUC first used the word *geology* in 1778.[168] The idea that the Earth was around 6,000 years old was still commonplace; seashells found far from the oceans seemed to be evidence of the biblical Flood. With new methods and technologies, geologists began to understand that the Earth was much older than they had thought. The notion of deep geological time developed during the 1780s. Geologists James HUTTON and Abraham

---

166 HOCHADEL 2003.
167 MCNEILL 2010: 348.
168 DELUC 1778–1780.

Gottlob WERNER, on opposite sides of the "Plutonist vs. Neptunist" debate, contributed tremendously to the understanding of this concept. They believed that Earth was not formed by a few catastrophes over a few thousand years but through processes that took place over extremely long periods of time.[169]

Philosopher Robert FRODEMAN compares HUTTON and WERNER's role in the discovery of deep geological time to the role astronomer Nicolaus COPERNICUS played in the human understanding of extraterrestrial space. FRODEMAN also emphasizes that the idea of deep time, and the change it brought about, is often underappreciated.[170] Over the next two and a half centuries, geology changed: almost every new generation discovered that the Earth was even older than the last generation had estimated until radiocarbon dating in the early twentieth century revealed that the planet was 4.56 billion years old.[171] Geologists must think in deep geological time, which is hard to comprehend for almost anybody who is not an Earth scientist.

This particular environmental history will start in deep geological time, with the formation of the Iceland mantle plume 130 million years ago. Using this unique timescale, I have incorporated other vital aspects of the story, including the convergence of the mantle plume and the Mid-Atlantic Ridge. The necessary precursors to the narrative happen within this time frame; this is an exceptionally long period for a study of history but a relatively short one for geology. It is about the length of a Wilson cycle, a model that describes the breakup of a continent, the subsequent opening and closing of an ocean basin, and the forming of a new continent.[172] Iceland's volcanism is unique due to its location at the converging point of a subaerial mid-oceanic ridge and a mantle plume. Ever since settlers first set foot on Iceland in the late ninth century, the "fires" in their new country became apparent, as did their potentially devastating effects.

In deep geological time, events like the Laki eruption are common. Colossal flood basalt events gave rise to large igneous provinces, the remnants of which are dotted around the globe. Even if our frame of reference is only the Holocene (about the last 11,700 years), the Laki eruption pales in comparison to larger flood basalt events such as those produced by the Katla and Bárðarbunga volcanic systems. This eruptive style was virtually unknown in 1783; today, armed with extensive data, perhaps we can prepare ourselves for a future event on the scale of Laki or larger.

History is part of the humanities and geology is part of the natural sciences. In environmental history, a combination of the two allows for precisely the kind of interdisciplinary approach needed to ask and answer new questions. Indeed, history and geology have much in common. For most historians and geologists, it is impossible to

---

169 GROTZINGER, JORDAN 2017: 192.
170 FRODEMAN 1995: 960. For the context of this debate, see also GOULD 1987; RUDWICK 2008; RUDWICK 2014.
171 RUDWICK 2005: 250–253.
172 GROTZINGER, JORDAN 2017: 262–267.

make direct observations of the events and processes that they study. Perhaps these events took place deep inside Earth and occurred over the course of millions of years, impossible to witness, or maybe they are veiled by the passage of time, an even greater impediment to observation. Both disciplines use a hermeneutic process: as historians work with written documents, geologists may work with outcrops, for example. Just as an image or a piece of text needs interpreting, so too does an outcrop. Due to the time that has elapsed between the event we study and the present, sometimes not all the data we need is available; both historians and geologists have to fill the gaps with knowledge and reasonable assumptions. While historians assume that humans in the past thought, felt, and acted as we would today – albeit within their historical context – geologists assume that the geological processes of the past were similar to those that can be observed in the present; this is called the principle of *uniformitarianism*.[173]

Teleconnections

In order to study transcultural relationships, historians have previously worked with concepts such as "entangled histories," *histoire croisée*, and connected histories. More recently, historians have adopted the concept of "teleconnections."[174] The term *teleconnection* comes from meteorology; it refers to the synchronicity of weather phenomena in different parts of the globe. The term was first used in 1935. It is often used in the context of atmospheric oscillations that have global effects, such as El Niño-Southern Oscillation.[175] Prior to the discovery of teleconnections, there was a missing link in the understanding of different climate patterns around the globe.[176] The concept of teleconnections first emerged in the late nineteenth century and was systematically verified by statistical analysis in the early twentieth century.[177]

In the early twenty-first century, the concept of societal teleconnections emerged.[178] Geographer Susanne MOSER and oceanographer Juliette FINZI HART assert, "[s]ocietal teleconnections link activities, trends, and disruptions across large distances, such that locations spatially separated from the locus of an event can experience a variety of impacts from it nevertheless." The idea is "to uncover distal vulnerabilities via a distinct

---

173 FRODEMAN 1995: 960–966; GROTZINGER, JORDAN 2017: 8.
174 BEHRINGER 2017: 27–28.
175 ANGSTRÖM 1935; GLANTZ 1996: 40–41.
176 BRIDGMAN, OLIVER 2006: 25–27.
177 BAUCH, SCHENK 2020: 17. For a more detailed history of the discovery of teleconnections, see GROVE 1997.
178 LIEBERMAN 2003–2009; ADGER, EAKIN, WINKELS 2009; CAMPBELL 2016; BAUCH, SCHENK 2020: 17.

focus on the connection itself."[179] The term is used in the study of the "direct and indirect causal links between historical phenomena of climatic and societal change."[180] Of course, the consequences of natural events vary; their impact depends on their scale and the societal and political circumstances of the affected region.[181] Did similar events happen before? What time of year did the incident take place? It would be deterministic to assume that climate directly leads history; however, the weather is still an important factor.[182] Climate is complex; many factors play a role. In the context of this interdisciplinary study, both physical and societal teleconnections triggered by the Laki eruption are relevant.

Teleconnections explain the time lag between events and their physical manifestations.[183] While the Laki eruption commenced on 8 June 1783, the dry fog it produced settled above Europe mid-month at the earliest. The climatic effects of the Laki eruption, a topic of much debate to this day, took more time to develop. The winter of 1783/1784 was severely cold, and several abnormal seasons followed. It took until 1787 for the weather to normalize. Some of the societal reverberations, in particular those that occurred after the realization of a connection between the eruption and the dry fog, took much longer to come about: some 100 years.

The concept of teleconnections also aids the interdisciplinary approach of environmental history, as it makes for more viable connections between concerned fields. Volcanic eruptions and their effects, like the realm of climate in general, are often abstract and difficult for laypeople to grasp. Nevertheless, we can find information hinting at these teleconnections in sources from 1783 and the following years.

Sources

The main sources for this book are an autobiography, official reports, newspapers, scientific publications, weather diaries, and travelogues. As is often the case, unusual or extreme weather is better documented than "normal" weather.[184] In some German newspapers from mid-July 1783, almost every report dealt with an extreme weather event.[185] Newspaper reports, scientific publications, and letters help reconstruct the phenomena of 1783, contemporary perceptions of them, and historical populations' ideas on the possible origins of said phenomena.

---

179 Moser, Finzi Hart 2015: 15.
180 Bauch, Schenk 2020: 17.
181 Hoffmann 2020: 281.
182 Fagan 2000: xiv.
183 Hoffmann 2020: 283–284.
184 Pfister 1999: 16–17; Fagan 2000: 51; Greyerz 2009: 42.
185 Examples are the issues of the *Königlich Privilegirte Zeitung* dated 17 July and 19 July 1783, both of which have numerous articles relating to the weather.

My sources for the eruption itself are chiefly the English translations of the auto-biography and "fire treatise" of Jón STEINGRÍMSSON (1728–1791), a Lutheran pastor in Kirkjubæjarklaustur, who observed the fiery columns of the eruption, the lava flows, and the effects of the eruption's ejecta on the vegetation and population first-hand. His "fire treatise," also called an *eldrit* in Icelandic, is a report about the eruption and its aftermath and has strong religious overtones. The other sources are by two Icelandic authors, Sæ-mundur Magnússon HÓLM (1749–1821) and Magnús STEPHENSEN (1762–1833): the former is a collection of letters and the latter is a report detailing observations made while on duty in Iceland for the Danish Crown.[186]

The invention of the printed newspaper, which Wolfgang BEHRINGER calls one of the most important media revolutions, occurred shortly after the so-called postal rev-olution. This created a perfect synergistic moment: news could now be printed more readily and distributed more efficiently.[187] Johann CAROLUS, a writer and former cor-respondent, established the first printed newspaper in Strasbourg in 1605.[188] Over the next two centuries, postal networks and roads improved, which meant, for instance, that the travel time of news between Hamburg and Augsburg was reduced from 30 days in 1615 to only five days by 1800.[189] The postal routes also dictated the order of the stories in newspapers: periodicals of the time printed news in the order that they received the reports. This meant the most important stories were not necessarily on the front page. Printed newspapers served three main purposes: first, to advertise and initiate trade and economic relationships; second, to maintain the political sys-tem; and third, to spread scholarly findings.[190] Newspapers also documented military successes and failures, diplomatic negotiations, theater, and other cultural affairs, as well as weather events and disasters.[191] A story would be printed simply if the editor considered it newsworthy.[192] As historian Margot LINDEMANN reminds us, the newspa-per was often no more than a cobbled-together collection of news and rumors.[193]

In the German Territories, the number of newspapers quadrupled to 200 between 1700 and 1800, with a particularly sharp increase in the last 25 years of that period.[194] Some of these ran daily, others only once a week. The frequency of the postal routes

---

186 STEINGRÍMSSON 1998; STEINGRÍMSSON 2002; HÓLM 1784a; HÓLM 1784b; STEPHENSEN 1785; STEPHENSEN, EGGERS 1786. For other Icelandic sources written around the time of the eruption, see THORDARSON et al. 2003b; GUNNARSDÓTTIR 2022.

187 For more information on this "media revolution," see BEHRINGER 2003: 680; BEHRINGER 2010: 51.

188 ARNDT, KÖRBER 2010: 20; WEBER 2005. For more information on the early days of the newspapers, see BEHRINGER 2005; MAUELSHAGEN 2005.

189 BEHRINGER 2003: 664.

190 WILKE 2010: 62.

191 BLOME 2010: 207.

192 STÖBER 2002: 159–160; ARNDT, KÖRBER 2010: 6.

193 LINDEMANN 1969: 34.

194 STEIN 2006: 222.

determined the frequency of newspaper publications: that is, if they received mail daily, they could publish daily.[195]

A few newspapers, such as the *Hamburgischer Unpartheyischer Correspondent* and the *Mercure de France*, were read in different regions and countries.[196] LINDEMANN states that the *Hamburgischer Unpartheyischer Correspondent* was, in fact, the most widely-read newspaper in Europe at the time. In 1789, each issue was printed 30,000 times. It was highly regarded because of its generally reliable reports. Because of this, it served as the news source for several other newspapers that did not have their own correspondents.[197] In 1783, newspapers primarily consisted of letters from anonymous correspondents: usually, the only information provided was where and when they had written the letter. The role of the newspaper correspondent has been misunderstood for a long time; it was assumed they worked on behalf of an authority and it has only recently become understood that financial incentives were motivating factors.[198]

In many territories, the state imposed a substantial degree of local censorship on newspapers; this meant that periodicals often forbore printing local news in favor of stories from other regions to avoid conflict with authorities.[199] To gather as much information as possible about different regions in the German Territories, I analyzed newspapers from Hamburg (*Hamburgischer Unpartheyischer Correspondent*), Berlin (*Königlich Privilegirte Zeitung, Berlinische Nachrichten*), Munich (*Münchner Zeitung*), Augsburg (*Augsburgische Postzeitung*), Göttingen (*Göttingische Anzeigen von gelehrten Sachen*), Breslau (*Schlesische Privilegirte Zeitung*), and others from Dessau, Hanau, and Vienna.[200] In addition to German newspapers, I also worked with British, French, and American periodicals.

A study conducted by historian Jürgen WILKE revealed that the *Berlinische Nachrichten* did not print any local news in 1736. By contrast, 60 years later, in 1796, local news made up ten percent of the paper. Over this period, most newspaper articles recounted news from other countries, often southern European countries; reports from the German Territories accounted for between one-fifth and one-third of all news. His study also showed that between 1736 and 1796, about half of the reports that appeared in the *Berlinische Nachrichten* also appeared in the *Hamburgischer Unpartheyischer Correspondent* or vice versa.[201]

---

195 BEHRINGER 2003: 667.
196 STOLLBERG-RILINGER 2011: 142.
197 LINDEMANN 1969: 163; STEIN 2006: 222.
198 ARNDT, KÖRBER 2010: 20.
199 See also LINDEMANN (1969: 111–123), to learn how different, larger territories dealt with censorship in the seventeenth and eighteenth centuries.
200 The newspapers from Hamburg and Berlin can be found on microfilm at the Staats- und Universitätsbibliothek Hamburg and the newspaper department of the Staatsbibliothek zu Berlin. All other newspapers have been digitized for the relevant years and can be found through ANNO and DigiPress.
201 WILKE 2002: 84.

At the end of the eighteenth century, the German Territories had an estimated 27.5 million inhabitants.[202] The sheer number of newspapers in that region prompts the question: who could read them? Literacy rates, which varied widely across Europe, substantially increased within the German Territories throughout the eighteenth century. Some territories introduced compulsory school attendance. Reading and writing were taught by family members, in the community, and through the Church. Lending libraries had been around since 1750 in the German Territories, 1717 in Switzerland, 1725 in England, and 1759 in France.[203] It is safe to assume that most people in cities – home to about 20 percent of the population – had access to newspapers. There is proof that handypersons and servants who worked in stately homes could read.[204] Generally, it was men rather than women who could read and write, as well as people from the upper classes; more people to the west of the Stralsund-Dresden line were literate than to the east.[205]

Newspapers were interested in selling as many copies as possible. Even as early as the seventeenth century, most had become interested in expanding their readership beyond the intellectual elite by targeting the "common man." Thus, by the mid-eighteenth century, newspapers were no longer the exclusive domain of the educated.[206] Even so, in 1783, newspapers often used technical terms that were not necessarily meant to be understood by everybody, perhaps in order to suggest authority.[207] Newspapers were not only accessible to the literate: often, they were read aloud within the family home, in coffeehouses, or in reading societies.[208] Verbal exchanges remained the most common form of communication in the early modern city.[209] Historian Peter STEIN estimates that in the 1780s, 300,000 German-language newspaper issues were printed weekly, reaching approximately three million readers (or listeners).[210]

The naturalists who influenced the journalistic and scientific discourse had different backgrounds; however, they shared the ideals of the Enlightenment.[211] Reasoned argument, the search for truth, and empirical evidence became of the utmost importance. In 1783 and the following years, many naturalists published papers about the different phenomena they witnessed and offered explanations. Scientific monographs were aimed at scholars and general audiences alike; some books directly addressed their readers. In contrast, articles in the journals of learned societies were written for a scholarly readership; these learned societies often financed themselves through

---

202 Hartmann 1995: 348.
203 Engelsing 1973; Stein 2006: 267–270.
204 Arndt, Körber 2010: 18. The size of the readership was estimated by Welke (1981: 163–166).
205 François 1989: 407–413.
206 Böning 2010: 236–237.
207 Georgi 2009: 68–69.
208 Puschner 2002: 194; Stein 2006: 265.
209 Georgi 2009: 71.
210 Stein 2006: 222; Lindemann 1969: 124.
211 Briese 1998: 15.

subscriptions to their journals. Scientific publications tended to be much more detailed than newspaper reports.[212]

Scientific books and articles were generally published in the common language – German in the German Territories, French in France, English in Britain, etc. This was part of the media revolution of the eighteenth century: in 1740, in the German Territories, Latin books made up 28 percent of the total number of published texts; in 1800, they made up only four percent of the total.[213] Neither newspapers nor scientific publications existed in a vacuum; frequently, they addressed one another. Often, newspapers referred to scientific findings and commented on them. Similarly, in some scientific publications, naturalists referenced rumors they had read in the newspaper and tried to correct inaccuracies.

One significant outlier in this trend was the *Ephemerides* of the Societas Meteorologica Palatina, a highly technical publication compiled and published two years after the data was gathered. Its Latin text made it clear that it was written for a highly educated and sophisticated readership. As the Society gathered its data from a network of around 30 weather stations in several countries, it made sense for its publication to use an international language. The *Ephemerides* also used many symbols that were not necessarily self-explanatory. During the summer of 1783, the Mannheim observatory, headquarters of the Societas Meteorologica Palatina, published a statement about the nature of the dry fog in several newspapers. That they chose to take such action underlines how urgently the readership of the newspapers wanted explanations for the unusual weather.[214]

This book mainly relies on weather diaries to reconstruct the summer and winter weather in Great Britain and North America. In the United States, almanacs were popular; here, the observer could record the daily weather and any other occurrences. Most weather observers described the weather simply, with words such as "fair" or "rainy." They also noted the readings from the instruments they had at hand, often thermometers, barometers, and hygrometers. Instrument readings from this time should be taken with a grain of salt, as the practices for using them had not yet been standardized. Temperatures were measured inside or outside, in the sun or the shade, without reference. Sometimes weather observers remarked upon the unreliability of their instruments and their frustration with them, such as when British naturalist Gilbert WHITE (1720–1793), in the winter of 1784/1785, complained that his thermometers were not up to the task of measuring the extreme cold.[215]

---

212 BARDILI 1783; FISCHER 1784.
213 MIX 2005: 283.
214 Münchner Zeitung, 10 July 1783: 422; Berlinische Nachrichten, 19 July 1783: 670: Report from the Mannheim observatory, 6 July 1783.
215 Gilbert WHITE, "The Naturalist's Journal," 1784, Add MS 31849, British Library, London, UK. Gilbert WHITE kept a diary for several years and, by means of close phenological observations, tried to decipher nature's pattern each year in order to "domesticat[e] climatic chaos into diaries of ecological self-awareness," as eco-critic Heidi SCOTT (2009) puts it.

I also conducted archival research in the United Kingdom and the United States to reconstruct the weather of 1783 and the following years. This subset of my study has three purposes. The first is to establish whether the dry fog appeared in North America, as Benjamin FRANKLIN famously claimed.[216] A few mentions of fog can be found in the sources from the United States for the summer of 1783; however, the fact that these are not further remarked upon is indicative of them being occurrences of "normal" fog. In contrast, sources from three Moravian settlements in Labrador, in today's Canada, reveal that the dry fog was visible there; it was perceived as a "smoke" that lingered for several months. The second is to reconstruct the weather during the summer in Great Britain, where the dry fog was also present. The third purpose is to analyze the winter climate in both regions, as both primary and secondary literature indicate that the winter was exceptionally cold in North America and Europe.

Travelogues were reports written by naturalists and private persons traveling to other countries. Often, they were aimed at a general audience. Iceland was an exotic destination and many were interested in reading about the "land of fire and ice." These travelogues included tales of the people, the landscape, and personal achievements, such as summiting Mount Hekla. Even the travelogues in the direct aftermath of the Laki eruption, which served the purpose of collecting information for the Danish king rather than entertaining readers, were translated into other European languages. Of particular importance for this book are the travelogues and other scientific articles authored by Sveinn PÁLSSON, Þorvaldur THORODDSEN, and Amund HELLAND. They play key roles in connecting the Laki eruption to the dry fog in Europe.[217]

Unless otherwise stated, I have translated all the quotes from the sources; the non-English texts can be found in the footnotes. Spelling in the original source texts differs from twenty-first-century spelling. The various newspaper reports and scientific publications are expressions of individuals, which reflect the complexity, plurality, and contradictions of the eighteenth century. This compilation of different responses to the dry fog helps to paint a picture of the collective atmosphere at the time.

Structure

This book has three distinct parts. In Chapter Two, I explore Iceland's geological formation and its volcanic activity during the Holocene. This chapter aims to illustrate the Icelanders' experience with volcanism from when they first set foot on the island until the late eighteenth century. Furthermore, I describe the Laki eruption itself and its consequences for Iceland. How did the people cope with this eruption? Given their

---

216 FRANKLIN 1785: 357–361.
217 PÁLSSON 1793a; PÁLSSON 1793b; PÁLSSON 1945; PÁLSSON 2004; THORODDSEN 1879; THORODDSEN 1925; HELLAND 1881; HELLAND 1882; HELLAND 1884; HELLAND 1886.

almost 900-year history in Iceland, could they have been prepared? What makes the Laki eruption so unique? For the most part, the sections on the geology and the history of Iceland are based on English-language secondary literature.

Leaving Iceland behind and effectively following the Laki eruption's gases where they lead us, in Chapter Three, I study the various real and imagined consequences of the eruption outside of Iceland, focusing on Europe, where the dry fog was most intense and lasted the longest. I first describe the various phenomena visible during 1783, including those unrelated to the Laki eruption. Naturalists at the time considered all the phenomena to be interconnected; therefore, an awareness of them is crucial to understand the story. After detailing the phenomena, I analyze how naturalists from different disciplines interpreted them. Influenced by the Enlightenment, the natural sciences experienced a transformation after 1750. Many disciplines became more specialized, and some were newly established. This period laid the foundation for the modern disciplines as we know them today. The main focus of this chapter is the summer of 1783, during which most of the Laki eruption's effects were observable. Toward the end, I briefly look at the cold winters that followed. The main research questions here are: what impact did Laki have on the Northern Hemisphere? How were the dry fog and other phenomena perceived? What explanation strategies were developed?

In Chapter Four, I trace the legacy of the Laki eruption. Legacy, in this case, means something transmitted from the past. The Laki eruption serves as a fascinating opportunity to trace the knowledge production of a meteorological and geological event that originated in a remote and sparsely populated country in the North Atlantic from the late eighteenth century to the present. In 1783, the discipline of geology had to compete against other spheres of knowledge and was itself characterized by internal disagreement. In this chapter, I study naturalists' and geologists' travelogues and letters, along with diaries from participants in European expeditions to Iceland between the 1790s and the early twentieth century. The principal question in this chapter is: when was the Laki eruption connected to the dry fog of 1783? Thus, I also analyze scientific publications that were compiled and published in the aftermath of the Krakatau eruption in 1883.

In Chapter Five, I make concluding remarks on my research and offer an outlook on the substantial changes that the fields of volcanism and geology have gone through since the Laki eruption. One such change was the widespread acceptance of the theory of plate tectonics in the late 1960s; this led to a major paradigm shift in geological theory. At present, the theory of plate tectonics explains most of the geological phenomena around the world, from earthquakes to volcanic eruptions.[218] Finally, I trace the perceptions of the Laki eruption in the present and discuss what scholars know about it today, what we have yet to discover, and how Icelandic volcanism will change in a period of anthropogenic climate change

---

218 MCKENZIE, PARKER 1967; MORGAN 1968. Xavier LE PICHON (1991) reconstructs what happened to the discipline of geology between 1967 and 1968.

2 A Volcano Comes to Life

Recently, Iceland has seen a rapid transformation from a country on the periphery of people's mental landscapes to a major tourist destination. This transformation is taking place in a country that is newborn, geologically speaking, having risen from the ocean floor a mere 24 million years ago.[1] Iceland plays host to an eruption every three to five years; in 2010, the volcano Eyjafjallajökull reminded the world of this country's volcanism.[2]

In this chapter, I explore the geology and history of volcanic eruptions on the island. In Iceland, historical time commenced at the beginning of the Norse settlement in the ninth century CE. Which significant volcanic eruptions occurred in Iceland before 1783? What hazards do the island's volcanic eruptions pose to Iceland and the world beyond? In addition to answering these questions, I also analyze descriptions of the Laki eruption and its direct aftermath within Iceland.

The Geological Formation of Iceland

Studies of history rarely consider geological timescales, which cover deep history, the rise of mountain ranges, and the birth and death of oceans.[3] That said, we must consider them, as a volcanic eruption is at the center of this book. Eruptions can wreak havoc on flora and fauna, and thereby local food supplies and livelihoods. They can also influence the weather, atmosphere, and even the climate over a surprisingly broad area.

130 million years ago, in the early Cretaceous, the last age of the dinosaurs, the Iceland mantle plume came into existence at the base of the lithosphere, where the Earth's immobile uppermost mantle meets its convecting mantle (Figure 4). A mantle plume is a mass of relatively hot (and therefore less dense) material that rises from the Earth's lower mantle.[4] The lithosphere is composed of numerous tectonic plates that make up our planet's continents and ocean basins. Geologists assume that the mantle plume is fed from the lower mantle, just above the core-mantle boundary about 2,700 to 2,900 kilometers below the Earth's surface. The lifespan of a mantle plume is typically 100 to 150 million years.[5]

The Iceland mantle plume formed the basaltic Alpha Ridge in the Arctic Ocean, northwest of today's Greenland. The Alpha Ridge is considered the first expression of

---

1 THORDARSON, HÖSKULDSSON 2014: 1.
2 THORDARSON 2010: 285; SCHMIDT et al. 2014a.
3 FORTEY 2005: 28.
4 CONDIE 2001: 1–2, 67–68, 123; CONDIE 2016: 2–3; Science Direct, "Mantle Plume."
5 SAUNDERS et al. 1997; CONDIE 2001: 123.

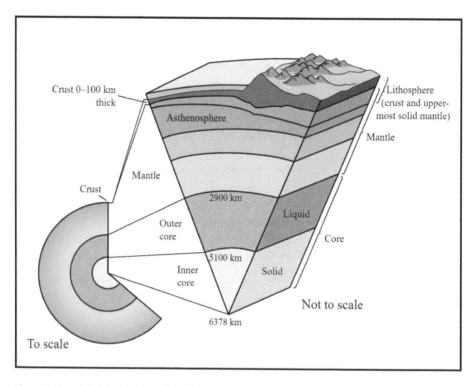

Figure 4: A model of Earth's internal density structure. The dynamic convection of the mantle takes place within the region illustrated by the three warm colors. The convecting mantle provides the energy for rising plumes and plate tectonics.

the Iceland mantle plume.[6] Mantle plumes require a continuous supply of hot materials from the lower mantle (Figure 5).[7] These buoyant materials move upward within a funnel-shaped zone toward the surface. The Iceland mantle plume is mushroom-shaped, with a narrow funnel and a head that is roughly 2,000 kilometers in diameter.[8] By definition, a hotspot is the surface manifestation of a mantle plume.[9]

Mantle plumes are planet-wide phenomena that occur in intraplate regions and at plate boundaries in both oceanic and continental crusts. If the magma within a

---

6 For a detailed study of the geological evolution of the North Atlantic region from the late Cretaceous to the early Eocene, see JOLLEY, BELL 2002. A recent study has discovered a geothermal heat flux anomaly running through Greenland from the northwest to the southeast, which MARTOS et al. (2018) interpret as the path that this plate took across the Iceland mantle plume between 80 and 50 million years ago.
7 SAUNDERS et al. 1997; CONDIE 2001: 123.
8 WHITE 1988: 8; WHITE, MCKENZIE 1989. Other scholars describe the Iceland mantle plume as disk-shaped, with a plume head of 600 kilometers in diameter that is fed by a vertical conduit, which is 50 to 100 kilometers in diameter; SAUNDERS et al. 1997: 52–53.
9 WHITE 1988: 3; CONDIE 2001: 1–2.

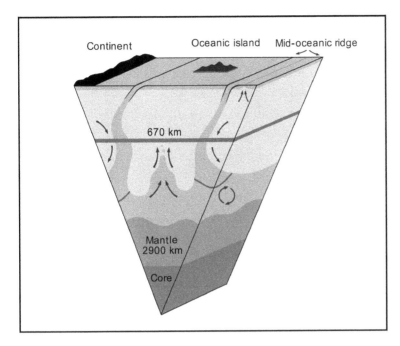

Figure 5: Dynamics of the possible effects of a dense layer in the lower mantle. Internal heating and heat flow along the core-mantle boundary drive the internal circulation of the lower mantle. Mantle plumes arise from local hotspots consisting of recycled slabs and other materials.

mantle plume breaches the surface, for a brief time, it is defined as lava before it cools and solidifies into volumes of igneous rock, the larger masses of which are called igneous provinces (Figure 6).[10] Igneous rock, also called magmatic rock, is one of three main rock types found on Earth, the other two types being sedimentary and metamorphic. The term "igneous" derives from the Latin word for fire, *ignis*.[11] In general, geologists assume that the large igneous provinces around the world formed as a result of mantle plume activity and not by the processes directly linked to plate tectonics.[12] The driver of both the mantle plumes and plate tectonics is convecting mantle.

The North Atlantic Igneous Province is one of the largest discovered so far, extending from Ellesmere Island in eastern Canada to the British Isles.[13] Between 62 million and 56 million years ago, igneous activity produced onshore magmatism in Canada's Baffin Island, Greenland, Scotland, and Northern Ireland, and offshore magmatism in eastern

10 CONDIE 2016: 48.
11 PARK 2010: x.
12 SAUNDERS et al. 1997: 48; CONDIE 2001: 54.
13 SAUNDERS et al. 1997: 45; LACY 1998: 13; CONDIE 2001: 56, 67; RICKERS, FICHTNER, TRAMPERT 2013: 39; BAR-NETT-MOORE et al. 2017: 251. For more information on oceanic spreading rates in the North Atlantic, see VIBÉ et al. 2018.

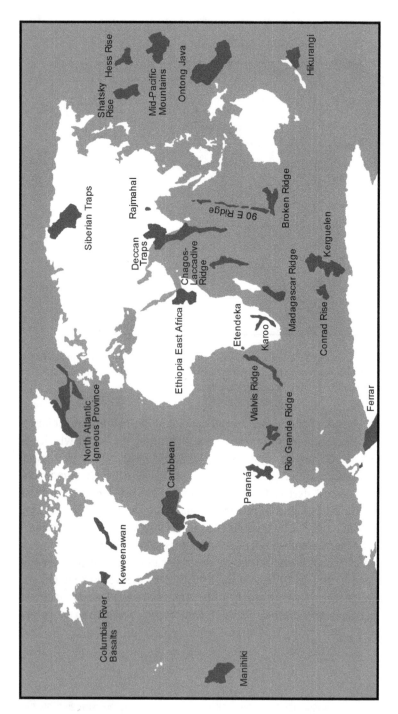

Figure 6: Large igneous provinces. The areas in black indicate the locations of large igneous provinces formed in the past 250 million years.

Greenland, the Greenland-Iceland-Faroe Ridge, and Iceland.[14] Even today, remains of large volumes of basaltic lava flows can be found in the depths of the North Atlantic.

The North American tectonic plate moved westward over time; therefore, the plume appears to have moved eastward, even though it remains stationary. These basaltic areas in Greenland and the North Atlantic trace the apparent trajectory of the mantle plume 30 to 40 million years ago.[15] Around 24 million years ago, the Iceland mantle plume and the Mid-Atlantic Ridge crossed paths and started interacting with one another, gradually forming the Iceland Basalt Plateau.[16] This plateau is more than 3,000 meters above the seafloor and has a crustal thickness ranging from 15 kilo-

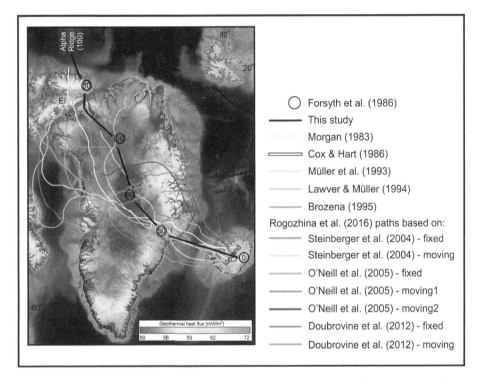

Figure 7: Proposed trajectories of the Iceland mantle plume over the past 100 million years. The numbers indicate the position of the Iceland mantle plume 80 million, 68 million, 60 million, and 50 million years ago to the present (0). The black line indicates the trajectory proposed by MARTOS et al. 2018, Figure 3.

---

14 SAUNDERS et al. 1997: 45, 59, 69, 71, 82. The age of these phases has been established through the reversal phases of the Earth's magnetic field; LUNDIN, DORÉ, 2005; THORDARSON, LARSEN 2007: 119.
15 Katla Geopark Project: 3–4.
16 THORDARSON 2010: 286.

meters in coastal areas to 40 kilometers in central Iceland, underneath the northwestern part of the Vatnajökull ice sheet (Figure 7 and Figure 8).[17]

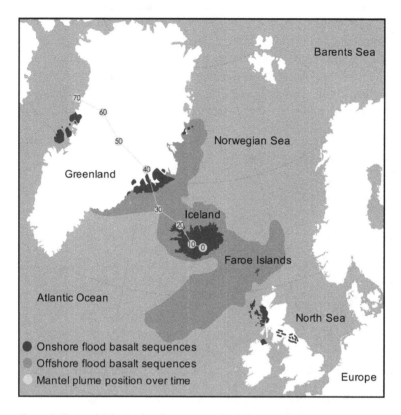

Figure 8: The North Atlantic's basalt structures. The dark colors show onshore (black) and offshore (gray) basalt formations created by the North Atlantic Igneous Province. The numbers within the gray circles show the present (0) and previous positions of the Iceland mantle plume (ten million to 70 million years ago).

This igneous activity is still ongoing. Today, the Iceland mantle plume sits beneath the Vatnajökull ice sheet. Although activity is more focused, magma production rates are two to three times lower than they were 56 million years ago when the large igneous province was formed.[18] Nevertheless, Iceland remains one of the most active volcanic regions on the planet above sea level.[19]

---

17 Katla Geopark Project: 3–4; FEDOROVA, JACOBY, WALLNER 2005: 123; SIGMUNDSSON, SÆMUNDSSON 2008. Vatnajökull literally translates as "Vatna glacier," but for the readers' convenience, at times, I use the full Icelandic name *Vatnajökull* followed by glacier or ice sheet.
18 SAUNDERS et al. 1997: 70–71.
19 THORDARSON, HÖSKULDSSON 2008: 197.

Volcanoes can form near subduction zones, which are convergent (or destructive) plate boundaries. The subduction zone west of South America, for example, is described as destructive because the lithosphere gets subducted into the mantle and is slowly destroyed. Oceanic crust is thinner and lighter than continental crust; thus, it gives way and is subducted underneath the continental plate. As a result, the continental crust is uplifted, forming mountain belts. The subducted slab can go as deep as 670 kilometers to the so-called Wadati-Benioff-Zone, the boundary between the upper and lower mantle. Volcanoes can form on the surface near these subduction zones. They can also form at divergent (or constructive) plate boundaries. The mechanisms of constructive plate boundaries form new crust. One such example is the Mid-Atlantic Ridge.[20] Sometimes volcanism takes place far away from a plate boundary, the manifestations of which are called hotspot or intraplate volcanoes.

Icelandic Volcanism in the Holocene

Let us jump forward almost 24 million years, from the formation of Iceland to the Holocene: our current geological epoch. It is challenging to establish pre-Holocene volcanic events because erosion caused by glaciation destroyed a great deal of evidence. During glaciations, ice caps grew to thicknesses of about 500 to 1,500 meters.[21] At the end of the Pleistocene, around 10,600 BCE, an extraordinary eruption occurred at Katla: the so-called Sólheimer eruption is the only known Plinian (VEI 6) eruption to have occurred at this volcano. It produced ten cubic kilometers of tephra and pyroclastic surges. Tephra is the name given to the solid particles ejected by volcanoes during eruptions. It consists of ash particles, rocks, and pumice.[22] The tephra layer from the Sólheimer eruption settled in Iceland and other Scandinavian countries and is called Vedde ash. Pyroclastic surges are waves of tephra and hot gas that sweep away from a volcano at speeds of up to 700 km/h and can reach temperatures of 1,000 °C.[23]

Prevailing winds can carry tephra great distances. When it finally rests, its layers vary in thickness, from millimeters to meters, and in color, from yellow to red, brown, gray, or black.[24] A single volcano can produce tephra of a similar chemical signature from separate eruptions occurring years apart. The uppermost tephra layers are from more recent eruptions; the layers below derive from eruptions that occurred further back in time. In the 1940s, Icelandic geologist Sigurður ÞÓRARINSSON established this chronology of tephra layers and invented the science pertaining to

---

20 GROTZINGER, JORDAN 2017: 22–34.
21 LACY 1998: 17.
22 LAMB 1970: 427.
23 ÞÓRARINSSON 1981; LARSEN 2000; ÓLADÓTTIR et al. 2005; ÓLADÓTTIR 2008; LARSEN 2010.
24 BYOCK 2001: 89.

them; tephrochronology.[25] Today, tephrochronology is widely applied in the natural sciences to date volcanic eruptions, model climate reconstructions, and corroborate archaeological findings.[26]

Perched upon the Eurasian and North American Plates, Iceland grows by about two centimeters per year: one centimeter to the east and one to the west.[27] This equates to roughly 20 kilometers of growth in one million years. Volcanism in Iceland occurs within axial volcanic zones, which cover a third of the country and are primarily located along the Mid-Atlantic Ridge (Figure 9). The volcanic zone that is most relevant here is the Eastern Volcanic Zone (EVZ). The EVZ is Iceland's most active volcanic zone and is home to eight volcanic systems, including the four most active: Grímsvötn, Bárðarbunga, Hekla, and Katla.[28] It is also Iceland's most productive volcanic zone in

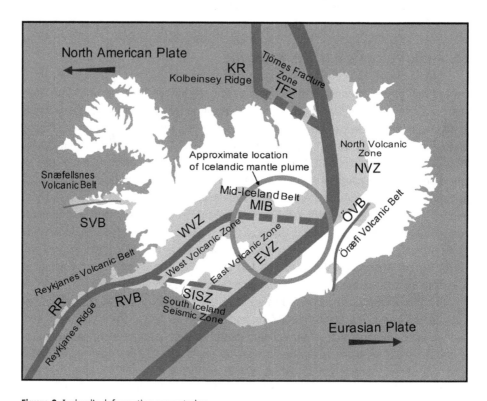

Figure 9: Iceland's deformation zones today.

25 Katla Geopark Project: 13; Þórarinsson 1944; Þórarinsson 1956b; Þórarinsson 1979.
26 Karlsson 2000a: 13–14; Dugmore, Vésteinsson 2012: 77, 81. For more information on tephrochronology in general, see Þórarinsson 1981; Dugmore et al. 2009a.
27 Pagli, Sigmundsson 2008: 4.
28 Thordarson, Larsen 2007: 118–121; Thordarson 2010: 286; Katla Geopark Project: 5.

terms of lava output. In historical times, 80 percent of Icelandic eruptions occurred in the EVZ. This zone has been active for two to three million years.[29]

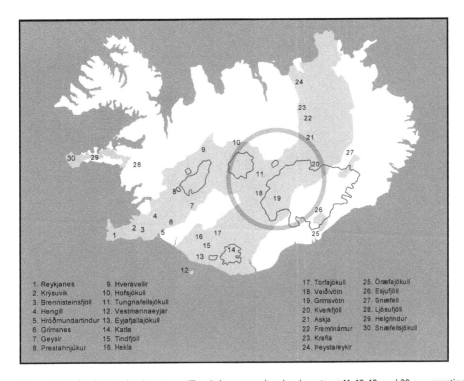

Figure 10: Iceland's 30 volcanic systems. The circle surrounds volcanic systems 11, 18, 19, and 20, representing the location of the Iceland mantle plume. The other lines indicate the locations of glaciers within Iceland. The dark gray area indicates the location of the spreading axis, the subaerial part of the Mid-Atlantic Ridge.

Iceland has 30 active volcanic systems (Figure 10). In this context, an active volcanic system is one that has had an eruption during the Holocene. A volcanic system consists of either a central volcano, a fissure swarm (a cluster of fissure volcanoes), or both. Activity close to the axial volcanic zones is connected to the phenomena of spreading and rifting and the creation of crust. It is not a continuous process; rather, it is episodic, with each phase traditionally referred to as a "fire," such as the Skaftá Fires. During such a rifting episode, the whole volcanic system is activated and expresses itself through one or more volcanic eruptions and recurring earthquakes.[30]

The Grímsvötn volcanic system (Figure 11) consists of two central volcanoes, Grímsvötn and Þórðarhyrna, and three calderas, 250 to 300 meters beneath the Vatnajökull ice sheet. A caldera is a large hollow that forms when a magma chamber collapses in

---

29 PASSMORE et al. 2012: 2594.
30 THORDARSON, LARSEN 2007: 121–123.

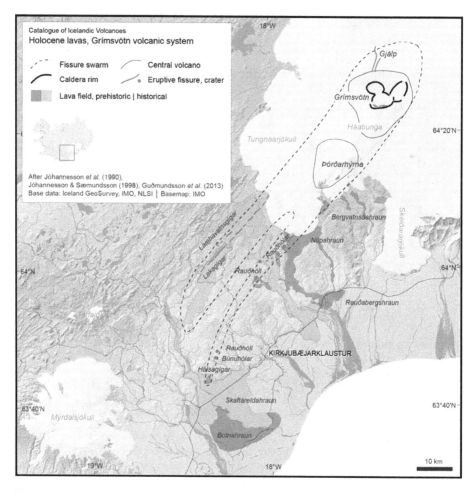

Figure 11: The Grímsvötn volcanic system. The caldera is underneath Vatnajökull (the glacier on the right), with subaerial parts of the Grímsvötn volcanic system represented by dotted lines. The pink color indicates the lava fields produced by Laki, while purple indicates older lava produced in pre-historical times.

on itself after an eruption.[31] Grímsvötn is a southeast-trending fissure swarm, some 80 kilometers long and ten to 15 kilometers wide, of which only the southernmost tip is ice-free. Four crater rows are visible here, all of which formed over the past 8,000 years: one of these rows is the Laki fissure.[32] Grímsvötn has seen more than 70 eruptions in

---

[31] For more information on subglacial lakes, see Bjӧrnsson 1974; Bjӧrnsson 1976; Bjӧrnsson 1992; Bjӧrnsson 2002; Russell et al. 2005: 163–164.

[32] Jakobsson 1979; Thordarson et al. 2003b: 13; Passmore 2012: 2594; Catalogue of Icelandic Volcanoes: Grímsvötn.

the last millennium.[33] It goes through a 60-to-80-year period of high activity consisting of 12 to 22 eruptions, followed by a 60-to-80-year period of low activity consisting of up to eight eruptions. Grímsvötn's geothermal activity frequently causes flooding on the plains around the Skaftá River.[34]

Iceland's 30 volcanic systems have produced roughly 2,400 eruptions and around 556 ± 100 cubic kilometers of lava since the last ice age. Scientists established this from tephra layers in Iceland and ice core records in Greenland, where many Icelandic eruptions deposit tephra.[35] Icelandic volcanoes range from majestic stratovolcanoes to earth-splitting fissures.[36] Subaerial, subglacial, and submarine eruptions occur here, which can erupt either effusively (most erupted materials are lava), explosively (most erupted materials are tephra), or with a mix of both.[37] Volcanologists classify eruptions, for the most part, as either wet or dry: most are wet (hydromagmatic or phreatic). Here, the magma interacts with water: a submarine eruption reacts with the ocean, a subglacial eruption reacts with the glacier and meltwater, and a subaerial eruption (which is not covered by water or ice) reacts with groundwater. The explosivity of wet eruptions depends on how much water is available. Wet eruptions constitute around 75 percent of all Icelandic eruptions. When magma does not react with water during an eruption, it is considered a dry eruption. When an eruption shows characteristics of being both wet and dry, it is called phreatomagmatic.[38]

An eruption that produces more than one cubic kilometer of lava is defined as a flood basalt event. Icelandic volcanologist Thorvaldur THORDARSON estimates that large flood basalt events in Iceland occur once every 300 to 1,000 years.[39] Flood basalt events are relatively common in the EVZ. Throughout the Holocene, many such events have taken place in this part of Iceland, sometimes with hundreds of years between each.[40] The largest of these was an eruption caused by the Bárðarbunga volcanic system around 8,600 years ago, which produced the Great Þjórsá Lava; this vast lava field consists of more than 22 cubic kilometers of lava and covers an area of 970 square kilometers.[41] This volume of lava would fill 8.8 million Olympic-sized swimming pools. As with the Laki eruption millennia later, the sulfuric gases released made this a "climatically significant eruption."[42]

33 THORDARSON, HÖSKULDSSON 2014: 141.
34 Catalogue of Icelandic Volcanoes: Grímsvötn.
35 THORDARSON, HÖSKULDSSON 2008: 197; ÁGÚSTDÓTTIR 2015: 1672.
36 THORDARSON 2010: 286.
37 ÞÓRARINSSON, SÆMUNDSSON 1979; ÞÓRARINSSON 1981; THORDARSON, LARSEN 2007: 118, 125–127; THORDARSON, HÖSKULDSSON 2008: 200; DUGMORE, VÉSTEINSSON 2012: 69.
38 THORDARSON, HÖSKULDSSON 2008: 207–209, 215.
39 THORDARSON, SELF 2003: 1; THORDARSON, HÖSKULDSSON 2008: 197.
40 THORDARSON et al. 2003b; THORDARSON, LARSEN 2007; PASSMORE et al. 2012: 2594.
41 THORDARSON et al. 2003a: 117; THORDARSON 2005: 205–209; Global Volcanism Program: Bárðarbunga. The Global Volcanism program dates this event at Bárðarbunga to 6650 BCE ± 50 years.
42 THORDARSON 2005: 218.

There were several huge eruptions at Grímsvötn between 10,400 and 9,900 years ago.[43] These eruptions formed the thickest tephra layer of the Holocene, the Saksunar-vatn layer; it is named after a lake on the Faroe Islands, its place of discovery. This layer is also present across almost all of Iceland. The amount of tephra released in total is esti-mated to be in the region of ten cubic kilometers.[44] One of the largest Holocene eruptions to date was Hekla 3, which occurred around 1000 BCE. Tree ring and stalagmite records indicate that this event significantly cooled the climate for the following decade.[45]

Icelandic Volcanoes in Historical Times

Iceland was the last country in Europe, and one of the last large islands in the world, to be settled (after Madagascar and before New Zealand).[46] While its people escaped the bloody and brutal conflicts of Europe and North America by virtue of their home-land's geographical location, they could not evade the hardships visited upon them by its geology.[47] Every generation of Icelanders has experienced volcanism and its im-pact on society and the environment. Nevertheless, the local perception of volcanoes greatly depends on their frequency and magnitude, as well as their proximity to set-tled areas of the island.[48]

Europeans had paid Iceland sporadic visits before its settlement. Archaeological ev-idence suggests that Irish monks landed there in the eighth and early ninth centuries. It may be that the Greek explorer Pytheas of Massalia (today's Marseille, France) landed there in the fourth century. In around 860 CE, Norsemen sailing from Norway to the Faroe Islands happened upon the island of Iceland, with settlers following in the 870s.[49] Over the next 60 years, more Norsemen and people from the British Isles trickled in.[50] No verified numbers regarding population for this time exist, but the Icelandic histo-rian Gunnar KARLSSON estimates that it had around 10,000 inhabitants in 930 CE, given

---

43 JOHANNSDÓTTIR, THORDARSON, GEIRSDÓTTIR 2006; THORDARSON, HÖSKULDSSON 2008: 219; BRAMHAM-LAW et al. 2013; Global Volcanism Program: Grímsvötn. The thickness of the Saksunarvatn layer indicates a very explosive volcanic eruption, perhaps even a super eruption (VEI 8). The tephra layer covered most of Iceland and can be found on both sides of the North Atlantic, even as far away as northern Germany. According to the Global Volcanism Program, the Saksunarvatn eruption was a VEI 6 event and the tephra can be dated to 8230 BCE ± 50 years.
44 Catalogue of Icelandic Volcanoes: Grímsvötn.
45 Global Volcanism Program: Hekla; BAILLIE 1989.
46 HJÁLMARSSON 1988: 14; VASEY 1996.
47 STARK 1994: 120–121; OSLUND 2011: 30.
48 THORDARSON 2010: 285, 290.
49 HJÁLMARSSON 1988: 13–15; LACY 1998: 76–77; KARLSSON 2000b: 9–15; VESTEINSSON, SVERRISDÓTTIR, YATES 2006. A study has been undertaken to carbon-date wood that was found in Reykjavík and is believed to have been from the early settlement; SVEINBJÖRNSDÓTTIR, HEINEMEIER, GUÐMUNDSSON 2004.
50 LACY 1998: 81; HELGASON et al. 2000; KARLSSON 2000a: 11–20; OGILVIE 2005: 257–258.

the population of 50,000 by 1100 CE.[51] Iceland's *Althing* was founded in 930 at Þingvellir and is one of the oldest parliamentary assemblies in the world.[52]

Reykjavík loosely translates to "steamy bay," so called due to the geothermal activity in the area.[53] During the initial years of settlement, two separate volcanic eruptions produced by one volcanic system, Bárðarbunga, occurred around the same time. As a result, a double layer of tephra exists called the "settlement layer," an archaeological term, as it contains the oldest settlement-related finds. It lends its name to an archaeological museum in Reykjavík, "The Settlement Exhibition Reykjavík 871 ± 2."[54]

The volcanic activity in Iceland during historical times remained the same as in the post-glacial and pre-settlement periods. In recorded history, Iceland has seen at least 213 volcanic eruptions (some of these are listed in Figure 12).[55] On average, Iceland sees 20 to 30 eruptions in a given century: these eruptions produce around eight cubic kilometers of lava in total.[56] The frequency of eruptions has varied over the past 11 centuries. It seems to have increased recently, though this is perhaps due to better documentation in the modern period.[57]

The first huge volcanic eruption, and the largest flood basalt eruption, to take place in Iceland in the last 2,000 years occurred just after the settlement period. From spring 939 to autumn 940, Eldgjá, "the fire gorge," formed in 30 eruptive episodes and produced 19 cubic kilometers of lava along a 75-kilometer-long fissure in a northeast-southwest direction extending to Vatnajökull.[58] The lava covered 780 square kilometers, an area about the size of New York City.[59] Katla, a volcanic system underneath the

---

51 KARLSSON 2000b: 15. Jesse BYOCK (2001: 9) also argues that the population levels must have reached "at least ten thousand people [by 930], and perhaps as many as twenty thousand, [. . .]." Terry LACY (1998: 77–79) believes Iceland might have had as many as 60,000 settlers by 930.

52 HJÁLMARSSON 1988: 27; KARLSSON 2000b: 9–13. *Thing* means assembly. For more information on the *Althing*, see also LACY 1998: 90–93; BYOCK 2001: 174–176; STEINGRÍMSSON 2002: 341.

53 The geothermal fields and steam might come from the Reykjanes or Hengill volcanic systems; THORDARSON, LARSEN 2007: 123–124.

54 The tephra layer with the lighter color, called Hrafntinnahraun lava, was produced by an eruption at Torfajökull. Global Volcanism Program, Torfajökull; Catalogue of Icelandic Volcanoes: Torfajökull. The layer with the darker color was produced at Vatnaöldur, which is an approximately 60-kilometer-long fissure and part of the Bárðarbunga volcanic system. The Global Volcanism Program estimates the eruption to have taken place around 870 CE, reaching VEI 4. Global Volcanism Program: Bárðarbunga; Catalogue of Icelandic Volcanoes: Bárðarbunga; THORDARSON, LARSEN 2007: 133. For more information on the "settlement layer," see also GRÖNVOLD et al. 1995; DUGMORE et al. 2009a.

55 THORDARSON, LARSEN 2007: 135–136. When this paper was published in 2007, Icelanders had seen 208 eruptions since the settlement. With the addition of the 2010 Fimmvorduhals, the 2010 Eyjafjallajökull, the 2011 Grímsvötn, the 2014–2015 Holuhraun (Bárðarbunga), and the 2021 and 2022 Fagradalsfjall eruptions (Krýsuvík-Trölladyngja), the number has grown to 214 volcanic events [as of January 2023].

56 GUÐMUNDSSON et al. 2008: 263; THORDARSON, HÖSKULDSSON 2008: 197.

57 THORDARSON 2010: 286.

58 OPPENHEIMER et al. 2018.

59 THORDARSON, LARSEN 2007: 142. New York City covers an area of 784 square kilometers.

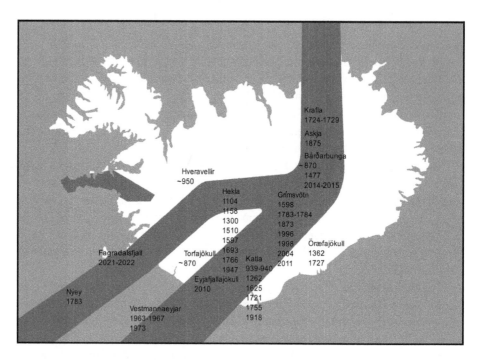

Figure 12: Map of eruptions in historical times featured in this book.

Mýrdalsjökull ice shield, was the source of this eruption.[60] The *Landnámabók*, the book of settlement composed in the twelfth and thirteenth centuries, briefly mentions this volcanic event.[61] The most obvious consequence of volcanic eruptions is lava flows that can continue for months or years, resulting in extensive property loss; this was the case during the Eldgjá eruption.[62] In addition, Eldgjá produced huge quantities of sulfur dioxide (about 220 megatons), making it the largest volcanic-pollution event in recorded history. A decade later, in 950, another flood basalt event, called Hallmundarhraun (*hraun* means lava field), produced around eight cubic kilometers of lava. This eruption was sourced by the Hveravellir volcanic system in the Western Volcanic Zone.[63]

In this heaving land they now called home, settlers dealt with the dramatic and the mundane. The foodstuffs they formerly benefitted from in their homelands were not necessarily available in Iceland; therefore, they had to change their diet according

---

60 Larsen 2000; Guðmundsson 2005: 140.

61 Thordarson 2010: 288.

62 Guðmundsson et al. 2008: 259.

63 Thordarson, Larsen 2007: 140–142; Dugmore, Vésteinsson 2012: 71–72.

to availability.[64] Icelanders used moss, a sort of lichen gathered in the mountains, to stretch flour. Seaweed was used for this purpose as well. At the beginning of the settlement, although plenty of fish and birds were available, there were no game animals or trees with fruit or nuts apart from a few kinds of bushes that bore wild berries.[65] While Iceland's climate is mild for its latitude, it was challenging to get by.[66] Agriculture was not easy because of the unstable weather patterns. These weather patterns are continually influenced by the Gulf Stream, which brings warm, damp air from the south, and the East Greenland Current, which brings cold, dry polar air.[67] The presence of sea ice lowered temperatures further.[68] Even at its warmest, during the Medieval Climate Anomaly between 870 and 1170, Iceland was never "self-sufficient as a grain producer."[69]

Many settlers brought livestock to Iceland, mainly sheep and cattle.[70] Grass became the main crop and was used primarily as winter fodder.[71] Early in the settlement process, Iceland's forests were cleared to create fields for grazing and to provide wood for fuel and charcoal production.[72] The lack of trees, the large-scale introduction of livestock, and frequent tephra deposits after volcanic eruptions made both erosion and desertification significant environmental challenges for the Icelanders.[73]

Fish, such as salmon, trout, and char, the occasional stranded whale, and birds' eggs supplemented the Icelandic diet. Until the late nineteenth century, fish were mainly caught in rivers and ponds or shallow waters close to the shore. The small row boats used by Icelanders at the time were inadequate for the open sea.[74] Iceland, cut off from Europe by a vast and boundless body of salt water, was, ironically, salt-deprived at all times. The resources vital for boiling saltwater were in short supply as well. Therefore, Icelanders smoked meat and air-dried fish to preserve them for the winter.[75] Seals, a staple for Icelanders, provided meat, oil, and leather.[76]

---

64 Karlsson 2000b: 44–51.
65 Rögnavaldurdóttir 2002: 1–3.
66 Lacy 1998: 21–24; Jónsson, Valdimarsson 2005: 81.
67 Bergþórsson 1969; Ogilvie 1984; Ogilvie 1986: 72; Ogilvie 2001; Steingrímsson 2002: 6.
68 Gunnarsson 1980; Ogilvie et al. 2000; Byock 2001: 26; Jónsson, Valdimarsson 2005: 81–82; Andrews et al. 2009; Ogilvie et al. 2009; Ogilvie, Hill, Jónsson 2011.
69 Rögnavaldardóttir 2002: 1 (quote). See also Byock 2001: 57; Steingrímsson 2002: 6: During the Medieval Climate Anomaly, temperatures were about 1 °C above the 1960–1990 average of 4 °C. As a comparison, the average annual temperature in Munich between 1993 and 2018 was 9.16 °C, based on weather data from Wetterkontor.de.
70 Rögnavaldurdóttir 2002: 2.
71 Ogilvie 2005: 257–258.
72 Dugmore, Vésteinsson 2012: 80.
73 Vasey 1991: 325; Lacy 1998: 43–48; Helldén, Brogaard 1999.
74 Gunnarsson 1983: 21–23; Rögnavaldurdóttir 2002: 2–3; Robertsdóttir 2008: 36; Oslund 2011: 11.
75 Rögnavaldurdóttir 2002: 1–4.
76 Pelly 2001: 5–6: Some place names in Iceland still reflect on the use of seals, such Seley ("seal island"), Kópavogur ("bay of baby seals"), and Urtusteinn ("female seal rock").

Most of what scholars know from the settlement period and afterward is from two written sources, the aforementioned *Landnámabók*, the book of settlement, and the *Íslendingabók*, the book of Icelanders. The latter was probably written down for the first time in around 1130 CE by the Icelandic historian Ari Thorgilsson (1067–1148). The so-called Saga Age was the period between 871, the settlement, and 1262, the year Iceland became a dependency of Norway.[77] In the Saga Age, maritime travel played a vital role. During this time, Icelandic seafarers landed in Greenland and, in around the year 1000, traveled on to North America, where Leifur Eiríksson set up camp at today's L'Anse aux Meadows in Newfoundland.[78]

The Scandinavian settlers continued to observe their polytheistic faith until the introduction of Christianity at the turn of the millennium.[79] The *Althing*, located at the boundary between the North American and Eurasian Plates, played host to the debate over whether or not to adopt this new faith. Records of this discourse reveal that some settlers had a surprisingly good understanding of volcanism. For example, when news arrived from the Reykjanes Peninsula about a fissure eruption that had produced enough lava to overrun the farm of a pagan priest, those present asked whether the Gods, "enraged" at the possible adoption of Christianity, were exacting punishment for this meeting. A pagan priest named Snorri pointedly remarked: "At what were the gods enraged when the lava on which we are standing formed [. . .]?"[80] Given that the settlers came from lands unstirred by volcanism – Scandinavia, Britain, and Ireland – it is impressive how quickly they came to understand their new environment.[81]

In 1104, for the first time since settlement, Hekla erupted (VEI 5).[82] According to geologist Sigurður Þórarinsson, the *Liber Miraculorum*, the Book of Wonders, contains the oldest known reference to this eruption. Chaplain Herbert of Clairvaux authored this text in 1180. He was a man of questionable geographical skills: he mistook Iceland for Sicily and called it "Hell's chimney."[83] Indeed, historian Terry Lacy suggests that the English expression "go to heck," meaning "go to hell," derives from Hekla's name.[84] The 1104 eruption produced extreme tephra fallout called Hekla 1 (H1) tephra. This tephra layer covered an area of 55,000 square kilometers (half of Iceland). Within

---

77 Hjálmarsson 1988: 13, 29, 45–51; Lacy 1998: 34–42, 80–81; Karlsson 2000b: 66–71; Kristjánsson 2007. Iceland was interchangeably called a province, a dependency, and a colony; Hálfdanarson 2014: 42.
78 Wallace 1991; Karlsson 2000b: 28–32; Oslund 2011: 12. For more information on the discovery of Greenland and its development during the Medieval Climate Anomaly and the Little Ice Age, see also Barlow et al. 1997; Lacy 1998: 123–133; Fitzhugh, Ward 2000; Ogilvie et al. 2000; Barrett 2003; Dugmore et al. 2009b; Lasher, Axford 2019.
79 Lacy 1998: 89–90, 100–101; Karlsson 2000b: 16–19, 33–43; Steingrímsson 2002: 2.
80 Thordarson 2010, 289.
81 Karlsson 2000b: 38–43; Vésteinsson 2000; Steingrímsson 2002: 2, 357; Thordarson 2010: 293.
82 Hjálmarsson 1988: 35–40; Thordarson, Larsen 2007: 141. According to the Global Volcanism Program, the eruption took place on 15 October 1104 ± 45 days; Global Volcanism Program: Hekla.
83 Þórarinsson 1956a: 5–6; Krieger 2019.
84 Lacy 1998: 108.

70 kilometers of the volcano, the tephra fall was more than 25 centimeters thick and resulted in the total abandonment of the farms in the area.[85] Tephra can fertilize the land, stimulating growth; in this case, however, too much of a good thing led to damaged vegetation, soil erosion, and barren surfaces that required decades or even centuries to recover. Fluorine is a common component in tephra's chemical make-up; this element can poison bodies of water and grazing lands, destroy crops and livestock, and pose a significant hazard for a community depending on pastoralism.[86] In 1693, another Hekla eruption caused ashfall and, with it, the death of trout populations in lakes 110 kilometers away. Sheep, cattle, and horses were also affected, particularly by dental lesions called "ash teeth."[87]

Until the thirteenth century, mentions of volcanic eruptions in Icelandic sources were brief, mostly one-line descriptions, such as "1158: second fire in Hekla." The chroniclers only mentioned eruptions if they impacted their immediate environment or the economy.[88] Several sources from mainland Europe (most of which were written some time after the fact) reported natural phenomena that might have been related to Icelandic eruptions, such as hazes, red sunsets, and famines.[89]

In the thirteenth century, a few prominent families gained most of the political and economic power in Iceland, which disrupted the previous balance. In 1262, the Icelanders accepted the influential Norwegian king as their leader to restore peace on the island. In a personal union with the king of Norway, the Icelanders agreed to pay taxes but were allowed to maintain control of their laws. Iceland remained under the thumb of a foreign power until 1944.[90]

Between 1210 and 1240, the Reykjanes Peninsula saw a series of lava flows overrun farms close to fissure systems.[91] Sometime before the thirteenth century, another flood basalt event called Frambruni, part of the Bárðarbunga volcanic system in the Eastern Volcanic Zone, produced around four cubic kilometers of lava.[92] Then, in 1262, Katla saw a remarkably strong (VEI 5) eruption; in 1300, another VEI 4 eruption occurred at Hekla.[93]

---

85 ÁGÚSTDÓTTIR 2015: 1674. Farmland is recoverable if it is under a tephra layer of 25 centimeters or less.
86 BYOCK 2001: 58–59; GUÐMUNDSSON et al. 2008; DUGMORE, VÉSTEINSSON 2012: 73–75; ÁGÚSTDÓTTIR 2015: 1673–1675.
87 PÉTURSSON, PÁLSSON, GEORGSSON 1984: 96.
88 DUGMORE, VÉSTEINSSON 2012: 67–68 (quote). FALK (2007) discusses the notable absence of volcanoes and volcanic eruptions in the Icelandic sagas, which otherwise is a genre praised for its realism.
89 WOZNIAK 2017: 736–738; OPPENHEIMER et al. 2018; EBERT 2019; EBERT 2021.
90 HJÁLMARSSON 1988: 52–60; LACY 1998: 139–148; KARLSSON 2000b: 72–95; STEINGRÍMSSON 2002: 3; OSLUND 2011: 12; HÁLFDANARSSON 2014: 40–44.
91 CLIFTON, KATTENHORN 2006; KEIDING et al. 2008; DUGMORE, VÉSTEINSSON 2012: 71–72; Catalogue of Icelandic Volcanoes: Reykjanes.
92 THORDARSON, LARSEN 2007: 140; DUGMORE, VÉSTEINSSON 2012: 71–72; Global Volcanism Program: Bárðarbunga.
93 Global Volcanism Program: Katla.

The mid-to-late thirteenth century also marked the beginning of the Little Ice Age in Iceland.[94] It was noticeable earlier in this part of the world than elsewhere: harsher weather, rougher seas, and sea ice around the coast of southern Iceland made shipping in the North Atlantic difficult and fishing even more dangerous.[95] This climatic change also impacted the Icelanders' diet.[96] They took to raising livestock instead of growing crops because, in emergencies, animals could sometimes be relied upon to fend for themselves. The flexibility of subsistence farming and the use of wild resources allowed the population to deal with setbacks brought on by cold winters or volcanic hazards; this was crucial, as they were almost always left alone to deal with the consequences.[97] In Iceland, the Little Ice Age lasted until the 1920s.[98]

In 1362, Öræfajökull, a 500-meter-deep caldera full of ice, sprang to life. It would be the most explosive eruption on the island since its settlement, reaching a VEI level of 6.[99] Lasting roughly from June to October 1362, it covered the surrounding meadows and fields with a thick layer of tephra and terrorized those nearby with glacial outburst floods. These torrential outbursts are a major threat during subglacial eruptions and are internationally referred to by their Icelandic name, *jökulhlaups*. In Iceland, they occur mainly at Mýrdalsjökull and Vatnajökull and can reach peak discharge rates of 300,000 cubic meters per second, flooding areas up to 400 square kilometers.[100] They contain debris such as ice blocks and other volcanic ejecta.[101] After the Öræfajökull eruption, at a distance of 15 kilometers from the volcano, the tephra was more than one meter thick.[102] The eruption also produced pyroclastic surges, which consumed 30 farms near the volcano.[103] In Iceland, this phenomenon is not a common occurrence; however, the 1362 eruption at Öræfajökull and the 1875 Askja eruption serve as reminders that Iceland could see pyroclastic surges again in the future.[104]

In 1380, Oluf II of Denmark inherited the Kingdom of Norway and all its territories, which included Iceland, when his father, Haakon VI of Norway, died; thus, Iceland became a Danish dependency.[105] From 1380 onward, Denmark ruled Iceland

---

94 For a debate on the length of the Little Ice Age, see DEGROOT 2018a: 2.
95 HJÁLMARSSON 1988: 12–13; OGILVIE, JÓNSSON 2000: 11–12.
96 RÖGNAVALDARDÓTTIR 2002: 1–2.
97 GUÐMUNDSSON et al. 2008: 263; DUGMORE, VÉSTEINSSON 2012: 71.
98 LACY 1998: 24.
99 Global Volcanism Program: Öræfajökull; here, the eruption is said to have reached VEI 5.
100 THORDARSON, HÖSKULDSSON 2008: 217; DUGMORE, VÉSTEINSSON 2012: 72.
101 THORDARSON, LARSEN 2007: 134.
102 GUÐMUNDSSON et al. 2008: 256.
103 ÞÓRARINSSON 1958; BYOCK 2001: 61–62; GUÐMUNDSSON 2005: 140–141; GUÐMUNDSSON et al. 2008: 259; DUGMORE, VÉSTEINSSON 2012: 71.
104 THORDARSON, LARSEN 2007: 133; GUÐMUNDSSON et al. 2008: 257–259.
105 LAURING 1960: 105–109; HJÁLMARSSON 1988: 64–68; LACY 1998: 166–177; KARLSSON 2000b: 2–3, 100–105; STEINGRÍMSSON 2002: 3; OSLUND 2011: 12.

from its central administration in Copenhagen.[106] Iceland was twice the size (103,000 square kilometers) of Denmark proper (42,933 square kilometers) and far away, which initially allowed for a certain amount of autonomy.[107]

In the first few years of the fifteenth century, the Black Death reached Iceland: one-third of the population perished from the disease. Surrendering to the plague and a worsening climate, some Icelanders gave up their farms and sold them to other landowners or the Church.[108] Iceland was not spared strong volcanic eruptions in the fifteenth century: in 1477, an unusually explosive eruption (VEI 5 to 6) occurred in the Icelandic highlands at Veiðivötn, producing ten cubic kilometers of tephra and small lava flows. Intermittent epidemics followed this, which thinned out the population yet more, killing around one-quarter of the hardened inhabitants.[109] Hekla erupted again in 1510 and caused yet more devastation.[110]

In 1523, the Kalmar Union, which had joined the kingdoms of Denmark, Sweden, and Norway under a single monarch, came to an end. Consequently, Iceland came under the direct rule of the Danish Crown.[111] In the same year, Denmark became Protestant, as did Iceland in 1550. The Protestant Reformation increased the Danish king's authority and landholdings as it allowed him to seize all the land formerly owned by Catholic monasteries.[112] From the sixteenth century onward, Lutheran Mass was conducted in Icelandic rather than Latin, and preaching and teaching became more important; however, much religious terminology was left unchanged.[113]

In 1602, the Danish king, Christian IV, gave privileges regarding all Icelandic trade to approximately 25 Danish merchants. This marked the beginning of the Danish trade monopoly, which lasted until 1787 and prevented the Hanseatic Trading League from getting a foothold in the Icelandic market. Each privileged merchant was allowed to send one or two ships to a certain trading post.[114] Usually, the ships left Denmark in the spring, arrived in Iceland in June or July, traded throughout summer, and returned to

---

106 OGILVIE 2005: 272–274.

107 ROBERTSDÓTTIR 2008: 40–41.

108 KARLSSON 1996; KARLSSON 2000b: 111–117; BYOCK 2001: 353; HUFTHAMMER, WALLØE 2013; CALLOW, EVANS 2016.

109 LARSEN 1984; ZIELINSKI et al. 1997; THORDARSON, LARSEN 2007: 133; THORDARSON, HÖSKULDSSON 2008: 219–220; STREETER, DUGMORE, VÉSTEINSSON 2012: 3664–3665. The Global Volcanism Program estimates that this eruption reached a VEI 6; Global Volcanism Program: Bárðarbunga.

110 Global Volcanism Program: Hekla. For a discussion of the climate in Iceland from 1500 onward, see OGILVIE 1991; OGILVIE 1992. Another eruption occurred at Hekla in 1597.

111 LACY 1998: 106–107; OSLUND 2011: 12.

112 HJÁLMARSSON 1988: 70–77; HASTRUP 1990: 37; KARLSSON 2000b: 128–137. In Denmark itself, the Reformation was relatively uneventful compared to what took place in other European countries; LAURING 1960: 141–144.

113 Today, "an ordained minister of the Icelandic Lutheran Church is called a *prestur* (priest in English), and the regular liturgical church service is called a *messa* (mass in English)." STEINGRÍMSSON 2002: 4, 357.

114 KARLSSON 2000b: 123–127; STEINGRÍMSSON 2002: 7; OSLUND 2011: 36–38.

Denmark in autumn. This seasonal rhythm was necessary because the journey to Iceland was all but impossible in the winter.[115] To maintain social stability, the merchants were prohibited from staying in Iceland during the winter or employing Icelanders.[116] These arrangements were disadvantageous to the Icelanders, to say the least: the Danes were now the only way Icelanders could get supplies and goods from outside the island. The privileged merchants could name their price and sell the Icelanders rotten goods, ensuring a profit.[117] Their trade was mainly made by barter, as very little money was in circulation on the island.[118]

In 1625, another eruption at Katla took its place in the litany of Icelandic disasters.[119] A day-to-day account of this eruption and its consequences survives until the present: the author of this was Þorsteinn MAGNÚSSON. His notes inspired others to document large eruptions in greater detail; publications of this genre became what is known as *eldrit* ("books of fire"). Those who authored the texts were usually government or Church representatives.[120] Sources became more plentiful from 1600 onward, allowing for more detailed climate reconstructions.[121]

Before the Laki Eruption

Iceland would suffer yet more hardships throughout the eighteenth century. In periods of plummeting temperatures, the Icelandic people could count on volcanic activity to worsen their already precarious situation. With two decades of reasonably mild weather in the 1760s and 1770s, the worst weather of the century came in the 1780s.[122] A colder climate and wildly unpredictable weather patterns made surviving difficult. The population fell victim to malnutrition and disease: life became a case of adapt or perish.

Denmark kept Iceland as a dependency for reasons of prestige rather than profit. The revenue from Iceland for the Danish king was small, and he was frequently called upon to assist Icelanders in times of crisis. In 1702, the Danish king, Frederick IV, began exploring ways to improve the Icelanders' situation. This endeavor was not as philanthropic as it may have seemed, for the king was motivated by the notion that

---

115 BYOCK 2001: 266.
116 ANDRÉSSON 1984: 232; GUSTAFSSON 1994; KARLSSON 2000b: 138–142; ROBERTSDÓTTIR 2008: 41–42.
117 HJÁLMARSSON 1988: 77–78; GUNNARSSON 1983: 11–12.
118 STEINGRÍMSSON 2002: 7; AGNARSDÓTTIR 2013: 13.
119 GUÐMUNDSSON 2005: 140.
120 THORDARSON 2010: 289, 293.
121 The data available shows that the Medieval Climate Anomaly was not uniformly mild, and the Little Ice Age was not uniformly cold. Reality, as it often is, was more complex. OGILVIE 1986; LUTERBACHER 2001: 30; OGILVIE 2005: 283–286; DEGROOT 2018a: 2, 32–37.
122 OGILVIE 1986: 67, 72; OGILVIE 2005: 283–284.

each part of the Danish realm should do its part to contribute to the well-being of the entire kingdom.[123]

A general census in 1703 revealed that Iceland had a population of 50,358.[124] In the eighteenth century, almost all Icelanders lived on farms in rural areas. It is anachronistic to describe these Icelandic settlements as villages or towns, the lack of which was the most notable difference between Iceland and western Europe at the time. There were no chartered towns in Iceland until 1786. The most significant settlements were the two bishoprics: Skálholt (85 people) in the south and Hólar (93 people) in the north. Reykjavík was a mere farmstead, just like a thousand others across the country.[125] Industry, such as tanning, wool spinning, knitting, and rope making, slowly emerged to complement farming and fishing during the eighteenth century. These crafts were usually carried out on farms when animal husbandry was not the sole focus, particularly during winter.[126]

As most of the farmers (roughly 96 percent) did not own the land they worked on, they frequently fell into poverty. Tenant farmers cared for the livestock they rented with the land. There was no insurance: if an epizootic disease or a volcanic eruption with poisoning ashfall struck, the farmer was liable for the cost of replacing the animals.[127] The Danish Crown and the Church owned almost half the land (48 percent), and private landowners, who made up around four percent of the population, held the remaining 52 percent.[128] In Iceland, there were around 180 parishes; most had one priest and a congregation of fewer than 500 people.[129]

Apart from a small elite on the island, Icelanders were relatively equal. Even members of this small elite – officials and clergy, such as district governors, sheriffs, bishops, and priests – farmed to provide for themselves and their families.[130] Thus, they also depended on good weather during the growing season to produce a good harvest. Volcanoes did not discriminate; they could affect all parts of Iceland and all

---

123 ROBERTSDÓTTIR 2008: 37–38, 368; OSLUND 2011: 13.

124 The National Archives of Iceland, Census Database: The Icelandic Census of 1703. The number of people in Iceland in 1703, as given on the manntal.is website, is 50,958, which is slightly higher than the above-mentioned figure. However, at least 400 inhabitants were counted twice, as people were moving around in the six-month period during which the census was taken. On the matter of population development, see also: KARLSSON 2000b: 161–168; STEINGRÍMSSON 2002: 358; OSLUND 2011: 22.

125 KARLSSON 2000b: 182–185; STEINGRÍMSSON 2002: 335–336; OSLUND 2011: 64; AGNARSDÓTTIR 2013: 13–16, 31. In the aftermath of the 1784 earthquake, Skálholt's population was relocated to Reykjavík; LUCAS 2009: 76–77; WIENERS 2020: 10. Icelandic towns only started to grow by the end of the nineteenth century; KARLSSON 2000b: 248–249. By 1920, 31 percent of the population lived in towns; GUNNLAUGSSON 1988; BYOCK 2001; AGNARSDÓTTIR 2012.

126 OSLUND 2011: 77.

127 VASEY 1991: 325; STEINGRÍMSSON 2002: 8.

128 ROBERTSDÓTTIR 2008: 35; AGNARSDÓTTIR 2013: 12.

129 MAGNÚSSON 2010: 33–34.

130 AGNARSDÓTTIR 2013: 12.

levels of society.[131] The threat that volcanic eruptions or adverse weather conditions posed to the harvest and livestock was the Icelanders' constant company. Unfavorable terms of trade imposed by the Danish Crown, coupled with the geographic distance from Denmark, a significant source of aid, created political and social vulnerability on the island.

In the 1680s, the office of the governor of Iceland (*stiftamtmaður*) was established. This office was the highest representative of the Danish king in Iceland for almost two centuries. Until 1752, the governors were only Danish; after that date, Icelanders, too, could serve in this role.[132] Beneath the governor, there were four district governors, called *amtmenn*, one for each of the four districts of Iceland (north, east, south, and west) (Figure 13).[133] Iceland was further divided into 23 counties called *sýslur* (singular, *sýsla*) (Figure 14), which, from the seventeenth to the nineteenth century, were

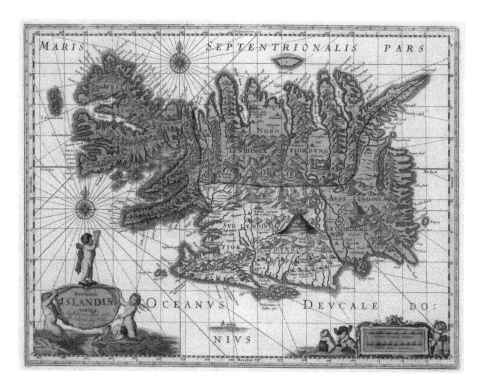

Figure 13: A map of Iceland, ca. 1700. This map by Peter SCHENK and Gerard VALK is titled *"Novissima Islandiæ Tabula,"* and it shows the four districts of Iceland in different colors. Hekla is depicted as erupting.

131 THORDARSON 2010: 293.
132 STEINGRÍMSSON 2002: 340; OGILVIE 2005: 272–274.
133 MAGNÚSSON 2010: 33–34.

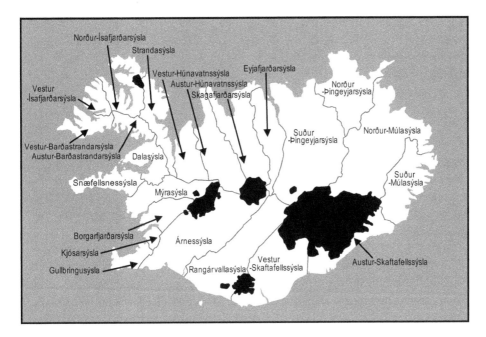

Figure 14: A map of Iceland's counties. Vestur-Skaftafellssýsla is in the south, near Vatnajökull, Iceland's largest glacier.

administered by sheriffs, local officials called *sýslumenn*.[134] Each *sýsla* was subdivided into approximately eight to ten regions, called *hreppar* (singular, *hreppur*). A *hreppur* was run by an unsalaried official, often an influential local farmer, aided by a regional committee consisting of other farmers and the parish priest; together, they were responsible for offering relief to people in poverty, which included relocating paupers to farms within the *hreppur* to prevent vagrancy.[135]

Death and disease spread throughout Iceland in the country's pre-modern period. The living conditions of most Icelanders, although not entirely to blame, did not help. Their houses, built from turf and stone for lack of timber, were ramshackle buildings that leaked and were poorly ventilated. These factors, together with the inevitable dampness and cold, left people susceptible to disease.[136] An epidemic of smallpox occurred between 1707 and 1709, which coincided with a famine; these events wiped out over a quarter of the population, leaving only 37,000 souls on the island.[137] A strong

---

134 OGILVIE, JÓNSDÓTTIR 2000; OGILVIE, JÓNSSON 2000; OGILVIE 2005: 272–274; AGNARSDÓTTIR 2013: 12.
135 STEINGRÍMSSON 2002: 7; MAGNÚSSON 2010: 33–34.
136 STEINGRÍMSSON 2002: 8; MAGNÚSSON 2010: 48–58.
137 HJÁLMARSSON 1988: 87–88; VASEY 1991: 346; KARLSSON 2000b: 177.

eruption (VEI 5) occurred at Katla from May to October 1721.[138] From 1724 to 1729, the Mývatn Fires in northern Iceland produced a lava flow that buried two farms and several farmhouses.[139] Öræfajökull also erupted from 1727 to 1728 and caused a *jökulhlaup*, killing three people.[140]

In the eighteenth century, the level of education in Iceland was relatively high compared to elsewhere in Europe. There were no elementary schools. Instead, the parents and the clergy were responsible for a child's education.[141] In the 1740s, around 50 percent of children were "at least minimally literate," and by the 1790s, this number increased to approximately 90 percent.[142] Industrious Icelanders traveled to Europe and beyond during the eighteenth century; some went to Copenhagen to study at the university, while others moved further afield to learn trades like spinning or weaving.[143] New ideas were brought back to Iceland by letter and return voyage. Icelanders were not as culturally isolated as they are often portrayed.[144]

The 1750s saw sea ice travel as far south as Vestmannaeyjar, a group of islands south of the Icelandic mainland. It made fishing all but impossible and was the first in a series of misfortunes that transpired during this decade. Bouts of scabies and lung disease were commonplace.[145] From 17 October 1755 until 13 February 1756, Katla saw another strong (VEI 5) eruption, which caused one direct and two indirect fatalities due to lightning at a farm 30 kilometers away. In addition, it caused severe damage from tephra fall and a *jökulhlaup* in the south.[146]

Farmers imported sheep to compensate for the loss of livestock; however, these sheep were highly susceptible to disease and infected the native animals, further depleting the Icelandic reserve.[147] A famine inevitably followed, peaking in the spring of 1757.[148] After this calamity, Iceland's population stood at 43,000.[149]

The Danish central administration introduced reindeer from Finnmark in an effort to combat the almost constant threat of famine. In 1771, the first shipment of 13 animals arrived; only about half survived the first year.[150] Thus, further shipments of reindeer followed in 1777, 1784, and 1787. In Iceland, they were allowed to live in the wilderness. Soon this invasion was met with resistance, as the reindeer fed on the

138 Global Volcanism Program: Katla.
139 GUÐMUNDSSON et al. 2008: 259; DUGMORE, VÉSTEINSSON 2012: 71–72.
140 Global Volcanism Program: Öræfajökull; GUÐMUNDSSON et al. 2008: 261.
141 MAGNÚSSON 2010: 88–89.
142 STEINGRÍMSSON 2002: 5, 326.
143 AGNARSDÓTTIR 2013: 25.
144 MAGNÚSSON 2010: 69–70.
145 OGILVIE 1986: 65, 72; KARLSSON 2000b: 177; OSLUND 2011: 71; DUGMORE, VÉSTEINSSON 2012: 71.
146 Global Volcanism Program: Katla; GUÐMUNDSSON et al. 2008: 257.
147 OSLUND 2011: 69.
148 VASEY 1991: 346.
149 KARLSSON 2000b: 177; DUGMORE, VÉSTEINSSON 2012: 71.
150 OSLUND 2011: 69–71.

lichen the Icelanders used and competed for pastureland with the surviving sheep. Farmers complained about the reindeer and requested permission to hunt them, which Copenhagen granted in 1798. Today, about 7,000 reindeer live in the wild in Iceland, descendants of that initial population.[151] In a nutshell, the plan to use reindeer to supplement animal husbandry and reduce farmers' dependency on sheep had failed.

In 1766, Christian VII became the king of Denmark. He would reign until his death in 1808. He was, however, only the nominal king, as he suffered from mental illness. Thus, from 1772 to 1784, his half-brother, Ove Høegh-Guldberg, served as regent and de facto prime minister. From 1784 onward, Christian VII's son (who would later be king), Frederick VI, served as the unofficial regent.[152] After the epizootics and famine of the 1750s and 1760s, Christian VII sent the first royal land commission (*landsnefndin fyrri*), consisting of three men, to Iceland in 1770 to gather information and (among other things) research ways to bolster the Icelandic economy.[153] The first royal land commission concluded that the office of the governor should be in Iceland rather than in Copenhagen and that the official ranks should consist of more Icelanders. Skúli Magnússon (1711–1794) became the first Icelandic treasurer. His task was to make the Danish merchants treat the Icelandic people better by insisting on fair prices and sufficiently fresh products.[154] Despite the criticisms leveled against it by the island's people, the Danish trade monopoly remained in place.[155]

By June 1783, Icelanders had had a wealth of experience with volcanic eruptions acquired over several generations. Thus, the economy could deal with sudden disasters such as bad weather and, importantly, could adapt to the effects of volcanic eruptions.[156] Historian Gunnar Karlsson believes that prior to the Laki eruption, Iceland had recovered from the famines of the 1750s and 1760s and had a population of around 50,000.[157]

Geoscientist Andrew Dugmore and archaeologist Orri Vésteinsson argue that every Icelandic volcanic eruption's impact must be judged on a case-by-case basis within its own historical and geographical context as the eruptions display great diversity in type and magnitude. Important factors to consider include the remoteness and strength of an eruption, the landscape around it, its temporal proximity to another large eruption, and the health of the human and animal populations. Usually,

---

151 Oslund 2011: 71–73; "Reindeer Warning in East Iceland." Iceland Review, 6 January 2018.
152 Lauring 1960: 178–188.
153 Lacy 1998: 188; Agnarsdóttir 2013: 14–15.
154 Hjálmarsson 1988: 89–91; Agnarsdóttir 2013: 25.
155 Oslund 2011: 36–38, note on 181.
156 Agnarsdóttir 2013: 13–14.
157 Karlsson 2000b: 178.

major eruptions only cause devastation if other unlucky circumstances exacerbate their negative impacts. Throughout their history, the Icelandic population and economy repeatedly recovered, showing tremendous fortitude.[158]

Given their history, one could argue that the Icelandic population could have been better prepared for the events of 1783 and their consequences; perhaps the Icelanders could have stored more non-perishable food and seeds for the seemingly inevitable unlucky times that they frequently suffered. However, Iceland was a poor country with extremely limited resources whose population lived from season to season.[159] The agricultural output had always been stifled by the country's high latitude, cold weather, and eroded soils, and by 1783 livestock numbers had been greatly thinned out by various epizootics.[160] Furthermore, the recurrence period of strong volcanic eruptions is quite long: approximately 30 to 50 years. Large flood basalt events are rarer still, which makes planning for them difficult.[161] To substantiate these claims, the following subchapter will analyze descriptions of the eruption, the complex factors that came together in 1783, and the eruption's aftermath.

The Laki Eruption

Kirkjubæjarklaustur and the Fire Districts

The Laki eruption took place in the remote highlands of south-central Iceland, above a rift zone between Mýrdalsjökull and Vatnajökull, also known as the Fire Districts.[162] Several large fissure swarms cut through this area.[163] These unique ridges were formed by subglacial eruptions during the last glaciation. Although the individual ridges are of different ages, they run in the same southwestern-northeastern direction, parallel to one another, as they formed on the spreading axis, the subaerial part of the Mid-Atlantic Ridge (Figure 15).[164]

Kirkjubæjarklaustur is a village on the southern coast of Iceland, nestled at the foot of a steep, 200-meter-tall cliff in the Síða region, not far from the *sandur* plains and just west of Vatnajökull.[165] Glacial rivers, such as the Skaftá and Hverfisfljót,

---

158 DUGMORE, VÉSTEINSSON 2012: 68–69.
159 MCCALLAM 2019: 217–218.
160 ANDRÉSSON 1984: 232; HÁLFDANARSSON 1984: 162.
161 HJÁLMARSSON 1988: 93; DUGMORE, VÉSTEINSSON 2012: 76.
162 THORDARSON, HÖSKULDSSON 2014: 128–131; WITZE, KANIPE 2014: 59–60.
163 GUÐMUNDSSON et al. 2008: 265.
164 Katla Geopark Project: 12.
165 THORDARSON et al. 2003b: 13.

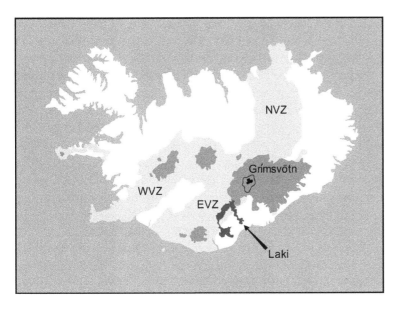

Figure 15: The location of the glaciers (light gray), the spreading axis (gray), and the location of the Laki lava (dark gray).

formed the *sandur* plains, which are covered by lava fields from other Holocene erup-tions.[166] The Skaftá River is named after Skaftárjökull, part of Vatnajökull, its source. From there, it flows down from the highlands and into the Atlantic Ocean, not too far from Kirkjubæjarklaustur.[167]

The cliff face north of Kirkjubæjarklaustur marks the beginning of the highlands. Kirkjubæjarklaustur, is often abbreviated as "Klaustur," which means "church farm cloister," and refers to a Benedictine convent that existed there from 1186 until the Reformation.[168] In 1783, Klaustur consisted of a church and a farmstead. The Síða re-gion is in Vestur-Skaftafellssýsla ("the county to the west of [the mountain] Skafta-fell"). Vestur-Skaftafellssýsla also encompasses Mýrdalsjökull and the western parts of Vatnajökull. The Síða region is one of the six parts of the Fire Districts.[169] It was com-parable to most Icelandic regions at the time: its inhabitants lived on farmsteads scat-tered throughout the surrounding area. Grímsvötn, situated around 70 kilometers from Klaustur and covered in a thick sheet of ice, is Iceland's most active volcanic system: it erupts every two to seven years.

---

166 THORDARSON, SELF 1993: 234.
167 ÞÓRARINSSON 1984: 35.
168 HERRMANN 1907: 105–106; BYOCK 2001: 340.
169 The other Fire Districts are Álftaver, Meðalland, Skaftártunga, Landbrot, and Fljótshverfi; THOR-DARSON et al. 2003b.

The danger posed by a flood basalt event was probably the last thing on anyone's mind in the spring of 1783 because it had been eight centuries since the last one had occurred in that area. Vestur-Skaftafellssýsla had around 2,200 inhabitants in early 1783.[170] Klaustur and its surrounding farmsteads, with their 613 inhabitants living only 32 kilometers from the Laki fissure, would take center stage in the unfolding drama. The region's best-known resident at the time was Jón STEINGRÍMSSON (1728–1791), the local priest. Like everyone else, STEINGRÍMSSON farmed to provide for his family.[171] He was married to Þórunn HANNESDÓTTIR (1718–1784), and together they had five daughters. The family lived in nearby Prestbakki, five kilometers north of Klaustur.

STEINGRÍMSSON was born in 1728 and was originally from northern Iceland. He studied at the diocesan school in Hólar. After his schooling, he took a position as a church deacon in Reynistaður. There, he fell in love with Þórunn, whose abusive husband died under dubious circumstances shortly after STEINGRÍMSSON's arrival. In 1753, they married and had a child soon after. In Reynistaður, in addition to his priestly duties, STEINGRÍMSSON served as a self-educated physician to his parishioners.[172] When the rectory for the parish of Kirkjubæjarklaustur fell vacant in 1778, STEINGRÍMSSON successfully applied for the position. The family moved to Prestbakki; their first few years in their new home were happy and prosperous.[173]

By the late eighteenth century, Icelanders had a general understanding of volcanic eruptions and their nature. STEINGRÍMSSON himself was aware of at least two other eruptions prior to Laki: the Mývatn Fires of 1724 to 1729 – documented by another priest by the name of Jón SÆMUNDSSON in a text STEINGRÍMSSON was familiar with – and the Katla eruption of 1755, an event that he himself had witnessed.[174] Jón STEINGRÍMSSON authored both an autobiography and a fire treatise (*eldrit*). His was one of the first Icelandic autobiographies. He wrote the first part of it during his first few years in Prestbakki before 1783 and the rest after the Laki eruption. Most of it was written in 1788, only a few years before his death in 1791. STEINGRÍMSSON never intended to publish this text, having written it exclusively for his daughters and their children. The fact that the autobiography was supposed to remain private allows for the assumption that STEINGRÍMSSON wrote honestly and openly about his feelings, fears, and struggles.[175] His *eldrit*, on the other hand, was intended for publication so that all Icelanders "born and yet unborn" could remember the Laki eruption.[176]

---

170 GUÐBERGSSON, THEODÓRSSON 1984: 112–115.
171 STEINGRÍMSSON 1998: 80; STEINGRÍMSSON 2002: 10.
172 STEINGRÍMSSON 1998: 7; STEINGRÍMSSON 2002: xvi, 10–15.
173 STEINGRÍMSSON 1998: 19; STEINGRÍMSSON 2002: 11–15, 178.
174 THORDARSON 2003: 2.
175 STEINGRÍMSSON 2002: 1–2, 18–22, 359. His sister's son, Steingrímur JÓNSSON, the bishop of Iceland between 1824 and 1845, received the manuscript and liked it so much that he decided it had to be published. For this reason he kept it safe and, in 1916, the full autobiography was published posthumously for the first time.
176 STEINGRÍMSSON 1998: 15. This text was originally published in 1788.

The two texts that Steingrímsson penned are the reason we know the exact start and end dates of the eruption and of the many incidents in between. Steingrímsson wrote daily for the first three months of the eruption; after this, he wrote less frequently and gave fewer details.[177] His writings provide descriptions of the events and illustrate how he and others interpreted them.[178] He distinguishes between his direct observations and his theological interpretations of the events. Steingrímsson confronted several phenomena that he had never encountered and, in all likelihood, never heard of before. Yet, he still found the language to describe them, allowing the modern reader to visualize what happened.[179] Examples include his descriptions of Pèle's hair, which he calls "volcanic hair," and passages that detail tephrochronology.[180] He was aware that several ash layers make up the soil, some thicker than others.[181] Steingrímsson showed great interest in observing nature and his surroundings. The values of the Enlightenment lived in Steingrímsson and throughout Iceland, just as in the rest of Europe.[182]

There are ten known *eldrit* that describe the fire columns of the Laki eruption.[183] Steingrímsson's is by far the most thorough. Icelandic volcanologist Thordarson has studied all ten and, in 2003, used them to analyze the seismic activity, explosivity, lava surges, and eruptive episodes of the eruption. Thordarson argues that Steingrímsson decided to write such a detailed account because he knew a description of the course of events and a list of damages would help when Icelanders negotiated disaster assistance relief with the Danish central administration.[184]

The Fire Priest

1783 had started promisingly in the south of Iceland, in contrast to the rest of the country, with the winter and spring being reasonably mild.[185] The previous year had been quite cold across the island.[186] After a few arduous seasons, a good harvest and

---

177 Thordarson 2003: 2.
178 Steingrímsson 2002; Thordarson, Höskuldsson 2014: 134.
179 Steingrímsson 1998: 7.
180 For Péle's hair, see Steingrímsson 1998: 27.
181 Thordarson 2003: 1–2.
182 Steingrímsson 2002; see also Demarée, Ogilvie 2001: 223; Witze, Kanipe 2014: 25, 106.
183 For a detailed list of all the *eldrit* written about the Laki eruption, see table 1 in Thordarson 2003: 3.
184 Thordarson 2003: 1, 4, 10; Thordarson et al. 2003b. This almost daily record of the eruption has greatly helped the modern understanding of flood basalt events and large igneous provinces.
185 Ogilvie 1986: 63: The winter of 1782/1783 was severe in the north and east, reasonable in the south, and cold in the west. The spring of 1783 was variable in the north, mild in the south, cold in the east, and reasonable in the west. Astrid Ogilvie (1992; 2005) has done much work on weather and climate history in Iceland. For this period, she used several Icelandic weather diaries.
186 Ogilvie 1986: 67–69.

enough fodder for the animals in the coming winter were essential. In mid-May 1783, the ominous occurrences began: the residents in the Síða region noticed many weak earthquakes.[187] By 1 June 1783, the earthquakes increased in number and size. Their intensity became so strong that at some farmsteads in Vestur-Skaftafellssýsla, people slept outside in tents because they feared their houses might collapse. The stronger earthquakes in that sequence were perceptible in other parts of southern Iceland, from Mýrdalur in the west to Öræfi in the east; the latter is some 75 kilometers away from the Laki fissure.[188] This phenomenon is called an earthquake swarm, a precursor to some volcanic eruptions in Iceland.

Residents of the Fire Districts could feel these tremors – which probably reached a magnitude greater than 4.0 on the Richter scale – up to 80 kilometers away from the fissure's eventual location.[189] The earthquakes between May and early June 1783 were the strongest to occur between 1783 and 1785; this was due to pressure in the magma chamber reaching critical stress levels. The subsequent inflation of the magma chamber led to the bending, fracturing, and weakening of the crust above.[190]

On roughly 20 May 1783, a Danish ship called the *Torsken* was skirting the southern Icelandic coast when the sailors noticed fires in the mountains or glaciers north of the Síða region.[191] These fires may have been activity at Grímsvötn, which would indicate volcanic activity prior to June 1783. This was the beginning of the volcano-tectonic episode at the Grímsvötn volcanic system.[192]

We can imagine the likely scene: on that fateful Sunday morning, STEINGRÍMSSON and his parishioners walked to church or rode on horseback. The sun was long up.[193] It was a clear day with blue skies and calm weather. Attending Mass and meeting other parishioners was a welcome break from the toil of farm work.[194] Whitsunday, which in 1783 fell on 8 June, celebrates the descent of the Holy Spirit upon Jesus' disciples. That said, in this case, it will always be remembered as the beginning of Iceland's descent into the worst disaster in its almost 1,150-year history, as this was the day the Laki fissure violently roared to life. At around 9:00 a.m., something unusual caught the parishioners' eyes: dark clouds appeared above the mountains just behind Klaustur and

---

187 The earthquakes were first noticed by a farmer by the name of Jón EIRÍKSSON at Ljótarstadir in the Skaftártunga district; THORDARSON et al. 2003b: 19.

188 ÞÓRARINSSON 1984: 35; THORDARSEN et al. 2003b: 19.

189 THORDARSON et al. 2003b: 27.

190 GUÐMUNDSSON 1989; THORDARSON, SELF 1993: 259.

191 SCARTH 1999: 107. The *Torsken* sailed from Denmark toward Hafnarfjörður, a harbor town around ten kilometers south of Reykjavík.

192 THORDARSON et al. 2003b: 19, 26, and their translations of the earthquake sources titled B1 on 40. The eruption mentioned might refer to Nýey, but it is not clear.

193 Klaustur is located close to the Arctic Circle; on 8 June, the sun rose at 1:30 a.m. and set at 9:56 p.m.; Time and Date website.

194 STEINGRÍMSSON 2002: 10.

steadily filled the sky.[195] The faint sun became a menacing red behind thick clouds of ash, tephra, and gas from an atmospheric eruption plume that reached as high as 15 kilometers and was visible across the length and breadth of Iceland.[196] The tephra fallout from the plume resulted in complete darkness up to 100 kilometers from Grímsvötn.[197] Although sunset was 11 hours away, the ash fall, or "black haze of sand" as STEINGRÍMSSON describes it, "caused darkness indoors."[198] Later the same day, rain fell the color of "black ink."[199] It irritated the eyes and skin.[200] Inhabitants heard thunderous noises from the highlands.[201] When the wind blew the ash clouds away, fires were distinct against the darkness in the distance behind the mountains.[202] Strong earthquakes occurred throughout the night.[203] The Skaftá Fires, *skaftáreldur*, had begun.

> Now comes the beginning of the Lord's chastisements and of the fresh calamities that befell me and others, as I shall now relate. And yet they came upon us with greater forbearance and leniency that we had deserved. On Pentecost, [8 June 1783], an eruption of molten rock took place coming out of the mountains of the highland pasturage, whose effects were to destroy farmland, men, and beasts far and wide.[204]

So writes Jón STEINGRÍMSSON of the onset of the Laki eruption in his autobiography. When he talks of divine forbearance, he is likely referring to the ominous dreams that had plagued him before the eruption; they had warned him of something awful that was about to happen. STEINGRÍMSSON also feared that his parishioners had lived carelessly and wastefully and that this, therefore, might be their just punishment.[205]

From 8 to 11 June 1783, the strong earthquakes continued. STEINGRÍMSSON states that "however much the earth and houses trembled and shook, thunderclaps crashed, and fireballs flew," he felt no fear.[206] On 10 June, the waters of the mighty Skaftá River evaporated, leaving an empty riverbed in their wake.[207] The next day the river flowed with lava instead. On 11 June, amazingly, Klaustur experienced "a snowstorm,

---

195 ÞÓRARINSSON 1984: 35.
196 THORDARSON, SELF 1993: 233. For more information on how to calculate the height of the fire fountains and the eruption columns, see THORDARSON, SELF 1993: 257–258.
197 PÉTURSSON, PÁLSSON, GEORGSSON 1984: 97; LACY 1998: 16; STEINGRÍMSSON 1998: 26–27; THORDARSON, HÖSKULDSSON 2014: 131.
198 STEINGRÍMSSON 1998: 25.
199 ÞÓRARINSSON 1984: 35.
200 PÉTURSSON, PÁLSSON, GEORGSSON 1984: 97; THORDARSON, HÖSKULDSSON 2014: 134.
201 ÞÓRARINSSON 1984: 35. According to the Catalogue of Icelandic Volcanoes: Grímsvötn, the "groundwater levels are high in many parts of the ice-free area [of the Grímsvötn system], making the likelihood of an initial phreatomagmatic eruption likely."
202 ÞÓRARINSSON 1984: 35.
203 STEINGRÍMSSON 1998: 25.
204 STEINGRÍMSSON 2002: 180.
205 STEINGRÍMSSON 1998: 21–22; STEINGRÍMSSON 2002: 178–179.
206 STEINGRÍMSSON 2002: 181.
207 ÞÓRARINSSON 1984: 35; STEINGRÍMSSON 1998: 25.

which came from the black cloud," with the snow cover lasting for five days.[208] Thereafter, the summer of 1783 was cold throughout Iceland, except in the north, where the weather was more variable.[209]

The first fissure segment opened on 8 June. Initially, the eruption was very explosive; this was unusual as fissure eruptions tend to be rather effusive. In total, there were ten eruptive episodes, and with each one, a new fissure segment opened (Figure 16). These fissure segments are *en echelon*, meaning they are arranged diagonally, and consist of 140 craters, vents, and cones.[210]

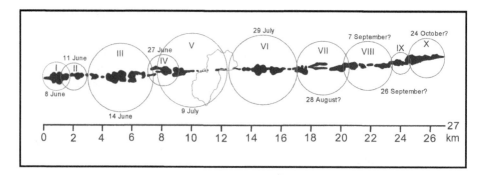

Figure 16: The different segments of the Laki fissure. Each segment was formed during a different eruptive phase, the start date of which is indicated. The southwestern part is on the left, Vatnajökull is on the right, and Mount Laki is almost in the center of the fissure (between segments V and VI).

The eruption began in the Úlfarsdalur valley, near the Hnúta Mountain, and continued in a northeast direction in an almost straight line toward the pre-existing Mount Laki. The mountain divides the fissure into two nearly equal parts. Eventually, the fissure continued past Mount Laki, reaching all the way to the tip of Vatnajökull.[211] The fissure, which may extend further underneath the glacier, measures 27 kilometers over its entire subaerial length. It sits at an altitude of roughly 600 meters above sea level, and its highest craters reach 100 to 120 meters above the surrounding landscape; by contrast, Klaustur is only 35 to 40 meters above sea level.[212]

208 STEINGRÍMSSON 1998: 26.
209 OGILVIE 1986: 63. In the spring of 1784, Sveinn PÁLSSON wrote a short treatise on the effects of the Laki eruption on northern Iceland. It contained some information on southern Iceland based on a letter from a friend, who lived there; for more information on the letter, see THORDARSON et al. 2003b: 20, source C7 therein; GUNNARSDÓTTIR 2022: 80.
210 THORDARSON, SELF 1993; THORDARSON et al. 2003a; PASSMORE et al. 2012: 2594; THORDARSON, HÖSKULDSSON 2014: 131, 136–137.
211 Katla Geopark Project: 10–11.
212 EINARSSON, SVEINSDÓTTIR 1984: 47.

The fissure segments to the southwest of Mount Laki had five eruptive episodes. Each episode began with seismic activity and a short-lived but explosive phase of Strombolian to sub-Plinian type (VEI 1 to 4), each lasting from half a day to four days.[213] The water table was very high at the site, as this area was quite boggy; the interaction of magma with the high groundwater resulted in this initial phreatomagmatic phase of the eruption.[214] These mild explosions created rootless vents and pseudocraters that are still visible today.[215] Once the availability of water dwindled, the next eruption phase began, with lava production becoming more effusive. The lava fountains reached 800 to 1,400 meters above the fissure and were visible from the lowlands around Klaustur. Within a few days, the first lava surge reached the lowlands, having traveled there at a pace of three to four kilometers per day.[216] Luckily, lava fields from previous flood basalt events had hindered its progress.

Each initial eruptive episode precipitated a lava surge that appeared in the gorge of the Skaftá River.[217] "[E]pisodic increase[s] in lava production at the fissures and hence in the magma discharge" were the source of these lava surges.[218] The second fissure segment opened on 11 June, and the third on 14 June 1783.[219] In the lowlands of Vestur-Skaftafellssýsla, the second lava surge came on 16 June; this lava did not stay in the riverbed but overran pastures and farms. STEINGRÍMSSON notes in his autobiography on 15 June 1783: "The uproar to the north of the mountains of the Síða region now grew violently, with such sounds of breaking and crashing, fire and smoke, and earthquakes, that no one knew whether the settlements here were in danger or not."[220] Many were frightened about the prospect of the fires spreading further and causing complete devastation. That same day, an exploratory party of three farmers walked north into the highlands and to Kaldbakur, the highest scalable mountain in the vicinity. When they returned, they told of as many as 27 fires that extended in a southwest direction from Mount Laki. These "fires" were likely lava fountaining from the 27 different vents and cones on the first fissure segments.[221]

The third lava surge came on 18 June 1783, the severity and speed of which forced people to flee; it carried some large rocks "tumbling about like large whales swimming,

---

213 THORDARSON, SELF 2003: 3. PASSMORE et al. (2012: 2595) estimate it lasted half a day to one full day.

214 LARSEN, THORDARSON 1984: 65; THORDARSON, SELF 1993: 233; THORDARSON et al. 2003b: 13. For more information on explosive lava-water interactions at Laki, see also HAMILTON, FAGENTS, THORDARSON 2010.

215 Katla Geopark Project: 12. These pseudocraters usually measure 50 to 60 meters across and are less than ten meters in height. For more research on rootless cones and vents, see also BOREHAM, CASHMAN, RUST 2018.

216 THORDARSON, SELF 1993: 233; THORDARSON 2003: 6; THORDARSON, SELF 2003: 3; PASSMORE et al. 2012: 2596; THORDARSON, HÖSKULDSSON 2014: 131.

217 PASSMORE et al. 2012: 2595.

218 THORDARSON et al. 2003b: 30.

219 PASSMORE et al. 2012: 2615, Figure 14a.

220 STEINGRÍMSSON 1998: 30.

221 ÞÓRARINSSON 1984: 35; STEINGRÍMSSON 1998: 35–36.

red-hot and glowing."[222] Molten lava dammed up former tributaries to the Skaftá River, causing flooding (of water this time).[223] The contact between the lava and these tributaries caused terrifyingly loud noises.[224] STEINGRÍMSSON asserts, "All that day and night, the thunderous crashing was so great that everything shuddered and shook, and the earthquakes made every timber crack and crack again."[225]

The first three fissure segments produced 6.1 cubic kilometers of lava between 8 and 21 June; this was 41.5 percent of the total volume of lava produced during the entire eruption. At its peak, the fissure ejected magma at a rate of 5,000 to 6,600 cubic meters per second. An Olympic-sized swimming pool contains 2,500 cubic meters of water; thus, at this rate, the discharge of the Laki eruption could have filled two to two and a half Olympic-sized swimming pools per second.[226] Throughout the eruption, lava was produced on average at a rate of around 580 cubic meters per second, meaning it would have taken (on average) about 4.3 seconds to fill an Olympic pool.[227]

The fourth fissure segment opened on 25 June 1783; the fourth lava surge in the Síða region followed soon after, on 29 June. The lava flow had already formed three branches, with which it terrorized the lowlands. It continued to burn down farmsteads and churches.[228] The fifth fissure opened on 9 July 1783; shortly after, on 14 July 1783, the fifth lava surge cascaded down the Skaftá Gorge. The lava mainly flowed within the Skaftá riverbed en route to Klaustur. By the time the fissure reached Mount Laki, 61 percent of the total magma volume had been ejected.[229] Utterly terrifying noises had accompanied each surge:

> [. . .] with boiling sounds, cracking and smashing and such quaking underground, as if everything were likely to break apart. [. . .] Flashes of fire were everywhere around us both indoors and out, but no one was killed outright by them. All week long, neither sun nor sky could be seen for the thick clouds of fumes and smoke which blanketed the area.[230]

After the initial surges, the lava continued to creep menacingly forward, seemingly unstoppable. To STEINGRÍMSSON and his parishioners, it now seemed inevitable that it would move through Klaustur and destroy everything in its path, including the church. On 18 July 1783, a Friday, STEINGRÍMSSON "imagined nothing but collapse and destruction." He expected none of his parishioners to sleep soundly in these tumultuous times.

---

222 STEINGRÍMSSON 1998: 33.
223 ÞÓRARINSSON 1984: 36.
224 STEINGRÍMSSON 2002: 15.
225 STEINGRÍMSSON 1998: 33.
226 THORDARSON, SELF 2003: 5.
227 EINARSSON, SVEINSDÓTTIR 1984: 48.
228 ÞÓRARINSSON 1984: 36; THORDARSON, SELF 2003: 5.
229 THORDARSON, SELF (2003: 5) and PASSMORE (2012: 2615) give slightly different dates for the openings of the fissure segments; here, I use the ones that THORDARSON and SELF (2003) gave.
230 STEINGRÍMSSON 1998: 47.

Locals could not drive their animals out of harm's way, so they released them to roam freely instead.[231]

When it was time for worship on Sunday, 20 July 1783, five weeks after Trinity Sunday, the lava was already alarmingly close to the church.[232] As well as the immediate danger posed by the lava, STEINGRÍMSSON and his parishioners were threatened from above by a sky overcast with "hot vapors and fog" accompanied by thunder and lightning, and from below, where they could feel "rumbling and thudding."[233] It must have been a truly terrifying day; even so, the faithful attended church. When they arrived, the air was so full of fumes and smoke that the outline of the church "could only be hazily seen."[234] In his *eldrit*, STEINGRÍMSSON mentions his despair at the thought that this might be the last service held in this church.[235] He notes that the church was "shaking and quaking from the cataclysm that threatened it from upstream" during his sermon.[236] "Claps of thunder were followed by such great flashes of lightning, in series after series, that they lit up the inside of the church and the bells echoed the sound, while the earth tremors continued unabated."[237] The parishioners had shown genuine commitment by attending, and he believed that they, like himself, felt no fear. Thus, he claims that they "were contented and prepared to accept whatever God would send."[238]

With lava approaching, the sermon commenced. When it ended, the congregation stepped outside and realized the lava flow had stopped; to their surprise, it had not moved since the sermon began. They believed that STEINGRÍMSSON's service had stopped the flow of lava.[239] "Instead [of advancing further] it piled itself up in a heap, layer upon layer. In addition, all the lakes and rivers came flooding down upon the heaped-up lava and violently quenched it. To God alone be the glory!"[240]

Because of these events, the church service of 20 July 1783 is known as *eldmessan*, the fire sermon, and STEINGRÍMSSON as *eldprestur* or *eldklerkur*, the fire priest: the man who stopped Klaustur's church from being consumed.[241] The lava surges in the Skaftá riverbed ended with the fire sermon. As the lava had piled up, tributaries of the former Skaftá River had started pouring over it, smothering the fire.[242]

---

231 STEINGRÍMSSON 1998: 48.
232 STEINGRÍMSSON 2002: 182, 380. Jón STEINGRÍMSSON mentions that the fire sermon was held on the fourth Sunday after Trinity. However, he also gave the date of 20 July, which is the fifth Sunday after Trinity. The date 20 July is believed to be the correct date of the fire sermon; GROTEFEND 2007: 203.
233 STEINGRÍMSSON 1998: 48.
234 STEINGRÍMSSON 1998: 49.
235 STEINGRÍMSSON 1998: 48.
236 STEINGRÍMSSON 2002: 182.
237 STEINGRÍMSSON 1998: 49.
238 STEINGRÍMSSON 2002: 182.
239 HENDERSON 1818: 279–286; KARLSSON 2000b: 178; WITZE, KANIPE 2014: 87–88, 153.
240 STEINGRÍMSSON 2002: 16.
241 STEINGRÍMSSON 2002: 1.
242 STEINGRÍMSSON 1998: 50.

Thereafter, the southwestern part of the Laki fissure continued to produce lava and gases until September 1783, although not quite with the same explosivity as it had in early June.[243] STEINGRÍMSSON was certain that the lava would remain visible "until the end of the world. [. . .] It will be for all eternity a source of the greatest wonder, that any living thing should have survived at all here in Síða."[244] Indeed, to this day, the Eldhraun lava field is still visible near Klaustur.

On 29 July 1783, the people in southern Iceland heard more frightening noises, this time from further northeast. These rumblings announced the opening of a new fissure segment, the sixth, located on the northeastern side of Mount Laki. Another dark cloud descended upon Klaustur and, once again, day turned into night.[245] Over the following three months (29 July to 30 October 1783), the sixth to tenth fissure segments opened.[246] The last two fissure segments on the east, the ninth and tenth segments, look slightly different due to erosion caused by the movement of the glacier and the flow of the Skaftá River since the eruption.[247]

On 31 July 1783, a familiar pattern repeated itself. While the Skaftá River runs by the west of Klaustur, another river runs to the east: this river is called Hverfisfljót. Two days after the parishioners had heard the noises from the northeast, the Hverfisfljót River started to heat up until it eventually evaporated. On 6 August 1783, the sixth surge, the first at Hverfisfljót River, cascaded down the empty river trunk and breached its banks, burning or engulfing the farmsteads that lay in its way. On 1 and 10 September 1783, the seventh and the eighth surges followed.[248] From Prestbakki, the fire made the sky glow behind the mountains. "Whenever the sun or the moon could be seen on the part of the sky where the fire vapours swirled about, each appeared red as blood."[249] The ninth surge commenced in late September. The tenth and largest surge in the Hverfisfljót River came on 25 October and continued until November 1783.[250]

Frightening sounds once again announced the lava surges and were accompanied by an "intolerable reek and odor [. . .] as if burning coal had been doused with urine or other acrid substance."[251] Over the summer of 1783, STEINGRÍMSSON harvested as much hay as he could. He used some as fodder for his one remaining cow; the unfortunate

---

243 ÞÓRARINSSON 1984: 36.

244 STEINGRÍMSSON 1998: 56.

245 STEINGRÍMSSON 1998: 54.

246 THORDARSON, SELF 2003: 3. The dates for the respective fissure segment openings are as follows: Fissure segment VI: 29 July to 9 August; fissure segment VII: 31 August to 4 September; fissure segment VIII: 7 to 14 September; fissure segment IX: 24 to 29 September; fissure segment X: 25 to 30 October.

247 THORDARSON, SELF 1993: 236–237. Detailed descriptions of the events during the individual eruptive episodes can be found here: THORDARSON, SELF 2003: 238–244.

248 For a discussion on the slightly differing dates of the drying of the Hverfisfljót River and the surge of the lava in the riverbed, see THORDARSON 2003: 6.

249 STEINGRÍMSSON 1998: 60.

250 ÞÓRARINSSON 1984: 36; STEINGRÍMSSON 1998: 63; THORDARSON, SELF 2003: 3.

251 STEINGRÍMSSON 1998: 56.

beast soon after perished. STEINGRÍMSSON later used this hay for warmth. In his *eldrit*, he claims that "it smoked and flamed like sulphur" when thrown into the fire.[252] One of STEINGRÍMSSON's strategies to mask this presumably sulfuric smell was to burn pieces of bark and juniper wood in his home.[253]

As the lava slowly crept closer from the east, the people in the Síða region were concerned that they might soon become trapped.[254] To the north lay the highlands, the very place the lava had originated; to the west, the smoldering remains of the first surges; to the east, the molten lava of the latest surges; and to the south, the coast; those who were still able to do so fled the area.[255]

The magma output decreased substantially after November 1783.[256] In December of that year, STEINGRÍMSSON wrote that "all the flames and glare in the sky began to decrease."[257] The new year of 1784 began with milder and calmer weather, with spells of sharp frost and northerly winds. Nevertheless, a strange odor still lingered in the air.[258] Those in the vicinity observed the fires of the Laki fissure for the last time on 7 February 1784; this is the official end date of the eruption. By February 1784, the lava measured 14.7 cubic kilometers and covered an area of 599 square kilometers, which is approximately the size of the Isle of Man.[259] Although the lava flows did not kill anyone directly, they did lay waste to eight farms and damaged 29 severely; two parishes remained almost entirely uninhabitable for two years.[260]

Volcanic and seismic activity in the Grímsvötn system, in the subglacial parts underneath Vatnajökull, did not cease on 7 February 1784.[261] The residents of the Síða region continued to hear "rumbling noises [. . .] from under the glacier."[262] Two large and foul-smelling *jökulhlaups* occurred in the spring of 1784.[263] At least four eruptive episodes occurred at the Grímsvötn volcano; two before the end of the Laki eruption and two after. Although they did not precipitate further lava surges in the lowlands, they did curse the surrounding area with yet more tephra fall.[264] The *jökulhlaups* and noises from Vatnajökull indicate that volcanic activity continued at the Grímsvötn volcano underneath the ice until 1785.[265] The eleventh eruptive episode occurred on

---

252 STEINGRÍMSSON 2002: 182.
253 STEINGRÍMSSON 1998: 83.
254 HERRMANN 1907: 96; STEINGRÍMSSON 1998: 51–52; THORDARSON, HÖSKULDSSON 2014: 135–137.
255 STEINGRÍMSSONS 2002: 182–183.
256 ÞÓRARINSSON 1984: 36.
257 STEINGRÍMSSON 1998: 63.
258 STEINGRÍMSSON 1998: 65.
259 More information on calculating the lava volume, see also THORDARSON, SELF 1993: 251–256.
260 HERRMANN 1907: 94–97.
261 STEINGRÍMSSON 1998: 65.
262 ÞÓRARINSSON 1984: 36.
263 ÞÓRARINSSON 1984: 36.
264 THORDARSON et al. 2003b.
265 THORDARSON, SELF 1993; THORDARSON et al. 2003a; PASSMORE et al. 2012: 2594.

24 November 1783, the twelfth episode from January to February 1784, the thirteenth episode from April to August 1784, and the last one, the fourteenth episode, from 4 to 26 May 1785.[266] The end date of this volcano-tectonic episode at Grímsvötn was 26 May 1785.[267] Two more *jökulhlaups* occurred in May 1785 and November 1785: the origin of the second *jökulhlaup* remains obscure.[268]

The seismic activity of the Grímsvötn system began in May 1783 and ended in May 1785. The Laki eruption took place within this timeframe.[269] The most prominent volcanologist working on the Laki eruption, Thorvaldur THORDARSON, has concluded that the simultaneity and the synchronization of the activity of the Laki fissure and Grímsvötn is a sign that both eruptions were part of the same volcano-tectonic event, caused by the same regional stress at the Grímsvötn volcanic system.[270]

In the past, volcanologists have wondered about the origin of the huge volumes of lava that engulfed the Síða region. Possible sources include a magma chamber located underneath the fissure or one within the caldera of Grímsvötn, probably located at a depth of two to five kilometers. The latter is unlikely, as the caldera would have collapsed from the outpouring of such a large volume.[271] Another possibility is a magma chamber underneath a *nunatak*, which is a mountain summit that protrudes out of a glacier.[272] Thorvaldur THORDARSON believes it is likely that a deep-seated magma chamber located at the crust-mantle boundary fed the eruption.[273] The crust-mantle boundary, in the northwest of Vatnajökull, is 40 kilometers deep.[274] This theory would explain the continuous magma flow for the entire eight-month period of the eruption.[275]

After May 1785, there was a protracted cessation of activity at Grímsvötn; the next eruption was in 1823, 38 years later.[276] In general, a period of rest is not unusual for volcanic systems after eruptions that produce extremely large volumes of lava. The same phenomenon occurred at the Katla volcanic system after Eldgjá erupted in 939: this eruption was followed by a quiescence of 240 years.[277]

---

266 The episodes' numbering and dates come from THORDARSON et al. 2003b: 25, table 2.
267 THORDARSON et al. 2003b: 26. The dates of the eruption come from THORDARSON, SELF 1993: 239–240.
268 STEINGRÍMSSON 1998: 87. *Jökulhlaups* at Grímsvötn can happen without volcanic activity; on average, they occur every two to three years; ÞÓRARINSSON 1984: 36.
269 THORDARSON, SELF 1993: 258; THORDARSON 2003: 1.
270 THORDARSON et al. 2003b: 33.
271 BJÖRNSSON, BJÖRNSSON, SIGURGEIRSSON 1982; GUÐMUNDSSON 1987; THORDARSON, SELF 1993: 258–259.
272 For the magma chamber underneath the fissure, see GRÖNVOLD 1984: 57. The lateral movement of magma from a crustal magma chamber similar to the Krafla Fires was suggested by SIGURÐSSON and SPARKS (1978); THORDARSON 2010: 288. For information on the magma originating from the Grímsvötn caldera, see Catalogue of Icelandic Volcanoes: Grímsvötn.
273 SIGMARSSON et al. 1991; THORDARSON et al. 1996: 215–216.
274 FEDOROVA, JACOBY, WALLNER 2005: 123, 132; ALFARO et al. 2007.
275 THORDARSON, SELF 1993: 259.
276 THORDARSON, SELF 1993: 259 THORDARSON et al. 2003b: 13.
277 THORDARSON, LARSEN 2007: 142; RUSSEL, DULLER, MOUNTNEY 2009; ÁGÚSTDÓTTIR 2015: 1674.

Pollution and Environmental Impact of the Laki Eruption in Iceland

The cessation of the outpouring of lava did not bring an end to the suffering of those in the Síða region or Iceland in general. Help was slow to arrive: Icelanders were left to deal with the situation with limited means. During the first few days of the eruption, it became apparent that the volcanic gases had "thickly contaminated" the air; STEINGRÍMS-SON states that he was unable "to breathe in fully, and hardly went outside [. . .], all that year [1783] and the next."[278] As early as 10 June 1783, he noticed that the "bitter rain [. . .] caused almost unbearable soreness to the eyes or bare skin, as well as a sense of dizziness."[279] Over the next two weeks, the eruption would show its deadly force:

> more poison fell from the sky than words can describe: ash, volcanic hairs, rain full of sulphur and saltpeter, all of it mixed with sand. [. . .] All the earth's plants burned, withered, and turned grey, one after another. [. . .] the first to wither were those plants which bore leaves, [. . .] and the horsetails were the last to go.[280]

The smell of the air was foul: according to STEINGRÍMSSON, it smelled as "bitter as sea-weed and [was] reeking of rot for days on end."[281] He laments the fact that those already plagued by pre-existing respiratory problems could no longer take a deep breath and remarks that "Indeed, it was most astonishing that anyone should live another week."[282] A combination of volcanic gases, together with lung lesions caused by PM 2.5 – particulate matter with a size smaller than 2.5 micrometers – was probably to blame for the breathing difficulties these people endured.[283] Between June and September 1783, there were at least 15 incidents of tephra fall up to 40 kilometers from the fissure.[284]

STEINGRÍMSSON also noticed how rapidly the eruption impacted his livestock. On Saturday, 7 June 1783, his cows and ewes had given eight buckets of milk. One week later, on Saturday, 14 June, they offered only six and a half. The animals' "[f]lesh and body were ravaged at the same time."[285] On 10 June, he detailed how newly shorn sheep received scorch marks on their skin when rain fell.[286] The Laki eruption ejected large amounts of fluorine, in addition to other gases. While fluorine is beneficial to humans (particularly their teeth) in small doses, in high concentrations it leads to dental and skeletal fluorosis. Dental fluorosis is a condition that causes deformation and loosening of the teeth, while skeletal fluorosis is a bone disease resulting in pain and

---

278 STEINGRÍMSSON 2002: 180.
279 STEINGRÍMSSON 1998: 25.
280 STEINGRÍMSSON 1998: 41.
281 STEINGRÍMSSON 1998: 41.
282 STEINGRÍMSSON 1998: 41.
283 PÉTURSSON, PÁLSSON, GEORGSSON 1984: 97.
284 LARSEN, THORDARSON 1984: 65.
285 STEINGRÍMSSON 2002: 180.
286 STEINGRÍMSSON 1998: 26.

damage to the bones and joints.[287] Icelanders had known that volcanic ash could lead to "ash teeth" and death in animals for at least a century.[288] In hindsight, STEINGRÍMSSON wrote, they should have slaughtered the livestock there and then while they still had some meat on them.[289]

Although people were afraid that the foul-smelling sheep meat was poisonous, they "nevertheless tried to dress it, clean it and salt it as best they knew how or could afford to."[290] STEINGRÍMSSON outlines how his parishioners tried to save their livestock but that most efforts were in vain.[291] His dairy animals, and those of his neighbors, all perished over the next few weeks: they had either been poisoned by pollution or starved for lack of fodder.[292] Between 12 August 1783 and 24 June 1784, his family had no dairy food at all.[293]

As early as 14 June 1783, STEINGRÍMSSON writes of birds fleeing and leaving their eggs behind, which could be collected but "were scarcely edible because of their ill odor and sulphurous taste."[294] With fish he fared no better as the ponds and the rivers nearby were all poisoned by the eruption's ejecta.[295] The trout, pipits, wrens, and white wagtails became disoriented at first and then died.[296] Field mice were similarly affected.[297] In his writing, STEINGRÍMSSON notes that "[a]ll the mice in this county and the next one to the west, which had often caused great damage to our lyme grass, grain and other stores, were killed and there has been no sign of them since."[298]

The meat the parishioners ate and the water they drank were contaminated.[299] Since the beginning of the eruption, every drop of water had been characterized by "its bad flavor and bitter taste in the mouth," which made it unpotable.[300] Nevertheless, STEINGRÍMSSON remarks that his family continued to drink the water for lack of an alternative. He writes: "We became so used to drinking [the] water that it tasted to us like sweet whey. But it was polluted and brought in its train more disorders than I care to mention."[301] This brief mention of "disorders" most likely refers to waterborne diseases such as dysentery.

---

287 The *Lakagígar* eruption was particularly rich in fluorine; THORDARSON et al. 1996: 205–225.
288 PÉTURSSON, PÁLSSON, GEORGSSON 1984: 96.
289 STEINGRÍMSSON 1998: 28.
290 STEINGRÍMSSON 1998: 76.
291 STEINGRÍMSSON 2002: 180.
292 PÉTURSSON, PÁLSSON, GEORGSSON 1984: 97.
293 STEINGRÍMSSON 2002: 180.
294 STEINGRÍMSSON 1998: 27.
295 HERRMANN 1907: 96–97; STEINGRÍMSSON 1998: 73.
296 STEINGRÍMSSON 1998: 27.
297 PÉTURSSON, PÁLSSON, GEORGSSON 1984: 97.
298 STEINGRÍMSSON 1998: 93.
299 STEINGRÍMSSON 2002: 180.
300 STEINGRÍMSSON 1998: 67.
301 STEINGRÍMSSON 2002: 182.

Shortly after the beginning of the eruption, the grass on the south-central coast withered and the harvest subsequently failed. This was unlucky, as sometimes haymaking is possible as early as July in the warmer regions of Iceland.[302] Hay was used as feed for the livestock during the winter months when the animals could not remain in the fields due to heavy snowfall.[303] Without this additional hay, Icelanders knew most of their animals would perish over the winter.[304]

The fine layer of tephra that blanketed most of the country led to the harvest failing almost everywhere (Figure 17).[305] In his *eldrit*, STEINGRÍMSSON concludes that "the other effects of this fire spread over the entire country, [are] the withering of the grass and the ensuing famine among men and animals alike, [. . .]."[306] The people in the Síða region quickly ran out of food: illness followed their hunger. People suffered from pain in

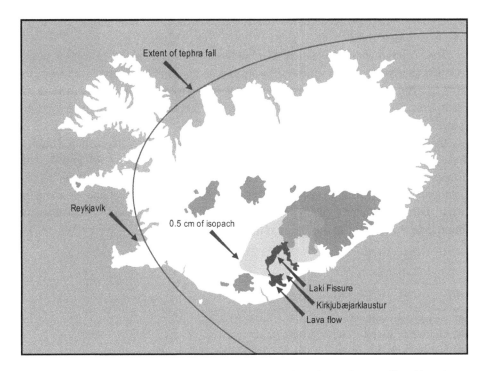

Figure 17: Tephra fall during the Laki eruption. The dark gray color indicates the area affected by at least 0.5 centimeters of tephra (7,200 square kilometers); the light gray color indicates the area covered in fine dust (200,000 square kilometers). Within this area, 60 percent of the livestock perished.

---

302 VASEY 1991: 333–335.

303 HÁLFDANARSSON 1984: 161.

304 VASEY 1991: 324, 333–336.

305 PÉTURSSON, PÁLSSON, GEORGSSON 1984: 96–97; OGILVIE 1986: 70; ÁGÚSTDÓTTIR 2015: 1674.

306 STEINGRÍMSSON 1998: 71.

their mouths, swelling of their gums, and toothaches.[307] These symptoms are typical signs of fluorosis; however, new research on the skeletal remains of animals in the region has called into question the extent of the problem in this area.[308] In addition, many suffered from scurvy, and otherwise relatively harmless infections became deadly.[309] As their health deteriorated, many people lost their hair, died of dysentery, or starved to death.[310]

Fish was a staple food for Icelanders. Yet, for most of 1783, even fishing in the shallow coastal waters was impossible due to the thick, blinding haze. To add to the Icelanders' bad luck, the East Greenland Current brought much sea ice to Iceland between 1781 and 1784, which lowered temperatures further and prevented many merchant vessels from reaching the Icelandic coast.[311] In early August 1783, making use of every possible opportunity to find additional sources of nourishment, STEINGRÍMSSON gathered a group of people to hunt seals at the river mouth of Hverfisfljót; they managed to club 26 of them.[312]

The eruption and bad weather also hindered Icelanders' access to trading posts. Iceland's southeastern and southern coasts had no harbors; the nearest trading posts were in Djúpivogur in the east and Eyrarbakki in the west (Figure 18). There were no roads from these posts and few bridges over the rivers; even under normal circumstances, journeys across the land by horse were difficult.[313]

In autumn 1783, STEINGRÍMSSON traveled to Skálholt to ask the bishop, Hannes FINNSSON (1739–1796), for his help. The trip disappointed STEINGRÍMSSON greatly: the bishop only gave him 20 *ríkisdalir*, a currency of silver coins, to buy food for his parishioners. Under normal economic circumstances, one *ríkisdalur* bought a sheep, seven *ríkisdalir* bought a cow, and eight *ríkisdalir* bought a horse.[314] STEINGRÍMSSON found little charity during this frigid autumn journey.[315]

---

307 STEINGRÍMSSON 1998: 78.

308 GESTSDÓTTIR, BAXTER, GÍSLADÓTTIR 2006: 32–33.

309 PÉTURSSON, PÁLSSON, GEORGSSON 1984: 97; THORDARSON et al. 1996: 205–225.

310 HERRMANN 1907: 94–97; STEINGRÍMSSON 1998: 78.

311 DAMODARAN et al. 2018: 520.

312 STEINGRÍMSSON 1998: 55.

313 VASEY 1991: 324. It was only in the late nineteenth century that roads were built to transport carriages. During this time, two bridges were built over the two largest rivers in southern Iceland; KARLSSON 2000b: 248–249.

314 GUNNLAUGSSON 1984: 213–214. This result was found by a special commission that had been initiated in February 1785 to investigate the conditions in Iceland; GUNNARSSON 1983: 144–147; OSLUND 2011: 36–38. Price series of Icelandic currency only go back to 1849 and, therefore, an accurate estimate of the value of *ríkisdalur* from 1783 is impossible. Table 12.25 in JÓNSSON, MAGNÚSSON 1997: 637; personal correspondence with Prof. Guðmundur JÓNSSON, University of Iceland, 16 February 2020.

315 STEINGRÍMSSON 2002: 180–181; see also OGILVIE 1986: 63.

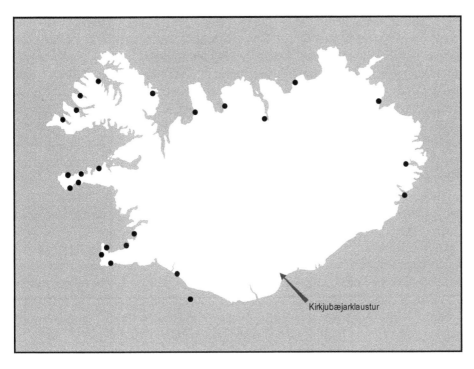

Kirkjubæjarklaustur

Figure 18: Danish trading posts in Iceland during the time of the Danish trade monopoly, 1602 to 1787. A lack of trading posts in the south and southeast around Kirkjubæjarklaustur becomes apparent.

Móðuharðindin, the Famine of the Mist

The situation in and around Klaustur was dire from the beginning of the Laki eruption, only growing worse with time. Winter was fast approaching, and there was no sign of help from Denmark. Yet more people left for the relative safety of the less affected parts of Iceland. Those who could not flee witnessed a seemingly interminable period of death and decay.

Usually, the trading posts in Iceland held only a meager surplus of cereals. By autumn 1783, however, these stations housed large amounts of fish, the main export product from the southwest and the west, as well as mutton from the north and east. All this was gathered for ships to take back to Denmark as part of the trade monopoly. The governor, district governors, and sheriffs of Iceland could prohibit the export of any Icelandic foodstuff if famine threatened. The officials felt uncomfortable making this decision without consent from Copenhagen: so it was that considerable amounts of food were exported from Iceland to Denmark on the eve of the most disastrous winter in Icelandic history.[316]

---

316 ANDRÉSSON 1984: 233.

While the mortality rate for Iceland in 1783 was actually below average, it seemed inevitable that all this would soon come to a head.[317] Early in the new year of 1784, people began to die in droves.[318] With no feed for animals or available winter pastures, famine struck.[319] In Iceland, the aftermath of the Laki eruption has a name: (*reykur-*) *móðuharðindin*, which loosely translates as "the famine of the mist." *Reykur* means "smoke," *móða*, "haze or mist in the air caused by the eruption," and *harðindi*, "hardship induced by a bad harvest."[320] The eruption's effects on mortality were not uniform in all parts of Iceland. Some regions, such as south-central Iceland and those north of the fissure, such as Skagafjarðarsýsla, Eyjafjarðarsýsla, and Suður-Þingeyjarsýsla, were hit harder than others.[321]

In his autobiography, STEINGRÍMSSON describes the state of affairs in early 1784. "I now had to travel on foot all the time, and to lean on my only horse for bringing dead bodies to the Church, since there were no other means of transportation."[322] Miraculously, one horse was able to survive on the contaminated hay; thus, perhaps it should rightly be called the Fire Horse. In the winter months, it was difficult for STEINGRÍMSSON to find able-bodied men to dig holes in the cemetery's frozen soil, which meant some people were buried in mass graves of up to ten people. However, all 76 people who died in STEINGRÍMSSON's parish in 1784 were given individual coffins.[323] The lack of fuel made people resort to burning furniture and even parts of their homes.[324] From STEINGRÍMSSON's description of what his parishioners ate during this time, it becomes apparent how hungry and desperate people were becoming. They "put their mouths on many things they never before thought to taste," even their robes and shoes.[325]

> All those who knew how, lived as sparely as they could, stretching what food they had, cooked what [. . .] ropes they owned, and restricted themselves to the equivalent of one shoe piece per meal, which was sufficient if soaked in soured milk and spread with fat.[326]

317 VASEY 1991: 328–329.
318 STEINGRÍMSSON 1998: 79.
319 VASEY 1991: 335–336.
320 HERRMANN 1907: 94–97; KRISTINSDÓTTIR 1984: 186.
321 VASEY 1991: 328–329.
322 STEINGRÍMSSON 2002: 183–184.
323 STEINGRÍMSSON 1998: 79; STEINGRÍMSSON 2002: 182–184.
324 STEINGRÍMSSON 1998: 80.
325 PÁLSSON 1945, vol. II: 597–598. "Hallæri það, sem á féll, neyddi vesalings fólkið, sem þraukaði heima í eldsveitunum, til að leggja sét til munns marga þá hluti, er það hafði aldrei áður látið sér til hugar koma að bragða á."
326 STEINGRÍMSSON 1998: 81–82.

His desperate parishioners mixed hay with their porridge and collected fishbones from the seashore to crush and eat. As a last resort, some people ate horsemeat; still, even in this wretched situation, not everybody could bring themselves to eat their horses.[327]

With the spring of 1784 came some hope: plants began to grow, the roots and leaves of which locals could use for food. The pastures, however, were still polluted and the grass produced little hay.[328] A study by Icelandic scholar Páll Bergþórsson demonstrated that a decrease of 1 °C in mean annual temperatures leads to a 30-percent reduction in hay production. Several cold spells throughout the eighteenth century amounted to just that. This led to severe shortfalls, time and time again, in fodder for livestock over the winter.[329]

In 1784, Steingrímsson mourned a loss of his own: on 4 October 1784, his wife of 31 years, Þórunn Hannesdóttir, passed away. After her death, he suffered from depression. The famine and general hardship he and his parishioners endured worsened his condition:

> The period from the autumn of 1784 until the spring of 1785 was the most dismal that I have ever lived through. When I lost my wonderful wife, everything, so to speak, collapsed around me. [. . .] I had no fuel for the lamps and so had to languish in constant darkness. My people had to be in the cowshed because of our milking stock, the thieves were on the prowl; so I had to lie or sit there in a state of anxiety. [. . .] The house was now empty and horribly cold. And yet I found pleasure in sitting often by the familiar bed and reading God's word. But then my hands and my feet became swollen by the frost so that I could not take up my pen for diversity. At Christmas, I injured my arm so that for five weeks, I could hardly get dressed. But the church services were not canceled, for it was – and still is the greatest delight to worship God in His house.[330]

Steingrimsson's parish, once 613 members strong, was reduced to just 93 souls by the end of 1784. Many had left in the early summer, settling in the fishing stations to the west. These refugees left most of their possessions, which were soon stolen if not left with a trustee.[331]

Weather-wise, the autumn of 1784 was reasonable everywhere in Iceland.[332] By contrast, the winter that followed was horribly cold again, resulting in another surge in mortality that lasted until the spring of 1785.[333] Bishop Hannes Finnsson estimated infant mortality of 30 percent for the year 1784, which he partly blamed on a lack of breastfeeding.[334]

---

327 Vasey 1991: 323–324; Steingrímsson 1998: 82.
328 Steingrímsson 1998: 68, 83.
329 Bergþórsson 1985: 117; Bergþórsson 1987: 396; Vasey 1991: 347.
330 Steingrímsson 2002: 187–188.
331 Steingrímsson 1998: 80.
332 Ogilvie 1986: 63.
333 Vasey 1991: 332–333, 346.
334 Lacy 1998: 187, 226; Magnússon 2010: 99.

The year 1785 began with inclement weather, including a very sharp frost. STEIN-GRÍMSSON observed how, on 23 January 1785, "a whole pint bottle of communion wine which stood on the altar during the service that day turned to slush."[335] Given that wine freezes at temperatures between 0 and −6 °C, it is hard not to admire the stead-fastness of all those at church that day. In the autumn of 1785, Iceland experienced a smallpox epidemic that started in Reykjavík and quickly spread in all directions.[336] This epidemic was relatively mild; nevertheless, given the population's weakened state, it killed around 1,500 people.[337]

Scholars estimate that the moss in Iceland took three years to recover; the inland trout fisheries and the offshore fishing grounds took about the same length of time. Only in 1786 did the food supplies slowly begin to recover.[338] Between 1783 and 1785, about 76 percent of Iceland's horses, 79 percent of the sheep, and half of the cattle perished.[339] Cattle were kept alive as a priority, rather than sheep or horses.[340] Farm-ers generally culled sheep and expected their horses to fend for themselves.[341]

Roughly 15 percent of all farms inhabited before the eruption were abandoned. Overall, the eruption initiated a movement from the inland areas to the coast and so diversified the economy, making it less dependent on livestock and introducing fish-ing and foreign trade.[342] Guðmundur PÉTURSSON and his colleagues estimate that Ice-land had a population of 48,884 before the eruption. This is based on extrapolated data from parish registers for births, deaths, and pastoral visitations, which date back to as early as 1668.[343] The hardships in Iceland were severe. The famine lasted until the spring of 1785, by which time 20 to 25 percent of the population had perished; around 10,000 people lost their lives.[344] In 1786, Iceland had a population of 38,363.[345] Vestur-Skaftafellssýsla lost about 44 percent of its population (from 1,964 to 1,072 peo-ple) between 1783 and 1785. 472 people from this region died and, during the same time period, 80 children were born; the remaining 500 or so fled (temporarily) to

335 STEINGRÍMSSON 1998: 86.
336 HÁLFDANARSSON 1984: 162.
337 PÉTURSSON, PÁLSSON, GEORGSSON 1984: 97; VASEY 1991: 333.
338 BJARNAR 1965: 416; ÞÓRARINSSON 1979: 152; VASEY 1991: 323–324, 333–336; OSLUND 2011: 36.
339 HENDERSON 1818: 275; FISHER, HEIKEN, HULEN 1997: 170; OPPENHEIMER 2011: 286.
340 PÉTURSSON, PÁLSSON, GEORGSSON 1984: 96–97; RAFNSSON 1984a: 178.
341 VASEY 1991: 324.
342 RAFNSSON 1984a: 178. Although many eruptions tested the Icelanders, no district in Iceland was ever permanently or entirely deserted; ÞÓRARINSSON 1979; THORDARSON 2010: 293.
343 More information on parish registers here: HÁLFDANARSSON 1984: 161; VASEY 1991: 326–327. The Ge-nealogical Society of Utah microfilmed these for their Genealogical Data Communications. Daniel VASEY (1991) analyzed them to ascertain the fate of most Icelanders prior to and after the famine.
344 This equates to 19 to 22 percent of the population at the time; BJARNAR 1965; DEMARÉE, OGILVIE 2001: 224. Other estimates put the figure at around 22 percent, such as THORDARSON, SELF 1993: 261. VASEY estimates it to be around 22.4 percent, VASEY 1991: 343; VASEY 2001.
345 PÉTURSSON, PÁLSSON, GEORGSSON 1984: 96–97.

other regions.[346] Historian Ólöf GARÐARSDÓTTIR has found a distorted sex ratio in the live births in Iceland in the aftermath of the Laki eruption, with a higher-than-average number of girls being born.[347]

The Danish Response to the Crisis

Copenhagen is around 2,000 kilometers from Iceland; the only way to send a message between these two places in the eighteenth century was via Danish merchant ship. Over the summer of 1783, the Danish merchants who were in Iceland witnessed the volcanic eruption and its dire consequences. When they returned to Copenhagen in late August 1783, they brought this news with them. Their accounts were subsequently printed in Danish newspapers and elsewhere in Europe. Most of the first newspaper reports were in large part based on a detailed letter written by Danish merchant C. J. SÜNCKENBERG in Stykkishólmur[348] on 24 July 1783; all the early reports contain the same inaccuracies and misspellings that were present in the letter.[349]

In 1783, the Icelandic governor was Lauritz Andreas THODAL (1718–1808), a Norwegian whom historians regard as a competent official. At the time of the Laki eruption, THODAL was ill.[350] He lived in the vicinity of Reykjavík, far from the Síða region and the people who suffered the most. THODAL had heard about the eruption but wanted to confirm these reports before he passed them on to the Danish central administration. On 16 September, he duly sent his report, which took over two months to arrive.[351]

In response to the dire accounts that trickled in, the Danish central administration sent two representatives to investigate the disaster and the woes it had inflicted upon the Icelandic economy and society. The representatives were Hans VON LEVETZOW

---

346 GUNNLAUGSSON 1984: 128.

347 Ólöf GARÐARSDÓTTIR gave a paper about this topic at the ESEH conference in Tallinn, 2019. In Sweden, this eruption seemingly had effects on perinatal health. In 1784, fewer males were born than before, which resulted in a distorted sex ratio. In 1785, both male and female infant mortality rates were unusually high; CASEY et al. 2019.

348 In the letter, the settlement is referred to as Holmershafen, which is located on Iceland's west coast and is one of the trading posts of the Danish trade monopoly. The letter was addressed to the directors of the Royal Monopoly Company in Iceland; THORDARSON et al. 2003b: 37. The letter can be found in GUNNLAUGSSON et al. 1984: 269–270.

349 Kjøbenhavns Adresse-Contoirs Efterretninger, no. 170, 5 September 1783; I thank Karl-Erik FRANDSEN for his assistance in obtaining this newspaper report. GUNNLAUGSSON 1984: 213.

350 WIENERS 2020: 2.

351 WIENERS 2020: 6–9. Claudia WIENERS makes a good argument here: In the summer of 1783, it was difficult for governor THODAL to predict that a severe, deadly famine would result from this eruption based on his information. In fact, it is difficult even in the present to foresee the magnitude of famines in certain parts of the world.

(1754–1829), a German-Danish nobleman, an official of the Danish central administration and later governor of Iceland (1785–1789), and Magnús STEPHENSEN (1762–1833), an Icelander conducting his law studies in Copenhagen at the time. The latter would later become chief justice of the high court of Iceland.[352]

The ship set off in autumn; unfortunately, adverse weather made it impossible to dock anywhere in Iceland and forced them to winter in Norway. In spring 1784, they set sail again, landing in Iceland in April.[353] STEPHENSEN and VON LEVETZOW began their survey in the summer of 1784, coming eventually to Vestur-Skaftafellssýsla. Unfortunately, they were unable to obtain the precise information needed for their survey, as there was no way to prove whether somebody had died or simply moved elsewhere.[354]

The delayed arrival of this reconnaissance mission was not the only thing that went wrong in the Danish response to the crisis. In addition, another ship carrying supplies for the expedition was wrecked off the coast of Meðalland, just to the west of the Síða region. Immediately after that, whoever had the means could buy flour, hemp, and iron from the survivors of the wrecked ship.[355]

In the spring of 1784, STEPHENSEN and VON LEVETZOW were able to give some relief to the people in southern Iceland as direct financial aid was made available for those who fled Vestur-Skaftafellssýsla to resettle elsewhere. People were also allowed to receive supplies and new livestock from the Crown's stores in Iceland. The privileged merchants and Icelandic officials administered this aid until 1785.[356] Each farmer was given roughly eight *ríkisdalir* to buy livestock, a task made difficult due to higher prices and a dwindling supply.[357] At this time, around 20 to 30 percent of farms were in dire need of animals.[358] The problems faced by the farmers in the Síða region were compounded by the demands of their landlords to pay rent and settle any other debts that had gone into arrears.[359]

During the first half of 1784, the central administration in Denmark published several royal decrees regarding measures to tackle famine and loss of livestock in Iceland. They knew, from what little information they had, that the Icelanders required urgent aid. However, as the gravity of the situation was unclear, the suggested measures were far from sufficient.[360] In early 1784, fundraising activities for the Icelanders began in Copenhagen. Despite this, merchants traveled to Iceland in the spring of 1784

---

352 STEINGRÍMSSON 1998: 85. Magnús STEPHENSEN's father, Ólafur STEPHENSEN, was the governor of Iceland from 1790 to 1806.

353 GUNNLAUGSSON 1984: 213; WIENERS 2020: 6.

354 STEINGRÍMSSON 1998: 85–86.

355 ANDRÉSSEN 1984: 233; STEINGRÍMSSON 1998: 85.

356 GUNNLAUGSSON 1984: 213.

357 STEINGRÍMSSON 1998: 84.

358 GUNNARSSON 1983: 144–147; GUNNLAUGSSON 1984: 213–214; OSLUND 2011: 36–38.

359 STEINGRÍMSSON 1998: 89.

360 GUNNLAUGSSON 1984: 213.

without additional food supplies and still exported their regular fish quota: for this thoughtless action, they were heavily criticized.[361]

Throughout 1784, ships returning from Iceland arrived in Copenhagen with updates on the grim situation. On 21 July 1784, a royal decree ordered the shipment of 3,000 to 4,000 barrels of rye to Iceland to be distributed amongst those in need. In the end, 5,300 barrels were shipped. The decree also ordered that fish from western and eastern Iceland be shipped to the north and south; in addition, extra cargoes of timber were sent to aid the effort to rebuild.[362]

Few of the proposed initiatives to help the situation in Iceland were quite as ambitious as the plan to relocate the poor, elderly, ill, and orphaned to the moors of Jutland in mainland Denmark.[363] Many Danes had left Jutland due to the area's problems with soil erosion.[364] This plan of resettlement, however, was never realized.[365] Moreover, new research and close examination of written protocols call into question whether the planned evacuation was ever taken seriously in Copenhagen, instead suggesting that it had only been briefly and informally discussed.[366]

The collection of funds in churches in other parts of the Danish kingdom, such as Norway, only began in 1785. In total, the amount raised for Icelandic relief was 46,000 *ríkisdalir*. In addition, the royal trading company donated another 32,000 *ríkisdalir* to the campaign.[367] 78,000 *ríkisdalir* was an immense sum; unfortunately, according to German medievalist Paul HERRMANN (1866–1930), only a quarter of the donations reached the intended recipients. The rest was used for other purposes, such as measuring and mapping the coastline.[368]

A second royal land commission was initiated in February 1785 and ran until 1794. Its mission was to analyze the consequences of the *móðuharðindin* and to follow up on issues that the first royal land commission from 1770/1771 had not concluded satisfactorily. Some unresolved matters included police ordinance, commerce, free trade, fishing, grinding grain, and manufacturing, to name a few.[369]

Around 95 percent of the population's livelihoods depended on farming, animal husbandry, and fishing, all of which were severely disrupted by the volcanic eruption.

---

361 OSLUND 2011: 36–38.
362 ANDRÉSSEN 1984: 233; GUNNLAUGSSON 1984: 213; THORDARSON 2003: 1, 10.
363 For a history of Jutland, see OLWIG 1984.
364 KJAERGAARD 1994; OSLUND 2011: 70, 181.
365 A short overview of the matter can be found here: BJARNAR 1965: 421; GUNNLAUGSSON 1984: 213; OSLUND 2011: 70.
366 WIENERS 2020: 8.
367 GUNNARSSON 1983: 145–146; GUNNLAUGSSON 1984: 213–214; OSLUND 2011: 36–38.
368 HERRMANN 1907: 94–97. According to GUNNARSSON (1983: 145–146), a special fund was created from the donations that were collected in Denmark and Norway. Most of this money remained in Copenhagen to be used for the Icelanders in the future, should the need arise; in the 1840s, a high school was built in Iceland with the money.
369 ROBERTSDÓTTIR 2008: 418.

The short-term response of the distant central administration in Copenhagen was too slow, and the aid provided too meager to ensure the survival of many Icelanders and their livestock.[370]

The Road to Recovery

The *móðuharðindin*, the famine of the mist, cut down one in five Icelanders. That on which they depended, land and livestock, was swept from beneath their feet. In many ways, it is surprising that Icelandic society did not totally collapse. The societal structure remained intact: most Icelanders lived in multi-person households, within which the burden of risk was shared. Some abandoned farms had been pillaged, but one can understand people's desperation in the face of starvation and almost certain death. Icelandic households showed solidarity with one another and cared for their own as best they could. They rationed food and ate whatever they could find.[371]

Daniel VASEY found that the mortality rate during the *móðuharðindin* was almost twice that of the Irish potato famine of 1845 to 1849.[372] Whereas the Irish response was emigration, the Icelandic response was domestic relocation. Perhaps they did not have the means to travel abroad or simply did not want to leave their country.[373]

Over time, the Icelandic population crept upward. The birth rate rose sharply in the late 1780s and early 1790s, a common occurrence after a famine. In 1801, when the second national census took place, the population hovered around 47,200. However, it took until the 1810s for Iceland to reach pre-eruption population levels again.[374] The region of Vestur-Skaftafellssýsla needed longer still, only reaching pre-eruption population levels half a century later, in the 1830s.[375] Throughout the nineteenth century, the average age of marriage fell, as did infant mortality.[376]

---

370 VASEY 1991; KARLSSON 2000b: 180–181; DUGMORE, VÉSTEINSSON 2012: 76. For a detailed study of the Danish central administration's response to the Laki eruption, see WIENERS 2020.

371 VASEY 1991: 344, 348–349.

372 Ireland, just before the potato famine, however, had a population of eight million. One million people died from hunger and diseases; another million left the country. So, in numbers, the Irish potato famine was much worse, but in terms of percentage, the mortality rate was higher in Iceland during the *móðuharðindin*; VASEY 1991: 344.

373 Emigration from Iceland to Canada took place in the second half of the nineteenth century. In total, around 17,000 Icelanders (20 percent of the population) left for North America between 1870 and 1914. To this day, they are referred to as "West Icelanders" by native Icelanders; KARLSSON 2000b: 234–238; MAGNÚSSON 2010: 64–84; OSLUND 2011: 50, 185.

374 The National Archives of Iceland, Census Database: The 1835 Census. Historical Demographical Data of the Whole Country of Iceland between 1703 and 2050. The data from populstat.info reveals that Iceland had 49,000 inhabitants in 1810 and 50,000 inhabitants in 1815. See also VASEY 1991: 344; KARLSSON 2000b: 186–192; MAGNÚSSON 2010: 21.

375 GUNNLAUGSSON 1984: 128.

376 MAGNÚSSON 2010: 22, 198.

The Laki eruption transformed Iceland's landscape. The course of both the Skaftá and Hverfisfljót Rivers changed due to the enormous outpourings of lava. In addition, the eruption also raised the land by an average of 25 meters; in some parts, it was well over 100 meters.[377] The lava fields of the eruption remain to this day in the form of undulating expanses of igneous rock flecked with moss.

The *móðuharðindin* was a "major hunger catastrophe."[378] As we have seen from the history of Iceland up to 1783, it was by no means the first of its kind. The Icelandic economy, as well as its population numbers, stagnated throughout the eighteenth century. The trade monopoly did not allow for any competition, and – as a result – Icelanders remained, for the most part, cash-poor subsistence farmers.[379] In the eighteenth century, Iceland was prone to famines and epidemics; when they struck in tandem, they usually wiped out around 20 percent of the population. This was the case between 1707 and 1709, between 1754 and 1759, and again between 1783 and 1786. Resources were distributed unevenly within Iceland, and the poor were more likely to perish than the wealthy.[380]

Why was the *móðuharðindin* the last major hunger catastrophe in the history of Iceland?[381] What did the Icelanders and the Danes learn from this experience? The effects of the disaster were devastating. With the loss of the hay harvest in 1783, only a meager harvest in 1784, no access to supplemental food sources due to fluorine poisoning, and high livestock losses, should one not ask why *only* one in five Icelanders died?[382] Perhaps they had learned from the many hunger-induced mortality crises that came before the *móðuharðindin*. Or was it a mere coincidence that Iceland was not struck again by a large famine?

Throughout Iceland's history, the Laki eruption and the *móðuharðindin* have been interpreted differently, depending on the zeitgeist. In the late eighteenth century, the debate focused on the reaction to the eruption so that such a calamity could be mitigated in the future. Inspired by the Enlightenment, many Icelanders came to regard the hardships of 1783 as anthropogenic rather than heaven-sent.[383] Even religious men like Jón STEINGRÍMSSON eventually came to this conclusion. He was a keen observer of nature and sometimes felt conflicted between science and the dictates of his faith. STEINGRÍMSSON conducted some experiments, such as throwing small boulders into the lava. When they did not melt, he concluded that there was no way that the lava could destroy the mountains surrounding Klaustur as some had feared.

---

377 THORDARSON 2010: 290–291.
378 GUNNARSSON 1984: 242.
379 VASEY 1991: 344.
380 GUNNLAUGSSON 1984: 212–213.
381 GUNNARSSON 1984: 242.
382 VASEY 1991: 344–346. The severe weather between 1782 and 1784 certainly played a role in the loss of life, both human and animal; OGILVIE 1986.
383 GUNNARSSON 1984: 242.

STEINGRÍMSSON was both a man of the Enlightenment and a priest of his time. In another passage of his autobiography, he talks of two men living together on a farm near Klaustur. He believed the men to be homosexual and considered that God may have produced the volcanic eruption to drive them apart.[384]

The bishop of Skálholt, Hannes FINNSSON, wrote a treatise titled *Mannfaekkun af Hallærum* ("Loss of Life as a Result of Dearth Years") in 1786 and published it ten years later. The inspiration for this work was most likely the hardships that the bishop witnessed during the *móðuharðindin*. FINNSSON defines a "dearth year" as one in which famine and hunger-related deaths affect the whole country of Iceland. For climate historian Astrid OGILVIE, FINNSSON's treatise is the first Icelandic work in the then-unnamed genre concerning the "human dimension of climate change."[385] FINNSSON concludes that several factors can contribute to a dearth year, the most important of which are climatic in nature. He points the finger specifically at an accumulation of consecutive cold seasons, which can lead to poor harvests and famine. FINNSSON did not think that the climate varied throughout the time of his study; he complained about a lack of data from earlier times in Icelandic history but nevertheless found that there had been fewer years of hardship before 1280.[386] Today, given what we know about the early onset of the Little Ice Age in the North Atlantic Rim, we can appreciate the accuracy of these findings.

Another treatise focusing on the hardship caused by the so-called dearth years was written and published in 1790 by Stefán ÞÓRARINSSON, district governor of northern and eastern Iceland from 1783 to 1823. This treatise focuses on the years between 976 and 1783. According to ÞÓRARINSSON, four events could cause hardship and adversity in Iceland: the grass harvest failing, restricted access to fishing waters, severe and extremely long winters, or the delayed arrival of Danish merchant ships. He regards volcanic eruptions as another cause of hardship but concedes that nothing can be done to prevent them. Therefore, he suggests another course of action. The details of this plan are revealed by the title of his treatise: *Thoughts for Greater Consideration Regarding Dearth Years and their Effects, in addition to the setting up of Food or Grain Reserves for Severe Years.* Food reserves, in the opinion of ÞÓRARINSSON, would allow Icelanders the opportunity to rise above the status of helpless victims and mitigate future disasters.[387]

By 1783, Icelanders had lived in and struggled with their natural environment for almost 900 years. When the Laki fissure erupted, they made do with what they had. Throughout their history, they had fended for themselves in the face of volcanic eruptions and hazardous weather conditions and so learned how to attenuate their effects. Isolated from Europe by hundreds of kilometers of ocean, they developed a communal

---

[384] STEINGRÍMSSON 1998: 7–8, 32.
[385] OGILVIE 2005: 280.
[386] OGILVIE 2005: 280.
[387] OGILVIE, JÓNSSON 2001; OGILVIE 2005: 280–281.

resilience rather than a dependence on Copenhagen: help, in the shape of resources or refuge, often came from areas within Iceland that were less affected.[388] Creatively, they looked for internal solutions, such as clearing the tephra from fields so that the grass could grow and animals could graze again. Similarly, they relocated farms that were destroyed and created new paths for those cut off by lava.[389]

The Laki eruption precipitated the termination of the Danish trade monopoly in 1787. Now, within the realm of the Danish kingdom, Danes and citizens of other dependencies were allowed to trade with one another. In 1855, yet more restrictions were lifted: Denmark granted free trade to all its subjects. Now, Icelanders were allowed to contact foreign merchants directly without Denmark acting as a middleman.[390] In the nineteenth century, the interpretation of the eruption's aftermath changed when a distinct anti-Danish sentiment arose, with some blaming the Danes for their lackluster aid effort. In the twentieth century, blame shifted to the "fire and ice" of Iceland. Interestingly, in both explanations, Icelanders are painted as helpless victims.[391]

What if the trade terms for Iceland had been fairer or agricultural methods had been different? Would they have fared better had they not suffered from endemic poverty? How would they have coped had they worked their own farms rather than rented lands? The fact remains that the Laki eruption was of a size that only occurs on average once every 500 years. Icelanders could not have known an eruption like this was an inevitability. Even today, nothing can protect fields from being poisoned by fluorine. Daniel VASEY is confident that the Laki eruption would "probably have caused excess mortality under any pre-industrial circumstances."[392] It seems many of the deaths during this period were unavoidable.

Having analyzed the eruption and its aftermath, we now leave Iceland behind and follow the Laki haze toward Europe. The haze mesmerized and frightened many, inspiring a few to find explanations for what they were witnessing.

---

388 DUGMORE, VÉSTEINSSON 2012: 76.
389 THORDARSON 2010: 293–294.
390 KARLSSON 2000b: 243–244; AGNARSDÓTTIR 2013: 14–15.
391 GUNNARSSON 1984: 242.
392 VASEY 1991: 348.

3 Shaking the World

Whereas Icelanders had long since familiarized themselves with the peculiarities of their homeland, in mainland Europe, volcanism was still quite an obscure concept. Indeed, it was not the obvious and dramatic consequences of a volcano that occupied the continent that summer but the protracted presence of a caustic mist. Chapter Three analyzes how contemporaries in Europe and beyond reacted physically, emotionally, and intellectually to the Laki haze and the numerous other unusual phenomena that characterized 1783.

History of Geology

Volcanoes: A Topic of Meteorology

In the mid-eighteenth century, geology was still a fledgling branch of the sciences. Volcano itself was a neologism; in the literature of that period, there is scant mention of the term *vulcanus*, but rather "fire-spitting mountain" or "earth fire."[1] The term "volcano" only emerged in the sixteenth and seventeenth centuries when Europeans started traveling to other parts of the world, such as the Canary Islands, the Moluccas, and Central and South America. Volcanoes were popular attractions with educated Europeans on their Grand Tours. The well-traveled elite soon realized that fire-spitting mountains were a recurrent worldwide phenomenon. The first mention of the noun "volcano" in the English language is in the Oxford English Dictionary of 1613. Spanish explorers of the time referred to these "mountains of fire" (*montañas de fuego*) as *vulcan* or *volcan*, amongst other names, referencing the Aeolian Islands north of Sicily in the Mediterranean. Historian of geology Kenneth TAYLOR states that Vulcano and Stromboli were often confused, so their names were used interchangeably for a long time. Until the seventeenth century, Etna was referred to as *Vulcanus mons*. The term volcanology was only coined in the nineteenth century.[2]

Today it seems obvious that volcanoes are a central part of geoscience; however, in the eighteenth century, the exploration of volcanoes and earthquakes was considered part of meteorology. The Greek adjective *meteoros* means "uncertain or inconstant." In the seventeenth century, the term *meteora* was used to denote "a class of 'phenomena which surprise us.'"[3] The phenomena included subsurface revolution events associated with volcanoes and earthquakes, thunderstorms, northern lights, comets, and shooting stars. In summary, *meteora* were natural occurrences that took

---

1 "Vulcanus" in ZEDLER 1747, vol. 51: 1239. "Feuer=speyende Berge" in ZEDLER 1734, vol. 9: 768.
2 TAYLOR 2016: 3; MCCALLAM 2019: 4–6.
3 TAYLOR 2016: 4; note 5.

place at odd times and inspired awe. *Meteora* is also the title of Aristotle's treatise that dealt with the atmosphere, geology, and hydrology, among other topics.[4]

Kenneth TAYLOR argues that in antiquity, the Romans regarded Mount Etna and Vesuvius as local phenomena rather than expressions of the same underlying principle and made no effort to categorize them. The seemingly accidental nature of volcanoes was related to the notion of *meteora*.[5] But, of course, the awareness of volcanoes in antiquity extended far beyond just Mount Etna and Vesuvius. Aristotle's *Meteora* inspired Pliny the Elder's (23–79 CE) 37-volume *Naturalis historia*, in which he thought deeply about volcanic eruptions.[6] Other notable scholars who engaged with these ideas include Isidore of Seville (560–636) and Bede the Venerable (ca. 672–735).[7]

The knowledge produced about volcanoes from antiquity was known and discussed among scholars in the Middle Ages. During the thirteenth century, Aristotle's *Meteora* regained its allure. Important thinkers such as Thomas Aquinas, Albertus Magnus, and Thomas de Cantimpré made good use of the treatise's insights. In 1349, Konrad von Megenberg published the first German-language natural history encyclopedia; it became extremely popular and made Aristotelian knowledge accessible to a lay audience.[8] Knowledge from antiquity on volcanoes and the natural world did not disappear in the Middle Ages; instead, it was transferred.[9] The echoes of the ideas of antiquity struck a chord with those of a curious bent during the Enlightenment. The period's predisposition toward "novelties of nature" and faraway travels to the New World further sparked an interest in these mysterious fire-spitting mountains.[10]

In the eighteenth century, volcanoes were increasingly explored, researched, and discussed. Fieldwork and travels to Italy or Iceland stimulated debate among naturalists with a surge in interest between 1763 and 1792, a period of relative peace on the continent. This period, by happenstance, coincided with activity at Vesuvius and Etna and so offered an opportunity for naturalists to consolidate their theories on the causes of eruptions and the formation of the planet. Naturalists at the time leaned on other disciplines, such as chemistry, when considering the explosive reactions between gases, rocks, or metals and seawater, air, or fire, and later physics, when considering the role of electricity.[11] The discovery of Pompeii and Herculaneum in the eighteenth century – frozen in time by a pyroclastic density current from Vesuvius' eruption in 79 CE – served as a reminder of humanity's mortality. Geological timescales threaten human resilience and make huge events easy to forget. Volcanoes, at

---

4 TAYLOR 2016: 4.
5 TAYLOR 2016: 3–4.
6 Plinius 2013.
7 ROHR 2017: 49, 64–67.
8 Konrad von Megenberg 2003.
9 ROHR 2017: 49, 64–67.
10 TAYLOR 2016: 4.
11 McCALLAM 2019: 1–3, 234.

least in contrast to earthquakes, leave a visual reminder in the landscape: "Generations pass, the smoking peak endures."[12]

Athanasius KIRCHER (1602–1680) was a German Jesuit who observed nature and volcanoes extensively. On a journey to Italy in 1638, he visited Mount Etna, Vesuvius, and an erupting Stromboli. Thus, he gained first-hand experience with volcanoes at a time when they were a rather exotic topic.[13] Prior to its 1628 eruption, Vesuvius had been considered extinct. KIRCHER speculated about the mechanisms of volcanic eruptions and assumed a connection between the different burning mountains around the world, which he famously depicted in a copperplate print published in his *Mundus Subterraneus*. KIRCHER considered volcanoes part of a "system of terrestrial operations" and imagined a subsurface network of chambers and passages linked to a central fire within the Earth.[14] The volcanoes depicted by KIRCHER were cone-shaped stratovolcanoes; the iconic image included in Figure 19 influenced the public perception of the appearance of volcanoes for centuries.

Beginning in the 1790s, a debate concerning the origin of basalt divided the scientific community. Two schools of thought existed, Neptunism and Plutonism. Neptunism, also called Diluvialism as it referenced the biblical Flood, was the assumption that all rocks, including basalt, derive from ocean sediments.[15] Abraham Gottlob WERNER (1749–1817) and his students taught Neptunism, which had, for a time, a proponent in Johann Wolfgang VON GOETHE (1749–1832). Plutonism, also called *Volcanism*, was a theory developed by Scottish geologist James HUTTON (1726–1797), which assumed that a central fire inside the planet created the Earth's crust through volcanic activity.[16] At the end of the eighteenth century, HUTTON suggested that geologists should consider volcanoes a central concept in geology.[17] This debate is explored in greater depth in the fourth chapter of this book.

Eighteenth-century geologists developed many theories as to why earthquakes and volcanoes occurred. The discipline of geology strove to identify the laws behind observations of nature. Volcanoes played a marginal role as they would only erupt at odd, irregular times, making it nearly impossible to determine the laws behind their existence. As TAYLOR points out, volcanoes needed to become ordinary to be integrated into the thought-world of early geology.[18]

---

12 McCALLAM 2019: 233.
13 TAYLOR 2016: 6.
14 TAYLOR 2016: 46; PYLE 2017: 56.
15 RAPPAPORT 2007: 103, 111–114; McCALLAM 2019: 234–235. Michael KEMPE (2003a, 2003b) has studied the Swiss polymath Johann Jakob SCHEUCHZER and the theory of the biblical Deluge; see also BOSCANI LEONI 2010.
16 PÁLSSON 2004: 175.
17 HUTTON 1788: 274; TAYLOR 2016: 1.
18 TAYLOR 2016: 1.

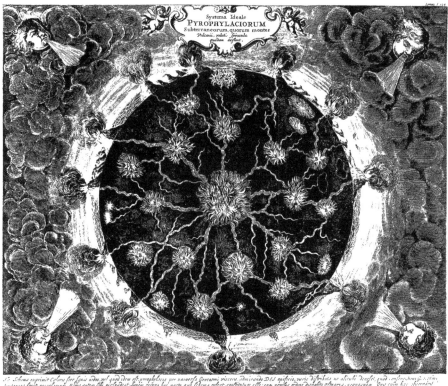

Figure 19: Athanasius KIRCHER, *Subterraneus Pyrophylaciorum*, 1665. This copperplate print shows the scheme of the Earth's fire canals, which KIRCHER believed connected all the volcanoes in the world.

Natural Sciences in the Late Eighteenth Century

"Nobody is so ignorant that they have never heard or read about fire-spitting mountains (volcanoes), even in the newspapers."[19] This statement by German philosopher Christoph Gottfried BARDILI (1761–1808) strongly suggests that volcanoes had a place within the collective consciousness in 1783. In the second half of the eighteenth century, volcanoes were considered sublime – terrifying and beautiful simultaneously.[20]

Today, the term "natural sciences," coined during the nineteenth century, is an umbrella term that encompasses the different fields of research that engage with facts and

---

19 BARDILI 1783: 9–10: "Niemand ist so unwissend, daß er nicht schon einmal etwas von feuerspeyenden Bergen (Vulkanen) gehört oder gelesen hätte, wäre es auch nur in den Zeitungen gewesen."
20 MCCALLAM 2019: 235.

natural processes. In the eighteenth century, the terms "science," "physics," "physical science" [*Naturlehre*], and "natural philosophy" were used synonymously.[21] Thus, natural scientists were referred to as scholars, naturalists, or natural philosophers.[22]

Toward the end of the seventeenth century, the sciences became increasingly institutionalized: universities, academies, and learned societies were founded. Local languages, such as German, replaced Latin as the language of science. In the second half of the eighteenth century, this process intensified in the German Territories and many centers of science arose.[23] The natural sciences underwent a transformative period that laid the foundation for the modern disciplines as we know them today. Many existing disciplines became more specialized. New knowledge led to the formation of new scientific fields. Throughout this differentiation process, knowledge production intensified.[24] The importance of this period cannot be overstated; the patterns of thought that emerged played a crucial role in the debates about the strange phenomena of the summer of 1783.

I will illustrate the rapidness of this change with examples from three scientific realms. The first is meteorology. By 1783, weather observers increasingly relied on scientific instruments to take measurements; these instruments had become more standardized, which made for more useful data.[25] The second is the invention of hot-air balloons; these aircraft offered a new perspective on the world and the chance to conduct weather observations at higher altitudes. Some of the first flights took scientific instruments onboard to take readings in previously inaccessible locations: the first step in airborne atmospheric research.[26] The third is research on electricity.[27] This will be expounded upon in a later subchapter on the public's engagement with lightning rods.

These examples were closely related to the advent of new technologies. A particular feature of the transformational period of the natural sciences in the second half of the eighteenth century was the concept of "public science." Naturalists were often able to demonstrate that their new technology was practicable and enlightened amateurs were interested in the spectacular and often surprising effects of scientific experiments.[28] Demonstrations took place at universities, learned societies, salons, shops, and even on the streets.[29] Science became increasingly experimental and empirical. It was more important than ever that scientific experiments were reproducible and that scientists were accountable. Although newspapers reported on new findings and discoveries, the sciences also

---

21 STEINLE 2009: 54, 56; "Natur=Lehre, Natur=Kunde, Natur=Wissenschafft, Physick" in ZEDLER 1740, vol. 23: 1147–1167.
22 STOLLBERG-RILINGER 2011: 183.
23 WEIGL 1987: 25; HOCHADEL 2003: 35.
24 D'APRILE, SIEBERS 2008: 74.
25 GOLINSKI 1999.
26 DE SYON 2002: 7–13; LYNN 2010; THÉBAUD-SORGER 2013.
27 HOCHADEL 2003.
28 HOCHADEL 2003: 17–19.
29 WEIGL 1987: 25–26; TRISCHLER, BUD 2018: 187.

had their own channels of communication in the shape of numerous journals and book publications.[30] These journals offered naturalists a new and direct way to engage with the public.[31] Unfortunately, they also highlighted the increasing gap between the knowledge of the educated and that of the "common people."[32]

Enlightened science in the eighteenth century had two foundational aims: the search for the truth and the glorification of God.[33] This was not perceived as a contradiction; it was possible to think about the natural cause of an extreme event and simultaneously believe that God could intervene.[34] Naturalists of the early modern period often viewed science as the search for divine truth; their God-given curiosity inspired their attempts to decipher creation. Although the search for natural laws was the primary focus, God remained the creator of those natural laws.[35] From a Christian perspective, research on volcanoes and what the findings implied was problematic: deep geological time, of which fieldwork produced more and more proof, stood in contrast to biblical chronology. Geologists began formulating ideas about the mechanisms behind volcanic eruptions and refuting the notion that they were God's punishment for human misbehavior.[36]

Understanding weather was crucial for agriculturally oriented societies in the premodern world; a successful harvest depended on it and so, throughout antiquity, the skies were carefully observed. Records of daily visual observations of the weather exist from as far back as the late Middle Ages.[37] In the following centuries, scientific instruments helped measure the weather empirically.[38] In 1592, Galileo GALILEI invented the thermometer. In 1643, Evangelista TORRICELLI invented the barometer, which measures changes in atmospheric pressure and therefore changes in weather.[39] From the 1660s, the barometer was widely used in observatories and from around 1720, they were available to anyone who had the means to purchase one. Well-off households were likely to own a barometer during this time.[40] In 1714, Daniel Gabriel FAHRENHEIT invented the mercury thermometer; at the end of the eighteenth century, these devices, although not yet fully standardized, were widely available. For many years, this meant that if three different thermometers were stationed in one place, each could offer a different reading.[41] In the eighteenth century, scholars realized that to gain insights into long-term

---

30 MÜNCH 1992: 496; HOCHADEL 2003: 28; ALT 2007: 12; STEINLE 2009: 69; BEHRINGER 2011: 203–204.
31 D'APRILE, SIEBERS 2008: 74.
32 SCHMIDT 1999: 227; STOLLBERG-RILINGER 2011: 186.
33 HOCHADEL 2003: 21, 33.
34 REITH 2011: 83.
35 RUPPEL, STEINBRECHER 2009: 14; REITH 2011: 92–93.
36 MCCALLAM, 2019: 235–236.
37 SCHMIDT 1999: 37–38; MALBERG 2007: 189–190.
38 KINGTON 1988: 3.
39 CAPPEL 1986: 15; GOLINSKI 2007: 144–145; BEHRINGER 2011: 25–26.
40 WEYER, KOCH 2006b: 85; GOLINSKI 2007: 121, 144–145.
41 KINGTON 1980: 8; WEYER, KOCH 2006b: 81.

climatic trends, it was necessary to observe the weather from different locations with standardized equipment.[42] While long-term trends are crucial to historical climatology today, first-hand accounts of individual days are also invaluable: they draw attention to extraordinary weather events that might otherwise remain invisible. Weather observations were taken conscientiously throughout the eighteenth century, so a lot of data is available for this time.[43]

The Laki eruption and the unusual weather of 1783, which inspired Louis-Sébastien MERCIER to deem the year an *annus mirabilis*, coincided with a period in which the Enlightenment had thoroughly captivated most of Europe's naturalists. Many institutions existed that were well-suited to engage with the unusual weather, such as meteorological networks, learned societies, and universities. Laypersons and naturalists alike looked skyward, recorded their observations, and published their findings.[44] Beyond the circles of academia, many well-heeled Europeans took an interest in the natural world; a great deal of them even purchased instruments and kept daily logs.[45] As well as non-standardized equipment, another problem existed insofar as scientists lacked a standardized vocabulary to describe the weather.[46]

The frenetic pace of discovery in the late Enlightenment and the whirlwind that this created left naturalists straddling two worlds, one of relative nescience and one of knowledge. They were only too keen to cast off the old world, but the limitations of this intermediary period meant that some discoveries lay just out of reach. A glorious sense of excitement in some left them keen to use any opportunity to produce knowledge but, consequently, left them equally as keen to pursue the next mystery. Never was there a clearer example of this culture than the dealings of those who were concerned with the sultry mist of 1783.

Historical Context

1783: The End of the American Revolution and "Ballomania" in Europe

In the 1760s, growing differences strained the relationship between Great Britain and its colonies in North America, which led to conflict. The American Revolutionary War began with open combat between British soldiers and the Massachusetts militia in

---

42 CAPPEL 1986: 15.

43 MÜNCH 1992: 136–137.

44 OPPENHEIMER 2011: 277. The CHIMES project, based at the University of Bern and led by Stefan BRÖN-NIMANN and Christian ROHR, as well as a conference on a global inventory of early instrumental measurements back to the 1760s, has produced the following studies: BRUGNARA et al. 2019; BRÖNNIMANN et al. 2019; L. PFISTER et al. 2019.

45 DEMARÉE, OGILVIE 2001: 223.

46 MAUELSHAGEN 2010: 45–46, 50.

1775. On 4 July 1776, the Thirteen Colonies declared independence as the United States of America.[47] The war, however, continued, with different European countries joining as co-belligerents on both sides.

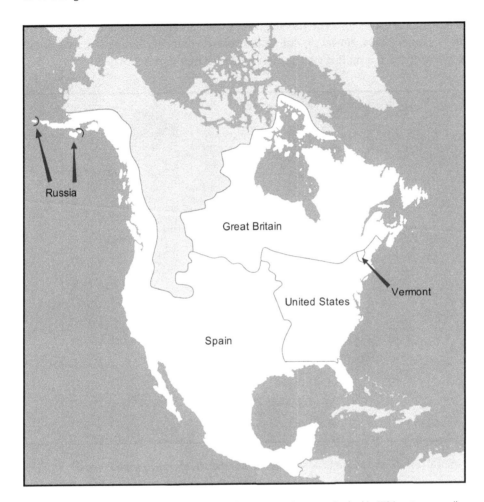

Figure 20: North America in 1784. This shows the claims non-native countries had in 1784, not necessarily the land they controlled.

Peace negotiations started in 1782. Four Americans, Benjamin FRANKLIN (1706–1790), John ADAMS (1735–1826), John JAY (1745–1829), and Henry LAURENS (1724–1792), were in Paris and had the authorization to negotiate a final peace treaty with Great Britain. On 3 February 1783, Great Britain acknowledged the independence of the United States of America and, soon after, on 15 April 1783, the Congress of the Confederation

---

47 GREENE, POLE 2000; BLACK 2001; FREMONT-BARNES, ARNOLD 2006; FERLING 2015.

ratified preliminary articles of peace. The Peace of Paris, a treaty between the United States and Great Britain, was signed on 3 September 1783; this formally ended the American Revolutionary War and granted the United States independence from Great Britain. Thus, the United States of America became a fully recognized and independent nation (Figure 20).[48] The US Congress of Confederation ratified the peace treaty on 14 January 1784. However, due to the severe winter in Europe (Figure 21) and North America, the ratified versions were not exchanged until 12 May 1784.[49]

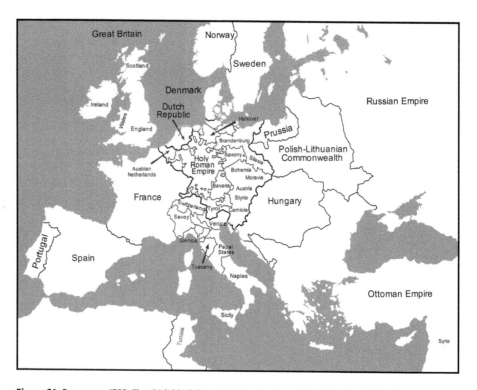

Figure 21: Europe ca. 1783. The thick black line marks the borders of the Holy Roman Empire.

During this period, George III was the king of Great Britain and Ireland (reigned 1760–1820). Louis XVI was the king of France (reigned 1774–1792); the last king of France before the abolition of the monarchy during the French Revolution. Joseph II, the brother of Louis XVI's wife, Marie Antoinette, was the Holy Roman Emperor (reigned 1765–1790). Frederick the Great ruled the Kingdom of Prussia (reigned 1740–1786).

---

48 FRANKLIN 2011: liv, lvi–lvii.

49 HOFFMAN 1986. For a detailed account of the adventurous journey undertaken by Josiah HARMER during the extraordinarily cold winter of 1784 to transport the ratified treaty from Annapolis, Maryland, to Paris, France, see SMITH 1963.

Catherine the Great was the empress of Russia (reigned 1762–1796). Christian VII was the king of Denmark and Norway (reigned 1766–1808). William V, Prince of Orange, was the last *Stadtholder* of the Dutch Republic (1751–1795).[50] According to the newspapers, Europe's main concerns at the time were the threat of invasion by the "Turks" and outbreaks of the plague.[51]

1783 also saw a milestone in human invention: in France, pioneers of air travel – like the MONTGOLFIER brothers – created hot-air balloons that could take off from the ground, fly to the skies, and (ideally) land safely. The term "ballomania," coined by Sir Joseph BANKS (1743–1820), adequately describes the enthusiasm for these "aerostatic globes" or "flying machines."[52] On 4 June 1783, the brothers Joseph-Michel MONTGOLFIER (1740–1810) and Jacques-Étienne MONTGOLFIER (1745–1799) demonstrated their first hot-air balloon, called a *Montgolfière*, to the public in their hometown of Annonay. The flight was uncrewed and lasted about ten minutes. It reached an altitude of about 2,000 meters, covered two kilometers, and landed safely in a nearby vineyard. News of this invention would dominate the headlines and communications among scientists for months. These "flying globes" inspired other inventors and would-be-balloonists to join the race to the skies.[53]

French inventor Jacques CHARLES (1746–1823), the only true competitor of the MONTGOLFIER brothers, also worked on a balloon of his own design; *Le Globe* was filled with hydrogen and made from silk and rubber rather than paper, like the *Montgolfière*. Using hydrogen instead of hot air meant *Le Globe* could be much smaller.[54] On 27 August 1783, at Champ-de-Mars in Paris, *Le Globe* saw its first uncrewed ascent. Benjamin FRANKLIN was in the crowd of 50,000 onlookers. These demonstrations were both scientific experiments and a kind of public entertainment that helped raise funds for the next experiment.[55] *Le Globe* traveled 21 kilometers but was destroyed with pitchforks and rocks by terrified residents in the village where it landed.[56] The competitive nature of this duel for the skies mirrored the nature of the debate regarding the meteorological phenomena of that year.

---

50 KINDER, HILGERMANN 2004: 276–287.

51 Münchner Zeitung, 17 July 1783: 439–440. The plague is caused by a bacterium called *Yersinia pestis*. After the Black Death in the mid-fourteenth century, the plague became endemic and recurred regularly until the nineteenth century.

52 GILLESPIE 1984: 242, 262.

53 GILLESPIE 1984: 249; FRANKLIN 2011: 394–396; HOLMES 2009: 125–128.

54 Hydrogen, a flammable gas lighter than air, had been discovered by English chemists Henry CAVENDISH and Joseph PRIESTLEY in 1766; HOLMES 2009: 125–131.

55 Journal de Paris, 28 August 1783. In a letter dated 30 August 1783 to Joseph BANKS, Franklin detailed his experience of the demonstration on 27 August; FRANKLIN 2011: 544, 547–555. FRANKLIN (2011: lvii, 393) kept the Royal Society in London apprised of the demonstrations of the balloon experiments, about which he was very enthusiastic. He used this opportunity to reestablish his connection to the Society. See also GILLESPIE 1984: 250–252.

56 FRANKLIN 2011: 548–551; FRANKLIN 2014: lvii.

On 19 September 1783, the MONTGOLFIER brothers demonstrated another *Montgolfière* in the front courtyard of the Palace of Versailles. The spectators included Louis XVI and Marie Antoinette. This aircraft had animal passengers: a sheep, a rooster, and a duck. The flight lasted eight minutes, traveled a distance of three kilometers, and reached an altitude of 460 meters. The animals managed to survive the journey.[57] Based on this positive

Figure 22: The view from Benjamin FRANKLIN's terrace in Passy on 21 November 1783. FRANKLIN was a keen observer of hot-air balloons. He also witnessed the first untethered journey of a *Montgolfière* hot-air balloon with a crew. FRANKLIN himself observed the ascent from the launching stage at the Château de la Muette.

result, the next step in the development of air travel was sending a person.[58] Contemporary engravings depict many of the test flights; significantly, for this book, none of these numerous depictions feature a hazy sky or a blood-red sun (Figure 22).

On 1 December 1783, the first-ever crewed journey in a hydrogen-filled balloon, *La Charlière*, named after Jacques CHARLES, took place in Paris (Figure 23). Benjamin FRANKLIN was also present at this event.[59] Five days after the experiment, FRANKLIN

Figure 23: The ascent of *La Charlière* in the Jardin des Tuileries in Paris on 1 December 1783. An estimated 400,000 spectators were present.

---

57 A report about the MONTGOLFIER demonstration to the Royal Family at the Palace of Versailles can be found in a letter to Joseph BANKS from 8 October 1783, FRANKLIN 2014: 81–83.
58 HOLMES 2009: 127–131; "Procès-verbal of the Montgolfier Balloon experiment," FRANKLIN 2014: 210–212; Letter to Joseph BANKS from 22–25 November 1783, FRANKLIN 2014: 216–220.

summarized the atmosphere surrounding the event in a letter to Henry LAURENS: "We think of nothing here at present but of Flying; the Ballons engross all Conversation."[60]

Several publications came out in and around 1783 that informed the interested reader about these "aerostatic balls."[61] Further hot-air balloon journeys followed in the next few years. Flying a hot-air or hydrogen balloon was a risky endeavor: fatal accidents were not uncommon. In 1785, the French aviator Jean-François PILÂTRE DE ROZIER crashed and died during an attempt to cross the English Channel from France to England.[62]

Earthquakes in Lisbon and Calabria

Three decades previous, a momentous event incited debate about science and God: the Lisbon Earthquake, which occurred on 1 November, All Saints' Day, in 1755. This devastating earthquake had its epicenter in the Atlantic, 200 kilometers southwest of Portugal, along the Azores-Gibraltar Transform Fault.

Geologists rely on descriptions and damage reports to estimate the intensity of historical earthquakes. The intensity scale used varies from region to region. These include the Modified Mercalli Intensity Scale (MMI), used in the United States, and the European Macroseismic Scale (EMS-98). Whereas magnitude scales indicate how much energy is released during an earthquake, intensity scales consider the level of destruction. Both intensity scales mentioned above have divisions between I ("not felt") and XII ("almost all structures destroyed").[63] The 1755 All Saints' Day earthquake in Lisbon was possibly Europe's largest and most destructive earthquake in recorded history.[64] It is estimated to have reached an XI on the MMI/EMS-98 scale.[65]

---

59 ROLT 1966: 29; HOLMES 2009: 131–133; letter from Benjamin FRANKLIN to Joseph BANKS on 1 December 1783, FRANKLIN 2014: 248–251.

60 FRANKLIN 2014: 264. A similar report was printed in the Münchner Zeitung, 19 September 1783: 586. "War and people were quickly forgotten and one did not talk about anything else other than the flying balloon," an anonymous correspondent snarkily remarked. "Es ist abscheulich zu sagen, was die Luft-kugel neulich hier für ein litterarisches Stiergeheze veranlasset hat. [. . .] Kaum war die Unterzeich-nung dann in Paris bekannt geworden, als schon Krieg und Frieden vergessen ward, und man sich von nichts mehr als dem fliegenden Ballon unterhielt. [. . .]."

61 For German translations of French books on the matter, see MURR 1784; LÜTGENDORF 1784; EHRMANN 1784.

62 GILLESPIE 1984: 249; HOLMES 2009: 153–155. For a brief overview of ballooning and flying after 1800, see HOLMES 2009: 159–162.

63 For more information on the Modified Mercalli Intensity Scale, see USGS, "The Modified Mercalli Intensity Scale." For more information on European Macroseismic Scale (EMS-98), see GRÜNTHAL 1998.

64 KÜBLER 2012: 14.

65 NOAA, Significant Earthquake Database, estimates that the Lisbon earthquake reached XI on the Modified Mercalli Intensity Scale. A recent study has suggested that the magnitude was not as large as

The earthquake and resulting tsunami affected several places in Portugal, Spain, and Morocco. The Portuguese capital of Lisbon was devastated. Between 30,000 and 70,000 people died during the earthquake and its fiery aftermath.[66] Furthermore, it was felt as far away as Finland and caused surface water waves (*seiches*) in Switzerland and Scotland.[67] Given the magnitude of the destruction, it was difficult for eyewitnesses to find words for the calamity. Often, they resorted to the tried and tested method of focusing on individuals' experiences, ordeals, and losses.[68]

The earthquake also had "intellectual aftershocks." It was interpreted as divine judgment as it destroyed almost all the churches in Lisbon, the conspicuous ruins of which left many feeling vulnerable.[69] The physical and intellectual consequences of this earthquake have been the subject of much scholarly attention. Many historians argue that the Lisbon earthquake profoundly influenced the European Enlightenment and fundamentally changed European culture and philosophy. The earthquake, and the death and destruction it left in its wake, led to a dialogue between French philosophers Voltaire (1694–1778) and Jean-Jacques Rousseau (1712–1778) on God's purported benevolence.[70]

German philosopher Immanuel Kant (1724–1804), residing in Königsberg, was fascinated by the reports he received from Lisbon; these reports prompted him to write three papers concerning earthquakes and volcanoes. Concerning volcanoes, Kant favored the hypotheses put forward by French naturalists Nicholas Lémery (1645–1715) and Georges-Louis Leclerc, Comte de Buffon (1707–1788), both of which suggested that subterraneous chemical reactions between sulfur and iron caused eruptions. This idea remained popular throughout the eighteenth century. It is unlikely that Kant ever experienced a significant earthquake himself. He had to rely on second-hand information gleaned from newspapers and other scientific publications for his writings.[71]

---

previously believed and the hypocenter might have been located closer to or even on the Iberian Peninsula; Fonseca 2020.

66 Brown 1991: 23. Jelle Zeilinga de Boer and Donald T. Sanders (2005: 94) suggest the earthquake reached a magnitude of 8.5 on the Richter scale. The USGS suggests a magnitude of 8.7; USGS, "Historic Earthquakes: Lisbon, Portugal, 1755 November 01." Another study suggests a magnitude of 8.5 to 9.0; Gutscher, Baptista, Miranda 2006: 154.

67 Demarée et al. 2007: 356. According to this text, a *seiche* is "a standing wave caused by seismic or atmospheric disturbance in an enclosed or partially enclosed body of water." See also Demarée, Nordli 2007.

68 Weber 2015: 360–361.

69 Löffler 1999; Frömming 2005: 155; Reith 2011: 82.

70 Dynes 1999. At the same time, Matthias Georgi (2009: 22, 163–168) states that he was unable to find indications of this clear change that is often assumed for the Lisbon earthquake in the English newspapers of the 1750s, nor in sermons and newspaper articles from Germany.

71 Reinhardt, Oldroyd 1983: 251–252; see also Kant 1755; Kant 1756a; Kant 1756b. All three of these texts by Immanuel Kant have been fully or partially translated by Reinhard and Oldroyd (1983).

Three and a half months after the 1755 Lisbon earthquake, another occurred in Düren, a town in western Germany. It probably reached a Richter magnitude of 6.2 ± 0.2 and precipitated in damage worthy of VIII to IX on the MMI/EMS-98 scale, making it one of the strongest earthquakes recorded in Germany. It killed two people, damaged several buildings, triggered a landslide, and brought home the destructive potential of earthquakes for those in the western parts of the German Territories.[72]

Gaston DEMARÉE and his colleagues describe the Lisbon earthquake "as a laboratory for new seismological concepts." Its consequences demonstrated the need for a better understanding of earthquakes and the hazards they pose.[73] Fatefully, a seismic sequence of five strong earthquakes that shook Calabria and Sicily some 30 years later would test the preparedness of Europe. Historian of science Deborah COEN highlights that the Calabrian earthquakes were the first to be described scientifically; *men of science* such as Scottish geologist Charles LYELL visited the affected regions in southern Italy, collected evidence, and documented the aftermath.[74]

The Calabrian earthquakes were caused by the subduction of the African tectonic plate under the Eurasian plate. The collision zone between the plates in southern Italy is called the Calabrian Arc. Stromboli and the Aeolian Islands show near-constant volcanic activity, and Mount Etna in Sicily is an active stratovolcano. These volcanoes are a product of the violent subduction of a tectonic plate. There were other tsunamigenic and deadly earthquakes in eastern Sicily in 1114, 1169, 1542, and 1693, and in the Messina Strait in December 1908.[75]

Some reports describe a "strange fog" that was visible before the first strong earthquake of the seismic sequence in Calabria (Figure 24), as early as 4 February 1783, which dispersed only to return for a few weeks over the summer.[76] An unknown author detailed his experience of the earthquake on 5 February 1783 (Figure 25) while on a ship in the Strait of Messina. Around noon, there was a terrible roar from the ocean and a loud noise permeated the air, which gave way to thick fog, similar to smoke, with a sulfuric smell. High waves added to the confusion, and the sailors feared that their ship would sink.[77] Given the nature of maritime weather, this "strange fog" could have simply coincided with the earthquake. Many Europeans

---

72 KÜBLER 2012: 64; 77. MEIDOW (1994) indicates an intensity of VIII to IX. LEYDECKER (2011: 48) states an intensity of VIII. A detailed analysis of the geology and the impact of the 1756 Düren earthquake, including an analysis of the question of whether it was related to the 1755 Lisbon earthquake, has been conducted by KÜBLER (2012).

73 DEMARÉE et al. 2007: 354.

74 COEN 2014: 15–16.

75 D'ANGELO, SAIJA 2002: 123, 133; GRAZIANI, MARAMAI, TINTI 2006: 1053; CARCIONE, KOZÁK 2008: 668. For the 1908 Messina earthquake, also see CARZIONE, KOZÁK 2008: 670; PARRINELLO 2015: 21–120.

76 KIESSLING 1988: 26; HOFF 1840, vol. 1: 108; KLEEMANN 2019a; KLEEMANN 2020.

77 ANONYMOUS 1783: 3.

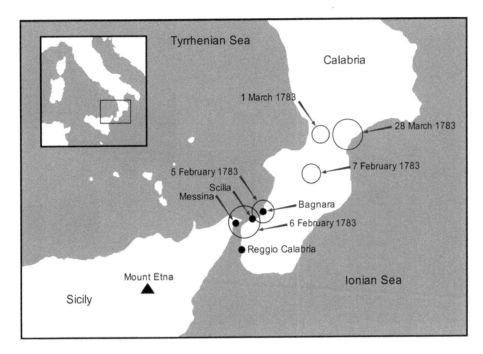

Figure 24: The locations of the five main earthquakes in Calabria in February and March 1783.

feared unusual fogs, as will become clear, believing them to be harbingers of misfortune. There had been a similar, ominous occurrence a mere two weeks before the All Saints' Day earthquake in Lisbon in 1755; a yellowish fog or smoke that was accompanied by a sulfuric smell appeared on 20 October 1755. An Icelandic eruption had caused this: Katla erupted from 17 October 1755 to February 1756 and released ash and gases that formed a haze, likely reaching Portugal just before the All Saints' Day earthquake.[78]

Figure 26 depicts the events of 5 February 1783. The tremors were probably severe enough to knock people off their feet. The artist portrays the wave-like ground movement that can occur during strong earthquakes. There are three different types of waves: the first to arrive are primary waves, which consist of push-and-pull movements. An up-and-down and side-to-side movement characterizes secondary waves. The third type of waves during strong earthquakes are surface waves, which can be subcategorized into Love waves, which travel sideways on a horizontal plane, or Rayleigh waves, which travel elliptically on a vertical plane (a reverse waveform).[79]

Furthermore, the print shows people falling as they try to run away; some are hit by the debris from collapsing buildings, while others are already trapped or buried

---

78 DEMARÉE et al. 2007: 337, 345, 349.
79 GROTZINGER, JORDAN 2017: 342–352, 372–382.

Figure 25: The Strait of Messina as seen from the north when the earthquake struck. On the left is the coast of Calabria, on the right is the harbor of Messina; in the distance, to the right, Mount Etna erupts (it is uncertain whether it erupted in 1783). Buildings, city walls, and lighthouses collapse, fires spread, and a tsunami (visualized as a whirlpool) causes trouble for the ships in the Strait.

under the rubble. Many buildings appear destroyed or severely damaged. Trees reportedly shook so violently that their tips touched the ground, and some were uprooted entirely. The ferocity of the earthquake was such that it even dislodged heavy pavement.[80] The print, however, is not a first-hand account: the artist lived in aseismic Paris where they imagined the poor of Calabria running around and the wealthy, equally terrified, being taken away in their horse-drawn carriages. Several other artworks of the earthquake and its aftermath depict deep cracks in the ground. According to geophysicist Jan KOZÁK, these are noteworthy as they indicate a macroseismic intensity of more than X.[81] It can be inferred that the earthquake on 5 February 1783 reached XI on the Mercalli-Cancani-Sieberg (MCS) scale, a predecessor to the MMI scale, which correlates with a Richter magnitude of around 7.0. This figure is based

---

80 DOLOMIEU 1784; JACQUES et al. 2001: 504.
81 KOZÁK, ČERMÁK 2010: 146–147.

Figure 26: The earthquake of 5 February 1783. Hand-colored copper engraving, ca. 1790.

on the reported destruction of almost all the built structures as well as occurrences of landslides and rockslides.[82]

Terrified, many spent the night outside; in Scilla, thousands slept on the beach, a fatal mistake. A second large earthquake occurred shortly after midnight on 6 February 1783, most likely reaching a magnitude of at least 6.5 on the Richter scale and X or XI on the MCS scale. A resulting rockslide near Scilla caused a tsunami, which swept away 1,500 of those seeking refuge on the shore. Both earthquakes also dammed rivers and created 215 new lakes. Another strong earthquake occurred on 7 February 1783 at 1:10 p.m., with an approximate Richter magnitude of 6.5. The next followed on 1 March 1783 at 1:40 a.m. This fourth earthquake was perhaps the smallest of the seismic sequence, reaching a Richter magnitude of less than 6.0. The fifth and last large

---

82 GRAZIANI, MARAMAI, TINTI 2006. A detailed description of the earthquakes and the damage they caused can also be found in TORCIA (1783).

earthquake occurred on 28 March 1783 at 6:55 p.m. with a Richter magnitude of 6.5 or greater. As it was felt at a greater distance, geologists believe it occurred at a greater depth than the others. The aftershocks lasted from 1783 to 1785, with more than 300 between February and May 1783.[83]

Many people had lost their homes. Destroyed infrastructure, such as bridges and roads, made transportation difficult. As is often the case with earthquakes, the spread of fire and disease caused problems. Landslides triggered by the tremors caused olive groves and cultivated flats to cascade hundreds of meters into the valleys below them, thus disrupting agriculture.[84] In June 1783, Ferdinand IV, the king of Naples and also the king of Sicily (reigned 1759–1816), established the *cassa sacra*, a governmental body to administer expropriated Church estates to aid with reconstruction.[85]

At the time, there was widespread interest among naturalists and scholars in this seismic crisis. Intellectuals, scientists, Italians, and foreigners all contacted their friends and family members in Calabria to get information.[86] News traveled, albeit slowly, via newspapers to other parts of Europe detailing this dreadful event. As early as 6 March 1783, there is a reference to the destruction caused by the Calabrian earthquakes in the *Münchner Zeitung*. The lengthy report covers almost two entire pages and is based on a letter from Rome dated 15 February 1783. The "terrible earthquake" was said to have destroyed 320 of the 375 towns and villages in Lower Calabria. Furthermore, the report states that "several terrible maws opened up and are now releasing thick smoke and sulfuric steam."[87] Other German-language newspapers published similar reports; only the numbers of existing and destroyed villages varied. News of each earthquake of the seismic sequence reached the German Territories within roughly one month of the event.[88] Descriptions of the earthquakes and tsunami must have sounded truly fantastic and petrifying to the readers: "The force of the volcano which caused all this was one of incomprehensible violence, as even the ships on the ocean were thrown up in the air, and all the elements and creatures felt His power."[89]

83 JACQUES et al. 2001: 504–506; D'ANGELO, SAJIA 2002: 126; GRAZIANI, MARAMAI, TINTI 2006: 1054–1059.
84 JACQUES et al. 2001: 503.
85 Das Wienerblättchen, 27 August 1783: 120.
86 PLACANICA 1985: 67.
87 Münchner Zeitung, 6 March 1783: 147–148: Report from Rome and Lower Calabria, 15 February 1783. "Im Messinesischen Grunde, und den umligenden Feldern ist die Erde an vilen Orten geborsten, und hat ungeheure Schlünde, aus denen diker Rauch, und Schwefeldampf emporquillt."
88 Hochfürstlich-Bambergische wochentliche Frag- und Anzeigenachrichten, 18 March 1783: 1–2 (News about the 5 February 1783 earthquake); Münchner Zeitung, 17 April 1783: 243–244. Giacomo PARRINELLO (2015: 219) states that the earthquakes flattened 182 towns and villages, 33 of which were later relocated.
89 Hochfürstlich-Bambergische wochentliche Frag- und Anzeigenachrichten, 15 April 1783: 2: Report from Italy, 24 March 1783. "Die Stärke des Vulkans welcher dies alles wirkte, muß von einer unbegreiflichen Gewalt gewesen seyn, weil sogar die Schiffe auf dem Meer in die Höhe geworfen worden, und also die Elementen und Kreaturen seine Wirkungen empfanden."

On 23 May 1783, William HAMILTON (1730–1803), the British ambassador based in Naples at the time, wrote to Joseph BANKS. His report, based on first-hand accounts of the affected region, is titled *An Account of the Earthquakes which happened in Italy, from February to May 1783*; it was read at the Royal Society on 3 July 1783 and published at the end of the same year.[90] To HAMILTON and others' surprise, the earthquake had not destroyed Reggio Calabria as they had expected. Indeed, it was in significantly better shape than Messina.[91] The *Hamburgischer Unpartheyischer Correspondent* published excerpts of HAMILTON's report on 30 July 1783. It is a terrifying account of the destructive power of earthquakes: "He [Hamilton] has also seen a house that had been thrown a quarter of an Italian mile from its initial location. A man and a woman had to lie under the rubble for four days until they were rescued." In addition, the report mentions that 40,000 people had been dug up from the rubble, but that 50,000 had perished.[92]

A study from 1935 by Guiseppe IMBÓ, director of the Catania Observatory, claimed that Mount Etna erupted on 17 February 1783 and that Stromboli erupted a few days later. Stromboli is a very active volcano: the term "Strombolian" is used to describe a volcano with near-constant activity. It is quite likely that Stromboli erupted around this time. However, as for Mount Etna, the Smithsonian Institution's Global Volcanism Program has only confirmed a VEI 2 eruption in March 1781 and a VEI 4 eruption in June 1787, but none in 1783.[93] Although contemporary newspapers shared stories about volcanic activity at Vesuvius, Stromboli, Vulcano, and Mount Etna, it remains unclear whether any of these volcanoes erupted. A point of certainty is that they did not affect the weather in any notable way.[94]

Nýey: A Burning "New Island"

On 1 May 1783, Jörgen MINDELBERG, the captain of a Danish fishing vessel called the *Boesand*, observed smoke rising from the sea southwest of the Reykjanes Peninsula. He noted this discovery in the ship's logbook at 3 a.m. On 3 May, the vessel returned to the area but found it impossible to inspect closely; within half a mile of the source

---

90 HAMILTON 1783.
91 JACQUES et al. 2001: 503.
92 Hamburgischer Unpartheyischer Correspondent, 30 July 1783. "Er hat auch ein Haus gesehen, das eine Italienische Viertelmeile weit von dem Platz, da es gestanden, geworfen worden. Ein Mann und eine Frau, die sich in dem Hause befanden, mußten 4 Tage unter dem Schutt liegen, ehe sie gerettet werden konnten." Michele TORCIA gave a detailed list of the towns and villages that lost people, he estimated the total to be 31,871 casualties; TORCIA 1783: 34–39.
93 IMBÓ 1935 (quoted after GRATTAN, BRAYSHAY, SADLER 1998: 26); Global Volcanism Program: Etna.
94 SIMKIN et al. 1981: 123; CAMUFFO, ENZI 1994: 32; CAMUFFO, ENZI 1995: 148. According to the "eruptive history" section of the Global Volcanism Program, Vesuvius was active from 18 August 1783 to 5 July 1784, VEI 3; Global Volcanism Program: Vesuvius. Stromboli and Vulcano were almost constantly producing magma.

of the smoke, the sulfuric smell became unbearable and MINDELBERG turned the ship around for fear his crew would faint from the stench.[95]

More famous were the reports by Danish fishermen aboard the *Torsken*. Captain Peder PEDERSEN and his assistant Gottfried SVENDBORG came across the burning island approximately 50 kilometers southwest of Reykjanes on 22 May 1783. Both wrote separate letters stating that the inhabitants of mainland Iceland had noticed smoke in the sea around Easter [20 April] without knowing the cause.[96] On 1 July 1783, an article was published in the Danish daily newspaper *Kjøbenhavns Adresse-Contoirs Efterretninger*, which stated that the island was surrounded by pumice, smoke, and fire, all of which was impeding sea travel in the area.[97] The *Königlich Privilegirte Zeitung* announced that the king of Denmark, Christian VII, had given the island the name *Ny-Oee*, which means "new island." In Icelandic, it was christened *Nýey*.[98]

A dramatic newspaper report detailed SVENDBORG's reaction upon discovering this island. When he first laid eyes on the smoldering outcrop, he feared that Iceland was lost.[99] However, upon circumnavigating the new island, SVENDBORG discovered it had a circumference of just one mile and reasoned that it could not be Iceland. He then had cause to travel further north, where he found the real Iceland the next day. He was relieved to find it intact, remarking on the welcome sight of the birds on the cliffs in their regular place. In a sign of things to come, a newspaper article remarked that this strange new island had emerged from the sea at the same time earthquakes had rocked Messina and Calabria.[100]

PEDERSEN and his crew were fishermen; they did not stick to the regimented sailing pattern of the merchants, who arrived in spring and departed in late summer or autumn. It seems that the fishermen left Iceland sometime in late May 1783, a few days before the earthquakes became stronger in the Síða region and the Laki eruption began. This explains why news of the new island reached Europe during the summer of 1783, but news of the Laki eruption only reached Europe in the autumn.

During the summer of 1783, Nýey received a lot of attention in the press; in particular, it fascinated naturalists and they speculated whether this newly emerging island might have produced the dry fog they were witnessing. This is unsurprising, given the

---

95 HALLDÓRSSON 2013: 20. "Before 1 May 1783" is also the time frame given for the eruption of Nýey by SIMKIN et al. (1981: 123).

96 HALLDÓRSSON 2013; STEINÞÓRSSON 1991: 136. These letters were sent to Professor HEINZE in Kiel and then they were translated from Danish into German; Hanauisches Magazin, no. 49 [first week of December 1783]: 449–450; DEMARÉE, OGILVIE 2016: 124; DEMARÉE, OGILVIE 2017.

97 Kjøbenhavns Adresse-Contoirs Efterretninger, 1 July 1783; World Data Center 1984: 12.

98 Königlich Privilegirte Zeitung, 12 July 1783: 688: Report from Copenhagen, 1 July 1783; WOOD 1992: 71; DEMARÉE, OGILVIE 2001: 230.

99 Hanauisches Magazin, no. 49 [first week of December 1783]: 450.

100 Hamburgischer Unpartheyischer Correspondent, 28 June 1783. A similar report can be found in the Berlinische Nachrichten, 10 July 1783.

temporal proximity of the appearance of the dry fog in Europe and the news about Nýey.[101]

In October 1783, prompted by further reports from the merchants who returned in early September, Christian VII ordered Magnús STEPHENSEN and Hans VON LEVETZOW to claim the island for Denmark.[102] This order was part of the grander expedition to inspect the extent of the damage caused by the Laki eruption in Iceland.[103] When they arrived in the spring of 1784, Nýey had vanished; it had succumbed to wave erosion.[104] Today, Nýey is a submarine crater, a submerged reef nine to 55 meters below sea level.[105]

Given what we know about other temporary volcanic islands that have emerged around Iceland in the past, it is unlikely that the Nýey eruption could have produced ejecta anywhere near the scale of that produced by the Laki eruption. Additionally, Nýey's eruptive activity took place in March, whereas the Laki haze only began in mid-June. Nýey could never have been responsible for the haze that mystified Europe.[106] Similar reactions had followed the news of newly emerging islands near Santorini in 1707, and in the Azores in 1720.[107]

The Weather of the 1780s

The 1780s were, in many regards, typical of the Little Ice Age. Temperature and precipitation extremes characterized the decade. The Maldà anomaly occurred from 1760

---

101 The news about Nýey reached Hamburg (28 June), Berlin, Dessau (3 July), Augsburg (7 July) and Munich (10 July), Bamberg, Zurich (11 July), Augsburg (14 July), Leiden and England (18 July), Vienna (3 August). Most of these reports were based on a letter from Copenhagen from 24 June. Hamburgischer Unpartheyischer Correspondent, 28 June 1783; Königlich Privilegirte Zeitung, 3 July 1783: 656; Dessauische Zeitung für die Jugend und ihre Freunde, 3 July 1783: 224; Augsburgische Postzeitung, 7 July 1783: 4; Münchner Zeitung, 10 July 1783: 423–424; Berlinische Nachrichten, 10 July 1783; Hochfürstlich-Bambergisches Wochenblatt, 11 July 1783; Zürcherische Freitagszeitung, 11 July 1783: 2–3; Gazette de Leyde, 18 July 1783: 6; Morning Herald and Daily Advertiser, 18 July 1783; Das Wienerblättchen, 3 August 1783: 5–6; FRANKLIN 2017: 293.
102 Hamburg Unpartheyischer Correspondent, 2 July 1783: Report from Copenhagen, 28 June 1783. "Die jüngst gedachte 7 Meilen von Island in der See empor gekommene Insel soll, auf Königl. Allerhöchsten Befehl an die Rentekammer, sogleich in Besitz genommen werden." The order, therefore, had already been given in June, but it was only executed later.
103 A letter from Copenhagen dated 8 October 1783 detailed Hans VON LEVETZOW's ("Le Chevalier de Levezau") mission to Iceland. Gazette de France, 4 November 1783.
104 WOOD 1992: 71.
105 World Data Center 1984: 12.
106 WOOD 1992: 71; Global Volcanism Program: Reykjanes (Nýey).
107 DEMARÉE and 2001: 230.

to 1800 and caused unusual weather patterns across southwestern Europe.[108] Weather was variable on a season-to-season and year-to-year basis, which caused concern. The Laki eruption heightened this variability from 1783 onward.[109]

The 1780s were a fascinating decade; several meteorological networks were founded that conducted systematic instrumental observations from multiple locations around the globe. In Europe, the most notable were the Societas Meteorologica Palatina (1780–1793) from Mannheim, the Society Royale de Médecine (1776–1789) in France, and the Bavarian Academy of Sciences and Humanities (1781–1789) (Figure 27). The dates speak to the fact that these networks did not survive the French Revolution and the Napoleonic Wars that followed. Outside of Europe, early instrumental weather records also existed in India (1784–1785), Iraq (1782–1784), and New South Wales (from 1788).[110]

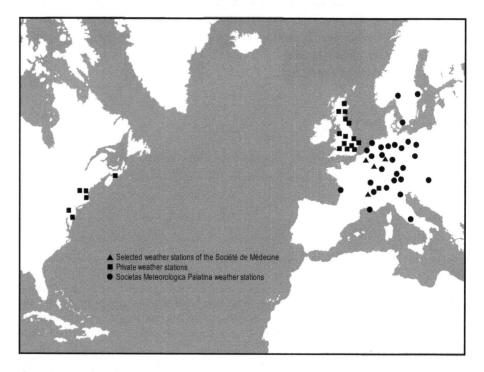

Figure 27: Map of weather stations in 1783 that featured in the research for this book.

---

108 Barriendos, Llasat 2003: 212; Michnowicz 2011: 9–11; Domínguez-Castro et al. 2012.
109 Lamb 1970; Kington 1988: 2; Barriendos, Llasat 2003: 201–202.
110 Damodaran et al. 2018: 517–518.

Charles Theodore, Prince-elector, Count Palatine, and Duke of Bavaria, founded the Societas Meteorologica Palatina. He had studied the natural sciences in university and gradually developed a keen interest in meteorology. His court chaplain, Johann Jakob HEMMER (1733–1790), was the Society's secretary.[111] At its peak, it boasted 39 weather stations in the Northern Hemisphere.[112] The Society's headquarters was in Mannheim in the Palatine region. Although most observatories were in Europe, some were in far-flung destinations such as the Ural Mountains, Greenland, and Massachusetts. Several of the weather stations were monasteries, such as the Andechs, Tegernsee, and Peißenberg stations in Bavaria and the Saint Gotthard Massif station in Switzerland. Each station received special instruments, including thermometers, barometers, and hygrometers, all of which had been calibrated in Mannheim by HEMMER. With these instruments came instructions to measure the temperature, pressure, and air humidity at three specific times during the day: 7 a.m., 2 p.m., and 9 p.m.[113] These measurements, along with general observations, were to be recorded in a standardized form. The records were then sent back to Mannheim every year, where HEMMER edited and published them in Latin as annual *Ephemerides*, roughly two years after they were recorded.[114] The Societas Meteorologica Palatina was unique because of its size and systematic approach to conducting weather observations.

HEMMER passed away in 1790. The Society struggled along, but the network of stations became less reliable, and its financial resources dwindled.[115] Then, in 1795, the French Revolutionary Wars brought chaos to Mannheim. The last volume of the *Ephemerides Societatis Meteorologicae Palatinae* was published on that same year, and the Society was dissolved soon after; Charles Theodore died in 1799.[116]

---

111 A list of private weather observers and a network of roughly 80 meteorological stations that made weather observations in Europe in the 1780s (either for the Societas Meteorologica Palatina or the Société Royale de Médecine) can be found in KINGTON 1988: 6–11.

112 CAPPEL 1986: 19; FLEMING 1998: 37; SCHMIDT 1999: 39; CAMUFFO 2002: 12.

113 KINGTON 1980: 12–15; KINGTON 1988: 14; BAUER et al. 2010.

114 Societas Meteorologica Palatina 1783: 56–61: The *Ephemerides* of 1783 were published in 1785. They covered the fog (*vaporis*) of 1783 extensively, they reported sightings of the fog in Mannheim and studied the relation of the fog to rain and wind, as well as the visibility of the stars. See also CAPPEL 1986: 19, 23; MÜNCH 1992: 137; MALBERG 2007: 190–191; PAPPERT et al. 2021.

115 CAPPEL 1986: 19, 25; KINGTON 1988: 12.

116 KINGTON 1988: 14; BAUER et al. 2010.

The Summer of 1783

The Extraordinary Dry Fog of 1783

Characteristics and Names of the Dry Fog

During the summer of 1783, a peculiar fog blanketed Europe. Peculiar because of its longevity and its dryness.[117] This haze, or smoke in some cases, had a blue hue.[118] On 17 June 1783, a weather observer named PREUS reported that a fine, smoke-like fog had started to appear around Sagan in Silesia.[119] The hygrometer reading indicated that there was very little moisture in the air, which was highly unusual in the presence of morning fog.[120] Icelandic poet and scholar Sæmundur Magnússon HÓLM (1749–1821) and French naturalist Jacques Antoine MOURGUE DE MONTREDON (1734–1818) both suggested that this dryness was due to sulfur within the mist. Later, and in a similar vein, Dutch mathematician and botanist Jan Hendrik VAN SWINDEN (1746–1823) judged sulfuric acid to be the cause of the dryness in his 1785 report in the *Ephemerides*.[121] In the eighteenth century, sulfuric acid was known as *oil of vitriol*, a byproduct of alchemy and metallurgy processes.[122]

PREUS observed this smoke-like fog at his weather station in varying intensities each day from 17 June to 13 September 1783. At first, it appeared as a fine mist, then on 19 June 1783, he observed thick fog for the first time. On 21 June 1783, he again described the fog as smoke. For the next three months, it remained, whether thick or thin, except for a few days in September. It reappeared, though more sporadically, through October and November.[123] Other similar reports came from the weather stations near the Saint Gotthard hospice in southern Switzerland and Padua in northern Italy.[124]

The unusual dry fog of 1783 is a significant focus of that year's 694-page-long *Ephemerides*; the preface, several special treatises, and the observations from most of their

---

117 Berlinische Nachrichten, 19 July 1783: Report from the Mannheim observatory, 6 July 1783: 670; PFISTER 1972: 23–24.

118 TITIUS 1783: 206–207. According to François VERDEIL (1747–1832) in Lausanne, Switzerland, the haze was bluish, sometimes red; PFISTER 1972: 24.

119 "Vapor t[enuis] quasi fumus," Societas Meteorologica Palatina 1783: 339. Sagan is today's Żagań in Poland.

120 Societas Meteorologica Palatina 1783, 17 June 1783: 359.

121 STOTHERS 1996: 82–83: It is described as sulfuric acid (by Jan Hendrik VAN SWINDEN in Societas Meteorologica Palatina 1783: 679–688) and volcanic sulfur gases (HÓLM 1784b; MOURGUE DE MONTREDON 1784).

122 KARPENKO, NORRIS 2002; MALILA 2018: 2. VAN SWINDEN calls it "vitriolic-acidic air" in Societas Meteorologica Palatina 1783: 688. "Aciditas, pondus, effectusque hujus *Gas* efficiunt, ut credam, ipsum ad naturam illius Gas, quod *aër-acidus-vitriolicus* dicitur, accessisse."

123 Societas Meteorologica Palatina 1783: 359–368.

124 Societas Meteorologica Palatina 1783: 186; 573.

weather stations mention the dry fog and detail its effects, such as the red coloration of the sun or moon. Most of the weather stations recorded the dry fog in their *meteora* column, referring to it sometimes as a *vapor*, with descriptions such as dry (*siccus*), thick (*spissus*), or fine (*tenuis*). On days with no fog, the observer wrote *nullus*. A symbol resembling the five dots on a die, followed by an asterisk, was also used to indicate thick fog.[125]

The time of year that the fog appeared was also curious. VAN SWINDEN remarked that between 1774 and 1783, he had never observed any fog in June. Typically, a storm, the sun, or winds dissipate a fog; neither rain nor storm seemed capable of dispersing this anomalous haze.[126]

Due to its density, this fog reduced the optical visibility on both land and sea. Richard STOTHERS calculated that in some parts of Europe, the dry fog reduced the visibility to two kilometers. On a clear day, from a high enough viewpoint (because of the curvature of the Earth), the human eye can see as far as 20 kilometers.[127] Today fog and mist can be separated by their density: fog allows for visibility of less than 1,000 meters, while mist allows for visibility greater than 1,000 meters.[128] Both fog and mist consist of tiny water droplets suspended in a cloud. If visibility is reduced due to dry particles, it is referred to as haze. Another term we use today is smog, which is a mixture of smoke, gases, and chemicals.[129] In 1783, these distinctions were not so clear.

During the summer of 1783, the dry fog was an almost pan-European occurrence and received different names in different regions: for instance, in England, the dry fog was primarily referred to as a *haze*; in France, it was called *vapeur* (vapor) or *brouillard sec* (dry fog); in Sweden, it was called *sol-röken* (sun smoke); and in Italy, it was called *caligine* (haze). In Iceland, the preferred term was *móða* (mist).[130] The German sources have several names for the dry fog, including *Dunst* (haze), *Duft* (smell, but in 1783, a synonym for haze, steam, or fog), *trockener Nebel* (dry fog), and *Höhenrauch* (high smoke).[131] Other German terms in circulation included *Heerrauch* and

---

125 Societas Meteorologica Palatina 1783: 57. A good explanation and a legend of the different symbols applied in the *Ephemerides* by the Societas Meteorologica Palatina can be found in KINGTON 1988: 24.
126 VAN SWINDEN 2001: 73; GLASER 2008: 234.
127 STOTHERS 1996: 82; GRATTAN, BRAYSHAY, SCHÜTTENHELM 2002: 100. From a high enough viewpoint, on a clear dark night, the human eye can see a candle flame as far as 48 kilometers away.
128 MALBERG 2007: 105; AHRENS 2009: 113–115.
129 GIBBONS 2018 on weather.com.
130 COTTE 1783; LAPI 1783; OPPENHEIMER 2011: 277; CASEY et al. 2019.
131 "Höhenrauch" in GRIMM; GRIMM 1877 (1984): 1711. "Duft" in KRÜNITZ 1776/1785, vol. 9. For a further debate on the names *Haarrauch*, *Heerrauch*, *Höhenrauch*, and *Moorrauch*, see DEMARÉE 2014.

Hahlrauch, which was a reference to the fog's dry and "smoky" quality.[132] VAN SWIN-
DEN, writing in Latin, dubbed it *nebula*, which can be translated as "cloud, fog, smoke,
mist, or haze."[133]

One consequence of the fog's thickness was the reddish appearance of the sun
and the moon, particularly at a few degrees above the horizon. At the Berlin weather
station, on 23, 24, and 26 June 1783, as well as 10 and 17 July 1783, "the sun set in the
color of blood," as a weather observer by the name of BÉGUELIN notes in the *meteora*
column of his journal.[134] On other days during this time, he describes the sun as
red.[135] As early as 18 June 1783, with the first appearance of the fog, the sun appeared
red at the weather station in Göttingen. The last time the sun and moon were described as red there was 3 September 1783.[136] At the Tegernsee weather station, at a
lake in Upper Bavaria, the weather observer P. DONAUBAUER writes in his annotations
that the sun and the moon at rising looked very similar to glowing iron.[137] Descriptions of the sun detail how its rays had lost all intensity, and it was either a bluish-
white hue or appeared as an iron-like red globe.[138] Many observers note that, at
times, the sun was blood-red or cherry-red, as was the moon.[139]

Planets and stars, previously visible to the naked eye from 20 to 40 degrees in the
sky, were at times rendered invisible.[140] The 20 brightest stars in the night sky are
called first-magnitude stars. When the fog was at its densest, these stars were hidden
from view in the lower parts of the sky from Scandinavia all the way to Italy.[141] At
times, the sun seemed to disappear entirely below ten degrees, such as in Copenhagen, Geneva, and Narbonne. During this period, parhelia, bright spots to the left and
right of the sun, and paraselenae, a similar occurrence involving the moon, were also
observed.[142]

---

132 Bayerische Akademie der Wissenschaften 1783, 44; "Hal" in Pfälzisches Wörterbuch, CHRISTMANN,
KRÄMER 1980: 591.
133 THORDARSON, SELF 2011: 66.
134 "Sol occidit colore sanguineo." Societas Meteorologica Palatina 1783: 108–110.
135 "Sol ruber" (on 11 and 18 July 1783), Societas Meteorologica Palatina 1783: 108–110.
136 Societas Meteorologica Palatina 1783: 663–673.
137 Societas Meteorologica Palatina 1783: 299.
138 STOTHERS 1996: 83–84.
139 PFISTER 1972: 24.
140 STOTHERS 1996: 82; GLASER 2008: 234; BEHRINGER 2011: 213. A description from Ofen, Hungary, indicates that the "vapor-filled air" (*dunstige Luft*) obscured the moon at times, such as on 31 July, 25 August,
and 4 November 1783, as stated in a report by the observers WEISS and BRUNA from the Royal Observatory of Ofen in the Berliner Astronomisches Jahrbuch for 1787; BODE 1784: 182–185.
141 Societas Meteorologica Palatina 1783: 60; HÓLM 1784a; HÓLM 1784b; TOALDO 1784; STOTHERS 1996:
82–84.
142 KIESSLING 1885; STOTHERS 1996: 82–83.

The Beginning of the Dry Fog

On 10 June 1783, the dry fog reached the Faroe Islands, 450 kilometers to the southeast of Iceland, only two days after the volcanic activity at the Laki fissure commenced (Figure 28).[143] Around the same time, the fog also appeared above the western coast of Norway and northern Scotland.[144] These first offshoots of the fog were faint. Ashfall was reported in the Faroe Islands and in Trondheim, Norway.[145] According to the 1882 geology textbook by Scottish geologist Archibald GEIKIE (1835–1924), in Caithness, northern Scotland, 1783 is famously remembered "as the year of the ashie," when a fine but persistent ash fall damaged crops and vegetation.[146] Some ash, although substantially less, fell in the Netherlands, Denmark, and northern Germany.[147]

Around 14 June 1783, faint manifestations of the dry fog appeared above the European continent, growing thicker over the next few days. Westerly winds carried yet more ash and dust eastward, and by the end of the month, a continent-wide veil of fog enshrouded the land.[148] On 30 June, the dry fog reached Aleppo in modern-day Syria; one day later, it was visible above Baghdad and the Altai Mountains in western China. The latter is approximately 7,000 kilometers from Iceland.[149]

Jan Hendrik VAN SWINDEN documented how the Norwegian Sea and the North Sea were covered by the dry fog from at least 25 to 30 June 1783. On the last two days of the month, the fog "was so dense, that it nearly removed all view."[150] VAN SWINDEN gathered this information from a logbook that covered the journey of a ship that left Norway on 19 June and reached Groningen on 2 July 1783. From this, VAN SWINDEN concluded that the dry fog must have come to the Dutch Republic from a northern region.[151]

Jens Jacob ESCHELS (1757–1835), a seaman originally from the North Frisian Island of Föhr, wrote his memoirs in 1831 and published them in 1835. He started going to sea in 1769. On 1 May 1783, he traveled from St. Thomas in the Caribbean to Europe

---

143 There are some descriptions of earlier appearances of the dry fog, such as a sighting on 24 May 1783 in Copenhagen and sightings on 6 and 7 June 1783 in La Rochelle, which were mentioned by KIESSLING (1888, vol. 1: 27–28). This might be either a transcription mistake or occurrences of a "normal" fog. Alternatively, it is possible that these were, in fact, thin offshoots of a dry fog that was produced by activity at Grímsvötn in May 1783 and before the beginning of the Laki eruption.

144 KIESSLING 1888; FIACCO et al. 1994; THORDARSON 1995; THORDARSON, SELF 2001: 66; GLASER 2008: 234.

145 HÓLM 1784; THORODDSEN 1914; THORODDSEN 1925; THORDARSON, SELF 1993: 249. For ashfall in Scandinavia, see NORDENSKIÖLD 1875; ÞÓRARINSSON 1981.

146 GEIKIE 1882: 219; VASEY 1991: 327; STOTHERS 1996: 80. A detailed list of translated historical sources that mention the first appearance of the Laki haze in Europe can be found in THORDARSON, SELF 1993: 20–24.

147 HÓLM 1784a; HÓLM 1784b; THORODDSEN 1914; THORODDSEN 1925; THORDARSON, SELF 1993: 249.

148 STOTHERS (1996: 80–82), OMAN et al. (2005), and THORDARSON and SELF (2003: 21–24) include detailed descriptions from the historical sources of when the dry fog appeared and where.

149 RENOVANZ 1788; ÞÓRARINSSON 1979; STOTHERS 1996: 80–81.

150 VAN SWINDEN 2001: 76 (all translations herein are by Susan LINTLEMAN).

151 VAN SWINDEN 2001: 76.

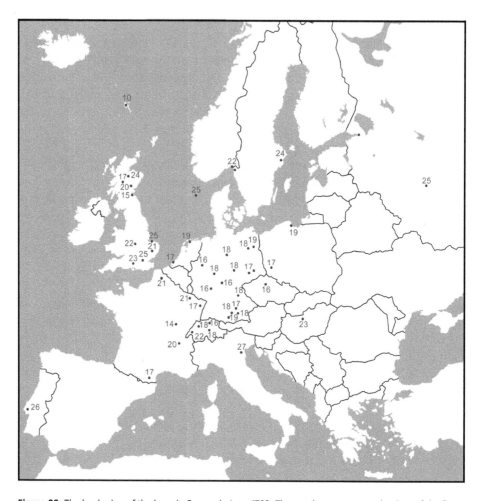

Figure 28: The beginning of the haze in Europe in June 1783. The numbers represent the date of the first mention of the dry fog in different cities around Europe in June. The dates are based on my sources (see list of illustrations for more detail).

and reached Heligoland on 6 July. When his ship reached the Azores – presumably in mid-June 1783 – he noticed the presence of fog which, according to ESCHELS, lasted until August 1783.[152] He details in his writings how the fog was visible above the Atlantic Ocean, the North Sea, and the Baltic Sea; for this reason, "the old seafarers referred to the year 1783 as the mist year [*das Mistjahr*]." Most of these "old seafarers" believed that earthquakes in Messina and Iceland caused the fog. He writes: "The fog was

---

152 ESCHELS 1928: 5–13, 128–130.

terrible for sailing as one was not able to see very far, but thank God I found my way! Once we reached the English Channel, we had 14 days of easterly wind and thick fog."[153]

Over the Baltic Sea, the dry fog was also visible, hampering travel. The *Frankfurter Staats-Ristretto* reported that travelers coming from St. Petersburg to the Lower Elbe region by ship had noticed a "great fog [. . .] everywhere" that made passage difficult.[154]

The Mediterranean was affected as well. On 6 July 1783, reports from Italy reached the German Territories stating that on the coast of the Adriatic Sea, the fog was so thick that ships had to signal each other with cannon fire in an effort not to collide.[155] On the Italian west coast, ships fared no better; it was reportedly impossible to navigate without the aid of a compass.[156]

The Fog in the Alps

The fog curtained the Alps from many a disappointed traveler that year; in Munich, one observer complained of being denied the spectacular vista to which they had become accustomed.[157] John Thomas STANLEY (1735–1807), a British peer and later politician, traveled through Europe on a Grand Tour in 1783; on or around 28 June 1783, he arrived in Switzerland near Bienne at Lake Biel. Unfortunately for STANLEY, this "splendid & beautiful scenery was concealed [. . .] for a considerable time [. . .] by a fog which had spread itself over a great part of Europe. It was of a peculiar kind, having no apparent moisture."[158]

This description corroborates the German newspaper reports that state that the Alps and Lake Geneva were barely visible.[159] Swiss weather observers, including Sigmund Gottlieb STUDER (1757–1834) from Bern, note a reduction in visibility; at times,

---

153 ESCHELS 1928: 129. "Als wir während dieser Reise auf der Länge der Azorischen Inseln waren, kamen wir in einen Nebel, der dieses Jahr den ganzen Sommer, bis im Augustmonat, im Ozean und in der Nord- und Ostsee war, weshalb das Jahr 1783 von den alten Seeleuten das Mistjahr genannt wird. Dieser Nebel ist der Vermutung nach von dem Erdbeben in Messina (welche Stadt meist unterging) und zu gleicher Zeit in Island entstanden, und weil diesen Sommer die meiste Zeit stiller Wind war und gar kein Sturm wehte, so konnte er nicht weggehen, sondern blieb so lange stehen. Es war für die Seeleute sehr schlimm zu fahren, weil man nicht weit vor sich hinsehen konnte; doch gottlob! ich fand meinen Weg. Als wir in den Englischen Kanal kamen, hatten wir vierzehn Tage Ostwind und dicken Nebel."
154 Frankfurter Staats-Ristretto, 25 July 1783: 486: Report from the Lower Elbe, 16 July 1783. "Reisende die von Petersburg kommen, berichten, daß der grosse Nebel sich auf der See ebenfalls überbreitet hat, und daß sie ihre Reise deswegen mit der größten Gefahr zurückgelegt haben."
155 Hamburgischer Unpartheyischer Correspondent, 22 July 1783: Report from Italy, 6 July 1783.
156 Frankfurter Staats-Ristretto, 4 August 1783: 508: Report from Naples, 15 July 1783.
157 Bayerische Akademie der Wissenschaften 1783: 45.
158 John Thomas STANLEY, MS, JRL 722, John Rylands Library, University of Manchester, Manchester, UK: 95–96.
159 Berlinische Nachrichten, 31 July 1783: 706: Report from Switzerland, 22 June 1783.

STUDER had difficulty spotting the Gurten hill, just south of Bern. Swiss priest Johann Jakob SPRUENGLI (1717–1803), who made his weather observations from Gurzelen, noticed that the Gurnigel mountain, only 6.5 kilometers away, and the Stockhorn mountain range, which was eight kilometers away, were at times completely obscured from view. The Swiss meteorologist and geologist Jean-André DELUC (1729–1812), from his vantage point in Geneva, could scarcely make out Mont Salève, a mere 6.5 kilometers away from him; the Jura Mountains, which were 16 kilometers away, were a faint silhouette.[160]

Mountains afforded naturalists the opportunity to see just how high the dry fog reached: DELUC climbed Mont Salève (1,379 meters above sea level) intending to reach the upper limit of the fog, but even at the summit, he found that the landscape remained veiled.[161] The French botanist, physicist, geologist, and meteorologist Robert Paul DE LAMANON (1752–1787) summited Mont Ventoux (1,912 meters) with similar intentions; at the peak, the fog still surrounded him.[162] Alpine herdsmen, who tended to their animals at around 2,000 meters above sea level, reported a mist (*Dunst*) that shrouded the highest plateaus. Swiss chamois hunters observed the dry fog on all the mountains and peaks they worked.[163] Between 1781 and 1789, Capuchin priests Pater Onuphrius and Pater Laurentius carried out observations at regular times at the hospice on the Saint Gotthard Massif for the Societas Meteorologica Palatina (2,469 meters).[164] During the day, they observed the dry fog at the hospice and the nearby summits of 2,700 to 2,900 meters; at night, the upper limit of the dry fog was said to have fallen to as low as 2,350 meters.[165] Even though the fog was perceptible at 3,000 meters above sea level, it became apparent that it thinned with altitude.[166]

The Peak of the Dry Fog

Modern volcanologists have estimated that the Laki eruption released 122 megatons of sulfur dioxide. This volume is equivalent to the emissions of 12,000 coal-fired power plants over one year.[167] 95 megatons of sulfur dioxide were released into the polar jet stream and traveled toward Europe; the lava emitted the remaining 27 megatons, which mainly affected southern Iceland (Figure 29).[168] The sulfur dioxide (SO_2) reacted with moisture in the atmosphere and formed the approximately 180 megatons

160 PFISTER 1972: 24.
161 PFISTER 1972: 24.
162 LAMANON 1799: 82–83.
163 PFISTER 1972: 24.
164 PFISTER 1975: 86.
165 PFISTER 1972: 24.
166 LAMANON 1784a; SENEBIER 1784; STOTHERS 1996: 82–83.
167 WITZE, KANIPE 2014: 133–134.
168 GRATTAN et al. 2005.

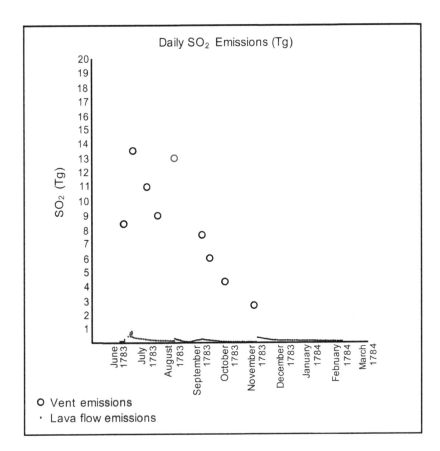

Figure 29: The Laki eruption's emissions of sulfur dioxide.

of sulfuric acid aerosols (H_2SO_4) that saturated the dry fog of 1783.[169] Additionally, seven megatons of chlorine and 15 megatons of fluorine were released.[170] Other gases released by the eruption include hydrogen sulfide (H_2S), ammonia (NH_3), and fluorine (F).[171] The Laki eruption's initial eruptive phases were the most powerful; the first three eruptive phases released as much as 40 megatons of sulfur dioxide over ten days.[172] Between 8 June and 8 July 1783, the first month of the eruption, the fissure released roughly 60 percent of the total volume of gases it would emit.[173] Thorvaldur THORDARSON and Stephen SELF estimate that by 26 June 1783, the dry fog enveloped almost all of Europe

169 THORDARSON, SELF 2003: 9; GRATTAN et al. 2005; THORDARSON 2005: 211.
170 GRATTAN et al. 2005.
171 DURAND, GRATTAN 1999.
172 GRATTAN et al. 2005.
173 GRATTAN, BRAYSHAY 1995.

with varying densities on a regional level. The thickest manifestations lasted roughly from 20 June to 23 July 1783.[174]

THORDARSON and SELF calculated that the conversion of sulfur dioxide into sulfuric acid aerosols at the altitude of the polar jet stream takes about one to two weeks; this correlates well with the time that passed between the onset of the eruption, the arrival of the dry fog, and the first descriptions of a sulfuric smell in Europe.[175] After 23 June 1783, episodic gaseous bursts, where the volcanic gases within the fog seemingly intensified, lasted a few minutes to a few days. These short-term, extremely poisonous episodes damaged human health, withered vegetation, and even corroded metal surfaces. The first of these bursts struck England and France on 23 June 1783 and the Dutch-German border region one day later; they then moved further south and eventually dispersed.

In England, several sources describe a "severe frost" event that occurred during the night from 23 to 24 June 1783.[176] On 23 June, the British naturalist Gilbert WHITE (1720–1793), in Selbourne, southern England, noted not only the first appearance of the dry fog in his area but also that "[t]he blades of wheat in several fields are turned yellow & look as if scorched with the frost." On 24 June, he wrote, "[. . .] Sun, sultry, misty & hot. [. . .] This is the weather that men think injurious to hops."[177] James WOODFORDE (1740–1803), an English clergyman based at Weston Longville, Norfolk, also notes "a smart frost this evening" in his diary for 23 June and "[a] smart Frost again this Night" for 24 and 25 June. These descriptions are surprising, considering the previous days had been quite hot, according to his weather diary.[178]

J. FENTON authored a similar report in Nacton in Suffolk, where he kept a weather diary; on 24 June 1783, he wrote that a particular sale at a market "began notwithstanding the close heat in the day, it was commonly reported that there was a sharp freezing rhyme this morning, which on the succeeding day caused the leaves to drop from the Trees."[179] The *Sherborne Mercury* also reported severe plant damage from this "frost" in the eastern parts of England:

---

174 DURAND, GRATTAN 1999: 371. THORDARSON, SELF 2003: 8, Figure 5. This figure compares the Laki haze in Grund, Iceland, and Mannheim, Germany. The haze's thickness in Europe corresponds well with the eruptive episodes that caused it.
175 THORDARSON, SELF 2001: 68–70.
176 GRATTAN, BRAYSHAY 1995. John GRATTAN and F. Brian PYATT (1999) compiled many descriptions of how the vegetation fared during the presence of the dry fog in 1783.
177 Gilbert WHITE, "The Naturalist's Journal," 1783, Add MS 31848, British Library, London, UK: 128.
178 James WOODFORDE, Weather Diary for Weston Longville, Norfolk, vol. 10: 1782–1784, MET/2/1/2/3/467, National Meteorological Library and Archive, Met Office, Exeter, UK.
179 J. FENTON, The Weather, etc. from June 27, 1779 to December 30, 1786 (at Nacton, Suffolk), Bib. No. 178296, National Meteorological Library and Archive, Met Office, Exeter, UK.

Throughout most of the eastern counties, there was a most severe frost in the night between 23 and 24 June 1783. It turned most of the barley and oats yellow, to their very great damage; the walnut trees lost their leaves, and the larch and firs in plantation suffered severely.[180]

Reverend John CULLUM (1733–1785), a clergyman from Hardwick, Suffolk, writes of the unseasonable frost on these nights with the following words: "All these vegetables appeared exactly as if a fire had been lighted near them, that had shriveled and discoloured their leaves."[181] Thomas BARKER (1722–1809) was a weather observer from Leicestershire in England and brother-in-law to naturalist Gilbert WHITE. BARKER noticed a strange smell on 30 June 1783; it was an evening with "thick smoaky air & smell of fens," with winds from the north to the east.[182] The "smell of fens" was probably hydrogen sulfide, a colorless gas that smells of rotten eggs.[183] On 24 and 25 June 1783, the sulfuric haze also crept north to Scotland.[184]

Although parts of England reportedly suffered episodes of "severe frost" and "thick ice" between 26 and 31 May 1783, in general, a great heat characterized the summer. It is unlikely that the episode on 23 June 1783 resulted from frost.[185] A report from an anonymous correspondent in the *Norwich Mercury*, printed on 19 July 1783, opined "that the late blast which affected the progress of the vegetation was not a frost," but rather an air "impregnated with sulphurous particles" due to the recent earthquakes in Messina and other places.[186] Modern natural scientists, such as John GRATTAN and F. Brian PYATT, have come to the conclusion that the damage described was too selective for a midsummer frost. However, it was "typical of damage by acids and halogens. [. . .] The shedding of leaves is a classic response to concentrations of fluorine, sulphur dioxide and hydrofluoric acid, and charring is typical of damage caused by a sulphuric acid aerosol."[187] Charged by the first three eruptive episodes, it seems fog rather than frost attacked the vegetation.

On the other side of the Channel, on the same evening, the "frost" was remarked upon in Arras, Pas-de-Calais, in northern France. Monsieur BUISSART, a weather observer for Société Royale de Médecine, notes this event not in the "daily observations" column of the standardized sheet but instead in the "special observations" field.[188]

180 Sherborne Mercury, 14 July 1783 (quoted after GRATTAN, PYATT 1994: 242).
181 CULLUM 1784: 417.
182 Thomas BARKER, Private Weather Diary for Lyndon Hall, Leicestershire, MET/2/1/2/3/227, National Meteorological Library and Archive, Met Office, Exeter, UK.
183 United States Environmental Protection Agency 1980; POWERS 2004; GREENWOOD, EARNSHAW 2008.
184 DAWSON, KIRKBRIDE, COLE 2021: 7.
185 Gilbert WHITE, "The Naturalist's Journal," 1783, Add MS 31848, British Library, London, UK: 124.
186 Norwich Mercury, 19 July 1783 (quoted after THORDARSON, SELF 2003: 14). A credible argument refuting the notion of an overnight frost during the hot and hazy summer of 1783 can be found in GRATTAN and PYATT (1994).
187 GRATTAN, PYATT 1994: 245.
188 Meteorological Observations for Arras, Pas-de-Calais, France, June 1783, by the Société Royale de Médecine. BUISSART also highlights that he had based this information on hearsay.

Further north, the botanist Eugène-Joseph D'OLMEN, Baron DE POEDERLÉ (1742–1813), based in Brussels and Saintes, then in the Austrian Netherlands, first describes a sulfuric smell in his notes for 24 June 1783.[189] The dry fog most likely reached a peak in sulfuric acid aerosol concentrations here on or around 24 and 25 June 1783. These peak concentrations became particularly apparent in the border regions between the Dutch Republic and the German Territories.

The *Ephemerides* of the Bavarian Academy of Sciences and Humanities reported on the plight of the people of the Dutch Republic, who had not seen daylight for half a week and had to cover their faces with sheets when outside to keep the foul odors at bay.[190] Dutch botanist and physician Sebald Justinus BRUGMANS (1763–1819), then studying in Groningen, wrote and published a book about this dramatic turn of events. He writes of a fog that materialized before 20 June 1783 in various locations across the Dutch Republic. In some areas, such as in Holland and Utrecht, it had no apparent sulfuric odor, but in others, such as in Gelderland and Overijssel, there was "a sulfuric smell [. . .] admixed to the fog."[191] BRUGMANS states that, on 24 June 1783, the fog was so intense that one could taste it with each breath. Although the smell began to dissipate over the following days, the dry fog remained; by the morning of 28 June 1783, the sulfuric smell had vanished entirely.[192]

Jan Hendrik VAN SWINDEN observed the sulfuric odor in Franeker. His description suggests that the town was redolent with a stench that crept through houses and clung to everything. According to VAN SWINDEN, those with "delicate lungs" had trouble with this odor; this was most likely a reference to those with pre-existing respiratory diseases. He pronounced that with each breath, he had to stifle a cough. This unfortunate situation followed him from the city in his retreat to the countryside. A growing number of people complained about headaches and respiratory difficulties, particularly asthma.[193] The connection between pollution and asthma was virtually unknown in the eighteenth century.[194]

In Groß Hesepe, in the neighboring Emsland region, this first wave of particularly sulfurous fog appeared on 24 June 1783, Saint John's Eve, and lingered until the next day. A few days prior, a smell that a local chronicler compared to "heated hay" had already manifested itself. A smell similar to burnt gunpowder and decay became apparent.[195] The comparison to burnt gunpowder strongly suggests that what they

---

189 DE POEDERLÉ 1784: 336.
190 Bayerische Akademie der Wissenschaften 1783: 45. A similar report can be found in the Frankfurter Staats-Ristretto, 15 July 1783: 465.
191 BRUGMANS 1783, foreword.
192 BRUGMANS 1783: 5.
193 VAN SWINDEN 2001: 73, 75. For a modern description of health problems that can be triggered by volcanic gases, see HANSELL, OPPENHEIMER 2004.
194 DURAND, GRATTAN 1999: 372.
195 SANTEL 1997: 108–110.

smelled was indeed sulfur dioxide, which is known to have a sharp smell similar to that of burnt matches. Locals "felt a sulfuric or saltpeter-like taste in their mouth on 24 June 1783 and the days after; some complained of itchy throats or breathing difficulties. The water in some rain tanks also started to have a saltpeter-like taste."[196]

Relatively low concentrations of sulfur dioxide can cause health problems after only a few minutes of exposure.[197] An itchy or sore throat and breathing difficulties are typical symptoms (Figure 30).[198] The people of Groß Hesepe saw, smelled, and even tasted the aftermath of a volcanic eruption that had occurred almost 2,000 kilometers away.

The Emsland area was no stranger to fog or smoke. Every year, the nearby bog colonists burnt part of the peat bog to generate fertile ash, in which they would plant seeds of buckwheat. This great wave of sulfuric fog was something else.[199] It was dramatically described as a "poisoning thaw," which withered everything it touched and covered all the land and ocean.[200] The *Frankfurter Staats-Ristretto* got wind of the unfolding story and published a report from a correspondent in Münster stating that the "smell of the steam that covers our soil is unhealthy in our region." This "smell" decimated the vegetation on the banks of the Ems River, where lush green rotted to postautumn brown overnight.[201]

The locals did not mistake this particularly poisonous dry fog for frost in this region.[202] On the morning of 25 June 1783, VAN SWINDEN remarked on the extensive

---

196 SANTEL 1997: 109–110. "Viele Menschen empfanden am 24. Juni und noch einige Tage danach einen schwefel-bis salpeterartigen Geschmack im Munde, der bei manchen ein Kratzen im Halse und sogar Atmungsbeschwerden hervorrief. Das Wasser in manchen Regenwasserbecken nahm ebenfalls einen salpeterartigen Geschmack an."

197 "Sulfur dioxide (SO_2), Air quality fact sheet," 2005. The threshold beyond which humans can perceive the taste of sulfur dioxide is 0.35 parts per million (ppm) – lower than the threshold to smell this pungent odor, which is possible at concentrations beyond 0.67 ppm; National Research Council (US), Committee on Acute Exposure Guideline Levels, 2010.

198 MEYER 1977.

199 LIER, TONKENS 1792: 263–265; STOCKMAN 1984: 220; DEMARÉE 2014.

200 SANTEL, SANTEL 1992; SANTEL 1997: 108: "vergieftender Thau."

201 Frankfurter Staats-Ristretto, 21 July 1783: 477: Report from Münster, 30 June 1783. "Der Geruch des unsere Erde bedeckenden Dampfes ist hier zu Land ungesund, und wirket auch stark auf die Blätter und Pflanzen. An den Ufern der Emse hin ist in einer Nacht alles Grüne verschwunden, als wenn es schon späte im Herbst wäre. Die Landleute befürchten alles, auch für das Hornvieh, von diesem Nebel, wenn sich nicht bald ein gesunder Regen einstellt." It is possible that not only the Ems region was affected but also Hamburg, located further to the northeast, a similar incident of all trees losing their leaves over night is reported in the Frankfurter Staats-Ristretto, 1 August 1783: 501. "In der Gegend von Hamburg sind alle Bäume in einer Nacht bey dem neulichen Erddampfe entblättert worden."

202 This poisonous dry fog, however, also affected areas further east: Plön in Holstein, for instance, experienced damage to vegetation on the night from 24 to 25 June 1783. Here, the event was compared to "rime" or "frost," as we have already seen for England or France; KUSS 1826: 170; KINDER 1904 (1976): 304–307.

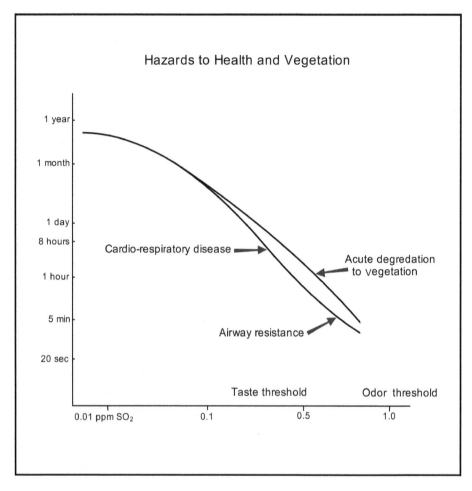

Figure 30: The hazards of sulfur dioxide to health and vegetation. The severity of the impact on human health and the environment depends on concentration and exposure.

damage that it had brought. Leaves had dropped from the trees, and grasses and plants withered. For the same day, BRUGMANS, too, remarked on the damage to plant life: he paid specific attention to their color, commenting on the fact that all around him, the vegetation appeared as it would in mid-winter.[203] He rejoiced that not all plants had been affected in this way. Some of the sturdier ones remained intact and seemed to have survived unscathed. The sudden change from lush green to the

---

[203] VAN SWINDEN 2001: 74–75. Similar reports come from Germany: In Jena, Johann WIEDEBURG (1784: 65–66, 83–84) noticed the orchard fruit grew very little and ripened unevenly. In Schleswig-Holstein, fruit trees were badly affected by the dry fog, bushes lost their leaves, and the fields were also affected; KUSS 1826: 171–172.

autumn colors of brown, black, gray, or white was highly unusual. While it is normal for plants and grass to turn yellow during a heat wave or drought, here, the plants changed their appearance overnight – the very night that the region was affected by the sulfuric smell. Fortunately, most vegetation seemed to recover well from their premature June withering, bearing leaves and fruit again later in the year. Some locals believed this happened because the fog had fertilized the soil with its saltpeter-like substances.[204] Animals living on the trees, mainly insects, were also affected: "[. . .] this haze made a great slaughter of insects, especially fleas, which settle on leaves of trees; when the leaves themselves were damaged, the insects of the trees, which were not injured by the haze, were killed [. . .]."[205]

The fog affected both living organisms and inanimate objects; the chemicals within the fog reacted with cloth, such as wet linen, and metal objects, particularly iron surfaces, which during this night turned green in some areas and rusty in others. "Even some soldiers, who kept watch in Coevorden [in the Dutch Republic, 33 kilometers to the west from Groß Hesepe], noticed a green coating on their gun barrels. Linen laid out on the grass to be bleached gained rusty stains that could not be removed."[206] These effects are similar to that of the vog – volcanic fog – in Hawai'i, where volcanic eruptions are rich in sulfur dioxide. Vog, unsurprisingly, has been linked to cases of asthma and bronchitis in addition to a reduction in agricultural output.[207] The Hawai'ian vog contains sulfuric acid droplets comparable to the corrosive chemicals found in battery acid.[208] Scientists doing fieldwork in Hawai'i have noticed that vog damages their metal equipment, causing it to corrode.[209]

With a few exceptions, these events were limited to England, northern France, the Low Countries, and northwestern Germany between 23 and 25 June 1783 (Figure 31). THORDARSON and SELF argue that this was due to the locations of two pressure systems. Between 21 June and 5 July 1783, while a low-pressure system was stationary above

---

204 SANTEL 1997: 108–110; VAN SWINDEN 2001: 74–75.
205 VAN SWINDEN 2001: 75.
206 SANTEL 1997: 109–110. "Der grüne Ausschlag wurde zuerst von wachhabenden Soldaten in Coevorden an ihren Gewehrläufen festgestellt. Auf dem Rasen zum Bleichen ausgebreitetes Leinen erhielt rostartige Flecke, die sich oft nicht wieder auswaschen ließen."
207 LONGO et al. 2010; ELIAS, SUTTON 2017; USGS, "What Health Hazards are Posed by Vog (Volcanic Smog)?" usgs.gov.
208 USGS, "Does Vog (Volcanic Smog) Impact Plants and Animals?" usgs.gov. "Corrosive" is a word that some contemporary sources use in their descriptions of the impacts on vegetation; Königlich Privilegirte Zeitung, 19 July 1783: 708: Report from Thuringia, 4 July 1783. "Viele Bäume, sowohl in Gärten als Holzungen, wurden seit den 23. Juny fast mit einemmale wie vom Feuer versengt, und verlieren ganz, oder zum Theil die Blätter, welche mit einer corrosiven Feuchtigkeit besprüht zu seyn scheinen, wovon sie ein schwarzes oder braunes Ansehen gewinnen, zusammenschrumpfen und verdorren."
209 BROENDEL et al. 2019, Third Pod from the Sun podcast.

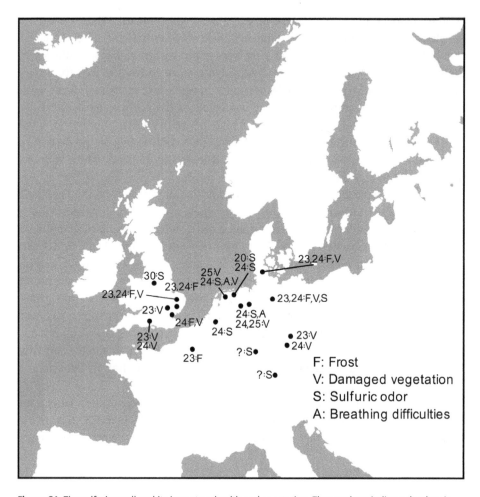

Figure 31: The sulfuric smell and its impact on health and vegetation. The numbers indicate the date in late June 1783 of observations of frost (F), damaged vegetation (V), a sulfuric odor (S), or breathing difficulties (A). (See the list of illustrations for more detail.) Other regions likely experienced these phenomena as well.

Iceland, a long-lasting high-pressure system hovered above northwestern Europe (Figure 32).[210] The polar jet stream transported volcanic aerosols from the Laki eruption in Iceland to Europe; then, the high-pressure system funneled the volcanic aerosols to ground level over the continent in a spiral-like movement (Figure 33). In the lower parts of the atmosphere, the aerosols reacted with the moisture, which caused

---

210 The locations of low-pressure and high-pressure systems have been reconstructed for every day of the 1780s by John KINGTON, based on contemporary meteorological descriptions; KINGTON 1988.

23 June 1783 24 June 1783 25 June 1783

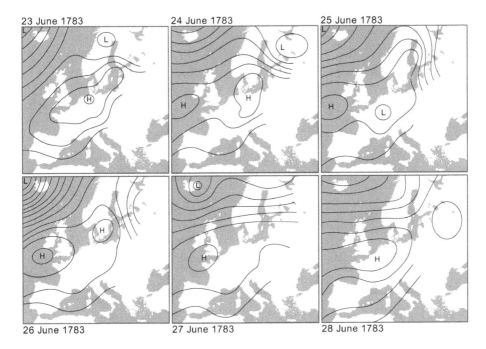

26 June 1783 27 June 1783 28 June 1783

Figure 32: Synoptic weather maps, 23 to 28 June 1783. The maps are based on data from John KINGTON's *The Weather of the 1780s Over Europe*. Reproduced with permission of Cambridge University Press through PLSclear.

the fog to become denser and the air to become very dry.[211] A particularly caustic blast of sulfuric-laden gases, transported by this pressure system, descended upon northwestern Europe on the days in question.

Human Health During the Haze's Peak

When the dry fog descended, it made life difficult for those with pre-existing respiratory problems.[212] The following excerpt from the *Münchner Zeitung* states as much: "In the evening, the fog returned after the thunderstorm, and blanketed the mountains again, and caused a shortness of breath [. . .]."[213] Amidst the gasps for breath

---

211 THORDARSON, SELF 2011: 70.

212 WITZE, KANIPE 2014: 110–111.

213 Münchner Zeitung, 11 July 1783: 425. "Am Abend nach dem Donnerwetter trat aber der Nebel wider ein, und verhüllte die Gebirge von neuem, macht Engbrüstigkeiten, und vermehrt die Vorurteile der gegenwärtigen Zeit." See also "Engbrüstigkeit" in KRÜNITZ 1777, vol. 11: 11.

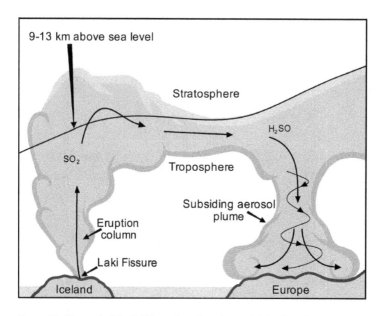

Figure 33: Dispersal of the Laki haze, based on the model developed by THORDARSON and SELF. The Laki eruption emitted ash, sulfur dioxide, and other gases to altitudes of nine to 12 kilometers during its first eruptive episodes between 8 and 14 June 1783. In the jet stream, the sulfur dioxide (SO_2) reacted with moisture and transformed into sulfuric acid aerosols (H_2SO_4), which at this altitude takes about one to two weeks. This "delay" corresponds very well to the time lag between the release of sulfur dioxide in Iceland and the appearance of the dry fog above Europe.

and fits of coughing, a multitude of people began to suffer from eye conditions, likely conjunctivitis.[214]

The poisonous wave of dry fog in the last week of June 1783 was extreme, rare, and perhaps singular in the early modern period. Undoubtedly, the arrival of noxious volcanic gases befouled the air in Europe.[215] Although the gases that arrived in Europe were weaker than they were at their Icelandic point of origin, they nevertheless left their mark.[216] The bitter cocktail consisted primarily of sulfur dioxide, hydrogen sulfides, and fluorine. Sulfur dioxide is detectable from concentrations as low as 0.35 parts per million; asthma worsens when the concentrations are higher than 0.572

214 Königlich Privilegirte Zeitung, 17 July 1783: 700: Report from Dillenburg, 2 July 1783. "In einigen benachbarten Ortschaften sind ganze Familien mit dem Uebel böser Augen behaftet." A similar report, describing people suffering from "bad eyes," presumably eye infections, can also be found in Frankfurter Staats-Ristretto, 11 July 1783: 453: Report from Dillenburg, 2 July 1783; CHRIST 1783: 28–29.

215 GRATTAN, BRAYSHAY 1994; GRATTAN, BRAYSHAY, SCHÜTTENHELM 2002: 92; GRATTAN, DURAND, TAYLOR 2003; GRATTAN et al. 2005.

216 DURAND, GRATTAN 1999: 375; SCARTH 1999: 114.

parts per million (ppm).[217] Hydrogen sulfide concentrations above ten ppm can cause eye irritation, and concentrations above 50 ppm can lead to severe eye damage. Both hydrogen sulfide and ammonia can cause paralysis of the sense of smell within a few minutes, which creates the illusion that the danger has passed even though it has not.[218] Fluorine irritates the eyes and respiratory system when concentrations reach 25 ppm.[219] Other health consequences of exposure to volcanic air pollution include headaches and a loss of appetite.[220]

The question remains: Did the Laki eruption cause excess mortality in European countries outside of Iceland? The mortality rate during June and July 1783 was not exceptional. The extreme spikes occurred in August and September 1783 and in the early months of 1784, which are discussed below.[221] Sources reveal that in June 1783, France reported cases of intermittent fever, possibly caused by malaria, typhoid fever, measles, or smallpox. The following month, dysentery and diarrhea were also reported.[222] Wilfried PITEL and Jérémy DESARTHE compared documents from several French cities for the months between June and September 1783. They identified that La Rochelle suffered only slightly higher mortality than usual, whereas, in Créteil, the number of burials was almost double the average for the same time in 1774 and 1789.[223] There is no evidence that this particular spike in mortality was directly related to the dry fog.

Often, in the sources, it is not clear whether heat or pollution was to blame for the plight of the people. In his journal from Selbourne, Gilbert WHITE details the deleterious effects of the weather on the town's laborers. For 11 July 1783, he writes that the temperature was "74 degrees [Fahrenheit]!" or 23.3 °C. The sun on this day was sultry: "[. . .] the heat [overcame] the grassmowers & [made] them sick."[224] Unfortunately, WHITE does not elaborate further on the nature of their sickness, so it is difficult to pinpoint its exact cause.[225] In Switzerland, a striking report from Graubünden on 24 June 1783 mentions that the fog was present for eight days and that many had fallen ill and some had died.[226] Just as in WHITE's report, the nature of the illnesses remains unclear.

---

217 WELLBURN 1994; BAXTER 2000; WITHAM, OPPENHEIMER 2004: 23; National Research Council (US), Committee on Acute Exposure Guideline Levels, 2010.
218 United States Environmental Protection Agency 1980; POWERS 2004; GREENWOOD, EARNSHAW 2008.
219 KEPLINGER, SUISSA 1968; LIDE 2004.
220 WELLBURN 1994; POPE, DOCKERY, SCHWARTZ 1995; PETERSEN, FISHER, TIMPANY 1996; DURAND, GRATTAN 1999: 373; GRATTAN, DURAND, TAYLOR 2003, 401; BEHRINGER 2011: 213; OPPENHEIMER 2011: 44–46; USGS, "Volcanic Gases and Their Effects."
221 HARREAUX 1858: 30–31; DURAND, GRATTAN 1999: 373; GRATTAN et al. 2003: 20.
222 Société Royale de Médecine 1782/1783: 19–22.
223 PITEL, DESARTHE, "Les Brouillards d'Islande événements extrêmes et mortalities"; GARNIER 2011.
224 Gilbert WHITE, "The Naturalist's Journal," 1783, Add MS 31848, British Library, London, UK: 130.
225 GRATTAN, DURAND, TAYLOR 2003: 411.
226 Königlich Privilegirte Zeitung, 19 July 1783: 708: Report from Graubünden, 24 June 1783.

The *Ephemerides* of the Bavarian Academy of Sciences and Humanities notes that most children and adults who died in 1783 passed between March and May, and most elderly people during June and July. The most common complaints and illnesses during the season of the dry fog were headaches, vertigo, general exhaustion, and strokes.[227] The Societas Meteorologica Palatina published a statement in the newspapers declaring that the nature of the fog had not given them the slightest indication that it had malignant effects: "Quite the opposite, diseases have been reduced, and the grapes are doing very well."[228]

The *Königlich Privilegirte Zeitung* printed an article wherein an author of a local magazine suggested that the "mist does not come from the Earth but comes from the higher regions of the air" and "causes a disadvantageous fermentation in the plants: [. . .] when you eat some portions of it, it causes violent, crampy stomachache." It also gave some practical advice for surviving the dry fog: livestock should remain inside; fodder should be washed outside during heavy rain; and vegetables should be adequately soaked before eating. The anonymous author further recommended smoking tobacco diligently.[229]

The dry fog's concentrations varied from place to place and from time to time. In some areas, the particulates within the fog might have been so great as to induce excess mortality, perhaps in conjunction with one of the many other unfortunate illnesses of the time. Some areas experienced a slow burn and would only play host to a mortality crisis some months later. These are probably the areas in which the fog had lower concentrations of small particulate matter for less time or areas where fever and disease were not so prevalent.

The Fog Days of Summer

In the German Territories, the first mention of the fog in the newspapers was on 5 July 1783 in the *Wiener Zeitung* from Vienna: "Because of the frequent rain that was followed by flooding along the rivers, the hot sun and the completely still air, vapor came up from the Earth and enveloped our horizons these days with a fog, which has

---

227 Bayerische Akademie der Wissenschaften 1783: 104–109. The sources describe strokes as *Schlagflüsse*, which is a term used to describe a sudden death when the cause of death cannot otherwise be determined; METZKE 2005: 62.

228 Berlinische Nachrichten, 19 July 1783: 670: Report from the Mannheim observatory, 6 July 1783. "Im Gegentheil haben sich die Krankheiten gemindert, und die Trauben haben ein herrliches Gedeihen."

229 Königlich Privilegirte Zeitung, 24 July 1783: 721–722: Report from Hanau, 18 July 1783. "Er glaubt, dieser äußerst subtilisirte Dunst komme nicht aus der Erde, sondern sey aus der höhern Region der Luft, durch die daselbst herrschende Winde und durch die Sonne niedergedrückt. Er bringe in die Röhren der Pflanzen, ziehe sie zusammen, hemme dadurch den Umlauf der Säfte, und verursache eine nachtheilige Gährung in denselben: der an den Baumblättern hängende starke Honigsaft, der, wenn man ihn in einiger Menge verzehre, heftiges Leibreissen verursache, bestätige dieses."

been particularly visible during sunrise and sunset in the morning and evening."[230]
On 8 July 1783, a report from Adorf from 26 June 1783, written by an anonymous correspondent, was printed in the *Berlinische Nachrichten*: "Also in our area, like in several areas in Germany since 15 June 1783, we observed a fog that looked similar to the air weighed down by the smoke of a burning forest."[231]

This quote makes it clear that the correspondent was aware of the supra-regional presence of the dry fog. Undoubtedly, most initial observers of this fog assumed it was a local phenomenon until news poured in from other places about its extent.[232] As early as 12 July 1783, the widespread nature of the fog was known in Austria: "The weather here [in Styria] is the same as in the whole of Europe, according to different reports."[233]

On 12 July 1783, the *Hamburgischer Unpartheyischer Correspondent* published a report from Paris dated 4 July: "For 14 days now, we daily had thick fogs [. . .] they deny us all the sunrays. Our naturalists now concern themselves with the discovery of the cause of this unusual appearance."[234] Given that many parts of Europe were denied "all the sunrays," it is unsurprising that there existed, among a few, the uneasy apprehension that they were witnessing "a harbinger of Judgement Day."[235]

---

230 Wiener Zeitung, 5 July 1783. "Die nach einem häufigen Regen, und die dadurch ausgetrettenen Flüsse, bey einer heißen Sonne und gänzlichen Windstille von der Erde aufsteigenden Ausdünstungen umgaben dieser Tagen unsern Horizont gleichsam mit einem Nebel, der des Morgens und Abends bey dem Aufgange und Niedergange der Sonne am meisten sichtbar war."

231 Berlinische Nachrichten, 8 July 1783: 629: Report from Adorf, 26 June 1783. "Auch bey uns herrscht, wie in mehrern Gegenden Deutschlands seit dem 15. dieses [Juni], ein Nebel, welcher einem von der Luft niedergedrückten Rauch eines brennenden Waldes ähnlich sieht [. . .]." There are several Adorfs in the German Territories and it is, unfortunately, unclear which Adorf this report refers to. Similar reports stating an awareness of the supra-regional character of the fog exist from elsewhere, such as in the Hamburgischer Unpartheyischer Correspondent, 11 July 1783: Report from the Lower Elbe, 10 July 1783.

232 CAMUFFO, ENZI 1995: 139; STOTHERS 1996: 85. A similar report can be found in the Wiener Zeitung, 9 July 1783: Report from Ofen, 2 July 1783; here, the fog first appeared around 20 June 1783. Either the correspondent from Ofen was aware of the presence of the fog in several parts of Germany, such as Munich, Dresden, and Regensburg, and parts of Hungary – or perhaps this detail was added later by the newspaper's editor.

233 Wiener Zeitung, 19 July 1783: Report from Graz in Styria, 12 July 1783. "Das Wetter ist bey uns eben so, wie es nach verschiedenen Berichten in ganz Europa herrscht. Doch auch wir haben davon nicht im geringsten eine üble Folge verspürt, es wäre denn die heftigen Ausbrüche des Donners, welche in unseren Gegenden geschehen, und hie und da nicht unbeträchtlichen Schaden anrichten."

234 Hamburgischer Unpartheyischer Correspondent, 12 July 1783: Report from Paris, 4 July 1783.

235 Anonymous ("E. R.") 1783: 405–406.

The Extreme Heat of Summer

The summer of 1783 was sweltering in northwestern and central Europe, with the peak of the heat wave arriving in early August. Many a task was left undone as temperatures left people exhausted, so much so that they were unable to work.[236] One report from Thuringia, which appeared in the *Königlich Privilegirte Zeitung* stated, "Also in this area, we noted a strong haze since 17 June 1783, which still continues with unbearable heat and strong northeastern winds."[237] Although the haze affected a much larger geographic area and lasted much longer than the heat wave, in the midst of this sultry spell, many presumed some connection.[238]

In Mannheim, a "great heat" struck on 2 and 3 July 1783; it peaked on the second day with a temperature of 27 ½ °Ré [34.4 °C] in the shade at 2 p.m.[239] England experienced heat and drought conditions as well; Gilbert WHITE recorded a temperature of 74 °F [23.3 °C] on 11 July 1783, and an atmosphere that was close and dark. Drought-like conditions followed: "There was not rain enough in this village [Selbourne] to lay the dust."[240] On 16 July 1783, as if answering WHITE's prayers, there was "a fine refreshing rain," and indeed, this had been the first rain in Selbourne since 20 June 1783.[241] James WOODFORDE from Norfolk kept a diary in which he frequently described the heat throughout the summer: for example, on 13 July 1783, he documented the moment that "one poor Woman by name Hester Dunham fainted in Church."[242] Stories about the heat in Europe, and countermeasures against it, even made it into North American newspapers. One such report told of people in Wales camping in the mountains to find some relief from the heat.[243]

August brought with it another fierce heat that peaked on the second and fourth day of the month. In Selbourne, for 2 August 1783, Gilbert WHITE writes: "Dew, cloudless,

---

236 Königlich Privilegirte Zeitung, 19 July 1783: 708: Report from Graubünden, 24 June 1783.

237 Königlich Privilegirte Zeitung, 19 July 1783: 708: Report from Thuringia, 4 July 1783. "Auch in hiesigen Gegenden wird seit den 17. Juny ein heftiger Nebel bemerkt, welcher mit unerträglicher Hitze, bey scharfwehenden Nordostwinden noch immer fortdauert." A similar report from the Lower Rhine area also mentioned that the haze was carried by winds from the north and northeast; Koblenzer Intelligenzblatt, 18 July 1783: Report from the Lower Rhine, 10 July 1783.

238 GRATTAN, SADLER 1999: 169; SCARTH 1999: 118–119; GLASER 2001: 205.

239 Berlinische Nachrichten, 19 July 1783: 670: Report from the Mannheim observatory, 6 July 1783. René-Antoine Ferchault DE RÉAUMUR introduced the Réaumur temperature scale in 1730; it was widely used in the German Territories and France. 0 °Ré is the melting point of ice and 80 °Ré is the boiling point of water.

240 Gilbert WHITE, "The Naturalist's Journal," 1783, Add MS 31848, British Library, London, UK: 130.

241 Gilbert WHITE, "The Naturalist's Journal," 1783, Add MS 31848, British Library, London, UK: 131.

242 James WOODFORDE, Weather Diary for Weston Longville, Norfolk, vol. 10: 1782–1784, MET/2/1/2/3/467, National Meteorological Library and Archive, Met Office, Exeter, UK.

243 The Connecticut Journal, 5 November 1783: Reprint of news from London, 22 [?] July 1783, reprint of a letter from Salon, Provence.

sultry, red evening. Burning sun. Workmen complain of the heat. Gardens burn."[244] James Woodforde noted the excessive heat on 2 and 3 August 1783 in Norfolk and his relief when, on 4 August, a little rain came in the afternoon, followed by thunder in the evening.[245] Across the water, in Franeker in the northwestern Dutch Republic, Jan Hendrik van Swinden noticed a particular heat after the fog had vanished, around 28 July 1783, when the thermometer measured 33.4 °C. He further remarked on yet another increase in temperature in August 1783 (Figure 34).[246]

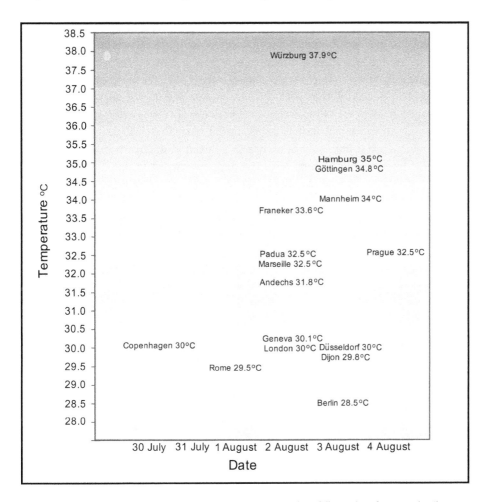

Figure 34: Peak temperatures of the August 1783 heat wave (see list of illustrations for more detail).

244 Gilbert White, "The Naturalist's Journal," 1783, Add MS 31848, British Library, London, UK: 133.
245 James Woodforde, Weather Diary for Weston Longville, Norfolk, vol. 10: 1782–1784, MET/2/1/2/3/ 467, National Meteorological Library and Archive, Met Office, Exeter, UK.
246 Van Swinden 2001: 80.

Data from temperature records based on 26 weather stations in Europe and three in North America within the Societas Meteorologica Palatina network reveal that temperatures in the summer of 1783 varied significantly between different regions. In north, west, and central Europe, the 1783 summer mean temperatures were 1 to 3 °C above average. However, in southern Europe and North America, temperatures remained within the normal range. Inland regions in Eurasia, from Poland to China, experienced unstable and cold weather.[247] Istanbul, then known as Constantinople, also had a very cold summer: "It is not yet sufficiently warm to put on our summer clothes. The inhabitants from Ankara and Izmir make the same complaints."[248]

Recent climatological studies show that the summer of 1783 was one of the warmest summers of the past three centuries in central and western Europe (Figure 35).[249] The July temperatures in northwest Germany were 2 °C above average for 1768–1798, and in central Germany, they were 1 °C above average for the same period.[250] In central England, July 1783 had an average temperature of 18.8 °C, which made it the hottest month of an entire temperature series, lasting from 1659 to 1973.[251] In fact, it took until 1995 for a hotter July to come around in England.[252] In Copenhagen, July 1783 remained the warmest month until 1893. Circulation dynamics in the atmosphere most likely supported dry and hot conditions above central and western Europe; this produced a rare anticyclone, which is a quasi-stationary high-pressure system, above Europe. High-pressure systems are associated with clear skies. Without clouds to reflect sunlight, the temperature increases during the day. The suffocating heat only compounded the sense of panic and hysteria brought about by the unwelcome fog.[253] As would be expected, the temperatures cooled down once the high-pressure system vanished.[254]

247 THORDARSON 2005: 214–215.
248 Journal historique et littéraire, 15 September 1783: 113 (translated by DEMARÉE, OGILVIE 2001: 227).
249 1783 was the warmest summer in western Europe in three centuries; HOCHADEL 2009: 45. In England, 1783 was the hottest summer in recorded history between 1659 and 1983; GRATTAN, SADLER 1999: 64.
250 SCARTH 1999: 117.
251 MANLEY 1974: 395, 398.
252 THORDARSON 2005: 214–215.
253 GRATTAN, BRAYSHAY, SADLER 1998: 29 (quote); see also GRATTAN, SADLER 2001.
254 GRATTAN, SADLER 1999: 169; SCARTH 1999: 118–119; GLASER 2001: 205; THORDARSON, SELF 2001: 68–70; GLASER 2008: 235. For more information on atmospheric pressure, see AHRENS 2009: 193–210.

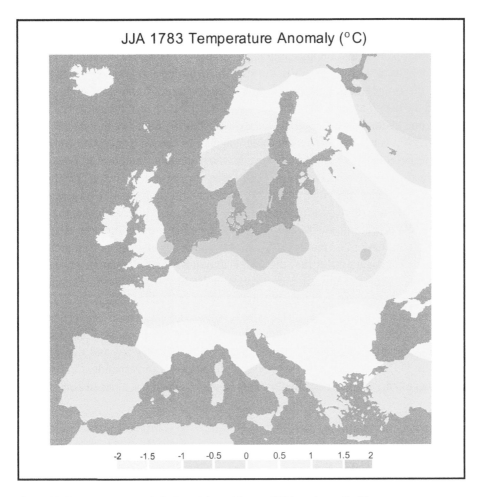

Figure 35: Temperature anomaly for June, July, and August 1783 based upon the 31-year mean, 1770–1800. Central, western, and northern Europe experienced warming, whereas southern Europe and the Mediterranean region experienced cooling.

Thunderclaps and Lightning

Thunderstorms frequently occur during spring and summer in the tropics and mid-latitudes; thus, they are not unusual in central Europe.[255] That said, during the summer of 1783, an alarmingly high number of severe thunderstorms occurred across much of the continent. A report from Zweibrücken dated 6 September 1783, printed in

---

255 SCARTH 1999: 117–118; WITZE, KANIPE 2014: 113.

the *Berlinische Nachrichten*, said: "Since time immemorial, this country has not experienced more damage from thunderstorms than in the present year."[256] A newspaper report from Mannheim that appeared in the *Königlich Privilegirte Zeitung* left readers in no doubt as to the ferocity of the storms: "The lightning and thunder, which accompanied the thunderstorm, cannot be described terrifyingly enough."[257] The storms reached the higher-altitude Westerwald region: "[. . .] during one thunderstorm, we did not see any lightning nor the usual storm clouds, but the numbing thunder was even stronger."[258] A consensus was forming. More and more, the word "unprecedented" was on the tips of tongues.[259] The thunderstorms were frightening and deadly; both people and animals were stunned, injured, and even killed by the accompanying lightning.

The prevalence of thunderstorms throughout the summer of 1783 was exceptional. Even though they began prior to the Laki eruption, it is possible that the significant volume of volcanic particles injected into the atmosphere influenced circulation patterns and therefore increased the frequency of the thunderstorms.[260] Rudolf BRÁZ-DIL and his colleagues attest that the Laki eruption had three "weather effects": the dry fog, the redness of the sun and moon, and the heavy thunderstorms that were sometimes characterized by a lack of rain.[261] Having studied the arguments of the natural scientists, Oliver HOCHADEL concludes that a causal connection between the dry fog and the thunderstorms is possible, even likely.[262] While scientists today have not yet firmly established a definite connection between the dry fog and the thunderstorms, in 1783, some could not be convinced otherwise. A correspondent in Brieg, Silesia, suggested: "The extraordinarily foggy weather, which has amazed almost all of

---

256 Berlinische Nachrichten, 25 September 1783: 890: Report from Zweybrücken, 6 September 1783. "Seit undenklichen Jahren hat dieses Land nicht mehr Schaden von Donnerwettern erlitten, als im gegenwärtigen."
257 Königlich Privilegirte Zeitung, 10 July 1783: 679: Report from Mannheim, 29 June 1783. "Die Blitze und Schläge, welche das Gewitter hat, sind nicht erschrecklich genug zu beschreiben; denn sie waren von den gewöhnlichen ganz unterschieden, fürchterlich, und dem Menschen fast unterträglich, die Blitze weißflammend, die unaufhörlichen Schläge rasselnd und knallend, als wenn stets tausend Granaten ins Kreuz und in die Quere auf einmal in der Luft zersprängen, wodurch die Grundfeste unaufhörlich zitterte, und sich alles bewegte."
258 Königlich Privilegirte Zeitung, 17 July 1783: 700: Report from Dillenburg, 2 July 1783. "Bey einem [Gewitter] sahe man keine Blitze auch keine gewöhnlichen Gewitterwolken, aber desto stärker war der betäubende Donner."
259 Königlich Privilegirte Zeitung, 12 July 1783: 686: Report from Mannheim, 1 July 1783. "Alle Nachrichten stimmen überein, daß das Knallen der Donnerschläge so erschrecklich gewesen, dergleichen noch nie gehört worden."
260 Schlesische Privilegirte Zeitung, 14 June 1783: 714: Report from Halberstadt, 1 June 1783. "Eines solchen Wetters können sich auch die ältesten Menschen bey uns nicht erinnern!"
261 BRÁZDIL et al. 2017: 148, 160.
262 HOCHADEL 2009: 56–57.

Germany, has also conquered the horizon here and caused thunderstorms on an almost daily basis."[263]

When comparing the dates of the thunderstorms reported at the different weather
stations of the Societas Meteorologica Palatina, it becomes apparent that either one
large storm lurched across Europe, affecting several areas, or many individual storms
occurred simultaneously in separate locations. Between May and September 1783, of all
the European cities with weather stations, Rome saw the most thunderstorms, with 40.
Padua saw the second most, with 39. Prague and Sagan drew for third, with 32 each.[264]
During the summer of 1783, Reverend Karel Bernard HEIN recorded 17 days with thunderstorms between June and August in the Moravian town of Hodonice, the highest
number among his observations during the 1780s.[265]

In England, Gilbert WHITE was delighted with the thunder and accompanying deluge of rain that fell upon Selbourne on 14 and 20 June 1783. This "growing weather"
undid the damage of the frost a month previous. The "well soaked" ground brought
welcome shoots of green to the withered landscape.[266] Soon, however, the color of his
reports changed. The storms passed, and the ensuing summer heat left Selbourne
parched. Even on 2 July 1783, when "Great thunder-shower[s]" descended upon Britain, Selbourne remained dry.[267] On 26 and 31 August 1783, WHITE wrote enviously
about "tremendous thunderstorm[s] in London."[268] His brother, Henry WHITE
(1733–1788), rector of Fyfield, also kept a journal. Throughout June and July 1783, he
witnessed and documented several powerful thunderstorms.[269]

Reports about damage caused by lightning strikes were plentiful throughout the
summer of 1783.[270] In some German newspapers, almost every article dealt with the
weather in one form or another.[271] Near Mannheim, trees had been "cut to bits."[272]

---

263 Königlich Privilegirte Zeitung, 2 August 1783: 749: Letter from a friend in Brieg [today's Brzeg in
Poland], 23 July 1783. "Die außerordentlich neblichte Witterung, welche fast ganz Deutschland in Ver-
wunderung gesetzt, hat auch den hiesigen Horizont eingenommen, und fast täglich Gewitter verursacht." David McCALLAM (2019: 214) has argued that the volcanic gases increased the humidity and air
temperature and thus triggered violent electrical storms.
264 Societas Meteorologica Palatina 1783.
265 BRÁZDIL et al. 2017: 153.
266 Gilbert WHITE, "The Naturalist's Journal," 1783, Add MS 31848, British Library, London, UK:
126–127.
267 Gilbert WHITE, "The Naturalist's Journal," 1783, Add MS 31848, British Library, London, UK: 129.
268 Gilbert WHITE, "The Naturalist's Journal," 1783, Add MS 31848, British Library, London, UK: 138–139.
269 Henry WHITE, Diaries of the Rev. Henry WHITE, Rector of Fyfield, county Southampton, brother of
Gilbert WHITE, 1783, Add MS 43816, British Library, London, UK.
270 For an overview of the damage done in different regions in Europe, see DEMARÉE, OGILVIE 2001:
621–622.
271 Such as this one: Königlich Privilegirte Zeitung, 17 July 1783: 700–702.
272 Königlich Privilegirte Zeitung, 12 July 1783: 686: Report from Mannheim, 1 July 1783. "Zu Eistadt
zerschlug ein Blitzstrahl einen Weidenbaum kurz und klein."

In Tönning, on the North Sea, a thunderstorm was briefly confused with an earthquake. In their newspaper report, the anonymous correspondent quickly concedes that "perhaps the houses have only been shaken by the thunderclaps and the whirlwind."[273] In Rendsburg, also in northern Germany, a severe thunderstorm and another unusual whirlwind occurred on 31 August 1783, during which lightning struck a church tower and a farm; 1,000 broken trees lay about the forest in its aftermath.[274]

The spires of the tallest buildings proved threatening to all and sundry that summer.[275] In the Westerwald region, lightning struck a church tower setting it aflame. Quick-thinking locals promptly doused the flames.[276] Unfortunately, people did not always display this presence of mind; in Dresden, a bolt of lightning set the castle on fire, which quickly spread to nearby buildings reducing them to ashes.[277] On 27 June 1783, Lautern in Swabia experienced "the most terrible thunderstorm," in which 46 sheep were killed as they huddled under a tree that was torn apart by lightning.[278]

Some of the most destructive and deadly lightning strikes were those that struck powder magazines; several newspapers tell of such incidents.[279] During a thunderstorm, a garrison and its surroundings were more or less sitting on a powder keg.[280] On 29 June 1783, in Klattau in Bohemia, lightning struck St. Adalbert's Church and set fire to the nearby armory, which naturally housed gunpowder. The place was blown to smithereens. Many were killed and injured.[281]

Less than a month later, on the afternoon of 19 July 1783, in Brieg, Lower Silesia, an artillery captain and a few of his men were on their way to transport "a large amount of gunpowder" into the newly built powder magazine when storm clouds began to gather in the sky. 350 *centner* [ca. 18 metric tonnes] of gunpowder already lay inside the

---

273 Königlich Privilegirte Zeitung, 19 July 1783: 708: Report from Holstein, 12 July 1783. "Zu Tönningen ist am 3. dieses [Juli], nachdem die Luft seit der Mitte des vorigen Monates sehr warm und mit vielen Dünsten angefüllt gewesen war, ein erschreckliches Gewitter mit Hagel, Wirbelwind gewesen, welches verschiedenemale eingeschlagen hat, und wobey einige eine Art eines Erdbeben bemerkt haben wollen. Vielleicht aber sind die Häuser durch die Donnerschläge und den Wirbelwind erschüttert worden."

274 Königlich Privilegirte Zeitung, 9 September 1783, 854–855: Report from Holstein, 31 August 1783.

275 Königlich Privilegirte Zeitung, 12 July 1783: 686: Report from Mannheim, 1 July 1783; Berlinische Nachrichten, 31 July 1783: 705: Report from Döbeln, 21 July 1783.

276 Königlich Privilegirte Zeitung, 17 July 1783: 700: Report from Dillenburg, 2 July 1783.

277 Hamburgischer Unpartheyischer Correspondent, 6 September 1783: Report from Saxony, 31 August 1783.

278 Königlich Privilegirte Zeitung, 12 July 1783: 686: Report from Lautern, 29 June 1783. "Am 27. dieses [Juni] Nachmittags um 2 Uhr hatten wir das fürchterlichste Gewitter, das man jemals hier gesehen hat." The thunderstorm in Lautern was also reported in Gazette van Gent, 17 July 1783: Report from Frankfurt, 11 July 1783.

279 SCHIFFER 2003: 184.

280 HOCHADEL 2003: 143.

281 Königlich Privilegirte Zeitung, 15 July 1783: 694: Report from Prague, 3 July 1783; BRÁZDIL, VALÁŠEK, MACKOVÁ 2003: 312. Klattau is today's Klatovy in the Czech Republic.

building.[282] As the sky darkened further, the men closed all the magazine's openings and retreated to a field some distance away. A lightning bolt struck the building but miraculously did not set the gunpowder on fire.[283] An explosion at a garrison could be enormous. On the Italian island of Gorgona, another powder magazine exploded during a lightning strike, which blasted off a part of the fortification; 35 kilometers away on the Italian mainland, residents of Livorno felt the shockwaves of this blast and thought it was an earthquake.[284]

Newspapers shared many stories of people who lost their lives during thunderstorms. Around 10 July 1783, an anonymous correspondent wrote: "In Pilsen, lightning struck a church tower and killed some of the people who, at the time, were ringing the bells."[285] A report from Langensalza on 31 July 1783 graphically described how a similar fate befell a child who was killed and left burnt, with a black tongue and blood running from his nose.[286] In the Holstein region in northern Germany, two female farm workers died during a thunderstorm on 31 August 1783.[287]

While it seemed all were at the mercy of the freakish weather, a pattern was emerging. During a thunderstorm in Hanau, a church tower was struck by lightning. Fortunately, those in the bell tower were only struck down and stunned. An anonymous newspaper correspondent expressed the following desire: "One wishes that the ringing of bells during thunderstorms would be prohibited and that only at the beginning of the thunderstorm would there be a sign for prayer given by the bell."[288] Church towers were often the highest point in the surrounding area and had metallic spires; lightning would strike here first. As early as 1745, German philosopher and theologian Peter AHLWARDT warned in his book, *Reasonable and Theological Considerations about Thunder and Lightning*, that one should not seek refuge in or near a church during a storm.[289]

So, why would anyone be in a bell tower during a thunderstorm? In 1783, in many regions in German-speaking countries, the custom of ringing church bells to avert thunderstorms was still widespread. In the German Territories, this practice

---

282 In Prussia, a centner was 51.448 kilograms (before 1818); VERDENHALVEN 2011: 65.
283 Königlich Privilegirte Zeitung, 2 August 1783: 749–750: Letter from a friend in Brieg, 23 July 1783.
284 Hamburgischer Unpartheyischer Correspondent, 5 August 1783: Report from Italy, 20 July 1783.
285 Hamburgischer Unpartheyischer Correspondent, 18 July 1783: Report from Prague, 10 July 1783. "Zu Pilsen traf es den Pfarrkirchthurm, und tödtete von 10 bey dem Gewitter läutenden Personen 6 auf der Stelle." Pilsen is today's Plzeň in the Czech Republic.
286 Berlinische Nachrichten, 12 August 1783: 745–746: Report from Langensalza, 31 July 1783.
287 Königlich Privilegirte Zeitung, 9 September 1783: 854–855: Report from Holstein, 31 August 1783.
288 Königlich Privilegirte Zeitung, 17 July 1783: 700: Report from Hanau, 8 July 1783. "Man wünschet, daß das Läuten unter den Hochgewittern abgestellt, u[nd] in Zukunft beym Anfange derselben das Zeichen zum Gebeth mit einer Glocke gegeben werde."
289 AHLWARDT 1745; SECKEL, EDWARDS 1984.

was referred to as *Wetterläuten*.[290] The belief was that if the bell ringer rang the bells furiously, ideally while praying, the sound would divert the thunderstorm. Many church bells from the fifteenth century onward were inscribed with the phrase *Vivos voco. Mortuos plango. Fulgura frango*, which translates as "I call the living. I mourn the dead. I break lightning." Compounding the problem was the fact that if a bell ringer died during a storm, the incident was concealed from the public, interpreted as divine punishment, or put down to neglect on the bell ringer's part.[291]

Ringing the bells and praying were the most common protection strategies employed to avert thunderstorms. While bell ringing was most prevalent among Catholics, it also existed in Protestant areas.[292] One real benefit of the practice was that it warned all within earshot of an approaching storm.[293] The idea that the ringing of a bell could avert a storm stemmed from the pre-Christian tradition of using it to drive off the God of thunder, Donar, or Thor.[294] The sacristan or members of the community that rang the church bells to prevent thunderstorms received a wage or grain for their services. The possibility of an additional income helped preserve the practice.[295]

The people of the early modern period interpreted thunderstorms in two ways. First, as a form of retribution from God. Thunder was His (angry) voice, and lightning His tool of divine punishment.[296] Second, in contrast to the first, as an expression of a good and loving father who blessed His people: storms cleared the air of haze and provided the fields and gardens with rain.[297] Ulrich Bräker, a farmer and writer from Switzerland, regarded thunderstorms as "God's punishment from beyond the clouds" and believed that they had a more significant influence on people than the words of "a thousand preachers."[298]

By the autumn of 1783, many were calling for the abolishment of "weather ringing." This movement was undoubtedly brought about by a broader awareness of the dangers. On 11 July 1783, the *Münchner Zeitung* posed the question: "Would a human

---

290 The practice of using noise to fend off evil goes back to the ninth century: A sermon delivered between 822 and 825 by the monk Hrabanus Maurus addressed the common belief that noise or the throwing of projectiles toward the moon during a lunar eclipse could prevent monsters from swallowing the moon; Haarländer 2006: 152–153. I thank Stephan Ebert for this information.
291 Rosinski 2006: 19.
292 Hochadel 1999: 143–144; Hochadel 2009: 51, 61.
293 Missfelder 2009: 88–89.
294 Dross 2004: 287.
295 Münchner Zeitung, 25 July 1783: 460; Rosinski 2006: 11–13.
296 Psalm 50:3–4: "Our God comes and will not be silent; a fire devours before him, and around him a tempest rages. He summons the heavens above, and the earth, that he may judge his people."
297 Schelhorn 1783: 2; Begemann 1987: 87; Briese 1998: 46–47; Missfelder 2009: 90.
298 Böning 1998: 108–109.

be unchristian if – after so many sad examples of the harmfulness of ringing the bells – this misuse would be put under tighter restrictions?"[299] Soon, a new invention would prove itself up to the task of providing protection for villages and towns across Europe and beyond from the perils of lightning.

Beginning in the late 1740s, Benjamin FRANKLIN endeavored to prove that lightning strikes were electrical discharges.[300] In May 1752, the first lightning rod, also called the Franklin rod, was realized in Marly-la-Ville, near Paris. This rod proved that lightning could be harnessed with technical means.[301] Had the news of this experiment reached FRANKLIN in time, he would not have had to conduct his famous and very risky kite experiment during a thunderstorm, which took place in the summer of 1752.[302] FRANKLIN's experimentations with electricity inspired many naturalists around the world to follow suit.[303]

Lightning rods were not an immediate success. In the German Territories, the first lightning rod, also referred to as a *Wetterstange* (weather stick), was installed on St. James' Church in Hamburg in 1770. Very few additional installations followed.[304] Historian Christa MÖHRING estimates that the widespread acceptance of the lightning rod in Germany and Europe occurred between 1780 and 1800.[305] In southern Germany, the Societas Meteorologica Palatina's secretary Johann Jakob HEMMER was the main proponent of the lightning rod; he personally installed as many as 150. He invented a lightning rod with five tips, the first of which he installed in 1776. In the same year, Charles Theodore ordered that all castles and powder towers in the Palatine region be equipped with the device. Over the next few years, HEMMER's five-tipped lightning rod became increasingly popular. When lightning struck it, the tips would fall off, and it had served its purpose. The idea behind the five-tipped design was to prove its functionality. Once lightning struck, an assessment was made: the first indicator was, of course, that the building had not been set on fire. The second

---

299 Münchner Zeitung, 11 July 1783: 428. "Wäre wohl der Mensch unchristlich, wenn man nach so vilen traurigen Beispilen von der Schädlichkeit dises Läutens dem lange der gewohnten Misbrauch engere Schranken bestimmte?"

300 FRANKLIN 1751; COHEN 1990: 82; HOCHADEL 2003: 145; SCHIFFER 2003: 161–162; MÖHRING 2005: 54–58. E. Philip KRIDER (2004: 1) puts 1746 as the year in which FRANKLIN and his colleagues started to experiment with electricity after they had seen some demonstrations of the phenomenon in Europe in the 1740s.

301 FRANKLIN 1751; COHEN 1990: 72–76; HOCHADEL 1999: 142; SCHIFFER 2003: 188; KRIDER 2004: 4–5; HOCHADEL 2005: 301; MÖHRING 2005: 12.

302 COHEN 1990: 67, 73; KRIDER 2004: 5; KRIDER 2006: 42.

303 WEIGL 1987: 13; BRIESE 1998: 22–23; MÖHRING 2005: 83–105; WEYER, KOCH 2006a: 93. In 1754, in Bohemia, Prokop DIWISCH had invented a lightning rod independent of FRANKLIN; BORNEMANN 1878.

304 WEIGL 1987: 12; MÖHRING 2005: 140–141 (Hamburg).

305 MÖHRING 2005: 123.

indicator was finding one of the splintered tips; this proved that the rod had protected the building by providing a low resistance path to the ground for the electrical current.[306]

When a rod successfully grounded a lightning strike, it often made the news.[307] Sometimes, lightning rods proved their worth right away. One such success story occurred on 28 June 1783 in Mainz, when a lightning rod, installed only six weeks earlier, saved a building during a "very terrible thunderstorm."[308] On that very same day, in Düsseldorf, a similar story unfolded: "The lightning launched onto the weather sticks [lightning rods] atop of two powder towers and happily ran down the lightning rods into the ground. Besides God, we have to thank these weather sticks for preserving the large barracks and the city as a whole."[309]

In France in 1783, many people awaited the outcome of a court case debated at the appeals court of Artois: Charles Dominique DE VISSERY de Bois-Valé, from nearby St. Omer, was appealing the decision of the local authority, which had ordered him to remove a lightning rod that he had installed on top of his chimney three years earlier. DE VISSERY, a lawyer and amateur physicist, complied but took the local authority to court. A lengthy three-year legal battle ensued. The junior lawyer Maximilien ROBES-PIERRE (1758–1794), a future politician, a proponent of the Enlightenment, and one of the best-known figures of the French Revolution, made a name for himself by representing DE VISSERY in court.[310] The case was important; its outcome would determine the relationship between science and legal authority for years to come. ROBESPIERRE argued for reason and science, to which humanity's welfare, he suggested, was indebted.[311] Finally, on 31 May 1783, the court ruled for the plaintiff and for the reinstallation of the lightning rod, which occurred two months later.[312]

This trial certainly brought the debate about lightning rods into the public eye. It did much to soothe the public's concerns about the safety of these instruments; even so, widespread adoption was not immediate. The ponderous pace of adoption meant

306 KLEMM 1969: 510; HOCHADEL 1999: 142, 146; HOCHADEL 2003: 145, 147; HOCHADEL 2009: 61.
307 Berlinische Nachrichten, 9 September 1783: 336–337: Report from Munich, 14 August 1783.
308 Münchner Zeitung, 7 July 1783: 414.
309 Königlich Privilegirte Zeitung, 17 July 1783: 700: Report from Düsseldorf, 28 June 1783. "[. . .] der Blitz stürzte sich auf die Wetterstangen zweyer Pulverthürme und lief an den Ableitern glücklich in die Erde herunter. Nächst Gott haben wir die Erhaltung der großen Kaserne und überhaupt der ganzen Stadt, diesen Wetterstangen zu danken."
310 DROSS 2004: 289.
311 WEIGL 1987: 30–31; RISKIN 1999.
312 FRANKLIN 2011: 352–353. A detailed account of this trial can be found in RISKIN 1999; RISKIN 2002.

many continued to suffer.[313] Severe thunderstorms hit Paris on 2, 3, and 15 July 1783, with 14 strikes in one region and four lives lost.[314] Even the Palace of Versailles was struck by a bolt; the Palace's roof was damaged, and falling debris killed some horses.[315] On 21 July 1783, Bethia ALEXANDER (1757–1839), daughter of merchant William ALEXANDER, reached out to Benjamin FRANKLIN, requesting his advice on how to protect her residence in France against a lightning strike. She implored him to hurry with his response; her urgency clearly sprung from the numerous severe thunderstorms terrorizing the country. She writes, "Do hurry, [. . .] because if you delay, and if a lightning bolt should lack the decency to wait until the answers arrive, you will be filled with remorse."[316] FRANKLIN did respond to the request and sent instructions on how to install a lightning rod or *paratonnere*.[317]

In Britain, buildings were struck by lightning and set ablaze, which often affected the crops on the fields and the livestock in the barns.[318] It seems that the much-needed rain overshadowed the destruction, and the storms were viewed more as relief from the excessive heat than as a threat; this despite the fact that the *Gentleman's Magazine* reported in July 1783 that "the thunder has been more alarming, and the lightning more fatal, during the course of the present month, than has been known for many years [. . .]."[319] The first lightning rod on an English church was erected in 1762, with one added to the spire of St. Paul's Cathedral in London in 1768.[320] In general, however, the rate of uptake was as slow in Britain as it was in France and the German Territories. The American declaration of independence in 1776 did not help matters; the "'Franklin rods' were more than ever abhorred by a multitude of persons, learned and unlearned."[321]

Many European sovereigns endorsed lightning rods, ostensibly for the welfare of their subjects; this endeavor also served to portray them as a monarch of the Enlightenment.[322] In July 1783, Mainz enacted a decree to this effect; Bavaria followed in August,

---

313 Meiningische Wöchentliche Zeitung, 26 July 1783: 120: Report from Frankfurt, 18 July 1783.
314 FRANKLIN 2011: 352–353. The Mercure de France, 28 December 1782: 188–189, reported on the installation of the first lightning rod in Paris.
315 Edinburgh Advertiser, 29 July 1783; BRAYSHAY, GRATTAN 1999: 183.
316 FRANKLIN 2011: 352: Excerpt from a letter from Bethia ALEXANDER to William Temple FRANKLIN.
317 FRANKLIN 2011: 354: Letter written some time after 21 July 1783.
318 BRAYSHAY, GRATTAN 1999: 183.
319 Gentleman's Magazine, July 1783: 621 (quote); James WOODFORDE, Weather Diary for Weston Longville, Norfolk, vol. 10: 1782–1784, MET/2/1/2/3/467, National Meteorological Library and Archive, Met Office, Exeter, UK, July and August 1783.
320 MÖHRING 2005: 272–273.
321 ANDERSON 1880: 42.
322 HOCHADEL 2009: 54.

Prussia in September, and Austria in November.[323] Paris followed in 1786.[324] In Spain, the practice of bell ringing continued until the 1850s, and in Upper Swabia, bells were still rung "to drive away the hail and prevent damage by lightning" as late as the 1860s.[325] It was the German Territories that paved the way for the lightning rod's breakthrough. It is unclear how many lightning rods were installed in 1783 here, but it was definitely no "average" year.[326]

Some regions had abolished the ringing of bells against thunderstorms before 1783. The many storms of the summer most likely accelerated legislative processes that had already been underway.[327] Newspaper reports often mention the presence of the sovereign or ecclesiastical officials during the installation of a lightning rod, almost certainly to bolster support for the invention.[328] Even after lightning rods were installed, people still rang the bells, both as a signal for prayer (at the beginning and during the thunderstorm) and (at the end) as a sign of gratitude.[329] The purpose of the bell ringing had changed but, unfortunately, the risks remained.

Earthquakes in Europe and Beyond

The powder magazine explosion on the Italian island of Gorgona that some thought an earthquake was not a singular occurrence. Throughout the summer of 1783, many phenomena were mistaken for earthquakes, be they thunderstorms, lightning strikes, or hailstorms; even the smallest tremor would be much discussed (Figure 36).[330]

On 1 July 1783, the *Berlinische Nachrichten* printed a report about a thunderstorm in the Silesian town of Schweidnitz on 22 June 1783 that some residents swore was an earthquake.[331] On 4 July 1783, the *Hamburgischer Unpartheyischer Correspondent* commented further on the Silesian region's woes with details of strong rainfall, damaged

---

323 HOCHADEL 1999: 143–144; BRÁZDIL, VALÁŠEK, MACKOVÁ 2003: 311–312; HOCHADEL 2003: 148; DROSS 2004: 293; HOCHADEL 2009: 54.
324 SECKEL, EDWARDS 1984.
325 COHEN 1990: 124.
326 DROSS 2004: 289–291; HOCHADEL 2009: 53–54. Lightning rods were much discussed in the media even before the dry fog of 1783 made an appearance in the newspapers: a rather detailed debate on the lightning rod can be found in the Frankfurter Staats-Ristretto, 28 June 1783: 425, and 1 July 1783: 433–434.
327 HOCHADEL 2009: 52.
328 HOCHADEL 1999: 147, 161, 167.
329 Münchner Zeitung, 7 August 1783: 485.
330 VERMIJ 2003: 234; REITH 2011: 81.
331 Berlinische Nachrichten, 1 July 1783: 605: Report, 25 June 1783. Similar reports can be found in the Königlich Privilegirte Zeitung, 1 July 1783 and the Hamburgischer Unpartheyischer Correspondent, 4 July 1783. Schweidnitz is today's Świdnica in Poland.

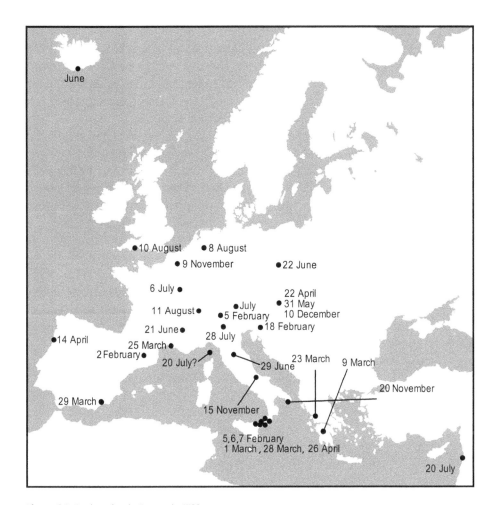

Figure 36: Earthquakes in Europe in 1783.

bridges, tumbled chimneys and cracked walls, and thunder so strong that it "was indistinguishable from an earthquake."[332] One week later, the same newspaper reported on another thunderstorm, this time near Glatz, detailed in letters they had received from Mittenwald. The letters also informed them about tremors at the mountains of Spitzberg and Schwarzberg, both in Lower Silesia. Miners in the coal pits near the Gottesberg "heard such a strong subterranean roar" that they were compelled to flee for fear

---

332 Hamburgischer Unpartheyischer Correspondent, 4 July 1783: Report from Schweidnitz, 25 June 1783. "Verwichenen Sonntag, den 22sten dieses [Juni], ist im Glatzischen ein so entsetzliches Donnerwetter gewesen, daß man es von einem Erdbeben nicht hat unterscheiden können."

of the consequences.[333] Today we know that these were most probably violent thunderstorms, misinterpreted. Seismologist Günter LEYDECKER's earthquake catalog for Germany between 800 and 2008 does not list any earthquakes between April 1783 and March 1784; that said, it is incomplete.[334] The seemingly frequent earthquakes of 1783 rattled contemporaries in more ways than one. The fear of earthquakes was real even if, in some cases, the earthquakes themselves were not.

The map (Figure 37) shows the probability of earthquakes occurring in different parts of Europe. Within the German Territories, it becomes apparent that some regions are more prone to earthquakes than others (Figure 38). The area of the Rhine Rift Valley from Basel in Switzerland to the Low Countries, particularly the Cologne Lowland area, has seen some strong earthquakes in the past. One example is the 1756 Düren earthquake. Other areas in Germany with increased seismic risk are those north of the Alps, around Lake Constance and Swabia, and the Vogtland region in eastern Germany. These intraplate earthquakes are caused by the Alpine orogeny, the formation of the Alps by way of the African Plate moving northward into the Eurasian Plate. Intraplate earthquakes, as the name suggests, take place some distance away from continental margins. They are less frequent and often weaker than interplate earthquakes, which occur at plate boundaries. Seismologists estimate that an earthquake with a magnitude of up to 6.4 on the Richter scale is theoretically possible for the Lower Rhine Graben.[335]

The news of the Calabrian earthquakes and their disastrous consequences still dominated the headlines during the summer of 1783.[336] Aftershocks continued throughout the

---

333 Hamburgischer Unpartheyischer Correspondent, 11 July 1783: Report from Silesia, 6 July 1783. "Eben so haben sich im Mittenwaldischen [Mittenwalde near Glatz, today's Kłodzko in Poland] an den Bergen Spitzberg und Schwarzberg Spuren einer Erderschütterung gezeiget, und in den Kohlengruben bey Gottesberg ist ein so starkes unterirdisches Getöse gehöret worden, daß die Arbeiter aus den Schachten, aus Furcht vor den Folgen, weggegangen sind." For more information on the geography of the region, see Weigel 1803, 105. A similar account can be found for miners in Düsseldorf, who complained that the "Earth vapors" (*Erddämpfe*) prevented them for working: the dry fog made it even darker in the mines, and they heard loud roars underground; Münchner Zeitung, 14 July 1783: 430: Report from Düsseldorf, 28 June 1783.

334 LEYDECKER 2011: 62–63: Between 18 February and 12 April 1783, there were six earthquakes with magnitudes between 4.0 and 5.0. It was recorded at Kraslice (Graslitz in German) in today's Czech Republic. For more information on earthquakes also felt in Bavaria, see GIESSBERGER 1922; GIESSBERGER 1924. John E. EBEL illustrates in his book that most earthquake catalogs not only list real historical earthquakes but also earthquakes that most likely never took place. Either the person who was responsible for the catalog misinterpreted a report, or they misinterpreted an old-style date with a new-style date. He points this out in his chapter "1658: The Earthquake That May Have Never Happened," EBEL 2019: 46–50.

335 HINZEN, REAMER 2007: Based on a 300-year catalog of earthquakes in the northern Rhine region, the authors of this study assume an earthquake with a maximum magnitude of 7.0 is possible here.

336 Even as late as 4 September 1783, the Münchner Zeitung reported that there were still "rather severe shocks of earthquakes" felt in Italy. These particular earthquakes had occurred on 29 and 30 July 1783; Münchner Zeitung, 4 September 1783: 550.

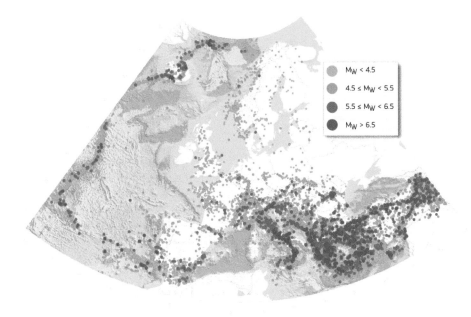

Figure 37: Earthquake history in Europe. This map shows the distribution of over 30,000 earthquakes with magnitudes larger or equal to 3.5 between 1000 and 2006.

summer, frightening locals and those who read the newspapers in faraway towns and cities.[337]

On 21 June 1783, an earthquake occurred in Belledonne near Lyon with a moment magnitude of 3.9.[338] In Florence, on 29 June 1783, at 4:30 a.m., "another weak earthquake" occurred.[339] In July 1783, there were reports about an earthquake in Austria.[340] On 6 July 1783, a notable earthquake was felt over a wide area in eastern France and Switzerland: specifically in Franche-Comté, Burgundy, the Jura, and Geneva. A letter from Dijon dated 7 July 1783 and printed in the *Gazette van Gent* shared some details about the earthquake and suggested a connection to the dry fog: "[. . .] while the atmosphere was covered with a thick fog, as was usual, an earthquake was felt from the east

---

337 Epp 1787: 30.
338 SHEEC: Belledonne Earthquake, 21 June 1783; Quenet 2005: 541. Today, moment magnitude is used to describe medium and large earthquake magnitudes. Smaller earthquakes are often measured in the local magnitude (Richter).
339 Hamburgischer Unpartheyischer Correspondent, 19 July 1783: Report from Florence, 30 August 1783; Königlich Privilegirte Zeitung, 24 July 1783: 723: Report from Florence, 30 June 1783. "Gestern früh um halb 5 Uhr haben wir hier abermals eine leichte Erderschütterung verspürt."
340 Königlich Privilegirte Zeitung, 31 July 1783: 742: Report from Austria, 19 July 1783.

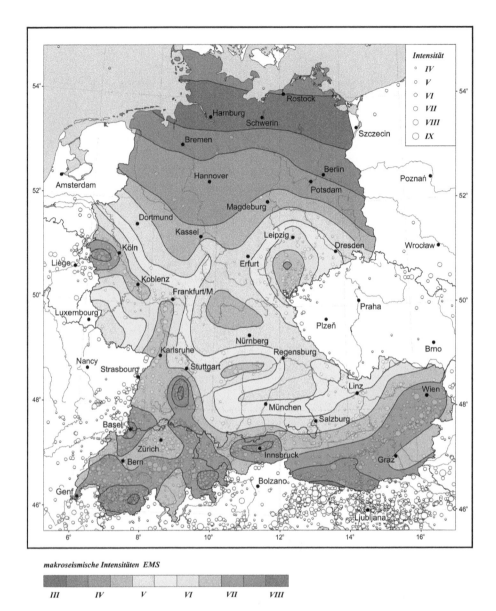

Figure 38: Earthquake risk in today's Germany, Austria, and Switzerland. GRÜNTHAL, Gottfried; MAYER-ROSA, D.; LENHARDT, W.: Abschätzung der Erdbebengefährdung für die D-A-CH-Staaten – Deutschland, Österreich, Schweiz. In: Bautechnik 75, no. 10 (1998), 753–767, here 764, Figure 6.

to the west, lasting three seconds."[341] In the affected areas, particularly in Beaune, glass windows were shattered, and "quite an amount of chimneys fell down."[342] Apart from these minor misfortunes, the area did not suffer much damage.[343] It is estimated that this earthquake originated in Vallée de l'Ouche, near Dijon, had a moment magnitude of 5.1, and reached an intensity of VI on the EMS-98 scale.[344] Another earthquake occurred in the early hours of 8 August 1783 at around 3 a.m. It shook western Germany, the Dutch Republic, and northern France, mainly between Aachen and Maastricht.[345] This month also saw mild tremors shake Devon in England.[346] On 11 August 1783, there was a 3.9-moment-magnitude earthquake in Lucerne, Switzerland.[347] A newspaper report concluded that "the whole of Europe is being haunted by this terror."[348] The sheer number of these rather weak earthquakes significantly increased the magnitude of fear throughout Europe.[349]

Outside of Europe, one significant seismic event occurred near Tripoli, Lebanon, on 20 July 1783.[350] News of this reached Europe in October 1783, with reports emphasizing the fact that this event in the Middle East had coincided with a dry fog.[351] With this news,

---

341 Gazette van Gent, 17 July 1783: Report from Dijon, 7 July 1783. "[. . .] terwyl de locht met eenen dikkeren mist, als na gewoonte, overdekt was, eene hevige aerdschuddinge van get oosten nae het westen had gevoelt, die dry seconden had geduert, [. . .]."

342 Münchner Zeitung, 25 July 1783: 459. "In Frankreich hat man kürzlich Erdbeben gespüret, wovon eine Menge Schornsteine abgefallen sind. Die Orte dises Vorfalls waren beinahe ganz Burgund, Bresse, Franchecomte, und das Genfer Gebiet, wo es am stärksten gespüret wurde" (quote). Damage in Beaune, particularly to glass windows and chimneys, was reported by the Gazette van Gent, 17 July 1783: Report from Dijon, 7 July 1783.

343 DEMARÉE, OGILVIE 2001: 225.

344 SHEEC: Vallée de l'Ouche Earthquake, 6 June 1783; QUENET 2005: 541.

345 DEMARÉE, OGILVIE 2001: 226; QUENET 2005: 541. That this earthquake occurred at 3 a.m. is reported in the Königlich Privilegirte Zeitung, 21 August 1783: 802: Report from Cologne, 10 August 1783. Note that this newspaper article falsely stated that the earthquake had occurred on August 9, not August 8. It is odd that this earthquake is not listed in the SHEEC database of the European Union.

346 Exeter Flying Post, 14 August 1783 (quoted after GRATTAN, BRAYSHAY 1995: 6). This earthquake might have had a magnitude of 3.6 (Richter) and was documented on 10 August 1783 in Launceston, Cornwall. It was felt in parts of Devon and Cornwall; British Geological Survey 1989.

347 SHEEC: Luzern [Lucerne] Earthquake, 11 August 1783.

348 Berlinische Nachrichten, 18 September 1783: 686: Report from Paris, 4 September 1783. "Briefe aus Portugall melden, daß man in verschiednen Gegenden dieses Königreichs, am 6. Julius ziemlich heftige Erdbeben empfunden habe; eben dergleichen wird auch aus England und selbst aus Island berichtet; also ist das ganze Europa von diesem Schrecken heimgesucht worden."

349 BRIESE 1998: 97.

350 Journal historique et littéraire, 1 November 1783: 363.

351 AMBRASEYS 2009: 613. The earthquake in Tripoli was reported on in the Massachusetts Spy; Massachusetts Spy, 8 January 1784: Report from Tripoli, Syria, 30 July.

Europe became increasingly anxious about the meaning of the earthquakes in Calabria.[352] There were additional vague reports about "terrible earthquakes" from further away, such as in the Antilles, China, Japan, and the Philippines.[353]

Fever and Mortality

As summer turned to autumn, the people of Europe's health deteriorated significantly.[354] In England and France, burial rates were average in June and July 1783, but spiked in August and increased again in September.[355] A study by Claire WITHAM and Clive OPPENHEIMER has identified 1783/1784 as a "mortality 'crisis year'" in England.[356] GRATTAN and colleagues define a "crisis year," in this respect, as when mortality is "greater than 10 percent in excess of the moving 51-year mean."[357] Mortality crises were not unusual in the early modern period due to poor sanitation and infectious diseases that commonly made the rounds, but this was different.[358] WITHAM and OPPENHEIMER identified two distinct periods that show excess mortality, with one lasting from August to September 1783 and the other lasting from January to February 1784. Together, these two periods are estimated to have caused excess mortality of around 20,000 people in England alone. Overall, between July 1783 and June 1784, the mortality in England was 16.7 percent above average.[359]

Sabina MICHNOWICZ analyzed burial records for Dorset, Cheshire, Yorkshire, Northumberland, London, Manchester, and Whitehaven. The records for these regions reflect the complex relationship between environment and mortality; they indicate "that different areas displayed differing sensitivity to environmental influences."[360] The mortality crisis was rather heterogeneous, affecting some regions more than others. Bedfordshire, Leicestershire, and Worcestershire were among the most severely affected in 1783/1784.

---

352 Journal historique et littéraire, 1 November 1783: 363. The Gazette van Gent, 17 July 1783: Report from Dijon, 7 July 1783 mentioned that the earthquake near Dijon "combined with the imagination of the Calabrian disasters had created a general fright among the inhabitants of Dijon." ("[. . .] welke aerdschuddinge, gevoegd by de verbeeldinge van de rampen in Calabrien, en by een alderschrikkelykste onderaedsch gedrys, eenen algemeynen schrik onder de Inwoonders van Dijon had verwekt.").
353 Berlinische Nachrichten, 2 August 1783: 715: Report from Paris, 19 July 1783 (Antilles); Berlinische Nachrichten, 26 August 1783: 791–792: Report from Versailles, 10 August 1783 (earthquakes in China, Japan, and the Philippines).
354 DEMARÉE, OGILVIE 2001: 237.
355 GRATTAN, DURAND, TAYLOR 2003: 406.
356 WITHAM, OPPENHEIMER 2004: 15.
357 WRIGLEY, SCHOFIELD 1989; GRATTAN, DURAND, TAYLOR 2003: 401 (quote).
358 GRATTAN, DURAND, TAYLOR 2003: 405.
359 WRIGLEY, SCHOFIELD 1989; WITHAM, OPPENHEIMER 2004: 15–16.
360 MICHNOWICZ 2011: 122. This topic was also researched by GRATTAN, MICHNOWICZ, RABARTIN 2007.

If these deaths were related to the volcanic emissions from the Laki eruption, it would mean that the eruption precipitated more deaths in mainland Europe than in Iceland.[361]

The mortality in the summer months of 1783 is highly unusual, as – under normal circumstances – summers are characterized by minimum mortality. Peak mortality can usually be observed in the spring, between March and April, with the lowest mortality generally occurring in the summer and early autumn months, depending on weather and harvest.[362] WITHAM and OPPENHEIMER pointed out that it was particularly striking that mortality peaked in September 1783, even exceeding that of January 1784 – the latter brought on by the severity of the winter.[363] The monthly mean temperature in England in July 1783 was 18.8 °C, about three degrees higher than average. The hot temperatures lasted roughly from 23 June to 20 July 1783, coinciding with the high-pressure system above Europe.[364] As became apparent during the summer of 2003, heat waves can be deadly in Europe. However, the heat wave was likely not responsible for the peak mortality in 1783 – as the onset of the mortality occurred after the heat wave had already passed.[365]

Several studies discuss the possible causes of the mortality crisis. One possibility is that the warm conditions might have raised the temperature of the soil and led to an increase in pests and so food contamination. Gilbert WHITE remarked that meat was inedible a day after the animal had been killed due to the heat and further complained about the presence of swarms of flies. The sultry conditions allowed mosquitoes to proliferate. Malaria was a problem in England at the time, mostly in areas with marsh and fenland. WITHAM and OPPENHEIMER argue that Kent, which usually saw the most deaths from malaria, did not experience a crisis in the summer of 1783, so perhaps blame cannot be laid on the mosquitoes.[366] Drainage schemes and better sanitation had made malaria less of a problem by the mid-eighteenth century.[367] Other diseases that might have caused the excess mortality include typhoid and dysentery, which spread relatively slowly and primarily affect babies and young children.[368] All of these diseases, together with the polluted air brought upon these people by the fog, likely exacerbated the problems caused by the harsh living conditions and so led to excess mortality.

In England, during August and September 1783, an "ague" or "fever" affected many people; the nature of this epidemic remains unspecified.[369] English poet William COWPER

---

361 MICHNOWICZ 2011: 118–119; 124.
362 WRIGLEY, SCHOFIELD 1989; GRATTAN, DURAND, TAYLOR 2003: 405.
363 WITHAM, OPPENHEIMER 2004: 18.
364 PARKER, LEGG, FOLLAND 1992; GRATTAN, CHARMAN 1994: 102; GRATTAN, SADLER 1999: 164–165.
365 WITHAM, OPPENHEIMER 2004: 19; CHARPENTIER 2011.
366 DOBSON 1980; DOBSON 1997; WITHAM, OPPENHEIMER 2004: 22–24.
367 EDWARDS 1998: xiv.
368 WITHAM, OPPENHEIMER 2004: 22.
369 WITHAM, OPPENHEIMER 2004: 20.

(1731–1800), who was based in Bedfordshire, wrote a letter on 7 September 1783 indicating that laborers suffering from fever were unable to work, which in turn affected the harvest. Naturalists from different parts of England commented on the "unhealthy season" and documented the several types of fever that were rampant.[370] WITHAM and OPPENHEIMER also argue that British soldiers returning from the American Revolutionary War might have brought diseases with them to which the English were not immune.[371] By September 1783, the dry fog had thinned; the eruptive phases that occurred throughout this month emitted much less sulfur dioxide than the earlier phases.[372] This comparatively small volume of sulfur dioxide, while not as harmful as the earlier blasts, could still have contributed to excess mortality, given that the people had been subjected to sulfur-laden air of varying potencies over the previous two months.

In Paris and the surrounding countryside, many suffered from different kinds of fever, which one anonymous correspondent assumed to be among the consequences of the extraordinary heat.[373] The records of the *Société Royale de Médecine* for 1783 reveal typhoid fever, diarrhea, dysentery, cholera, measles, pox, and sore throats were prevalent in August of that year.[374]

GRATTAN, RABARTIN, SELF, and THORDARSON conducted a study on mortality in France in 1783. They examined birth and burial records from four parishes in Loiret, 44 parishes in Seine-Maritime, and five parishes in Eure-et-Loir. They concluded that more people than usual died in France in August 1783, a number that increased further in September and October; during these months, the burial rate was 38 percent above average. Between August 1783 and May 1784, mortality was 25 percent above average; eastern France suffered the most.[375] Brittany in the northwest, and the regions of Eure-et-Loir in central France, achieved similar grim records. Normal levels of mortality were only seen again in May 1784.[376]

In the Austrian Netherlands, the Baron DE POEDERLÉ noticed that cases of dysentery greatly affected several provinces beginning in August 1783. By September, most had recuperated.[377] Some, including Mathias VAN GEUNS, a doctor from Harderwijk, Groningen, blamed the extraordinary heat and the dry fog.[378] The *Ephemerides* of the Bavarian Academy of Sciences and Humanities in Munich reported that during the dry fog, they did not hear of any increase in deaths or rampant diseases. Nevertheless, they "had to endure other menaces just like other kingdoms in Europe," referring to drought, heat,

---

370 GRATTAN, DURAND, TAYLOR 2003: 401, 405.
371 WITHAM, OPPENHEIMER 2004: 24.
372 THORDARSON, SELF 2003: 4, Figure 2.
373 Augsburgische Postzeitung, 11 October 1783: Report from Paris, 1 October 1783.
374 Société Royale de Médecine 1782/1783: 22–27; METZKE 2005.
375 GRATTAN et al. 2005.
376 SUTHERLAND 1981; RABARTIN, ROCHER 1993; GRATTAN, DURAND, TAYLOR 2003: 410; GARNIER 2011.
377 DE POEDERLÉ 1784: 342–344.
378 DEMARÉE, OGILVIE 2016: 141.

heavy downpours, and flooding.[379] Some cases of dysentery and typhoid fever affected people in the German Territories. Pox virus infections also presented a problem throughout the year, particularly in November and December 1783.[380] In Stuttgart, in September 1783, many of the city's young people contracted typhoid fever.[381] Reports of dysentery seem mostly absent from the German newspapers – perhaps due to local censorship.

Studies on excess mortality in the aftermath of the Laki eruption conclude that deaths rose by as much as 40 percent in England, France, and the northern parts of the Dutch Republic.[382] Further research analyzing birth and burial records for that time in other regions of Europe has yet to be conducted.[383] There can be no definite conclusion as to what might have caused this significant peak in mortality. While it is very likely the gases and particulate matter emitted by the Laki eruption were detrimental to the health of Europeans at the time, weakening the very young, the elderly, and the asthmatic, it is unlikely the sole cause of the widespread death that followed.

The End of the Fog

With some short interruptions, the dry fog lingered over Europe for much of the summer of 1783.[384] It was omnipresent and became conspicuous by its absence.

---

379 Bayerische Akademie der Wissenschaften 1783: 46. "Die Fruchtbarkeit war aller Orten überaus gesegnet. Alles Obst und Getreid hat außerordentlich gediehen. Man hat auch nirgend von Sterbefällen und grassierenden Krankheiten gehörtet. Doch mußten wir andere Landplagen erdulden, so wie andere Königreiche Europens."

380 Bayerische Akademie der Wissenschaften 1783: 104–109, "Kindsblattern" could refer to either smallpox or chickenpox.

381 Augsburgische Postzeitung, 30 September 1783: Report from Mannheim, 27 September 1783; Berlinische Nachrichten, 9 October 1783, 935: Report from Stuttgart, 21 September. Here the term *Gallenfieber* is used to describe typhoid fever; METZKE 2005: 60.

382 DEMARÉE, OGILVIE 2016: 142.

383 Recently, Geoffrey HELLMAN (2021) carried out research using 1,500 parish registers from the British Isles and beyond, which contradicts previous findings.

384 Münchner Zeitung, 10 July 1783: 422: Report from Berlin, 5 July 1783. "Last Tuesday, during the night there were two thunderstorms, the latter one burnt down a few houses nearby. A strong wind came up yesterday, which changed the weather." "Am verfl[ossenen] Dienstage folgten endlich in der Nacht zwei Donnerausbrüche aufeinander, wovon der leztere im nahe gelegenen Dorfe Reustädtl einige Häuser einäscherte. Seit gestern hat ein stärkerer Wind die Witterung geändert." Königlich Privilegirte Zeitung, 9 August 1783: Report from Mannheim, 27 July 1783: Prematurely, an anonymous correspondent remarked, "the long-lasting sad fog is now completely gone. It was lost twice after a thunderstorm before [. . .]." "Der langwierige traurige Nebel ist nun völlig verschwunden. Da er sich 2mal nach einem Gewitter verlohren [. . .]." The dry fog returned, however, in Mannheim, a *vapor tenuis* was last mentioned on 5 October 1783. Societas Meteorologica Palatina 1783: 16.

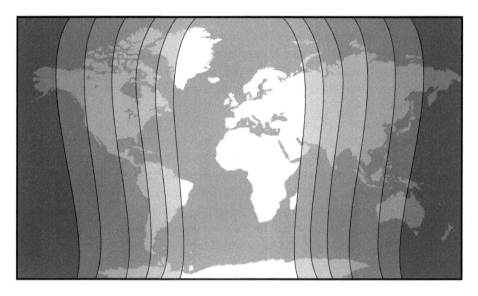

Figure 39: The total lunar eclipse of 10 September 1783. The white areas on this map show where the total lunar eclipse was visible, while gray areas illustrate partial visibility; in the dark areas, it was not visible.

Throughout Europe, a lunar eclipse was observed on 10 September 1783 (Figure 39).[385] That this lunar eclipse was visible is indicative of a significant decrease in the density of the dry fog by that time. In the German Territories, there was great interest in this celestial event.[386] In Berlin, Johann Esaias SILBERSCHLAG (1721–1791), a member of the Royal Academy of the Sciences in Berlin, observed and listed 10:40 p.m. as the time when the Earth's shadow first dimmed the moon and 11:39 p.m. as the moment when the moon was entirely eclipsed.[387] The *Münchner Zeitung* reported on the event, even though it was unobservable in Munich; this might have been due to cloud cover.[388] The eclipse was also witnessed in England and as far away as North America.[389] In Nain, a Moravian settlement in Labrador, Benjamin LA TROBE (1764–1820) observed it through a

385 National Aeronautics and Space Administration (NASA), Five Millennium Catalog of Lunar Eclipses, −2000 BCE to 3000 CE: A–153.
386 Göttingische Gelehrte Anzeigen 1784, register, 53; Göttingische Anzeigen von gelehrten Sachen, 8 May 1784: 738.
387 SILBERSCHLAG 1784: 134–147.
388 Münchner Zeitung, 18 September 1783: 581.
389 It was observed from York, PIGOTT 1786: 410; and from Selbourne: Gilbert WHITE, "The Naturalist's Journal," 1783, Add MS 31848, British Library, London, UK: 141; and from Kent, MACKAY 1793: 166.

"smoky sky."[390] Reverend Manasseh CUTLER also observed the eclipse from Cambridge, Massachusetts.[391]

It is difficult to ascertain whether the records from October onward refer to the dry fog or a typical humid fog when they mention this kind of weather. In Regensburg, the last "thick fog" occurred on 19 August 1783. There are mentions of fog three times in September and occasionally in November; however, this was likely "normal" fog.[392] In Sagan, the last definite mention of a "fog, like smoke" was on 11 October 1783, with an ambiguous entry in the last days of that month.[393] The last mention of the *vapor tenuis* (fine haze) in Mannheim was on 5 October 1783.[394] In Scotland, where foggy weather is commonplace, there is mention of "a very uncommon fog" in Belmont Castle, Perthshire, that lasted until 31 August 1783. A "thick mist" is recorded on the first two days of October 1783, then again on 13 and 18 October. Once again, the entries are ambiguous.[395] Gilbert WHITE observed some haze and red sunshine on the first few days of September 1783. On 18 and 27 September 1783, he noticed a wet fog, which can confidently be presumed to be of a different origin from the "smoky" mist of the summer.[396]

THORDARSON and SELF created a map containing the last reported observations of the Laki haze across Europe that they found in their sources. Some of these dates are surprisingly late; a reported sighting of the dry fog in Copenhagen exists from as late as January 1784.[397] However, certainty is not guaranteed; the later mentions could very well be fogs typical of the season (Figure 40).[398]

---

390 Meteorological Observations at Nain & Okak at Labrador, Hudson's Bay Company, MA/143, Archives of the Royal Society, London, UK, here: 10 September 1783.

391 The Continental Journal and Weekly Advertiser, 27 November 1783: 2.

392 KÖNIG 1783, here: Regensburg.

393 Societas Meteorologica Palatina 1783: 346.

394 Societas Meteorologica Palatina 1783: 16.

395 James Stuart MACKENZIE, Weather Reports Journal of James Stuart Mackenzie at Belmont Castle, Perthshire, 1771–1799, RH4/100, National Records of Scotland, Edinburgh, UK.

396 Gilbert WHITE, "The Naturalist's Journal," 1783, Add MS 31848, British Library, London, UK: 139–142.

397 THORDARSON, SELF 2003: 7.

398 One contemporary source from Austria mentioned that the fog was also present during phases of snowfall during the winter of 1783/1784, however, it is likely this was an occurrence of "normal" fog; STRÖMMER 2003: 208.

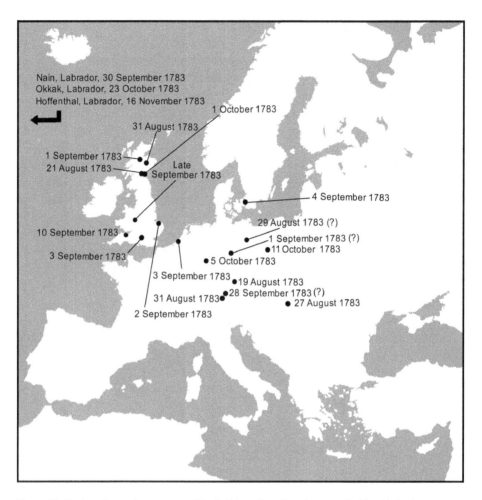

Nain, Labrador, 30 September 1783
Okkak, Labrador, 23 October 1783
Hoffenthal, Labrador, 16 November 1783

1 October 1783

31 August 1783

1 September 1783
21 August 1783
Late
September 1783

4 September 1783

29 August 1783 (?)

10 September 1783

1 September 1783 (?)
11 October 1783

3 September 1783

5 October 1783

3 September 1783 19 August 1783
28 September 1783 (?)
31 August 1783
27 August 1783

2 September 1783

Figure 40: The last observed occurrence of the Laki haze, based on data compiled for this book.

The Harvest

With leaves abscised and vegetation withered and damaged, prospects of a good harvest seemed dim.[399] Newspaper reports from England, the German Territories, Norway, Sweden, France, Italy, and the Dutch Republic remarked that the condition of plant life made it seem as though winter had arrived early.[400] However, what came to pass defied the pessimistic outlook.

---

399 Demarée, Ogilvie 2001: 233.
400 Grattan, Pyatt 1994; Grattan, Brayshay, Sadler 1998; Thordarson, Self 2003; Thordarson 2005: 214.

A report from late June found that the fields and vineyards around Vienna showed "such an abundance of grapes that we cannot remember such fertility in recent years."[401] Many surmised that the dry fog and the fertility were directly related. A correspondent from Stuttgart interviewed a vinedresser about the dry fog: "I don't know where it is from, the naturalists may decide that, but I can assure you that the weather is magnificent. [. . .] Thank the Lord for blessing us with such a good year." All was well, and the interviewer concluded, "May the populace in the cities and villages become smart from the old vinedresser's comments, be grateful to the Lord, and stop asking ridiculous and useless questions [. . .]."[402]

In mid-July 1783, German newspapers reported that "the field fruits are ripening, [. . .] it is known that the sublimated sulfur agrees with the vegetables."[403] Assumptions that the dry fog had beneficial consequences seemed to outweigh concerns, at least in print. On 19 July 1783, the *Berlinische Nachrichten* printed another calming report, which emphasized that vegetation, particularly the vineyards, were prospering.[404] Another report in the *Meiningische Wöchentliche Nachrichten* confirmed that the dry fog had not stopped the growth of plants: on the contrary, the fruit harvest seemed to increase.[405] Franz VON BEROLDINGEN (1740–1798), a Swiss-German autodidactic geologist and volcanologist, noticed that plants persisted despite the drought and the dry fog. Even his potatoes did well despite only growing small leaves and blooming early.[406]

A report from Mannheim on 19 September 1783 indicated that all regions with vineyards were thriving in the warmth. It further opined that those "who saw hunger when they saw the dry fog and who saw doom when they looked at the blood-red sun, may regret their superstition and now listen to the daily Enlightenment of nature's secrets, which will make them more confident in the future when trusting the

---

401 Hamburgischer Unpartheyischer Correspondent, 8 July 1783: Report from Vienna, 28 June 1783. "Die Getraidefelder und Weingärten in unserer Nachbarschaft versprechen die reichste Ernte. Es zeigen sich so viel Trauben, daß man sich seit vielen Jahren einer solchen Fruchtbarkeit nicht erinnert."

402 Königlich Privilegirte Zeitung, 12 July 1783: 686: Report from Stuttgart, 30 June 1783. "[. . .] woher dieser Dunst komme, das weiß ich nicht, das mögen die Gelehrten ausmachen, aber das kann ich ihn versichern, daß die Witterung herrlich ist. [. . .] Gott sey gedankt, der uns mit einem so guten Jahr segnet. – Möchte doch unser Pöbel in Städten und Dörfern von diesem alten Weingärtner lernen klug werden, Gott für den gegenwärtigen Segen danken, und alle fürwitzige und unnütze Fragen von Bedeutungen und vom Zukünftigen ununtersucht lassen."

403 Münchner Zeitung, 15 July 1783: 434. "Bei alle dem haben sich, Gott Lob, weder in unsren Gegenden, noch in irgend einem Orte Europens, wo diser trokne Nebel sich zeigte, die Krankheiten vermehret; noch die Fruchtbarkeit der Erde vermindert. Vielmehr scheinen sich gewisse, sonst in diser Jahreszeit herrschende Krankheiten verloren zu haben; und die Früchte reifen wenigst so gesegnet, als jemals, welches ebenfalls in der Vermutung starker Schwefelausdünstungen bestätiget, da es bekannt ist, das die sublimirten Sulphurteilchen dem vegetabilischen Reiche sehr gut bekommen."

404 Berlinische Nachrichten, 19 July 1783: 670: Report from Mannheim, 6 July 1783.

405 Meiningische Wöchentliche Nachrichten, 12 July 1783:114: Report from Heilbronn, 29 June 1783.

406 BEROLDINGEN 1783: 17–20.

Almighty."[407] Of course, these reports were probably put to print to assuage the fears of the public; however, given the sheer number of such stories, it is likely that the vineyards were genuinely expecting a bumper year.

These prognoses were correct: in the German Territories, the harvest was plentiful. The fruit harvests, particularly the wine grapes, were exceptional. Either the fog was not as injurious as suspected, or the plant life had made a remarkable recovery. In some circles, the fog was now considered a blessing.[408] Reports indicate that the German wine harvest from 1783 surpassed that of 1727 and 1766.[409] The abundance of 1783 extended to the smaller allotments and gardens in and around Munich and many other parts of the German Territories. "The remark of the naturalists seems almost to be confirmed; the long-lasting fog above Europe had a fertilizing power."[410] Mounting evidence suggested a clear connection between the dry fog and the plentiful harvest.

Britain had also profited from a particularly good harvest in 1783.[411] Despite the reports of crop damage in June and July 1783, no documentary evidence supports claims of large-scale harvest failure in England later that year.[412] Gilbert WHITE, however, casts a shadow, noting that some parts of the harvest had not proliferated, including hops and kidney beans.[413] In Scotland, on 17 August 1783, John ALVUS at Dalkeith recorded in his weather diary that "The Corn Harvest [began] & the crop [was] generally very good."[414] Another meteorological record from Scotland confirms that early September 1783 was "excellent for harvest work & cut down crops, tho[ugh] ill-suited for ripening late grain."[415]

The same was true for the Austrian Netherlands. The Baron DE POEDERLÉ wrote, for October 1783, that the vegetation was beautiful and healthy for the season and

---

407 Münchner Zeitung, 25 September 1783: 597: Report from Mannheim, 19 September 1783. "Die auf die Güte des Herrn mistrauischen Menschen, welche in dem Hehrrauche Hunger, und in der blutrothen Sonne Verderben zu ligen glaubten, mögen also ihren Irrwahn bereuen, den täglich heller leuchtenden Aufklärungen der Naturgeheimnisse mehr Gehör verleihen, und eben dadurch künftig zuversichtlicher auf die Wege der Allmacht bauen."
408 Berlinische Nachrichten, 16 October 1783: 956: Report from Vienna, 8 October 1783.
409 Berlinische Nachrichten, 30 September 1783: 902: Report from Saxony, 20 September 1783. A similar report can be found here: Königlich Privilegirte Zeitung, 30 September 1783: 911.
410 Münchner Zeitung, 14 October 1783: 642–643. A similar report can be found here: Berlinische Nachrichten, 16 October 1783: 956: Report from Vienna, 8 October 1783. "Fast scheint sich die Bemerkung der Naturforscher zu bestätigen, daß diese außerordentliche Fruchtbarkeit, dem so lange über Europa geschwebten Nebel zuzuschreiben sey."
411 Das Wienerblättchen, 31 August 1783: 5: Report from London.
412 WITHAM, OPPENHEIMER 2004: 24.
413 Gilbert WHITE, "The Naturalist's Journal," 1783, Add MS 31848, British Library, London, UK: 131–132.
414 John ALVUS, Weather Diary for Dalkeith kept by John Alvus, MET1/4/37, National Records of Scotland, Edinburgh, UK.
415 Scottish Meteorological Society, MET1, National Records of Scotland, Edinburgh, UK.

that the plant life was as lush as during September.[416] In Italy, the situation was similar. According to Italian priest and physicist Guiseppe TOALDO (1719–1797), the dry fog had had a detrimental effect on some plants but also "brought forth very great fertility in all fruits of the earth."[417] In France, the harvest occurred earlier than usual because of the hot summer. The grapes were said to be of a higher quality than usual and very sweet.[418] As in England, some plants suffered damage.[419] Historian of medicine Alain LARCAN remarked that municipal records from Paris and Lyon did not mention any subsistence crisis throughout the summer or autumn of 1783.[420] In Bohemia, there were reports of crop failure in late August 1783. Heavy rain was followed by extreme cold, and fog covered the region before the heat returned.[421] Here, the cost of rye, barley, and oats dropped between 1782 and 1786, whereas wheat prices increased between 1783 and 1785. Rudolf BRÁZDIL and his colleagues identified similar patterns in Prague and Brno.[422] In Sweden, however, the grain harvest failed in several parts of the kingdom in September 1783; the hay harvest had also produced very little, possibly pushing farmers to cull their livestock for lack of fodder.[423] In neighboring Norway, the vegetation suffered similarly.[424] The Laki eruption and its dry fog did not precipitate an agricultural crisis across western Europe in 1783; in many cases, it had the opposite effect.[425]

A Year of Awe

The dry fog was not the only unusual occurrence of 1783: several geophysical and meteorological events made this year remarkable. Five huge earthquakes devastated Sicily and Calabria; a heat wave scorched many parts of Europe; thunderstorms terrorized the continent with lightning; the sun turned blood-red; and a molten, smoking island appeared off the coast of Iceland. These events profoundly impacted Europeans throughout the year – all of these phenomena were mentioned in the newspapers, time and time again.[426] Contemporaries perceived the almost simultaneous occurrence of

---

416 DE POEDERLÉ 1784: 346–347.
417 VAN SWINDEN 2001: 77.
418 SOUKUPOVA 2013: 14.
419 RABARTIN, ROCHER 1993: 38–42.
420 Alain LARCAN's comment to GARNIER 2011.
421 Berlinische Nachrichten, 9 September 1783: 835–836: Report from Bohemia, 24 August 1783.
422 BRÁZDIL et al. 2017: 156.
423 DEMARÉE, OGILVIE 2001: 235.
424 HÓLM 1784a: 234; ÞÓRARINSSON 1981.
425 GRATTAN, BRAYSHAY, SCHÜTTENHELM 2002: 100.
426 DEMARÉE, OGILVIE 2001: 221.

these phenomena as extraordinary.[427] The year 1783 truly lived up to its designation: *annus mirabilis*, a year of awe.[428] The following subchapter considers how contemporaries responded to the unusual weather that characterized that year.

Reactions to the Unusual Weather

The Response to the Dry Fog

There seemed to be three different, not necessarily mutually exclusive, reactions to the unusual weather: some interpreted it as a harbinger of Judgment Day; some feared the unexplainable, a natural emotion in the face of something unfamiliar; and some were inspired to search for explanations within the scope of the physical world.

A report from Lautern in southwestern Germany from late June 1783 mentions the terrifying thunderstorms, deadly lightning strikes, and very thick fog that "left many people fear-stricken."[429] In Switzerland, the dry fog and other unusual phenomena created "great alarm" in the population.[430] The fog robbed the people of the ability to gaze upon a surety of life: "people almost doubt whether the sun and the moon still exist."[431] According to the *Preßburger Zeitung*, in Preßburg, in mid-July 1783, "everybody was in fear about those things, which might come." The fog was their main concern, and there was hope that the deluge of thunderstorms soaking Europe would disperse it. This fog, however, stubbornly remained, thwarting expectations.[432]

In a similar vein, in England, Gilbert WHITE remarked that the red sun, coupled with the news of earthquakes in Calabria, was enough to bother the enlightened mind.[433] David HIGGINS argues that WHITE was implying that "something ominous, perhaps even apocalyptic" was taking place on this "most portentous" summer.[434] In

---

427 Hamburgischer Unpartheyischer Correspondent, 15 August 1783: Report from Munich, 2 August 1783.
428 MERCIER 1784; STEINÞÓRSSON 1992; BRÁZDIL et al. 2010: 182.
429 Königlich Privilegirte Zeitung, 12 July 1783: 686: Report from Lautern, 29 June 1783. "Traurig sah das Auf- und Untergehen der Sonne aus, welche die Gestalt und Farbe einer rothglühenden Kugel hatte und viel Leute in Angst setzte."
430 Report from Zurich, KIESSLING 1888, part 1: 27–28.
431 Königlich Privilegirte Zeitung, 19 July 1783: 708: Report from Graubünden, 24 June 1783. "Unsere ganze Gegend ist seit 8 Tagen von Nebeln so verfinstert, daß man fast bezweifelt, ob Sonne und Mond noch existiren."
432 Preßburger Zeitung, 30 July 1783: Report from Hermannstadt, 14 July 1783. "Die Einbildung gab dem Nebel einen stinkenden Schwefelgeruch, und jedermann war hier in Furcht über die Dinge, die da kommen sollten." Preßburg is today's Bratislava in Slovakia.
433 MABEY 2006: 265.
434 HIGGINS 2019: 131–136.

Leicestershire, Thomas BARKER likened the summer of 1783 to "Virgil's description of the summer after Julius Cesar's death."[435]

France, too, was struck by "great fear," and gossip spread about the imminent end of the world.[436] Jacques Antoine MOURGUE DE MONTREDON got specific, stating that he had heard from some that the end would occur on 1 July 1783.[437] In his book, published in 1784, Johann Ernst Basilius WIEDEBURG (1733–1789), a German naturalist and astronomer based in Jena, suggests that 1783 had been one of the most baffling years for naturalists and gave them much reason to debate.[438] Generally, scientific inquiry took precedence over eschatological concerns as to the cause of all this; perhaps this is because a thick fog is not traditionally a harbinger of the end. In biblical stories, particularly in the Book of Revelation, the end is often announced by the appearance of comets.[439]

Generally speaking, the authors of the historical sources make no explicit mention of being panicked or fearful; during the Enlightenment, the discourse was devoid of such talk, with metaphor and rhetoric used in its stead.[440] Metaphors often play a compensatory role when dealing with one's fears of nature.[441] The newspapers found three different ways to hearten their readership. First, they relayed reassuring messages from naturalists. Second, they searched for precedents: the elderly were interviewed about their experiences with similar weather phenomena throughout their long lives, and old records and chronicles were scoured for mentions of similar events in the past. And third, they tried to appeal to their readers' pride by claiming that only the superstitious were panicking, whereas the Enlightened remained calm.

Many newspapers of the day concluded that the fog was nothing special. According to them, similar fogs had occurred in the past and, in all likelihood, they would happen again in the future. Although it is rarely explicitly stated, a certain level of fear and panic must have been in the ether, given the number of subtle and not-so-subtle attempts to spread calm.

The Voice of Reason

In the eighteenth century, it became apparent that individual weather phenomena could affect several different regions or countries.[442] Weather observers in 1783

---

435 Thomas BARKER, Private Weather Diary for Lyndon Hall, Leicestershire, MET/2/1/2/3/227, National Meteorological Library and Archive, Met Office, Exeter, UK: 213.
436 GARNIER 2011.
437 MOURGUE DE MONTREDON 1784: 764.
438 WIEDEBURG 1784: 1.
439 Revelation 12:1–17.
440 BRIESE 1998: 8–9; GRATTAN, BRAYSHAY, SCHÜTTENHELM 2002: 98.
441 BRIESE 1998: 13–14; SCHMIDT 1999: 297.
442 KINGTON 1988: 18–19.

quickly concluded that the phenomenon they were witnessing locally, the dry fog, also occurred in several other places around Europe. No remarks in the sources indicate whether the sheer scale of the fog made it more terrifying or less so. Perhaps its incredible expanse encouraged apocalyptic thoughts, or perhaps, to paraphrase fourteenth-century Italian historian Dominicus de Gravina, it brought comfort that others suffered too.[443]

The palpable uncertainty of 1783 prompted some naturalists to intervene. Members of the Societas Meteorologica Palatina of Mannheim felt compelled to draft a statement addressing the weather. The Society's publications were published entirely in Latin and thus remained mysterious to all but the highly educated; however, in early July 1783, they released a statement that made many of the European newspapers. This statement was lengthy and underlined the responsibility of the sciences to inform the public:

> [The] observatory in Mannheim gives us news that the fog started on 16 June and became increasingly thicker. [. . .] Within 15 days, it had covered a large part of Europe. [. . .] From the Réaumur hygrometer, we have learned that the atmosphere was extraordinarily dry, this vapor was no moist precipitation, unlike other fogs, but it consisted of dry, hard particulates, [. . .] which must have had their origin in the electric matter, in the opinion of the observatory, [. . .] Regarding the nature of this dry fog, it was not malignant at all. The mortality did not increase under it; it caused no new illnesses; the mortality rather sunk, and the fruits, particularly the grapes, thrived.[444]

Naturalists in Mannheim favored this positive interpretation of the dry fog, one of many in that vein circulating at the time. The report even went so far as to suggest that the dry fog might have even been beneficial to human health. This further clarifies the report's intention, which was to pacify the public and assure them that although this dry fog was unusual, it was not malevolent.

In his hastily published monograph, Christoph Gottfried BARDILI took a strong stand against superstitious interpretations of the dry fog. He saw no need for people to be

---

443 Dominicus de Gravina 2011.

444 Münchner Zeitung, 10 July 1783: 422. "[. . .] dise brave Sternwarte von Mannheim gibt Nachricht, das der Nebel am 16ten Brachm[onat]. Seinen Anfang nahm, sich allmälig verdikte, selbst die Donnerwolken den Augen entzog, und nach allmäligem Abnehmen eine Dauer von 15 Tagen durch eine grosse Streke Europens vollbrachte. Die Luft war dabei schwer und warm. Aus dem Reaumurschen Feuchtemesser nahm man ab, das die Atmosphäre ungewöhnlich troken, und also diser Duft kein feuchter Niderschlag, gleich anderen Nebeln war, sondern aus troknen, festen Teilchen bestand, die aus den Erdkörpern in die Luft mit grosser Kraft musten erhöhet worden sein, welche Kraft nach Meinung der Sternwarte blos der elektrischen Materie zugeschriben werden kann, deren ungemeine Auflösungskraft den Naturforschern bekannt ist. Was die Natur dises Hehrrauchs betrifft, so war er gar nicht bösartig. Die Sterbefälle haben sich dabei nicht gemehrt, keine Krankheiten sind darunter ausgebrochen; diese schinen sich sogar zu vermindern, und die Früchte, besonders der Weinstok hatte ein herrliches Gedeihen." Similar reports can be found in the Frankfurter Staats-Ristretto, 7 July 1783: 444: Report from Mannheim, 2 July 1783 and Berlinische Nachrichten, 19 July 1783: 760: Report from the Mannheim observatory, 6 July 1783.

worried; indeed, he reminded his readers that at this point, in early July 1783, there was too little data available to make grand assumptions about its possible negative conse- quences.[445] BARDILI's publication was aimed at both naturalists and the "uninitiated" (*Uneingeweyhte*), which presumably meant a non-educated, general audience.[446] Franz VON BEROLDINGEN conjectures that the fog was indeed harmful but that the cleansing na- ture of plant life would negate this over time.[447] This statement alludes to photosynthe- sis, an earlier discovery by Joseph PRIESTLEY (1733–1804), an English physicist and philosopher.[448] Nine years prior, PRIESTLEY had discovered that plants produced oxygen under the influence of sunshine at the cost of carbon dioxide, essentially filtering toxins from "unclean air."[449] In contrast to VON BEROLDINGEN, Johann WIEDEBURG predicted that the fog would ultimately prove detrimental to plant life but have no negative conse- quences for humans or animals.[450]

The red sun particularly upset people.[451] The *Königlich Privilegirte Zeitung* printed a calming explanation: "The current fog does not let any other of the seven colors of the sun through, other than the red one. That is the whole secret! No need to be scared!"[452] Indeed, the scattering of photons as the light traveled through the sulfur-dioxide-laden atmosphere caused the effect. Volcanically induced red sunsets usually occur about one hour after the actual sunset.[453]

On 15 July 1783, the *Königlich Privilegirte Zeitung* stated:

> There is no need to seek refuge in the earthquakes in Italy to explain our German fogs. As soon as you think the fog is from Italy, you feel your chest tightening, see glowing balls falling from the sky, and fear earthquakes. But [. . .] the ghost rumbles in our head.[454]

---

445 BARDILI 1783: 14–16.

446 BARDILI 1783: 1–9; 9 (quote).

447 BEROLDINGEN 1783: 30.

448 SCHIFFER 2003: 71.

449 STOLLBERG-RILINGER 2011: 168.

450 BEROLDINGEN 1783: 31; WIEDEBURG 1784: 66.

451 Königlich Privilegirte Zeitung, 12 July 1783: 686: Report from Lautern, 29 June 1783.

452 Königlich Privilegirte Zeitung, 15 July 1783: 695: Report from Stuttgart, 1 July 1783. "Der jetzige Nebel läßt uns von den sieben Farben der Sonne eben keine andere, als die rothe, zukommen. Das ist das ganze Geheimniß! Lassen Sie uns nicht bange seyn!"

453 AHRENS 2009: 532–534.

454 Königlich Privilegirte Zeitung, 15 July 1783: 695: Report from Stuttgart, 1 July 1783. "Auszug eines Schreibens aus Stuttgard, vom 1. July. Sie fragen mich über meine Meynung wegen der jetzt so lang anhaltenden Nebel. Ich halte sie für nichts besonders. [. . .] Wir haben also gar nicht nöthig, zu dem Erdbeben in Italien unsre Zuflucht zu nehmen, um unsere deutschen Nebel zu erklären. [. . .] Sehen Sie nur die Liste nach, so werden Sie finden, wie wenig Personen dieser Tage gestorben sind. Aber sobald man denkt, der Nebel komme aus Italien, so fühlt man seine Brust beklemmt, sieht glühende Kügeln vom Himmel fallen, und ahndet Erdbeben. Denn auch hier, wie in tausend andern Fällen, pol- tert das Gespenst in unserm Kopfe."

Evidently, the author of the publication believed a local explanation would satisfy the newspaper's readership more than a foreign one. News stories of the devastating earthquakes in Calabria had been printed in the same newspaper as the aforementioned quote for some months.[455]

A Search for Precedents

Naturalists and laypersons alike searched through chronicles and other records, hoping to find something that would lead to a better understanding of Europe's current predicament. The *Berlinische Nachrichten* printed a report from a named correspondent, an 84-year-old preacher called HöPPEL, who had found evidence of similar events in the past. While studying *Saur's Calendario Historico* from 1594, he came across mentions of strange weather in 1157, 1546, and 1571. Those years also featured a great heat, drought, and a thick fog that made the sun look like a fiery ball. The chroniclers did not mention any catastrophic aftermath, which suggested none was to be expected in 1783.[456] Johann Ludwig CHRIST (1739–1813), a German Lutheran pastor from Rodheim and an expert on fruit-growing and insects, wrote a book about the dry fog. In the book's appendix, he documents some evidence he found of a similar event. An old calendar revealed that the summer and autumn of 1652 were affected by a haze and blood-red sunsets and sunrises.[457] Karl Ludwig GRONAU (1742–1826), a German weather observer and Lutheran pastor, gives dates for precedents that were comparable to the reddish haze of 1783: they occurred on or between 22 and 25 April 1547; 15 February 1652; mid-July 1661; 5 and 6 August 1730; 6 to 8 June 1756; and 3 to 7 August 1766.[458] The monk and naturalist Dom Robert HICKMANN (1720–1787), based at the Saint-Hubert Abbaye in the Ardennes region, found records of similar fogs in 1746 and 1764; however, he admits that, unlike the current fog, those had only lasted a few days.[459]

Similarly, Jan Hendrik VAN SWINDEN "remembered without prompting" an event similar to 1783. He was born in 1746, so he must have been familiar with the records: "In 1721, over a vast tract of land, the sun on 1 June had been seen as white, destitute of rays."[460] The 1721 Katla eruption (VEI 5) in Iceland was perhaps the cause of this.[461] MOURGUE DE MONTREDON also looked to the past for similar events and reported on one such incident during the summer of 1721 in Persia, when a great fog enshrouded the

---

455 Berlinische Nachrichten, 17 July 1783: 664: Report from Naples, 24 June 1783.
456 Berlinische Nachrichten, 12 August 1783: 746: Report from Anspach, 23 July 1783.
457 CHRIST 1783: 3, 34–35. His book was published in July 1783: it was advertised as early as 19 July 1783 in the Frankfurter Staats-Ristretto, 19 July 1783: 473.
458 GRONAU 1785: 97.
459 HICKMANN 1783: 505–507.
460 VAN SWINDEN 2001: 73.
461 VAN SWINDEN 2001: 77–78.

land, and the sun had the color of blood. Both France and Italy experienced similar occurrences that same year.[462] Carl Friedrich HINDENBURG (1741–1808), a mathematician and professor of physics and philosophy, confirmed another of MOURGUE DE MONTREDON's assertations that a haze was observed in June 1721 over parts of France and Italy. Additionally, he came across mentions of a "sun smoke," whereby the sun took on a copper red or blackish appearance in Mecklenburg in July and August 1766. He thought it worth mentioning that there were no adverse effects observed.[463]

In his 1783 publication, *Vues sur la nature et l'origine du brouillard qui a eu lieu cette année*, Robert Paul DE LAMANON includes a long list of prior occurrences of an obscured sun dating as far back as Roman times.[464] Guiseppe TOALDO's publication mentions some of the same precedents: a darkened or reddish sun appeared in the year of Rome 291 [462 BCE], 542 [211 BCE], 552 [201 BCE], 554 [199 BCE], and 710 [44 BCE]. The last example, the year of Caesar's death, is known to be the year of a large volcanic eruption.[465] In the Common Era, the sun had an unusual color in 264, 396, 790, 937, 1020, 1104, 1154, 1206, 1227, 1263, 1383, and 1549.[466]

A correspondent from Hildburghausen interviewed a man who was – allegedly – very old. "A 102-year-old shepherd from nearby Fulda has confirmed that this is the third time he has experienced this great fog. In his experience, they were always followed by fertile years."[467] Many publications interviewed the elderly, using the perceived authority and wisdom that comes with old age to give credence to their point of view.[468]

One might readily question how far the "collective memory culture" of the contemporaries really stretched.[469] "Extreme weather memory" seemed to be fairly short in 1783. A weather event quickly became "the worst in living memory," even if written sources could prove that a comparable event had occurred only a few years before. Christian PFISTER reminds us to consider that the human capacity to remember the past, regarding the weather, can only reach back a few days to weeks.[470] Brian FAGAN comes to a similar conclusion: "The traumas of extreme weather events fade rapidly from human consciousness."[471] In 1783, it was difficult to compare past extreme weather events with one another; often, the descriptions were only qualitative, not quantitative, and exaggerations were commonplace. "The 'worst,' 'coldest,'

---

462 MOURGUE DE MONTREDON 1784: 763–764.
463 LUDWIG 1783: 219–221.
464 Journal de Paris, 13 August 1784: 964–965. DE LAMANON's publication was also translated into Dutch (1784b) and English (1799). More on DE LAMANON's life can be found in CARTWRIGHT 1997.
465 MCCONNELL et al. (2020) traced the latter eruption to 43 BCE and identified Alaska's Okmok volcano as the eruption's source.
466 TOALDO 1799: 421–422.
467 Königlich Privilegirte Zeitung, 31 July 1783: 742.
468 FLEMING 1998: 5.
469 MÜNCH 1992: 131.
470 PFISTER 1999: 36.
471 FAGAN 2000: xiii.

'hottest,' or 'wettest' weather in living memory seemed to crop up every few years."[472] In the zeitgeist of the Enlightenment, there was also much room for doubts, speculations, and creative ideas.[473]

John GRATTAN and Mark BRAYSHAY studied English newspapers published between June and September 1783 for mentions of the dry fog. They conclude that these newspapers seemingly accepted its unique nature. The British mindset was reflected upon in a letter by William COWPER to John NEWTON on 29 June 1783. First, COWPER describes the state of the weather: "So long, in a country not subject to fogs, we have been cover'd with one of the thickest I remember. We never see the Sun but shorn of his beams, the trees are scarce discernible at a mile's distance, he sets with the face of a red hot salamander, and rises with the same complexion." He continues to speak on the mindset of the people, stating that many had given the origin of these phenomena much thought. In his expressive epistle, COWPER seems fearful or anxious and wonders if he is witnessing the beginning of the end time.[474]

Most British newspapers that GRATTAN and BRAYSHAY studied were less inclined to mention Judgment Day but frequently used the adjective "violent," which appeared in 25 percent of the reports. GRATTAN and BRAYSHAY also caution the reader that the use of hyperbolic language might have been a rhetorical device used during the British Romantic period.[475]

The uniqueness of the dry fog was not generally accepted in France, perhaps because the search for historical precedents had been more successful.[476] The fact that French naturalists and newspaper correspondents considered the weather to be "nothing new" did little to persuade their British counterparts.[477]

Within the German Territories, reactions were more ambiguous than those of the English or French. Some newspaper reports concluded that the dry fog and the thunderstorms of 1783 were incomparable to past events. For instance, a report in the *Königlich Privilegirte Zeitung* remarked that the storm on 27 June 1783 at 2 p.m. was the most terrible that they had ever seen.[478] An article from Mannheim in the same newspaper issue asserted that "The oldest people cannot remember to have experienced something similar. [. . .] the thunder was so terrifying that something similar has never been heard before."[479] In mid-July, the *Münchner Zeitung* stated that "other chroniclers also

---

472 SCARTH 1999: 117.
473 SCARTH 1999: 117.
474 MENELY 2012: 477.
475 GRATTAN, BRAYSHAY 1995: 130–132.
476 ROBERJOT 1784: 399–400. L'ábbe ROBERTJOT nevertheless published a letter in which he titled the dry fog in France "un Phénomène singulier du Brouillard de 1783."
477 GRATTAN, BRAYSHAY 1995: 132.
478 Königlich Privilegirte Zeitung, 12 July 1783: 686: Report from Lautern, 29 June 1783.
479 Königlich Privilegirte Zeitung, 12 July 1783: 686: Report from Mannheim, 1 July 1783. "Der schon seit dem 17. [Juni] Tag und Nacht anhaltende trockene Nebel kann gewiß für eine ganz außerordentliche Erscheinung gehalten werden; den ältesten Leuten ist es nicht erinnerlich, dergleichen erlebt zu haben.

do not know of any such event. In Roman history, we find one that lasted for three summer months, but it was only present in Italy."[480] This confirms findings from a study conducted by atmospheric physicist Dario Camuffo and historian Silvia Enzi on the presence of dry fogs in Italy between 1374 and 1819.[481] In addition to all this, an unsettling connection was established: "some very old people remember that such hot and foggy summers preceded the very severe winters of the years 1709 and 1740. It is [therefore] necessary to [be prepared with] stocks of wood."[482] Mixed messages, and seeds of doubt in a generally reassuring overview, probably left the general public confused.

Blaming the Superstitious

Particularly in the fifteenth and sixteenth centuries, disastrous weather events were considered to be either divine intervention or portents of things to come.[483] The ecclesiastical authorities argued that the natural phenomena were either the expression of God's inscrutable will or manifestations of His righteous anger against the sinfulness of the world.

On occasion, in 1783, the Bible was used to interpret the possible punitive-theological meaning of weather events. Particularly, signs in the sky, such as the red sun and meteors, could be interpreted as bad omens.[484] It is not surprising that some could see their current situation reflected in biblical verses like these: "There will be signs in the sun, moon, and stars. On the earth, nations will be in anguish and perplexity at the roaring and tossing of the sea [. . .] When these things begin to take place, stand up and lift up your heads because your redemption is drawing near."[485]

In the source sample for this book, there were very few concrete mentions of the "end times." That the dry fog, thunderstorms, and flooding coincided with one another did – for some – give the impression of an event of biblical proportions. "In the evening, after the thunderstorm, the dry fog returned and enveloped the mountains again, [. . .]. The cries of the winds, the crying of the people, the banging of the thunder, and the rushing of the water were mixed! [. . .] The Day of Judgment seems to have started already."[486] In Lausanne, the dry fog was seen as the maw from the

---

480 Münchner Zeitung, 14 July 1783: 429–430: Report from Düsseldorf, 28 June 1783. "Die älteste Mensch erinnert sich nicht um dise Zeit keines so lange anhaltenden und so diken trokenen Duftes; auch unsre Kroniken wissen in älteren Zeiten nichts davon. In der Römischen Geschichte finden wir einen, der die drei Sommermonate hindurch gewährt, aber nur in den Italiänischen Gegenden."
481 Camuffo, Enzi 1995: 156.
482 Demarée, Ogilvie 2001: 239.
483 Schmidt 1999: 43.
484 Jakubowski-Tiessen 1992: 98–99.
485 Luke 21:25–28.
486 Münchner Zeitung, 11 July 1783: 425–426. "Am Abend nach dem Donnerwetter trat aber der Nebel wi[e]der ein, und verhüllte die Gebirge von ·neuem, macht Engbrüstigkeiten und vermehrt die

underworld, and its presence reminded those who paid attention of a passage in the Book of Revelation 9:2: "when he opened the Abyss, smoke rose from it like the smoke from a gigantic furnace. The sun and sky were darkened by the smoke from the Abyss."[487] In Switzerland, the authorities announced days of repentance, fasting, and prayer to prevent the worst and to prepare the population for the approaching end times.[488] In Pas-de-Calais in France, the bishop called for three days of prayers. In Antwerp, in the Austrian Netherlands, public prayers were ordered from 1 August 1783 onward in the hope that divine intervention would bring rain.[489]

There were other religious and political interpretations of the dry fog. Some believed that the "Enlightened" were responsible, as their thoughts and actions were in clear opposition to the doctrines of the Church. God might have used this natural event as a punishment to correct the "degeneration of the Enlightened zeitgeist." In his poem, *Hänts Leutel sagts mä do* (People, tell me), Austrian writer Peter Gottlieb LINDEMAYR (1723–1783) blamed the introduction of a new law: Emperor Joseph II had introduced the Patent of Toleration in 1781, which allowed Protestants and Jews to practice their religions freely. His Secularization Decree from 1782 also banned several monastic orders and liquidated a third of all monasteries; furthermore, he redefined marriage as a civil contract, forbade pilgrimage and processions, and cut priests' salaries. Needless to say, Catholics opposed all of these new laws and interpreted the fog as punishment for those turning away from Rome. LINDEMAYR wrote this poem in the very last days of his life; he died on 19 July 1783.[490]

Many viewed the fog as a blessing from God. The aforementioned bumper harvests were reason enough to come to this conclusion. For instance, a report from Stuttgart on 30 June 1783 – when the city was in the midst of the fog and under blood-red sunsets and sunrises – attested that the weather was "marvelous," citing the grape harvest as a positive consequence. The report reprimanded the "populace" for such bizarre ideas and suggested they thank God for sending this blessing.[491]

In the late eighteenth century, newspapers began to reach wider audiences due to the increasingly popular practice of reading them aloud in public places. Even so, the content of the reports makes it clear they were still squarely aimed at an audience of educated elites. They frequently ridiculed the irrational and paranoid beliefs of the so-called common folk. The diary of Reverend Henry WHITE, Gilbert WHITE's brother,

---

Verurteile der gegenwärtigen Zeit. [. . .] Die Menschen riefen sich von den Dachgibeln einander um Hilfe, ohne sich helfen zu können. Vieh, Bäume, Prüken, Zäune, Trümmer von Hütten, alles wurde unaufhaltbar dahin gerissen. [. . .] Das Heulen der Winde in das Heulen der Menschen, in das Knallen des Donners, und das Brausen des Wassers gemischt! – und denn die kläglichsten Auftritte für das Aug! Der jüngste Tag schi[e]n angebrochen zu sein."
487 Revelation 9:2; VERDEIL 1783; PFISTER 1975: 87.
488 KIESSLING 1888: 28; PFISTER 1975: 87.
489 DEMARÉE, OGILVIE 2001: 234, 237.
490 LINDEMAYR 2010: 132; BRÁZDIL et al. 2017: 159.
491 Königlich Privilegirte Zeitung, 12 July 1783: 687–688: Report from Stuttgart, 30 June 1783.

expresses similar sentiments. He noticed on 19 July 1783, "the air seems clearer from the late blue thickness which has been so very remarkable, that the Superstitious Vulgar in Town & Country have abounded with the most direful presages and prognostications."[492]

It may seem as though Europe was divided, with the fearless "enlightened" and the terrified "superstitious" on opposite sides, but the reality is more complex. Not all "common people" were petrified by the weather, and not all the "enlightened" were free from fear. Famously, Georg Christoph LICHTENBERG (1742–1799), a German naturalist and professor of experimental physics, was so frightened by thunderstorms that he canceled his lectures in Göttingen when inclement weather was approaching.[493] In a letter to an acquaintance, LICHTENBERG mentioned his fear but also relativized it by stating that his primary concern was for his instruments.[494] A goal of the Enlightenment was to reduce superstition and prejudices, the tool for which was rational thought and experimentation. Demonstrating an understanding of natural phenomena was a favorite pastime of proponents of the Enlightenment.[495]

In 1783, the lines between religion and science remained blurred. Physicotheology was a reform movement that started in England and France around 1700.[496] Its aim was to bridge the gap between the natural sciences and faith. The wonders of nature (*physis* in Greek) were evidence of God's existence.[497] For physicotheologians, both nature and the Bible were books of divine revelation.[498] The difference between the naturalist and the ignorant was, according to the *Münchner Zeitung*, that the first did not feel fear when confronted with natural occurrences. In contrast, "the ignorant populace ogles at these wonders of nature and interprets them as sad premonitions."[499]

---

492 Henry WHITE, Diaries of the Rev. Henry WHITE, Rector of Fyfield, county Southampton, brother of Gilbert WHITE, 1783, Add MS 43816, British Library, London, UK, 30 (19 July 1783).

493 PROSS, PRIESNER 1985; WEIGL 1987: 37.

494 LICHTENBERG 1985: 675, Letter 1123, Georg Christoph LICHTENBERG to Johann Andreas SCHERNHAGEN, Göttingen, 7 August 1783.

495 BEGEMANN 1987: 83–86; HOCHADEL 2003: 21.

496 ALT 2007: 34; REITH 2011: 92; BLAIR, GREYERZ 2020.

497 D'APRILE, SIEBERS 2008: 72.

498 JAKUBOWSKI-TIESSEN 1992: 84.

499 Münchner Zeitung, 30 June 1783: 397: Report from Munich. "Der unkündige Pöbel begasset nun di[e]se Naturwunder, als traurige Vorbedeutungen, und geht mit seinen unsinnigen Spekulationen so weit, das es ihm vom Herzen bange wird, was er von den 3 Geißeln der Menschheit Pest, Hunger, und Krieg für eine für sich herausklauben soll. Der Naturkundige, dem solche Erscheinungen nichts unerwartetes, nichts neues mehr sind, bauet auf Gottes Vorsicht, bewundert die Wunder der Natur, und bleibt ruhig."

Fears of the Fog

Given the sheer quantity of newspaper reports discussing the dry fog and the weather in the summer of 1783, we can infer that the topic interested the readership. Although fear must have been present across Europe, it was rarely explicitly expressed in popular discourse. From the newspapers' overly reassuring pronouncements, it is possible to reconstruct the uncertainty that was present at the time: health was of the utmost concern. On 17 July 1783, the *Münchner Zeitung* printed an article stating that Joseph Jérôme Lefrançois DE LALANDE (1732–1807), also known as DE LA LANDE, a French astronomer and member of the *Académie royale des sciences* in Paris, "has clearly shown, to put the common man's fears at ease, that there was similar weather in 1764, and it did not have any negative consequences for fertility or health."[500]

In the early modern period, fear of nature often amounted to much more than the fear of the phenomena themselves; it extended to a fear of God, His power, and what His actions implied.[501] Faith did not preclude naturalists from thinking about the earthly causes of extreme weather events or nature-induced disasters.[502] An analysis of a sample of newspapers from 1783 shows that contradictory interpretation patterns existed simultaneously.[503]

Although many discoveries throughout the eighteenth century shed light on the development and consequences of weather events, the knowledge of these new advances had not necessarily trickled down into the reasoning of the middle and lower classes. The newspapers attempted to establish a line of communication between the sciences and the general population. The explanations and interpretations, some of which were very detailed, guided the readers and assisted their efforts to make the correct assessment of natural processes.[504] The simultaneity and coexistence of changeable interpretation patterns were a mark of the Enlightenment; theological and magical interpretations coexisted with scientific explanations. New studies focusing on European religiousness in the late eighteenth century show that, particularly in times of crisis, the faithful relied upon a combination of different coping strategies, especially when religious doctrine could not explain the complexity of a crisis.[505]

---

500 Münchner Zeitung, 17 July 1783, 439–440. "Der berühmte Astronom de la Lande hat, um dem gemeinen Manne die Angst zu benehmen, klar gezeigt, das man im J[ahr] 1764 die ähnliche Witterung gehabt habe, ohne schädliche Folgen für Fruchtbarkeit und Gesundheit." For DE LALANDE, see DEMARÉE, OGILVIE 2016: 125.
501 BEGEMANN 1987: 71, 76; ROHR 2009: 21.
502 ROHR 2009: 20.
503 GLACKEN 1990: 505; SCHMIDT 1999: 47; MISSFELDER 2009: 83.
504 SCHMIDT 1999: 6, 191, 306.
505 HOCHADEL 2003: 29; REITH 2011: 83, 92.

The Speculation about the Cause of the Unusual Weather

The dry fog and the tumultuous weather throughout the summer of 1783 left an indelible mark on the continent. Richard STOTHERS remarked that it had a "profound psychological influence on Europe."[506] In 1783, the Enlightenment was well underway, and the summer's events had given naturalists across Europe something upon which to focus their attention.[507] The weather was studied, experiments were conducted, and findings were published in books, scientific journals, or newspapers. Not only did they try to explain where the dry fog might have originated, but they also attempted to connect the various unusual phenomena. In this subchapter, I share some of their compelling and diffuse explanations.

What was the reach of scientific publications? These texts were for the scholarly community and a small, educated audience with some pre-existing knowledge of science. Scientific publications had a small print run; they were fewer in number and more expensive than newspapers. However, even in 1783, there were lending libraries and book clubs. The biggest challenge was illiteracy. However, some books had an uneducated (perhaps illiterate) audience in mind. Johann Nepomuk FISCHER (1749–1805), a professor of mathematics from Ingolstadt, intended with his book "to convince the ignorant audience [*das ununterrichtete Publikum*] about the nature of thunderstorms and ringing the bells [. . .]."[508] It did not seem impossible or even improbable to him that an uneducated person might read or listen to his book. He particularly wanted members of the clergy to read his book to their parishioners. FISCHER's core message was that ringing the bells against thunderstorms was an injurious practice. Overall, the word choice in the scientific publications analyzed for this book was much less radical than in some of the very polemic newspaper articles: one example of the latter being a report in the *Münchner Zeitung* that explicitly called out the "stupid farmers."[509]

A passage from Franz VON BEROLDINGEN's book shows that the popular discourse was not entirely separate from the scientific discourse. He references German newspaper reports in the following excerpt: "You expect me to share my thoughts about

---

506 STOTHERS 1996: 79.
507 GRATTAN, BRAYSHAY 1995: 129.
508 FISCHER 1784: 14. "Also ist das Glockenläuten offenbar kein zuverlässiges Mittel gegen die Gewitter. Da aber meine Schrift die Absicht hat, das ununterrichtete Publikum von der Natur des Gewitters, und des Glockenschalles zu überzeugen, daß diese zwey Dinge einander gar nicht widersprechen; so muß es mir genug seyn, diese erwähnte Erfahrung zur Gewährschaft meiner Theorie einmal für allemal angeführt zu haben."
509 Münchner Zeitung, 26 August 1783: 530. "Unweit Altenburg hat die Errichtung eines Blizableiters eine Bauernrevolte (wer als dumme Bauern hätte wohl sonst wider Erfahrung und Vernunft sich empören können!) veranlasset."

the unique long-lasting dry fog, that did not just affect our region, but according to the newspapers has spread throughout Germany and most of Europe."[510]

The different interpretation patterns offer insight into how naturalists in the late eighteenth century tried to find explanations for extraordinary natural events. These hermeneutical devices show that the various disciplines were not yet fully defined. In some cases, earthquakes, lightning, diseases, harvest, and the dry fog, were explained by the presence of one another. The idea that the different phenomena were related dates back to the Renaissance, when the field of astrometeorology was established, which, at its core, was the belief that meteorological events were influenced by other, seemingly unrelated, natural occurrences.[511]

A Naturalist's Perspective

DE LALANDE authored one of the earliest scientific explanations for the presence of the strange fog: it was published on 2 July 1783 in the *Journal de Paris*.[512] While fog itself was perhaps nothing special, its duration was.[513] DE LALANDE argued that the vapors were merely the consequence of evaporation after a series of heavy rainfalls.[514] He found a description of a similar fog in 1764. Based on this, he believed the 19-year moon cycle might have influenced the weather.[515] The British and German press printed translations of his findings soon after the initial publication.[516]

Almost as quickly as DE LALANDE's findings were shared, criticisms of them arose. For amateur weather observers and naturalists, his argument served as the foundation for lively debate. Most took particular offense to his insistence that this was an ordinary fog; many naturalists debunked DE LALANDE's argument by systematically outlining the differences between this fog and regular, moist fog.

---

510 BEROLDINGEN 1783: 3. "Sie verlangen meine Gedanken über den so lange anhaltenen Nebel, der nicht nur in unsrer Gegend ungewöhnlich war, sondern sich auch laut der Zeitungsnachrichten in den meisten Gegenden Deutschlands, ja auch die meiste Provinzen Europens verbreitet hat."
511 DEMARÉE, OGILVIE 2001: 224.
512 Journal de Paris, 2 July 1783: 762–763; Journal de Paris, 9 July 1783: 789–790. The Berlinische Nachrichten quoted a report from Stuttgart, 1 July 1783, that outlined DE LALANDE's ideas. The Königlich Privilegirte Zeitung quoted a report from Paris, 8 July 1783. Berlinische Nachrichten, 15 July 1783: 654: Report from Stuttgart, 1 July 1783; Königlich Privilegirte Zeitung, 19 July 1783: 708–709: Report from Paris, 8 July 1783. In England, the news about DE LALANDE's explanation and his precedent findings appeared in the Morning Herald and Daily Advertiser, 15 July 1783; and in Felix Farley's Bristol Journal, 19 July 1783: Report, 4 July 1783. GRATTAN, BRAYSHAY 1995: 131–132; HOCHADEL 2009: 60.
513 STOTHERS 1996: 85.
514 Hamburgischer Unpartheyischer Correspondent, 19 July 1783: Report from Paris, 12 July 1783.
515 Münchner Gelehrte Zeitung, 29 August 1783: 60.
516 DE LALANDE in Felix Farley's Bristol Journal, 19 July 1783: Report, 4 July 1783. In the German Territories, his findings were printed in LALANDE 1783: 95–99.

Because of the avalanche of opposition to his findings, it is unlikely DE LALANDE managed to convince many.[517] The *Hamburgischer Unpartheyischer Correspondent* printed DE LALANDE's explanation right next to a counterstatement that openly doubted it. This counterstatement from Brest in France noted that the dry fog was visible at sea and hindered navigation, refuting DE LALANDE's theory. The correspondent concluded that "The fog definitely has a different cause than the heat after a lot of rain, as is believed by Mr. de la Lande in Paris."[518]

Another fact that refuted the evaporation theory quickly became apparent: the fog was present in regions that had not been affected by long-lasting rainfall and extreme flooding.[519] To their credit, newspapers did not simply collect information and blindly share it; instead, they made efforts to critique their sources. This ability to think and reflect critically is a central tenet of the Enlightenment, which had started to envelop all areas of life.[520]

Still, some naturalists agreed with DE LALANDE's findings. In Geneva, the encyclopedist Charles-Benjamin LUBIERES adopted DE LALANDE's explanation.[521] Many advocated for other arguments, of course. Dom Robert HICKMANN criticized Joseph DE LALANDE's theory; HICKMANN thought that a prolonged rainfall would naturally have created a wet fog rather than a dry one.[522] Franz VON BEROLDINGEN set about proving that the fog was not moist. He had observed a barometer when the sky was dull and hazy, and the reading remained high. This observation was proof that this was "a lighter fog, one that did not press on the mercury column."[523] VON BEROLDINGEN further argued that if the heat was to blame for this fog, it would appear every year, and concludes: "An extraordinary occurrence must have an extraordinary cause."[524] He also reasoned that sulfuric air had to come from inside Earth. VON BEROLDINGEN believed earthquakes, particularly those in Calabria and Sicily, were the source of the haze. He

---

517 HOCHADEL 2009: 50; WITZE, KANIPE 2014: 119.
518 Hamburgischer Unpartheyischer Correspondent, 19 July 1783: Report from Brest, 4 July 1783. "Der Nebel hat also gewiß eine andere Ursache, als Hitze nach vielem Regen, wie Herr de la Lande zu Paris glaubt."
519 Münchner Gelehrte Zeitung, 29 August 1783: 61.
520 WÜRGLER 2009: 43. The Münchner Zeitung was particularly good at signaling which reports sported editorial comments by placing them at the end of a report and highlighting them with a small asterisk at the beginning of the comment.
521 PFISTER 1972: 25.
522 HICKMANN 1783: 505–507.
523 BEROLDINGEN 1783: 3–4, 23 (quote). "Der Stand des Barometers war, wie die ganze Zeit über bey dem trüben, neblichten Himmel, hoch; genugsamer Beweis, daß die die Sonne verdunkelnden Dünste keine wässerigten Dünste, sondern von leichter Art gewesen seyn müssen, die nicht auf die Säule des Quecksilbers drückten."
524 BEROLDINGEN 1783: 9. "Ganz recht, mein Freund! eine ausserordentliche Erscheinung muß auch eine ausserordentliche Ursache haben."

felt the reddish color of the fog was another indicator that it was flammable air rather than a normal fog, which has a grayish tone.[525]

As VON BEROLDINGEN criticized DE LALANDE, so he too was criticized. In particular, Georg Christoph LICHTENBERG engaged with Franz VON BEROLDINGEN's publication. Although LICHTENBERG enjoyed his train of thought, he did not appreciate the description of the fog as simply "flammable air," which alone, he said, could not constitute a fog.[526] LICHTENBERG received so many letters regarding the fog and the unusual state of the atmosphere that he had trouble answering them. As a joke, he signed one of his letters with *"in nebula nebulorum,"* which means "in the fog of fogs."[527]

Reverend HILLIGER in Niedersgörsdorf, south of Berlin, noted a burning smell on 31 July 1783. He suggested that the dry fog originated in boggy peat areas or heathland regions and was then carried on the wind to the four corners of the continent. Given that the prevailing winds came from the northeast at the time, he deemed it possible that the smell came from Pomerania or Lüneburg.[528] Sebald Justinus BRUGMANS' book, published on 9 July 1783, argues that although the north had no shortage of potential sources that could produce a continuous sulfuric vapor, such as the fire-spitting mountain Hekla, he did not believe that this fog hailed from that location. According to BRUGMANS, either exhalations originating from the sulfuric innards of the Earth had produced the dry fog, or it had come from the upper atmosphere.[529]

In 1783, several theories on the origin of the fog were in circulation; earthquakes, volcanoes, and lightning rods were among the most prevalent. Whether the fog was singular in living memory and its very essence was up for debate. Some further hypotheses are analyzed in greater detail in the following subchapters.

Active Volcanoes, Active Imaginations

An extraordinary explanation unlike any other materialized that summer. This explanation stood in stark contrast to the reassuring tone of DE LALANDE's and involved at least four volcanic eruptions within the German Territories.[530]

Newspapers ran reports of fire-spitting mountains and one burning mountain within the German Territories. News of the "eruptions" featured alongside articles about national and international political and military affairs. While volcanism in

---

525 BEROLDINGEN 1783: 10–13.
526 LICHTENBERG 1985: 702, Letter 1144, Georg Christoph LICHTENBERG to Johann Andreas SCHERNHAGEN, Göttingen, 18 September 1783.
527 HOCHADEL 2009: 49.
528 HILLIGER 1783: 277–278.
529 BRUGMANS 1783: 54–55.
530 Parts of the subchapter, "Active Volcanoes, Active Imaginations," were originally published in *Global Environment*, see KLEEMANN 2022a.

Germany is not entirely extinct, the last known Holocene eruption took place in the Eifel region around 10,762 years ago (± 150 years), creating the Ulmener Maar.[531] In 1783, naturalists were aware of the German Territories' volcanic past.[532] It was known, for instance, that the landscape around Frankfurt am Main and the Laacher See was of volcanic origin, the latter described in at least one scientific publication as "an old crater."[533] The Laacher See is a caldera filled with water. The last eruption of this volcano took place around 12,900 years ago; it was a Plinian eruption that reached a six on the index of volcanic explosivity.[534] In 1774, the first researchers came to the Vulkaneifel to study its landscape. After the Napoleonic Wars, more geologists followed.[535]

The Cottaberg and the Gleichberg

The first mention of volcanic activity within the German Territories in my source sample was in an article published in the *Königlich Privilegirte Zeitung* on 8 July 1783. The story dealt with a report from the Meißen region in the Electorate of Saxony dated 1 July 1783: "For a few days, the Cottaberg has been throwing out burnt stones, which allows the assumption that in this area a fire-spitting volcano wants to come to life, too."[536] The *Hamburgischer Unpartheyischer Correspondent* ran with an almost identical account on 11 July 1783, as did the *Münchner Zeitung* on 24 July 1783.[537] Today, the Cottaberg is referred to as Cottaer Spitzberg due to its proximity to the town of Cotta. The formation is a 390-meter-tall basalt dome on the western edge of Saxon Switzerland, around 30 kilometers away from the city of Dresden. The hill is of volcanic origin, but the last eruption occurred 25 million years ago, long before 1783. During the Paleogene and the Neogene, two consecutive geological periods that began 66 million years ago and lasted almost 64 million years, the region was shaped by intense volcanism; individual intrusions of magma forced their way through the sandstone platform of the Elbe Sandstone Mountains. One such intrusion resulted in the formation of the Cottaberg.[538] It was a quarry in the nineteenth century, and today it looks more like a knoll. It still dominates the surrounding landscape and is a prominent landmark.

The Cottaberg was not the only eruption reported in the German Territories in the summer of 1783. Just a few days after the report about the Cottaberg, came news

---

531 Global Volcanism Program: West Eifel Volcanic Field; GRATTAN, GILBERTSON, DILL 2000: 307, 313; LUTZ, LORENZ 2013: 1. The term "maar" was first used in 1819.
532 BARDILI 1783: 16.
533 Göttingische Anzeigen von gelehrten Sachen, 17 July 1783: 1144.
534 RIEDE 2017; VOGRIPA: East Eifel Volcanic Field.
535 LUTZ, LORENZ 2013: 2.
536 Königlich Privilegirte Zeitung, 8 July 1783: 670: Meißen region, 1 July 1783.
537 Hamburgischer Unpartheyischer Correspondent, 11 July 1783; Münchner Zeitung, 24 July 1783: 453.
538 The Cottaberg is one hill in a line of volcanic inselbergs.

that the Gleichberg had sprung to life. On 12 July 1783, the *Augsburgische Ordinari Postzeitung* printed an extract of a letter from Hildburghausen, dated 24 June 1783. Unlike the report about the Cottaberg, this one was quite detailed. The letter declared that since Easter, the Gleichberg had been steaming intensely, which had created a thick fog between Römhild and Hildburghausen.

> All the forests in the area are white instead of green; the whole sky looks like chalk; the fog is true natural sulfur, which spoils everything that it touches; the sun and moon always set in a blood-red color. For eight days now, inside the mountain there has been a horrendous and frightening bashing, as if cannons were being fired; then, finally, the whole mountain opened up under the plumes of thick sulfuric smoke; and in the entire area you can hear a constant terrible roaring and rushing [*Sausen und Brausen*] from the opening.[539]

The report continued to detail how terrified locals were fleeing, how people in the churches of the surrounding towns were praying, and how "the whole mountain might collapse and create further disaster."[540] Extracts from the letter from Hildburghausen were printed in the *Frankfurter Staats-Ristretto* on 12 July 1783, the *Münchner Zeitung* on 15 July 1783, and the *Königlich Privilegirte Zeitung* on 22 July 1783.[541] Another report from 22 July 1783, printed in the *Berlinische Nachrichten*, quoted a letter from Frankfurt am Main from 12 July 1783 confirming that the Gleichberg seemed to be a fire-spitting mountain; the wording was very similar to the letter from Hildburghausen.[542]

In addition, an article in the *Münchner Zeitung* stated that a correspondent from Hildburghausen confirmed the Gleichberg had been erupting for three weeks and was

---

539 Augsburgische Postzeitung, 12 July 1783: Report from Hildburghausen, 24 June 1783. "Die Wälder in dieser ganzen Gegend sind alle weiß statt grün; der ganze Himmel wie aufgeflogener oder sublimirter Kalck; der Nebel ist wahrer natürlicher Schwefel, welcher alles was er berührt, verderbt; Sonne und Mond gehen immer bluthroth auf und unter. Seit etwa 8. Tagen that es in dem Berg so entsetzliche und fürchterliche Schläge als würden Kanonen gelöset; denn öffnete sich endlich der Berg ganz unter lauter dickem Schwefelrauch; und in der ganzen dasigen Gegend weit umher hört man aus seiner Oefnung ein beständig fortdauerndes entsetzliches Sausen und Brausen; in allen Kirchen werden Bethstunden gehalten; aus allen umliegenden Ortschaften haben sich die erschrockenen Einwohner bereits geflüchtet, da sie befürchten und vermuthen, der ganze Berg möchte endlich einstürzen, oder noch mehr anderes Unglück bringen. Das fernere Wichtige dieser Sache will ich Ihnen ebenfalls schriftlich melden. Also haben wir dennun in unserm lieben Deutschland auch einen feuerspeyenden Berg, den Gleichberg." A very similar report can be found in the Frankfurter Staats-Ristretto, 12 July 1783: 456: Report from Hildburghausen, 24 June 1783.
540 Augsburgische Postzeitung, 12 July 1783.
541 Frankfurter Staats-Ristretto, 12 July 1783: 456: Report from Hildburghausen, 24 June 1783; Münchner Zeitung, 15 July 1783: 433–434; Königlich Privilegirte Zeitung, 22 July 1783: 714: Report from Hildburghausen, 24 June 1783.
542 Berlinische Nachrichten, 22 July 1783: 677–678: Report from Frankfurt am Main, 12 July 1783.

like a coal oven, emitting steam and choking sulfuric smoke without any interruption all day and night.[543] Despite these descriptions, newspapers were calm and considered in their reactions. The report further speculated on the origin of all this chaos: "[. . .] the Earth has experienced an enormous overburden of subterraneous sulfur, which is being unburdened via flammable matter or harmless vapors."[544]

The identity of the correspondent(s) in question remains obscure: very little information about the network of correspondents for newspapers from this period is available. The wording of the articles suggests that several newspapers used the same source(s). What remains unknown is whether the correspondent(s) sent their letters to several publishers or whether one newspaper simply copied the other.

The Gleichberg, technically speaking, is made up of two hills of 679 and 641 meters in height, the *Großer* and *Kleiner Gleichberg*, located in what was then the county of Henneberg-Römhild.[545] As with the Cottaberg, the Gleichberg was known at the time to be of volcanic origin. The mountains are part of the *Heldburger Gangschar* system, which formed due to Cenozoic volcanic activity.[546] Today only a few of the volcanoes of this system are still visible at surface level. The Gleichberg is younger than the Cottaberg; the last eruption occurred as recently as 15 million years ago. Like the Cottaberg, it consists of volcanic basalt cones. The volcano cannot be classified as entirely dormant yet as there is still geothermal activity there today, which results in hot springs and could explain the steaming mentioned in the report.[547]

What is fascinating about this alleged eruption is how believable it seems – and the fact that it was reported on 24 June 1783, when the fog was at its most intense and a mere week after it first appeared over the German Territories. Importantly, this would have been before the naturalists and thinkers of Europe had had the chance to firmly establish the idea that the fog was of volcanic origin. The letters describe a blood-red sun and moon, a somewhat hazy sky, and thick dry fog accompanied by the

---

543 Münchner Zeitung, 15 July 1783: 433–434. Similar reports about thick sulfuric smoke spoiling everything it touched also appeared in Berlin; Berlinische Nachrichten, 22 July 1783: 677–678: Report from Frankfurt am Main, 12 July 1783; Königlich Privilegirte Zeitung, 22 July 1783: 714: Report from Hildburghausen, 24 June 1783.

544 Münchner Zeitung, 15 July 1783: 433–434. "[Der Ausbruch des Gleichbergs und der Nebel] macht einen nicht unbeträchtlichen Beitrag zu der Vermutung, das die Erde sich einer ungeheuren Ueberladung von lange Zeit über gesammeltem unterirdischen Schwefel, und anderen brennbaren Materien zu entbürden, und also sich teils durch unschädliches Ausdünsten, teils an Orten, wo den Ausdünstungen grösserer Widerstand geschieht, oder wo der auszuführenden Materia peccans zu vil ist, durch heftige Ausbrüche, oder Vulkane in die Atmosphäre auszuschütten habe. Ein Correspondent aus gedachtem Hildburghausen beteuert uns wirklich, das der vor 3 Wochen aus dem Gleichberge, wie aus einem innerlich brennenden Kolenofen, ringsum emporgestigene Dampf die vollkommene Farbe sowohl, als ganz den erstikenden Geruch des Schwefelrauchs gehabt habe."

545 KÖBLER 2007: 268.

546 The Cenozoic is the current geological era; it extends from around 66 million years ago to the present.

547 GRATTAN, GILBERTSON, DILL 2000: 307; HOFBAUER 2008: 71.

strange, "choking smell" of "sulfuric smoke." Whereas the dry fog can explain all the aforementioned phenomena, what can account for the rest?

The Gleichberg "eruption" was said to have sounded like cannon fire. The whole area could perceive a "constant terrible roaring and rushing."[548] These noises might have been thunderstorms, which we know had been plentiful in 1783; indeed, contemporary sources endow the thunder of that summer with an almost mystical quality, with noises the likes of which they had never heard before.[549] It is possible that these storms, which others so easily mistook for earthquakes, could also give the impression that a volcano was erupting. Although the *Ephemerides* of the Societas Meteorologica Palatina did not have a weather station close to the Gleichberg, other German stations did record several thunderstorms over the eight days in question.[550]

Thunderstorms consist of both thunder and lightning; however, it is possible that the lightning remained obscured by the fog. Without this visual component, the panicked locals could have easily jumped to conclusions. Descriptions from that year of thunderstorms with apparently no lightning are available: for instance, in Dillenburg in early July, one report went as follows: "With one [thunderstorm], one saw no flashes of lightning and no normal thunder clouds, but that made the thunder much more numbing."[551]

One element of the letter from Hildburghausen that remains mysterious is the mention that the beginning of the "eruption" is said to have been around Easter. Additionally, how the Cottaberg "eruption" threw out stones could have nothing to do with the Laki eruption and perhaps has more to do with the overactive imagination of the correspondent.

Other "Volcanic Activity": The Roßberg, the Gottesberg, and the Burning Mountain

More of this imagined volcanic activity followed. The newspapers reported on at least two more that year (for a map of the locations, see Figure 41). It is quite extraordinary that reports of dramatic eruptions in the German Territories occurred in a year when a very real eruption was influencing all of Europe.

---

548 Königlich Privilegirte Zeitung, 12 July 1783: 686: Report from Mannheim, 1 July 1783.
549 Königlich Privilegirte Zeitung, 12 July 1783: 686: Report from Mannheim, 1 July 1783. "Alle Nachrichten stimmen überein, daß das Knallen der Donnerschläge so erschrecklich gewesen, dergleichen noch nie gehört worden."
550 Societas Meteorologica Palatina 1783. For instance: Erfurt was plagued by thunderstorms on 15 and 20 June; Würzburg on 15 and 21 June; Ingolstadt on 14, 15, 16, and 24 June; Berlin on 15, 16, 20, and 22 June; and Sagan on 12, 13, 16, 20, and 21 June.
551 Königlich Privilegirte Zeitung, 17 July 1783: 700: Report from Dillenburg, 2 July 1783. "Während dem auch in hiesigen Gegenden und auf dem ganzen hohen Westerwalde ausgebreiteten Dufte, sind verschiedene sonderbare Gewitter bemerket worden. Bey einem sahe man keine Blitze auch keine gewöhnlichen Gewitterwolken, aber desto stärker war der betäubende Donner." A very similar report can be found in Frankfurter Staats-Ristretto, 11 July 1783: 453: Report from Dillenburg, 2 July 1783.

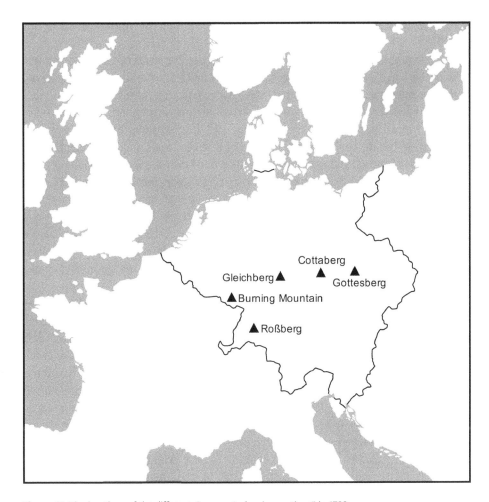

Figure 41: The locations of the different German "volcanic eruptions" in 1783.

The Roßberg near Tübingen, in the southwest of Germany, allegedly sprung to life on 18 July 1783: "The well-known Roßberg near Genkingen is producing a subsurface roar [*unterirdisches Getöse*]. For a few weeks, some have complained about eye infections here."[552] The Roßberg is located at the western margin of the Swabian Jura (*Schwäbische Alb*) and is about 100 meters higher than the surrounding area. As with the Gleichberg and Cottaberg, Roßberg is located in an area known to have had Neogene volcanic activity and is in the proximity of the Swabian volcano (active 17 to

---

552 Königlich Privilegirte Zeitung, 9 August 1783: 773: Report from Tübingen, 18 July 1783. "Man will versichern, der bekannte Roßberg bey Genkingen lasse auch ein unterirdisch Getöse von sich hören. – Seit einigen Wochen klagen hier verschiedene Personen über Augenkrankheiten."

11 million years ago). Geologists have identified more than 350 volcanic vents in this area.[553]

Although the Roßberg might be of volcanic origin, there was certainly no volcanic activity there in 1783. This newspaper report was much shorter and lacked the detail of the Gleichberg account. The "subsurface roar" could well have been loud thunder, once again mistaken for the rumblings of a volcano. It is worth noting that this area is one of the most seismically active in Germany.[554] It is possible that an earthquake, coupled with thunder and the dry fog, could have given locals the impression that the Roßberg was erupting. The accounts of eye infections mirror reports from other parts of the country, so it can be inferred that the dry fog was present in this area too.

Talk of a "subsurface roar" was probably a genuine attempt to give a local explanation for a frightening occurrence; however, the possibility remains that this element was fabricated. A similar "strong subterranean roar" was reported near the Gottesberg in Lower Silesia. That noise terrified local coal miners and prompted them to flee.[555] As with the Roßberg, this event might have been caused by an earthquake or severe thunder.

In the west of Germany, near Dudweiler, there was another "volcano." The *Allerneueste Mannigfaltigkeiten* reported on this event in great detail; descriptions of lava and unbearable smells color the report.[556] While the term "burning mountain" was synonymous with "volcano" in 1783, in this case, the name of the alleged source of the eruption actually is *Brennender Berg* (burning mountain). In the seventeenth century, this 356-meter-high mountain was a shale mine. A smoldering coal-seam fire started inside the mountain in the 1660s and continues until the present. Attempts to extinguish the fire were unsuccessful. Initially, a fiery glow was visible, which weakened over time. Today, smoke seeping from the mountain is occasionally visible.

Additional research may unearth even more false eruptions within the German Territories; furthermore, it would be interesting to study non-German sources to see whether people in other countries also applied this explanation strategy.

Retractions
The Gleichberg, Cottaberg, Roßberg, and Gottesberg did not erupt in 1783. The Burning Mountain is real; the eruptions were not. Despite best efforts, false stories sometimes made it to print.[557] Apparent corroborative evidence from multiple newspapers was anything but.

---

553 Geological Map of the Urach-Kirchheimer region, 2015.
554 Modern earthquake catalogs do not list any earthquakes in the region for 1783; however, these catalogs are not necessarily complete; LEYDECKER 2011.
555 Hamburgischer Unpartheyischer Correspondent, 11 July 1783: Report from Silesia, 6 July 1783; Münchner Zeitung, 14 July 1783: 430: Report from Düsseldorf, 28 June 1783.
556 Allerneueste Mannigfaltigkeiten, week 31 (late July/early August), 1783: 473–476.
557 GRATTAN, BRAYSHAY, SCHÜTTENHELM 2002: 101.

The first newspaper to dispute the story of the Gleichberg eruption was the *Meiningische Wöchentliche Zeitung* on 19 July 1783, the headquarters of which was only a short distance from the mountain. The letter in the *Frankfurter Staats-Ristretto* had come to the attention of the editor of the *Meiningische Wöchentliche Zeitung* in the days following its publication. The editor stated that it was "almost funny" to read and commented on the "strange influence of this evil fog on the heads of the people."[558] The editor found harsh words to criticize the spread of this story:

> One should be ashamed to terrify the superstitious people among the common folk. [. . .] Our eyes do not see, and our ears do not hear anything about the fleeing terrified residents, the threatening collapse of the Gleichberg, the roaring and bashing, the thick sulfuric smoke, [. . .]. One also does not taste anything of the sulfur on cherries or berries, of which there have been plenty this year.[559]

That sulfur was perceptible in the air was a particularly controversial topic. It seems those who did not experience it personally did not believe those who claimed to have smelled or even tasted it during the summer and thought it a fantastical and outlandish claim. The evidence suggests that the sulfuric odor was most potent in the Low Countries and some western parts of the German Territories during the last week of June 1783. Perhaps Meiningen, due to its location, was spared from this stench; that said, the previous newspaper issue from 12 July 1783 had referred to the dry fog as *der Duft* ("the smell").[560]

The editor of the *Meiningische Wöchentliche Zeitung* disproved most aspects of the story mentioned in the original "letter from Hildburghausen" and further mentioned that the dry fog and the unusual coloration of the sun and moon were not unique to the Gleichberg; this shows that the editor was aware of the supra-regional reach of the dry fog. As a closing thought, the editor expressed doubt as to whether the letter even originated in Hildburghausen and suggested instead that "the author of this untruth" must be located near Frankfurt.[561] The editor must have been unaware that the same

---

558 Meiningische Wöchentliche Zeitung, 19 July 1783: 116: Report from Meiningen, 18 July 1783. "Es ist fast lustig zu lesen und zu bemerken, welchen sonderbaren Einfluß der böse Nebel auf Menschenköpfe gehabt hat."

559 Meiningische Wöchentliche Zeitung, 19 July 1783: 116. "So würde er [der Verfasser des Schreibens aus Hildburghausen] sich schämen, abergläubische Personen unter dem gemeinen Volk in Furcht zu setzen. [. . .] Eben so ungegründet ist alles übrige Vorgeben. Vom Wegflüchten der erschrockenen Einwohner, vom angedrohten Einsturz des Gleichbergs, von dessen Sausen, Brausen, Krachen, dickem Schwefelrauch, der den Wäldern ein weißes Kleid angelegt haben soll; sehen unsere Augen und hören unsere Ohren nichts. Auch schmeckt man nichts von dem angeblichen wahren natürlichen Schwefel an Kirschen und Beeren, woran dies Jahr besonders reichhaltig ist."

560 Meiningische Wöchentliche Zeitung, 12 July 1783: 113: Report from Meiningen, 3 July 1783.

561 Meiningische Wöchentliche Zeitung, 12 July 1783: 117. "Man glaubt gar nicht, daß diese Legende in Hildburghausen ist verfertigt worden; sondern es ist vielmehr wahrscheinlicher, daß der Verfasser dieser Unwahrheit sich in der Nähe von Frankfurt aufhalten müsse."

letter was printed in Augsburg on the same day and in various other German newspapers shortly afterward.

Within a relatively short period, updates arrived from the affected regions, which revealed that they had, in fact, avoided devastation. On 25 July 1783, the *Hamburgischer Unpartheyischer Correspondent* printed a report from Bayreuth, dated 17 July 1783: "Travelers, who are coming from Hildburghausen, do not know of any change at the Gleichberg, nor of its uproar or fire-spitting. A fog prophet [*Nebelprophet*], of which there are many these days and who do not sense anything but bad luck, has probably spread the terrible news about this mountain."[562] On 28 July 1783, the *Münchner Zeitung* printed a report that stated: "The Cottaberg [. . .] and the Gleichberg [. . .] have cracked, but instead of hatching a volcano, they gave birth to a mouse. Someone has, we don't know where from, cooked up this stunning lie."[563] Later, on 31 July 1783, the *Königlich Privilegirte Zeitung* printed a letter from Hildburghausen, dated 16 July 1783. It reads: "at a steep wall, there is an outcrop, and one can see the basalt columns, which make up the inner part of the mountain, and they are still standing upright and look to be intact."[564] As the basalt columns were intact, the correspondent concluded that no recent volcanic eruption could have occurred there.

In September 1783, the editor of the *Dessauische Zeitung für die Jugend und ihre Freunde* admitted they had reported on the Cottaberg and Gleichberg eruptions "without reason." Around the time of their retraction, the editor had realized that stories about the mountains emitting "terrible steam and throwing out stones" might have agitated those with sensitive souls, for which they pleaded for the readers' forgiveness.[565]

---

562 Hamburgischer Unpartheyischer Correspondent, 25 July 1783: Report from Bayreuth, 17 July 1783. "Reisende, welche von Hildburghausen kommen, wissen von keiner Veränderung am Gleichberg, weder daß er tobe noch Feuer speye. Vermuthlich hatte ein Nebelprophet, dergleichen es jetzt viel giebt, und die nichts als Unglück wittern, die fürchterliche Nachricht von diesem Berge verbreitet."

563 Münchner Zeitung, 28 July 1783: 461. "Der Cottaberg im Meissnischen, und der Gleichberg bei Hildburghausen haben gekracht; aber anstatt einen Vulkan auszubrüten, eine Maus gebohren. Man hat uns, weis nicht woher, die schöne Lüge angebunden."

564 Königlich Privilegirte Zeitung, 31 July 1783: 742. "Dieser Berg [. . .] ist [. . .] vulkanischen Ursprungs. An einer steilen Wand desselben, die von Erde und Gesträuch entblößt ist, kann man die Basaltsäulen, die das Innere ausmachen, noch aufrecht stehend und unzerstückt erblicken." The same letter is quoted in the Frankfurter Staats-Ristretto: 22 July 1783: 480.

565 Dessauische Zeitung für die Jugend und ihre Freunde, no. 39, September 1783: 305: Report from Dessau, 20 September 1783. "Zum Schlusse dieses 5ten Quartals der Jugendzeitung sehe ich mich genötigt, den Lesern das offenherzige Geständnis zu tun, daß es mit dem Zeitungswesen nicht anders, als mit andern menschlichen Dingen beschaffen ist: überall schleichen sich Irtümer und Täuschungen neben der Wahrheit ein. Am meisten geschieht dieses alsdenn, wenn meine Herren Korrespondenzen nicht fleißig und prom[p]t sind, und mich in der Notwendigkeit lassen, andern öffentlichen Blättern zu folgen. Auf diese Art habe ich z.B. ohne Grund den Cottaberg (St. 30, S. 235) und den Gleichberg (32, 242) einen gewaltigen Dampf und eine Menge Steine auswerfen lassen, und dadurch vielleicht manche zärtliche Gemüter in Furcht und Schrecken versezt; weswegen ich sie hier inständig um Verzeihung bitte."

A Grand Hoax?

German climatologist Rüdiger GLASER calls the story of the Gleichberg one of the more obscure speculations in the discourse on the origin of the dry fog.[566] The tone of these stories is altogether more alarming and sensational than most others circulating at the time. Reports of several volcanic eruptions within the German Territories were terrifying, especially in the context of the many earthquakes that rocked other parts of Europe in 1783. How did these fallacious rumors find their way to print? Was it just a hoax fabricated by an unknown perpetrator to generate panic?

The news about a German volcanic eruption quickly crossed borders. The story of "Mount Gleichberg" was printed in the *Whitehall Evening Post*, the *Morning Herald and the Daily Advertiser* on 12 August 1783. John GRATTAN and his colleagues studied the case further, using English and German newspaper articles and geological information. Their aim was to determine whether it is possible to reconstruct geological events from historical documents, a technique they refer to as "excavating words." They concluded that it is very unlikely the Gleichberg erupted in 1783. Perhaps if the eruption was said to have taken place in a more distant period, in Roman times, for example, they might not have been so certain.[567]

Regarding the credibility of the accounts published by the newspapers, GRATTAN and his team admit that some of the statements are quite detailed and convincing.[568] The correspondent, who seemingly knew what to expect of a volcanic eruption, could have intended "to exploit the fear and panic that was becoming widespread in Europe from early July [1783]."[569] They conclude that "the original reports described above were either an elaborate hoax or the result of a genuine misunderstanding."[570]

It seems probable that the stories of these volcanic "eruptions" were ultimately unsuccessful attempts to explain the strange weather; it was known at the time that the Gleichberg, and the other mountains, were of volcanic origin, so it was not impossible to imagine that they might have come back to life. This, coupled with what felt like an urgent need for an explanation, might have forced the correspondent(s) to jump to certain conclusions. This was a mere eight days after the arrival of the dry fog in the German Territories, which meant whoever had written this account did not yet know that this was a country-wide, and indeed Europe-wide, occurrence. A local occurrence naturally would have a local source.

Newspapers at the time were subject to strict censorship regarding local news. To circumvent this censorship, they tended to print news from other regions.[571] The

566 GLASER 2008: 234.
567 GRATTAN, GILBERTSON, DILL 2000: 311–314. The principle of "excavating words" was established by Stefi WEISBURG (1985: 91–94).
568 GRATTAN, GILBERTSON, DILL 2000: 313.
569 GRATTAN, GILBERTSON, DILL 2000: 313.
570 GRATTAN, GILBERTSON, DILL 2000: 314.
571 GRATTAN, BRAYSHAY 1995: 127; WÜRGLER 2009.

correspondent must have known that the people who would read this letter would not be local. Thus, any motive for a hoax that depends on the correspondent's inclination to whip up some panic can be dismissed because the people in the correspondent's locality would likely never read the report. Additionally, as the correspondent put pen to paper, for all they knew, the towns in which their reports would be printed could have been basking in sunshine and unlikely to be shaken by news of a distant weather event.

The Gleichberg and Cottaberg reports appeared almost at the same time. The Cottaberg story was based on a letter from 1 July 1783 and printed on 8 July 1783 in Berlin; the Gleichberg story was based on a letter from 24 June 1783 and printed on 12 July 1783 in Augsburg. Both reports were printed for the first time within five days of one another in different parts of the German Territories. Newspapers almost always printed extracts from letters they received anonymously, only stating the letter's date and place of origin. As the correspondents in both cases were anonymous, it is possible that they were one and the same person. This assumes, however, that one correspondent would know about mountains of volcanic origin in different parts of the German Territories. Could it have been possible that it was two hoaxers working in tandem? Possibly, however, this interpretation works on the assumption that at least one of them had a basic idea of ancient volcanic activity within the German Territories and also had the will and the inclination to orchestrate a multi-person hoax. And then, the question that appears so obvious when we consider that two or more of these eruptions were part of a grand hoax is, of course, why would one report be so rich in detail and highly dramatic and the other less so?

And so, if we reason that these were independent incidents, with no element of human communication between them, to assume they were hoaxes would be to assume that several people, independently of one another and within a short space of time, conjured up the same plan. Therefore, I argue that the Gleichberg, Cottaberg, Roßberg, and Gottesberg eruptions were genuine attempts to explain and link unusual phenomena that were otherwise impossible to explain. A persistent dry fog, a sulfuric smell, breathing difficulties, sore eyes, and withered plants were all conveniently explained by the local volcano.

A Time of a Subsurface Revolution

Physician Christian LUDWIG (1749–1784) published a report in the *Leipziger Magazin zur Naturkunde, Mathematik und Oekonomie,* in which he and his friend, Carl Friedrich HINDENBURG, one of the editors of the journal, discussed the thick fog. LUDWIG calls this "the currently fashionable conversation" (*Modegespräch*) and notes the

contemporaneous occurrence of the dry fog and the Calabrian earthquakes; he believed both events to be connected.[572]

LUDWIG was not alone in his assessment. The idea that strong earthquakes in Calabria had caused the dry fog came to be one of the most popular theories regarding its
origin. The thinking behind the theory was that subterraneous winds and fermentation
processes had initiated violent chain reactions underground, which were then expressed by volcanic eruptions and earthquakes in Italy and beyond. These violent occurrences released flammable air through cracks in the Earth caused by earthquakes.
For proponents of this idea, questions regarding the origin of the dry fog could be put
to bed. With the matter seemingly settled, for some, there was no reason to look
elsewhere.

Based on newspaper reports, it is clear that this theory was in circulation as early
as July 1783. A few, like this correspondent, still appeared to be on the fence:

> Some naturalists want to assign the dry fog to the revolution caused by the earthquake in Mes-
> sina and Calabria, as subsurface roars [*unterirdisches Getöse*] have been heard in different parts
> of Germany and France. Others say that during the dry season of the year, these sort of fogs are
> normal.[573]

The last remark is an oblique reference to Joseph DE LALANDE's theory. Some newspaper reports hoped to debunk the theory that the Calabrian earthquakes caused the
dry fog by promoting DE LALANDE's theory: "Herein [de Lalande's theory] and not
within the earthquakes lies the cause this appearance."[574]

The idea that the Calabrian earthquakes had a connection to the unusual weather
was given further credence with the news that a similar fog had allegedly occurred during the first tremors in February 1783. Furthermore, the earthquakes in Calabria were
far from over; on 8, 11, and 12 June 1783, aftershocks occurred, and on 20 June 1783, another unusual fog appeared.[575] Johann WIEDEBURG remarked that the disappearance of

---

572 LUDWIG 1783: 211–212.
573 The term can be found as early as 11 July 1783: Hamburgischer Unpartheyischer Correspondent,
11 July 1783: Report from the Lower Elbe, 10 July 1783. "Einige Naturforscher wollen selbigen [den
Nebel] der durch das Erdbeben in Meßina und Calabrien verursachten Revolution zuschreiben, weil
man in verschiedenen Gegenden Deutschlands, auch Frankreichs, ein unterirdisches Getöse gehört
habe. Andere sagen, daß bey einer anhaltenden tröckenen Jahrszeit dergleichen Nebel gewöhnlich
wären." Hannoverisches Magazin, 15 September 1783.
574 Berlinische Nachrichten, 15 July 1783: 654: Report from Stuttgart, 1 July 1783. "Von den vielen Gewittern, Regen, Wolkenbrüchen, und Ueberschwemmungen, womit so manche Gegenden heimgesucht
worden, steigen eine Menge Dünste auf, die sich sonst in der Luft zu zerstreuen pflegen, aber jetzt
von der kühlen, herbstgleichen Witterung verdickt, und folglich daran verhindert werden. Hierinn
und nicht im Erdbeben liegt die Ursach dieser Erscheinung."
575 Berlinische Nachrichten, 17 July 1783: 664: Report from Naples, 24 June 1783; WIEDEBURG 1784:
63–64.

the dry fog at around the time of the cessation of the last earthquakes in Calabria and Sicily suggested a direct connection between them.[576] The *Berlinische Nachrichten* commented on an earthquake in France on 6 July 1783 with the following: "One has noticed that the local, strange fogs have ceased once the tremors began in Dijon and Besançon."[577] The report not only suggests a connection but goes further and suggests that the earthquakes had a role in dispersing the fog. That said, another report printed in the same newspaper one week later suggested the opposite: "The latest reports from Messina tell us that for some weeks, there has been such a thick fog here that one person can barely see another; since [this occurrence began] one has not felt any earthquakes."[578] These conflicting reports undoubtedly added to the confusion about the connection between the fog and the earthquakes.

Italian naturalist Michele Torcia (1736–1808) suggested that the dry fog was the thickest and most sulfurous in southern Italy, where the epicenters of the Calabrian earthquakes had been and where the aftershocks still rattled the Earth.[579] Guiseppe Toaldo also linked the dry fog in northern Italy to the Calabrian earthquakes, suggesting that winds from the south "may have carried with them a large mass of exhalations."[580] What struck him as strange, but did not push him to reevaluate, was that the dry fog did not seem to be an exhalation from below; rather, it seemed to have "proceeded downwards as if it had fallen from the atmosphere."[581] Robert de Lamanon postulated a slightly more complex set of events, which was necessary, he thought, as the Calabrian earthquakes had taken place in February 1783, and the dry fog only appeared in mid-June, "that is to say, [not] till more than four months after."[582] He believed that when the exhalations rose to the sky, they lingered until they became saturated and fell back to Earth, where they formed the fog. Robert de Lamanon believed Toaldo to be a reliable naturalist but suggested he might not have known that the dry fog was not unique to Italy.[583] This kind of critical interaction was becoming commonplace, as we have seen.

---

576 Wiedeburg 1784: 77–78.
577 Berlinische Nachrichten, 2 August 1783: 715: Report from Paris, 19 July 1783. "Man hat die Bemerkung gemacht: daß die hiesigen sonderbaren Nebel aufgehört haben, da man die Erdstöße von Dijon bis Besançon verspürte."
578 Berlinische Nachrichten, 9 August 1783: 739: Report from Naples, 20 July 1783. "Die neuesten Berichte aus Messina bringen mit: daß daselbst seit einigen Wochen ein so dicker Nebel vorhanden sey: daß ein Mensch den anderen kaum erblicken könne; seit dem habe man auch keinen Erdstoß mehr verspürt."
579 Torcia 1783b: 840; McCallam 2019: 230.
580 Toaldo 1784; Toaldo 1799: 419 (quote); Grattan, Brayshay, Sadler 1998: 26.
581 Toaldo 1799: 419.
582 Lamanon 1799: 86–87.
583 Lamanon 1799: 86–89.

Jacques Antoine Mourgue de Montredon and German naturalist Christoph Gott-fried Bardili developed surprisingly similar theories to explain the cause of the nu-merous earthquakes and volcanic eruptions occurring in Europe. Bardili suggested that Europe's current geological mood swings were down to a subterraneous fire. Fur-thermore, he believed an earthquake in one location could activate the fiery parts (*Feuertheilchen*) of another location, even over great distances.[584] In the autumn of 1783, Mourgue de Montredon wrote about these underground fires (*des feux souter-rains*) and traced the path they had taken. He followed this "subterraneous revolu-tion" from the newly emerging island off the coast of Iceland and the volcanic eruption near Mount Hekla (which later turned out to be the Laki eruption) through Europe, past the Gleichberg and Calabria, and all the way to the earthquakes in Trip-oli. This "fiery path" underneath Europe, according to Mourgue de Montredon, had left volcanic eruptions, earthquakes, sulfuric exhalations of flammable air, and thick fog in its wake.[585]

The *Münchner Zeitung*, too, opined that recent events seemed "to indicate a vio-lent revolution of planet Earth" and continued, "this leads to the presumption that planet Earth is attempting to discharge an enormous overload of subsurface sulfur and other combustible materials [. . .] into the atmosphere via strong eruptions or vol-canoes."[586] This idea was grand enough to fit with Franz von Beroldingen's assertion that an extraordinary phenomenon needed an extraordinary explanation.[587]

While news of the Calabrian earthquakes had been present in the media since the spring, reports of other earthquakes poured in over the summer and autumn, fur-ther stoking the (subterraneous) fire. The *Berlinische Nachrichten* summed up these concerns in a letter that concluded, "[. . .] the whole of Europe seems to have been plagued by this horror."[588] The idea of a subsurface revolution was much more than a fringe theory, for it had been seemingly corroborated by naturalists such as Mourgue de Montredon, Bardili, von Beroldingen, and Wiedeburg, among others.

In his book, Wiedeburg remarks that early "naturalists believed the occurrences of volcanoes and earthquakes to be the mere impact of mineral vapors and air that were locked in subterraneous cavities."[589] Wiedeburg is leaning on Aristotle's idea, which

---

584 Bardili 1783: 9–10. The preface in his book is dated to 9 July 1783.
585 Mourgue de Montredon 1784: 757–761.
586 Münchner Zeitung, 15 July 1783: 433. For the German original, see above.
587 Beroldingen 1783: 9–12.
588 Berlinische Nachrichten, 18 September 1783: 868–869: Report from Paris, 4 September 1783. "[. . .] also ist das ganze Europa von diesem Schrecken heimgesucht worden."
589 Wiedeburg 1784: 30. "Schon die ältesten Naturforscher hielten die Erscheinungen der Vulkane so-wohl als der Erdbeben vor eine blose Wirkung unterirdischer mineralischer verschlossener Dünste und der Luft in Höhlungen."

proposed that the Earth's interior consisted of gases and that earthquakes and volcanic eruptions were the consequences of subterraneous winds that released flammable materials.[590] Aristotle's *Meteora* and other knowledge from antiquity were still commonly referenced in the eighteenth century.[591] The idea of *subterraneous* fires that rage beneath our feet was not new. Athanasius KIRCHER had already postulated in 1665 that all volcanoes were connected by fire channels, as mentioned above; he also argued that chemical reactions in subterraneous passages and caverns caused earthquakes and volcanic eruptions.[592] In the eighteenth century, Nicholas LÉMERY, Georges-Louis LECLERC, Comte DE BUFFON, and Immanuel KANT promoted these ideas. The sheer number of unusual subsurface phenomena and their purported accidental nature overwhelmed many in 1783. The subsurface *revolution* hypothesis was a grand, overarching, and satisfying explanation. Historian of geology Rhoda RAPPAPORT distinguishes accidents and revolutions with the following statement: "Accidents are always local, revolutions usually so. But an accident is merely local and thus not very important, while revolutions are part of a recurrent, common pattern."[593]

To the contemporaries, it seemed that there was more seismic and volcanic activity than ever before, but this was not true. News of such events simply traveled faster and further than ever before.[594] It is also possible that people of the time suffered from a cognitive bias called frequency illusion, which shaped their perception of the news. In this case, after experiencing their first earthquake or reading sensational stories about such events, they were more likely to notice reports on earthquakes than ever before.

Elektrizitätstaumel: Are Lightning Rods to Blame?

The eighteenth century saw the spread of "a true 'electricity delirium' in Germany."[595] Electricity became the explanation par excellence; everything that had previously

---

590 ROHR, VLACHOS 2010: 467–474; CRAIG 2011: 62–64.

591 OESER 2003: 13–16; DEMARÉE et al. 2007; GLASER 2007: 793.

592 ZIEHEN 1783, vol. 1: 45–46.

593 RAPPAPORT 1982: 40–41.

594 Conevery Bolton VALENCIUS (2015: 197) found that there was "a sense of global interconnection" in the aftermath of the 1811–1812 New Madrid earthquakes: reports of the New Madrid earthquakes were mentioned together with other, seemingly simultaneous, events around the globe, such as earthquakes or volcanic eruptions, which were perceived as "terrestrial symptoms of deep and powerful disturbance."

595 CAPPEL 1986: 20.

been hard to explain became a byproduct of electricity.[596] The idea that lightning rods might even make cities earthquake-proof was still entertained. To some, it seemed too much of a coincidence that those cities severely affected by earthquakes, such as Lisbon and Messina, were those without lightning rods.[597] The technological optimism in the 1780s was boundless, and many thought rain and hail rods were conceivable.[598] At the height of the fever pitch, in the *Münchner Zeitung*, a perhaps exasperated correspondent commented: "We do not see why one always has to include electrical matter in everything [. . .]."[599] It is, therefore, unsurprising that naturalists also attempted to explain the presence of the Laki haze with hypotheses involving electricity.

In the late eighteenth century, lightning rods were installed in great numbers across Europe, changing how people viewed nature.[600] However, doubt remained and some questioned whether they were a step too far. The proponents of the lightning rods argued that their installation was an act of self-preservation, just as flood barriers were.[601] God had given humanity intellect, by which means they were able to invent the lightning rod in the first place. Would God have allowed the invention of the lightning rod if it was not part of His plan?[602]

The new technology elicited new questions.[603] Would the fertility of agricultural fields decline if lightning could not strike naturally? Would the lightning rod cause rainfall to decrease? Could the rods cause an imbalance that would precipitate earthquakes?[604] The severe thunderstorms of 1783 led to a sharp increase in the uptake of lightning rods; this, ironically, led to suggestions that the lightning rods had caused the many thunderstorms.

---

596 This also applies to the Calabrian earthquakes: Christoph Gottfried BARDILI (1783: 1–10) suggested the Calabrian earthquakes had increased the electricity on the Earth's surface, which attracted the fog and prevented it from dissipating. A similar theory was posited in the *Königlich Privilegirte Zeitung* on 9 August 1783; Königlich Privilegirte Zeitung, 9 August 1783: Report from Mannheim, 27 July 1783.

597 RISKIN 1999: 73; DROSS 2004: 301; HOCHADEL 2009: 50–51; BEHRINGER 2011: 203–204. There are exceptions to this train of thought. In 1755, after Boston and much of New England had been affected by the magnitude 5.9 Cape Ann earthquake off the coast of Massachusetts, it was debated whether the installation of a lightning rod was in itself a blasphemous act. The Franklin rod was a brand-new invention at the time. Benjamin FRANKLIN, who was originally from Boston, invented them. And in 1755, Boston had more lightning rods than any other town in New England – and, coincidentally or not, Boston seemed to have been "more deadfully shaken" than any other town, as claimed by the reverend Thomas PRINCE, who published about this; TILTON 1940: 85–97.

598 FISCHER 1784: 99–100; WIEDEBURG 1784: 46; BRIESE 1998: 23.

599 Münchner Zeitung, 10 July 1783: 422. "Wir sehen nicht, warum man immer die elektrische Materie überall ins Spil ziehen; oder ihr die Kraft eines Monstrums beilegen soll."

600 BEGEMANN 1987: 90.

601 CAPPEL 1987: 20; KITTSTEINER 1987: 20.

602 SCHELHORN 1783: 16–17.

603 KITTSTEINER 1987: 26; RISKIN 1999: 77; HOCHADEL 2003: 151.

604 Münchner Gelehrte Zeitung, June 1783: 44–48: "Von der Unschädlichkeit der Blitzableiter." WEIGL 1987: 13; HOCHADEL 2009: 50.

Newspapers reveled in the back-and-forth frenzy the rods had ignited. These questions formed the basis for a clash between the established religious authority and scientific rationalization. The latter was in its infancy and seemed a weak replacement for the traditional and time-tested religious explanations.[605] Although strong resistance to the installation of lightning rods was uncommon in the long run, they remained under fire from a vocal minority and were eyed suspiciously by those looking for quick answers.[606] "If one thinks how common the weather rods have become [. . .], one naturally has to think that this invention could have contributed a lot to the current foggy haze [*Nebeldünsten*] [. . .], which otherwise would have been absorbed by the lightning or the northern lights."[607] In this instance, the correspondent insists the rods gave rise to the fog and blames the Dutch, who happened to be fond of Franklin's new invention, for exacerbating this sad situation.

In 1783, there was rarely a distinction made between thunder and lightning.[608] German-language sources of the time indicate that a thunderclap was still considered dangerous.[609] Once there was a distinction, the lightning rod became a tool to disarm God.[610] Thunder, too, became a tool used as a means to determine the distance of a storm.[611] Thunderstorms were no longer random but rather one part of nature.[612] This realization led to a growing disenchantment with nature.[613] However, at the end of the eighteenth century, superstitious explanation strategies still had a strong foothold, partly because naturalists were not always successful in their efforts to convey new scientific ideas to the general public in a comprehensible manner.[614] Nevertheless, progress and reasoning led to a decline in apocalyptic proclamations in the late eighteenth century, as the lightning rod continued to protect both those who championed it and those who did not.[615]

---

605 BRIESE 1998: 20–21.
606 HOCHADEL 1999: 145; DROSS 2004: 284–294, 302.
607 Münchner Gelehrte Zeitung, 29 August 1783: 60. "Denn denkt man nach, wie sehr der Gebrauch der Wetterableiter seit einigen Jahren in Aengelland, und in den Niederlanden überhandgenommen, und fast zum Misbrauch geworden, so mus man natürlicher Weise auf den Gedanken fallen, das dise Erfindung viles zu den izigen Nebeldünsten beigetragen haben könne. [. . .] Was kann natürlicher daraus folgen, als das die aufgezogenen Schwefeldünste, welche sonst durch das elektrische Feuer oder Wetterleuchten verzehrt wurden, in der Luft zurükbleiben, und denjenigen trokenen Nebel verursachen, den uns die Niederländer bisher mit ihrem Winde so häufig zugeschickt haben."
608 MISSFELDER 2009: 91.
609 SCHELHORN 1783: 20.
610 BEGEMANN 1987: 79.
611 MISSFELDER 2009: 91–92.
612 KITTSTEINER 1987: 21, 25; GREYERZ 2009: 46; REITH 2011: 88–92.
613 WEBER 1993; GRAF 2006: 342; LEHMANN 2009: 9, 19–20.
614 SCHMIDT 1999: 310.
615 FLEMING 1998: 6–7.

The Fireball

In Britain, 18 August 1783 was another hot and sultry summer's day. It had been a "clear" day at Gordon Castle in the Scottish Grampians.[616] Thomas BARKER describes the weather as "sunny and hot & calm" in Leicestershire.[617] Similarly, William HUTCH-INSON of Liverpool remarks that it was "mostly hazy, but [with] some faint sunshine, warm and pleasant."[618] James WOODFORDE of Weston Longville, Norfolk, also writes of hot weather that day: "Morn very fair & very hot, Afternoon ditto."[619]

As this hot day drew to a close, something extraordinary happened. In the twilight hours, with many still out and about, William COOPER, at Hartlepool, near Stockton, notes:

> something singularly striking in the appearance of the night, not merely from its stillness and darkness, but from the sulphureous vapours which seemed to surround us on every side. In the midst of this gloom, and on an instant, a brilliant tremulous light appeared to the N.W. by N.[620]

Sometime between 9:15 p.m. and 9:30 p.m., a meteoroid entered the atmosphere somewhere over the North Sea to the north of Scotland.[621] This bolide – a bright meteor – shot over the eastern parts of Scotland and England at great speed. Next, it crossed the Channel, illuminating the skies over the Austrian Netherlands (today's Belgium) and northeastern France (Dunkirk, Calais, Ostend, Brussels, and Leiden). It continued toward southeastern France and northern Italy until it struck the Earth's surface, thus becoming a meteorite.[622]

The meteor was described as a fireball or globe of fire. In his painting (Figure 42), Henry ROBINSON, a schoolmaster from Nottinghamshire, calls it a *"Draco Volans or flying dragon,"* a shooting star.[623] At the time of the meteor's appearance, the sky was dark enough over the British Isles and mainland Europe for it to produce a stark

---

616 T. HOY, Private Weather Diary for Gordon Castle, Grampian, Scotland, MET/2/1/2/3/486, National Meteorological Library and Archive, Met Office, Exeter, UK.
617 Thomas BARKER, Private Weather Diary for Lyndon Hall, Leicestershire, MET/2/1/2/3/227, National Meteorological Library and Archive, Met Office, Exeter, UK, 18 August 1783: 201.
618 William HUTCHINSON, Private Weather Diary for Liverpool Dock, MET/2/1/2/3/230, National Meteorological Library and Archive, Met Office, Exeter, UK.
619 James WOODFORDE, Weather Diary for Weston Longville, Norfolk, vol. 10: 1782–1784, MET/2/1/2/3/ 467, National Meteorological Library and Archive, Met Office, Exeter, UK, 18 August 1783.
620 COOPER 1784: 116. It is important to note here that COOPER wrote his letter to BANKS on 19 August, the day after this event, so his impressions of the meteor were still very fresh.
621 The times the meteor was observed vary greatly, but most are within this 15-minute window. An asteroid has a size larger than one meter in diameter, whereas meteoroids have a diameter between one centimeter and one meter. Shooting stars are only a few millimeters in size. Meteoroids become meteors when they enter the Earth's atmosphere. Meteors become meteorites when they hit the surface of the Earth.
622 PAYNE 2011: 21. Evidence for Burgundy and Belgium: DEMARÉE, OGILVIE 2016: 142. Evidence for Calais, Dunkirk, and Ostende, BLAGDEN 1784: 203–204. Evidence for Brussels, Leiden, PAYNE 2011: 21.
623 BURKE 1986: 6.

contrast against the evening gloom.[624] British amateur astronomer and businessman Alexander AUBERT (1730–1805) estimated that the meteor's brightness equaled that of two full moons.[625] It traversed the sky from the northwest to the southeast.[626] The meteor was extraordinary because it was visible for quite some time, somewhere

Figure 42: A contemporary depiction of the Great Meteor. Henry ROBINSON, "An accurate representation of the meteor," as seen at Winthorpe, Nottinghamshire, England, on 18 August 1783. © The Trustees of the British Museum. Shared under a Creative Commons Attribution-NonCommercial-ShareAlike 4.0 International (CC BY-NC-SA 4.0) license.

---

624 At Windsor, England, the sun set at 7:17 p.m. and astronomical twilight ended at 9:36 p.m. At 9:17 p.m. or 9:23 p.m., by the time the fireball appeared, it was very dark. Although it was not proper night yet, most stars should have been visible. The moon was a little fuller than a half moon that night, almost in its third quarter. In Blair Athol, Scotland, the sun set at 7:46 p.m., nautical twilight (which is lighter than astronomical twilight) ended at 9:28 p.m., so the sky was illuminated by the sun a little bit more than in Windsor when the meteor was visible; based upon information from the Time and Date website.

625 AUBERT 1784: 274.

626 John ATKINS, "The Meteorological Journal for the Year 1783. Kept at Minehead in Somersetshire by Mr. John ATKINS. Presented at the Royal Society in London on 19 January 1786," MA/166, Archives of the Royal Society, London, UK; Thomas HUGHES, Private Weather Diary for Stroud, Gloucestershire, MET/2/1/2/3/410, National Meteorological Library and Archive, Met Office, Exeter, UK, August 1783.

between 12 and 30 seconds, and traveled an estimated 1,600 to 1,900 kilometers, which is an exceptional distance (Figure 43).[627]

A number of private weather observers noticed the meteor and wrote brief journal entries about it. Many were in awe; some were frightened.[628] The man depicted in Figure 42 seems to be taken aback by the sight, even raising his hands as if to take a defensive position while observing the meteor, which at this point has already broken up into one principal fireball and several smaller fireballs that make up its tail (Figure 44). Joseph BANKS, then the president of the Royal Society of London, made the timely decision to collect accounts of the meteor's passage.[629] In 1784, many of these accounts were published in the *Philosophical Transactions of the Royal Society*. Historian Noah MOXHAM views these efforts as an example of "crowd-sourcing" in eighteenth-century science.[630]

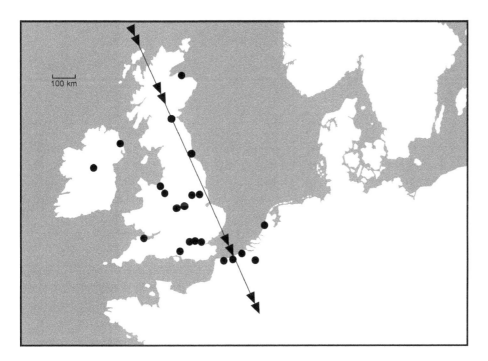

Figure 43: The approximate path of the meteor on 18 August 1783.

627 BEECH 1989: 131; DEMARÉE, OGILVIE 2001: 229; PAYNE 2011: 20–23; WITZE, KANIPE 2014: 115–116.
628 COOPER 1784: 116–117.
629 BEECH 1989: 130–132.
630 MOXHAM 2013.

Figure 44: Paul SANDBY's watercolor, "The Meteor of 1783, as seen from the East Angle of the North Terrace, Windsor Castle." The watercolor shows the procession of the meteor and portrays the most famous spectators of this event.[631]

In one contemporary account, Alexander AUBERT states that the haze obscured any stars below eight degrees; however, some naturalists put this figure at up to 20 degrees. AUBERT asserts further that even though the meteor "had got high enough to be quite out of the hazy part of the horizon, it was surrounded and accompanied in its whole course with a kind of whitish mist or light vapour."[632] This information is crucial; it shows that the Laki haze still lay thickly on the horizon. Observers wondered whether the omnipresent haze gave the meteor its bluish and later red color.[633] From Mullingar in Ireland, the meteor appeared in "the most vivid colours; the foremost part being in the brightest blue, followed by different shades of red."[634]

The *Gentlemen's Magazine*, a London-based monthly publication of letters and reports from Britain, continental Europe, and beyond, published several reports about the "uncommon meteor." An anonymous author stated that they presumed

> [it] may have been occasioned by some of the vapours issuing from the volcanoes upon the New Island lately sprung up in the ocean, about nine leagues to the S.W. of Iceland, or perhaps only from that profess exhalation of vapours occasioned by the excessive warm and dry weather we have experienced this summer.[635]

---

631 BEECH 1989: 130–134.
632 AUBERT 1784: 112–115, 113 (quote). Tiberius CAVALLO (1783: 108–111) estimated the stars at Windsor had set 18 to 20 ° above the horizon; STOTHERS 1996: 83–84.
633 CAVALLO 1784: 109.
634 EDGEWORTH 1784.
635 Gentleman's Magazine 1783: 711–713.

News about the fireball(s) spread. In the German Territories, the first reports appeared on 30 August 1783.[636] The *Münchner Zeitung* was critical of the British reaction. One correspondent argued thusly: "It seems every country has their prophets of doom, disadvantages, and fools." This statement referred to the fact that in Britain, all efforts seemingly focused on establishing whether this fireball was a bad omen.[637]

In 1783, it was thought outlandish to presume that rocks fell from the sky.[638] It was considered more likely that meteors were simply terrestrial exhalations; this idea persisted throughout the eighteenth century, even though British astronomer Edmond HALLEY postulated an extraterrestrial origin for these rocks in 1714.[639] The European intelligentsia was averse to the idea that meteors could possibly come from outer space. On the other side of the Atlantic, however, the discourse was markedly different: John WINTHROP, a professor of natural philosophy and astronomy at Harvard, wrote a paper on the extraterrestrial origin of meteors, which proved so controversial that the Royal Society refused to publish it.[640] With mounting evidence, the consensus shifted in the early nineteenth century and, gradually, the idea that extraterrestrial rocks came crashing into the Earth's atmosphere from space was accepted.[641]

1783 saw several other fireballs: sightings were reported on 26 September and 4, 19, and 29 October.[642] The most notable of these occurred on 4 October 1783: this fireball was visible above Britain, lasting three or four seconds. Descriptions include several discrepancies regarding its size, color, and trajectory.[643] Surprisingly, there was confusion over how to differentiate meteors from hot-air balloons; few people had seen either of these "flying balls" with their own eyes.[644]

---

636 Berlinische Nachrichten, 30 August 1783: 305: Report from Nieuwerkerk, 19 August 1783.

637 Münchner Zeitung, 8 September 1783: 559.

638 BLAGDEN 1784: 201. Even Aristotle believed comets to be meteors in the higher parts of the atmosphere. In the sixteenth century, Tycho BRAHE showed that comets were astronomical phenomena; BURKE 1986: 6.

639 BURKE 1986: 5–8.

640 MOXHAM 2013.

641 BURKE 1986: 11, 37–58; PRINCE 1986: 102–103; BEECH 1989: 132–134.

642 PAYNE 2010: 4; WITZE, KANIPE 2014: 117.

643 BLAGDEN 1784: 219–212: BLAGDEN was based in London. On 4 October 1783, the sun set there at 5:30 p.m. It was already fairly dark at 6:43 p.m., and the meteor occurred during astronomical twilight, the third phase of twilight that occurs just before it becomes proper night, based on the Time and Date website. Gilbert WHITE, "The Naturalist's Journal," 1783, Add MS 31848, British Library, London, UK: 143.

644 Parker's General Advertiser and Morning Intelligencer, 9 October 1783; ALEXANDER 1996; PAYNE 2011: 23. This is not only true for Britain, but also for Germany. In the Münchner Zeitung, a report about the "fire balls" in England was followed by an article about "artificial meteors," which referred to hot-air balloons; Münchner Zeitung, 8 September 1783: 559–560. Another report stated that these

In May 1784, Benjamin Franklin wrote a letter to his friend, the Manchester physician Thomas PERCIVAL (1740–1804), who read it at a meeting of the Manchester Literary and Philosophical Society on 22 December 1784. In his letter, FRANKLIN offered an explanation for the dry fog of the previous year: he suggested meteors might be to blame.[645] The following passage from FRANKLIN's letter is particularly interesting:

> The cause of this universal fog is not yet ascertained. Whether it was adventitious to this earth and merely a smoke proceeding from the consumption by fire of some of those great burning balls or globes which we happen to meet with in our rapid course round the sun, and which are sometimes seen to kindle and be destroyed in passing our atmosphere, and whose smoke might be attracted and retained by our earth.[646]

FRANKLIN was familiar with the American discourse on meteors and corresponded with American astronomer David RITTENHOUSE (1732–1796).[647] Thus, he was drawn to the idea that the Laki haze was in some way connected with the fireball that was visible in August, even though this fireball arrived two months after the dry fog's initial appearance.

Experiments on the Origin of the Dry Fog

Several naturalists were inspired to conduct experiments on the fog, some of which were physically demanding, even dangerous. Independently of one another, Johann WIEDEBURG in Jena and Benjamin FRANKLIN in Passy conducted tests that seemed to show that the fog weakened solar radiation. WIEDEBURG outlined how he tried to melt lead with the help of a magnifying glass. In previous experiments with the same equipment, the lead had begun to melt after ten minutes; in this instance, it resisted.[648] In his letter to Manchester, Benjamin FRANKLIN remarked upon his version of the experiment, stating that the rays of the sun "were indeed rendered so faint in

---

balls were "no terrible air phenomena, but only machines made from light canvas und covered by paper that cannot cause bad luck [. . .]"; Königlich Privilegirte Zeitung 18 September 1783: 882–883: Report from Paris, 5 September 1783. "Diejenigen also, welche dergleichen Kugeln, welche das Ansehen des verdunkelten Mondes haben, am Himmel entdecken sollten, werden hierdurch benachrichtiget, daß es gar keine fürchterliche Lufterscheinung, sondern nur von Taft oder leichter Leinwand gemachte und mit Papier überzogene Maschinen sind, die kein Unglück stiften können [. . .]."
645 PAYNE 2011: 23.
646 FRANKLIN 1785: 373–377.
647 RITTENHOUSE 1786: 173–176; BURKE 1986: 17–24; PAYNE 2010: 2–5; FRANKLIN 2017: 293.
648 WIEDEBURG 1784: 70.

passing through [the fog], that when collected in the focus of a burning glass [magnifying glass], they would scarce[ly] kindle brown paper."[649] FRANKLIN's results corroborate the findings of WIEDEBURG's experiment.[650]

Johann WIEDEBURG conducted other experiments, including collecting dew from the fog and – in the spirit of the Enlightenment – tasting it.[651] He was not the only one who used his body for the advancement of science. Georg Christoph LICHTENBERG, despite having a laboratory full of equipment at his disposal, used his body as a "trained instrument." In a letter, he humorously remarks, "my body is, as one can expect of the body of a physics professor, a never-disappointing barometer, thermometer, hygrometer, manometer, etc." He did not believe the dry fog had deleterious consequences, as he did not experience any.[652] Sebald Justinus BRUGMANS also tasted the dry fog; it is unclear if this was deliberate or whether, during the peak of the dry fog in the Dutch Republic, anybody could avoid tasting it.[653]

Many reported that they could look directly at the sun without any injurious effects. WIEDEBURG even peeked at the sun through a telescope with his naked eye without "punishment."[654] Jan Hendrik VAN SWINDEN, who first noticed the haze in Franeker on 19 June 1783, noted the "deep red" color of the sun, which he too dared to stare at, at midday no less, without apparent damage to his retina.[655]

VAN SWINDEN also detailed an account relayed to him by M. DU VASQUIER of Neuchâtel, Switzerland, who had waited for particularly foggy days and left presumably mediocre paintings exposed in a meadow for quite some time, which resulted in the discoloration of the canvas. "The red of the paintings first became orange, then purple when washed with water. The black paint was partly washed away. The purple lost its vivacity."[656] An attempt was made to reproduce these results using different acids with known chemical reactions; for this purpose, DU VASQUIER colored a canvas with red, violet, and black and then immersed half of it in diluted acid. A similar result was obtained, particularly with nitric and sulfuric acid. DU VASQUIER was convinced that these chemicals, or chemicals with similar properties, resided within the dry fog.[657]

---

649 FRANKLIN 1785: 259–361; PAYNE 2010: 3.

650 GLASER 2001: 204.

651 WIEDEBURG 1784: 70.

652 LICHTENBERG 1985: 640, Letter 1102, Georg Christoph LICHTENBERG to Gottfried Hieronymus AMELUNG, Göttingen, 3 July 1783. "Mein Körper ist, wie es sich für den Cörper eines Prof. Physices geziemt, ein nie versagendes Barometer, Thermometer, Hygrometer, Manometer pp. allein ich empfinde von diesem Nebel keine besondere Wirckung; wir haben ihn hier sehr starck, und bemercken auch allerley, aber nichts was sich nicht auch von einer trockenen Hitze erwarten liese. Man kan bey solchen Dingen nicht genug zweifeln."

653 BRUGMANS 1783: 58.

654 WIEDEBURG 1784: 70.

655 VAN SWINDEN 2001: 78–79.

656 VAN SWINDEN 2001: 78–79.

657 VAN SWINDEN 2001: 73, 78–80.

Christoph Gottfried BARDILI, upon reading reports from the German Territories about the smell of sulfur that was said to have accompanied the fog, became convinced that it must partially consist of sulfur. He suggested an experiment whereby a piece of silver would be left exposed to the haze. If the silver turned black – then a known result of a reaction of sulfuric particles with this precious metal – one could ascertain if the haze did indeed contain sulfuric particles. He remained doubtful whether the particles were large enough to cause a reaction and concluded that even if the silver objects did not turn black, there could still be sulfuric particles in the dry fog.[658]

In Paris, members of the local observatory attached meat to kites and guided them into the haze. The meat was "entirely corrupted," which was seen as proof that the haze had a deleterious effect on organic material and possibly human health.[659]

News from Iceland

As Iceland was a Danish dependency, news from the island was usually disseminated throughout Europe via Copenhagen.[660] The revelatory story of the Laki eruption was published in *Kjøbenhavns Adresse-Contoirs Efterretninger* on 5 September 1783. The report described a fire that broke out in the "Skaftefields district" on Whitsunday and dried up the "Skaptaa" River, destroying two churches and eight farmsteads. Smoke, ash, and sand filled the air, covering the whole country in a haze.[661] On 6 September 1783, this news item appeared in the *Hamburgischer Unpartheyischer Correspondent*, apparently

---

658 BARDILI 1783: 12–13.
659 GARNIER 2011: 1046; MCCALLAM 2019: 227–228.
660 DEMARÉE, OGILVIE 2001: 220.
661 Kjøbenhavns Adresse-Contoirs Efterretninger, 5 September 1783. The report is based on a letter from Iceland, dated 24 June 1783. A newspaper report in the Münchner Zeitung, 15 July 1783: 433, mentions: "The Danes give us news that Mount Hecla [sic] in Iceland has begun to spit fire and fill rivers with burning lava from its fiery maw. They add that several springs have started to run dry on this ice-covered island." "Die Dänen geben uns bei diser Gelegenheit Nachricht, das der Berg Hecla in Island vil Feuer auszuwerfen, und mit ganzen Strömen von brennender Lava aus seinem Feuerschlunde die Gegenden zu überschwemmen angefangen habe. Sie se[t]zen hinzu, das mehrere Quellen mit siedendem Wasser sich in verschidenen Gegenden diser beeisten Insel eröffnet haben." This news cannot be found anywhere else. It is unclear whether the newspaper really received a letter from a correspondent so early. Nobody else received it or followed up on it. Given that all the news about any unusual natural phenomena was widely shared, it is highly unlikely that the news of a Hekla eruption would have received so little attention. The lava flows in southeastern Iceland appeared on 12 June 1783; even if a report about these events had been written and sent aboard a vessel from Iceland this early, it is unlikely that the news would have reached Munich by mid-July. This is evidenced by the fact that news written on 24 July 1783 reached Copenhagen in early September.

based on a letter from Copenhagen dated 2 September.[662] There remains the possibility that one of the merchant vessels returned in late August rather than early September 1783. A report from Copenhagen, dated 26 August 1783, found its way into the *Berlinische Nachrichten* and was published on 18 September 1783. It included similar details to the previously mentioned report but described a haze that made the "sun gleam like a lump of fire." It sketched out how the "fires" hampered fishing efforts and affected grass growth and the production of milk.[663]

> Ships that have returned from Iceland yesterday relay the unpleasant news that in Skaptefields Syssel [sic], not far from Mount Hekla, several new volcanoes, including Myrdals Jökull [Mýrdalsjökull], have erupted steam and fire, and their lava flooded the Skaptaa [Skaftá] River, along a length of 15 miles and a width of 7 miles, like a stream of water; three churches and a monastery have fallen victim to the lava. The air itself has been filled ever since with a steam and fine dust, which darkened the sun and damages the fields. The newly formed island near Reickenäs [Reykjanes] is ever increasing, it burns and smokes incessantly.[664]

The report contains geographic inaccuracies: Mýrdalsjökull is a glacier that covers the caldera of Katla, which was not part of this eruption. Another glacier, called Vatnajökull, directly borders the Laki fissure. Unsurprisingly, the Laki fissure is not named as such. It is interesting that Hekla is highlighted here. The Laki fissure is closer to Katla than Hekla, but Hekla is singled out, being the more famous of the two.[665]

From September 1783, this report found its way onto the pages of many other newspapers, almost word for word.[666] The news was printed in Hamburg (6 September), Berlin, Stockholm (11 September), Breslau (15 September), Munich, Augsburg, The Hague (18 September), Vienna (20 September), Brussels, St. Petersburg (22 September), Bamberg, London (23 September), Florence (27 September), Paris (30 September), Luxembourg, Bern (1 October), Warsaw (8 October), Madrid (17 October), Lisbon

---

662 Interestingly, the German newspapers seem to refer back to a letter from Copenhagen on 2 September 1783, while the French newspapers refer to a letter from Copenhagen on 5 September 1783. Perhaps it was a typo or there simply were two letters.

663 Berlinische Nachrichten, 18 September 1783: 686: Report from Copenhagen, 26 August 1783. "[. . .] wenn sie [die Sonne] sich bey Auf- oder Untergange durch die schrecklich dicke Luft zeigt, wie ein Feuerklumpen schimmert."

664 Hamburgischer Unpartheyischer Correspondent, 6 September 1783: Report from Copenhagen, 2 September 1783. "Mit den gestern von Island zurückgekommenen Schiffen erfährt man die unangenehme Nachricht daß in Skaptefields Syssel, unweit dem Berge Hekla, verschiedene neue Vulcane, worunter auch der sogenannte Myrdals Jökull, mit Dampf und Feuer ausgebrochen sind, und mit ihrer Lava die ganze Gegend bey dem Fluß Skaptaa, die eine Strecke von 15 Meilen in die Länge und 7 Meilen in die Breite ausmacht, wie ein Wasserstrom überschwemmt, auch 3 Kirchen und ein Kloster zerstöhrt haben. Die Luft daselbst ist seitdem beständig mit Dampf und feinem Staube angefüllt, der die Sonne verfinstert und die Felder beschädiget. Die bey Reickenäs entstandene neue Insel nimmt immer mehr zu; sie brennt und raucht aber noch unaufhörlich."

665 FRÖMMING 2005: 104, 124.

666 Only the number of destroyed or damaged churches and farms varied slightly.

(24 October), and Barcelona (25 October), among other places.[667] News about the Skaftá Fires did not precipitate any revelatory conclusions concerning the dry fog, which had all but disappeared by this time.[668]

More detailed accounts would eventually make the rounds; for example, on 13 March 1784, the Austrian newspaper *Provinzialnachrichten aus den Kaiserlich Königlichen Staaten* published a four-page-long report about the eruption, with a follow-up on 17 March.[669]

Speculating about a Connection

Some naturalists, independently of each other, considered a volcanic eruption a viable explanation for the fog. It remains debated in the modern scholarly community who was the first to propose the idea.[670]

Dom Robert HICKMANN wrote a letter to the *Journal encyclopédique ou universel* on 29 July 1783, responding to and criticizing Joseph DE LALANDE's theory. HICKMANN's letter was published in the journal's issue on 15 September 1783.[671] He argued, no doubt inspired by Athanasius KIRCHER, that there was a permanent fire in the center of the Earth (*un feu permanent dans le centre de notre globe*) with more or less vertical subterraneous canals forming volcanoes from one pole to the other; some new, some old, and some extinct. He believed the fire moved from one volcano to another (via *canaux de communication*) and caused earthquakes, eruptions, and other convulsions. Concerning the dry fog, HICKMANN rightly assumed that the convulsions (*ces bouleversemens*) in Iceland were more significant than the Calabrian earthquakes. He attributes the "famous dry and sulfuric fog" to the Icelandic "convulsions," by which he meant the formation of Nýey.

---

667 Hamburg: Hamburgischer Unpartheyischer Correspondent, 6 September 1783; Berlin: Königlich Privilegirte Zeitung, 11. September 1783: 863; Berlinische Nachrichten, 11 September 1783, more detailed information followed in the issue published on 18 September 1783: 686; Munich: Münchner Zeitung, 18 September 1783: 583; Bamberg: Hochfürstlich-Bambergische wochentliche Frag- und Anzeigenachrichten, 26 September 1783; Augsburg: Augsburgische Postzeitung, 18 September 1783; Paris: Gazette de France, 30 September 1783: 345; Breslau [today's Wrocław in Poland]: Schlesische Privilegirte Zeitung, 15 September 1783: 1099. The other towns and dates have been mentioned by Gaston DEMARÉE and Astrid OGILVIE (2001: 239–240; 2016: 136).

668 WOOD 1992: 70; DEMARÉE, OGILVIE 2001: 239–240.

669 Provinzialnachrichten aus den Kaiserlich Königlichen Staaten, 13 March 1784: 330–333 (part 1), 17 March 1784: 349–350 (part 2).

670 SCARTH 1999: 116; THORDARSON, SELF 2003: 2; GRATTAN et al. 2005: 644.

671 That Dom Robert HICKMANN was one of the first, if not *the* first, to presume a connection between the fog and Icelandic volcanoes is detailed in MONGE et al. 1793: 232–236; HICKMANN 1783: 512. At the end of his report, HICKMANN gives the date as 29 July 1783.

HICKMANN admitted that it was strange that the "convulsions" in Iceland took place in March 1783 and the fog only became visible that June; he would not be drawn to pinpoint a fixed time for the fog's first occurrence and suggested that the newspapers did not sufficiently inform the public about the exact date that the dry fog descended upon Europe. He considered it possible that shifting winds had only eventually brought the dry fog to the continent after carrying it here and there.[672] As news from Iceland traveled slowly, HICKMANN remained unaware that another "convulsion" was rocking Iceland as he wrote.

Another naturalist who established the connection was Jacques Antoine MOURGUE DE MONTREDON. He presented his idea in front of the Société Royale des Sciences in Montpellier on 7 August 1783, about a week after HICKMANN had written down his findings.[673] MOURGUE DE MONTREDON's work was published in 1784 and was updated after August 1783. This is apparent because the published version includes details about events in Iceland that only reached European ears in early September.[674] MOURGUE DE MONTREDON states that the vapors in the atmosphere were a rare phenomenon and of the type that affected the educated reader just as much as the general population. He observed that during the last four days of June, strong northern winds did not disperse the vapors; on the contrary, they seemed to add to their density. Indeed, the fog had never been so thick nor the sun so red as when Boreas breezed by at the close of the month.

MOURGUE DE MONTREDON describes a great number of the unusual phenomena of the year, including the eruption of the Gleichberg. The earthquakes and other upheavals of the year led him to believe that the subterraneous fires (*les principaux foyers des feux souterrains*) existed and reached from Iceland to Syria via Calabria. He also acknowledges that he had heard about the emergence of Nýey in early June 1783 and comments on its extremely thick smoke and duration. He wonders rhetorically whether this new volcanic eruption could explain the vapors that hung over Europe. MOURGUE DE MONTREDON opined that exhalations from some volcanic eruption overloaded the atmosphere with flammable air, which led to the frequent thunderstorms that had spread across the continent. He also believed the interplay of the flammable air with the atmosphere had caused the "terrible and impressive meteors."[675]

Another naturalist who was very familiar with Icelandic volcanism and who was said to have established the island's connection to the dry fog was Christian Gottlieb KRATZENSTEIN. Originally from Germany, he was then a professor of physics at the University of Copenhagen, where, in the second half of 1783, he presented his hypothesis in a lecture. Sæmundur Magnússon HÓLM likely attended this lecture and put this idea to paper, acknowledging KRATZENSTEIN. HÓLM writes that when poisoning rain fell

---

672 HICKMANN 1783: 507–512.
673 MOURGUE DE MONTREDON (1784: 761) mentions "nouveaux volcans à peu de distance du mont Hecla."
674 His publication also included copies of his daily observations from 1 June to 30 September 1783.
675 MOURGUE DE MONTREDON 1784: 754–763.

in Iceland, it soon after fell in Norway, "burnt" the leaves on the trees, and turned the grass in the fields black. Furthermore, he points out that the Faroe Islands experienced sand and ashfall accompanied by a "sulfur vapor" when the winds came from the northwest; this also suggested that these gases and vapors could travel long distances. Hólm bolsters the idea further by commenting on several ships that sailed between Iceland and Copenhagen that were reportedly covered in black sand and other fine particles. He points to all this as evidence that the fog in Denmark was connected to an "earth/soil fire" (*Jordbranden* in Danish, *Erdbrand* in German, and *earth fire* in English) in Iceland. This remark can be found in Hólm's original Danish book, written in late 1783 or early 1784, as well as in the German translation.[676] Both books were published in Copenhagen in early 1784, with the foreword in the Danish version dated 25 February 1784.[677]

A naturalist called H. Guerin also noticed that southern winds thinned the fog, and northern winds made it thicker. He was drawn to conclude that "terrible volcanic eruptions in Iceland" had caused this "evil." Guerin blamed the eruptions for the frequent diseases of 1783, which he says were characterized by decay and inflammation.[678] This information, however, is only second-hand; Guerin's findings were published in the daily Swiss newspaper *Neue Zürcher Zeitung* on 5 November 1783 in the miscellaneous section. He presented these findings in the weeks before 5 November, either verbally or in writing; the original account remains missing.

A naturalist by the name of Johann Rudolf von Salis-Marschlins (1756–1835) wrote a report for the weekly Swiss newspaper *Der Sammler: eine gemeinnützige Wochenschrift für Bündten* titled, "Some remarks on the general vapor, which spread also in our area in June and July of this year," which was most likely published between 17 and 23 November 1783. The article includes his weather observations from June through August 1783 and references the news from Copenhagen about the eruption in Iceland, so it was likely written sometime in September. Von Salis-Marschlins observes that the "terrible eruption of the fire-spitting mountain in Iceland" occurred at the same time as the "vapor." Just as Guerin had done, von Salis-Marschlins notes the direction of the wind and details how this played a role in his assumptions.[679] He asks

---

676 Hólm 1784a: 52–53; Hólm 1784b: 66–67.
677 The foreword in the German version is a translation of the Danish foreword dated 25 February 1784.
678 Demarée, Ogilvie 2016: 135; Neue Zürcher Zeitung, 5 November 1783: 3–4. "Ein Naturforscher Nahmens H. Guerin hat beobachtet, daß die Nebel, so dieß Jahr Europa bedeckt haben, durch die Südwinde vermindert und hergegen durch Nordwinde verdickt worden seyen. Hieraus hat es die Folge gezogen, daß die schrecklichen vulkanischen Eruptionen in Island diese Nebel verursacht hätten. Dieser Eruption schreibt er auch die häufigen Krankheiten zu, die dieses Jahr geherrscht, und durch Symptome der Fäulung und Inflammation die Aerzte in Verwirrung über die Natur gesetzt haben, da die Kranken, welchen Blut gelassen worden, wie die Bemerkungen der medicinischen Fakultät zu Paris besagen, meist das Opfer dieses Irrthums geworden sind."
679 Salis-Marschlins 1783: 397.

rhetorically: "[. . .] Should the coincidence of these two natural phenomena make us think of a relationship between the two?"[680]

The editor of *Leipziger Magazin zur Naturkunde, Mathematik und Oekonomie*, Carl Friedrich HINDENBURG, commented on a letter published in his magazine written by Christian LUDWIG, which suggested a causal relationship between the Calabrian earthquakes and the fog. HINDENBURG supplemented LUDWIG's idea by listing other possible sources of the dry fog, referring to the "earth fire" in "Skaptaa Jokul."[681] It is uncertain when this was published; however, judging by its content, it was likely put to print between September and December 1783.

In early 1784, the Baron DE POEDERLÉ, a botanist, studied the effects of the dry fog on plants and trees in detail. *L'Esprit des Journaux François et Étrangers: Par une Société de Gens-de-lettres* published his study in May 1784. The text includes observations from January through December 1783, so it is safe to assume he wrote it sometime in early 1784.[682] DE POEDERLÉ, using reasoning similar to that of GUERIN and VON SALIS-MARSCHLINS, decided that the seemingly ceaseless eruption near Mount Hekla,[683] which filled the air with dust and caught the rays of the sun, was an integral part of the equation. DE POEDERLÉ did not think this was the sole cause of the dry fog; he also blamed the earthquakes in northern and central Europe. He thought it too much of a coincidence that when an earthquake occurred in Messina on 19 June 1783, the fog appeared. He mapped out more of his theory, suggesting that the earthquakes ceased because they had released all of their previously pent-up gases. He lauds 1783 as remarkable in several regards. He hoped that scholars of meteorology, agriculture, and medicine would later show an interest in the events he had witnessed.[684]

Benjamin FRANKLIN, in the same letter detailed above in which he addressed the issue of meteors, also discussed a possible connection between the fog and Iceland.[685] Regarding the fog's origin, FRANKLIN writes: "[. . .] whether it was the vast quantity of smoke, long continuing to issue during the summer from Hecla in Iceland, and that other volcano which arose out of the sea near that island, which smoke might be

---

680 SALIS-MARSCHLINS 1783: 393. "Wenn die Nachrichten von dem fürchterlichen Ausbruche eines feuer-speienden Berges in Island, der den 8 Junius erfolgt seyn sollte, zuverläßig sind, sollte die verhältnismä-ßige Uebereinstimmung der Zeit, nebst dem bemerkten Umstande, daß der Dampf sich beym Nordwind starker einfand, und hingegen vom Südwind einigermassen vertrieben wurde, uns nicht auf den Ge-danken eines Zusammenhanges zwischen diesen zwei Naturbegebenheiten bringen?"
681 LUDWIG 1783: 219–221.
682 DEMARÉE, OGILVIE 2016: 137.
683 Of course, it was Laki, undiscovered in the Icelandic highlands, that was ceaselessly erupting.
684 DE POEDERLÉ 1784: 336, 349.
685 FRANKLIN's letter is dated to May 1784. It is likely that FRANKLIN kept notes about the weather throughout 1783. FRANKLIN's friend Thomas PERCIVAL in Manchester read the letter titled "Meteorologi-cal Imaginations and Conjectures" in front of the Manchester Literary and Philosophical Society on 22 December 1784.

spread by various winds over the northern part of the world, is yet uncertain."[686] While DE POEDERLÉ argued very precisely that he believed the volcanic eruption that had commenced on 8 June 1783 *near* Mount Hekla was the culprit, FRANKLIN's conjectures were more speculative. He thought, as many naturalists did at the time, that the origin of the fog was not simply near Hekla but Hekla itself or perhaps Nýey.[687] This confusion is unsurprising; the newspaper reports often gave only vague descriptions of this new "fire" in the Skaftá region, which were often riddled with errors.[688] FRANKLIN may have been the first English-language author to write about the Icelandic connection; it seems, however, he was ultimately more convinced that comets had caused the dry fog.

Jan Hendrik VAN SWINDEN, based at Franeker in the Dutch Republic, observed the dry fog of 1783. The *Ephemerides* of the Societas Meteorologica Palatina published his insights in 1785. His article consists of three parts: a summary of his observations during 1783; the expert opinions of his friend and colleague Sebald Justinus BRUGMANS about the impact of the dry fog on plants; and VAN SWINDEN's own speculations as to the cause of the dry fog. He comments on the numerous earthquakes, the "dense haze," the coincidental nature of the appearance of the "haze" together with "a new mountain catching fire in Iceland" on 8 June 1783, and the new island that had appeared near Iceland. VAN SWINDEN noticed the simultaneity of the strange fog in Europe and the two volcanic eruptions in Iceland during the summer. He describes the aftermath of the eruption "in the middle of the mountains called Skaftan." He remarks that there was "ash, such that it obscured the air." VAN SWINDEN failed to see the similarities to the weather he himself had experienced. After considering all the evidence, he concluded that, in all probability, earthquakes caused the fog.[689]

The naturalists mentioned above established links between the volcanic eruptions in Iceland and the weather in Europe. Modern insights have proven that they were correct in their assumptions; however, for differing reasons, they failed to pinpoint the true cause of the dry fog.[690] The Laki fissure was unknown to naturalists at the time, so, unsurprisingly, there is no mention of it in their speculations. Hekla and Nýey were considered possible sources of the fog but were somewhat lost in the sea of other possible explanations, which included: too much or too little electricity in the atmosphere; aurora borealis; the smoke of meteors; vapors from a comet passing by the Earth; the smoke of peat fires; or – arguably the most popular explanation – vapors from deep within the Earth released during earthquakes.[691]

---

686 FRANKLIN 1785: 360.

687 DEMARÉE, OGILVIE 2016: 137.

688 OSLUND 2011: 180, note 10: "Benjamin Franklin commented on the dense fog experienced in North America in the summer of 1783. Like many of his contemporaries, he incorrectly attributed its cause to an eruption of Hekla rather than Laki."

689 VAN SWINDEN 2001: 78.

690 PAYNE 2011: 23.

691 STOTHERS 1996: 85.

The Summer of 1783 outside of Europe

Although it was soon common knowledge that the dry fog was more than a local phenomenon, the true extent was likely not known at the time; in the west, the dry fog reached North America. Reports from three different settlements in Labrador – today's Canada – also note a "smokey haze" that lasted up to five weeks. The dry fog may even have reached Alaska. Local weather observers in Labrador believed that Indigenous tribes "let some great woods on fire [as] they do so sometimes."[692] In the east, it was also visible as far away as the Altai Mountains on the western border of China and possibly triggered famine in India and Egypt (Figure 45).

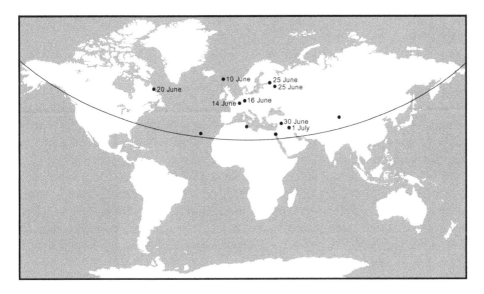

Figure 45: First appearance of the dry fog across the Northern Hemisphere. The dry fog appeared in the area north of the black line. For a detailed map of Europe, see Figure 28.

The Western Hemisphere

The United States of America
North America did not suffer the same tribulations as Europe. The *Continental Journal* from Boston printed the following on 5 August 1783: "we have not experienced finer weather for agriculture for some time past. From almost every part of the country,

---

692 Meteorological Observations at Okkak and at Hoffenthal, Hudson's Bay Company, Labrador, Newfoundland, 1782–1786, MA/144, Archives of the Royal Society, London, UK, 2–4 July 1783.

we have the most satisfactory information that all appearances indicate a year of astonishing plenty."[693]

In a letter written in May 1784, Benjamin FRANKLIN remarks that the dry fog had been visible over a "great part of North America."[694] Robert DE LAMANON argues in a footnote in his 1799 publication that he presumed the dry fog had appeared in North America after an eight-year-long drought, which was, in his eyes, a primary cause of the dry fog.[695] Prior to the English edition, DE LAMANON had published his text in French in 1784 with the same footnote at the end of the article.[696] It is very likely that Benjamin FRANKLIN read DE LAMANON's argument before he wrote his conjectures about the dry fog in his letter to the Manchester Literary and Philosophical Society and that this was the source of his information. Richard STOTHERS suggests that the dry fog was not visible in North America, based on weather diaries from Michigan, Massachusetts, and North Carolina.[697]

Other North American weather diaries suggest that the dry fog was not visible between New England and Delaware. According to the weather diary of Jacob HILTZ-HEIMER, based in Philadelphia, the summer of 1783 was mostly warm and pleasant.[698] William ADAIR from Delaware describes June 1783 as fair, with warmer spells and occasions of fog toward the end of the month. He gives no indication that these fogs were in any way unusual. He describes July and August 1783 as variable, with a mix of clear and cloudy weather.[699] Caleb GANNETT, who kept a daily weather log at Cambridge, Massachusetts, notes a mixture of fair and cloudy days between June and October 1783, with an occasional rainy day and no mention of fog.[700] Edward WIGGLESWORTH kept another daily weather diary in Cambridge, Massachusetts, and noted foggy weather on 21 June 1783, "hazey" weather on 9, 10, and 17 July 1783, and "hazey" weather again on 5, 6, and 9 August 1783. Otherwise, the weather during this time was, according to WIG-GLESWORTH, usually "fair."[701]

Cotton TUFTS, a physician, kept an annotated almanac at Weymouth, Massachusetts. According to his records, the weather during the summer of 1783 was mainly warm and hot. However, he notes occasional foggy and misty days, such as 16 June

---

693 The Continental Journal and Weekly Advertiser, 5 August 1783: Report from Hartford, CT.
694 FRANKLIN 1785: 357–361.
695 LAMANON 1799: 88–89.
696 LAMANON 1784a: 17.
697 STOTHERS 1996: 82.
698 Jacob HILTZHEIMER Diaries, vol. 13 (March 1783 to February 1784), Mss.B.H56d, American Philosophical Society, Philadelphia, PA, USA.
699 William ADAIR Meteorological Notebook, Meteorological observations taken at Lewes, Delaware, 1776–1788, Mss.551.5.Ad1, American Philosophical Society, Philadelphia, PA, USA.
700 Caleb GANNETT, A meteorological register kept at Cambridge 1783, MS Am 1360, bMS Am 1360 (7), Houghton Library, Harvard, Cambridge, MA, USA.
701 Edward WIGGLESWORTH, Meteorological journal. A.MS.s. Cambridge, 1780–1789, 1793, MS Am 1361 vol 33, Houghton Library, Harvard, Cambridge, MA, USA.

and 18 and 19 July.[702] Edward HOLYOKE, also a physician, describes June 1783 as very hot, with some thunderstorms, July and August as unusually rainy, and September as variable.[703] As these occasions of fog received little attention, it can be assumed they were regarded as normal. In Massachusetts, two weather diaries note frost on 10 August 1783. Reverend Joseph LEE at Concord, Massachusetts, annotates his almanac thusly: "This morning we got frost."[704] Ezra WHEELER, also at Concord, writes of 10 August and 4 September, "frost so as to kill things."[705] In Nova Scotia, Mary SEWALL, who fled from Massachusetts with her family during the war, describes June 1783 as a warm month. For 26 June 1783, she writes: "The weather is so exceeding[ly] warm I can hardly do anything."[706]

When, precisely, did the news of the Laki haze reach the United States? The first mention was in the *Pennsylvania Packet and General Advertiser* on 21 October 1783. The source was a letter from Salon in Provence, France, dated 11 July 1783: "For twenty days, a singular fog, such as the oldest man here has before not seen, has reigned [. . .] similar observations have been made at Paris and in many parts of Italy."[707] The same letter was also printed in the *Connecticut Journal* on 5 November 1783, in the *Massachusetts Spy* on 20 November 1783, and in the *Vermont Journal* and the *Universal Advertiser* on 3 December 1783.[708] The letter was written by Robert DE LAMANON, who also referred to his experience in Salon in his *Observations on the Nature of the Fog of 1783*.[709]

Other reports of a thick, dry fog in Europe appeared in American newspapers, too; on 25 October 1783, the *Royal Gazette* printed a report from Naples, where from 23 June 1783, "the atmosphere [was] loaded with a thick fog."[710] Likewise, on 2 December 1783, the *Connecticut Courant* printed a report about "an uncommon fog, [that]

---

702 Cotton TUFTS diaries, 1748–1794, Ms. N-1686, Massachusetts Historical Society, Boston, MA, USA.

703 Edward HOLYOKE, Medical Records, 1782–1788, Folder 2 (1783), Ms. N-1427, Massachusetts Historical Society, Boston, MA, USA.

704 Lee FAMILY Diaries [manuscript], 1783–1807, vol. 1 (1783), 271762, American Antiquarian Society, Worcester, MA, USA.

705 Ezra WHEELER Diary, 1783, Almanacs Mass. B490 1783a, American Antiquarian Society, Worcester, MA, USA.

706 Mary Robie SEWALL Diary, May–October 1783, call no. Ms. N-804, Massachusetts Historical Society, Boston, MA, USA.

707 The Pennsylvania Packet and General Advertiser, 21 October 1783: 2: Extract of a Letter from Salon in Provence, 11 July 1783.

708 The Connecticut Journal, 5 November 1783: 2: Reprint from news from London, 22 July 1783, reprint of Letter from Salon, Provence. The Massachusetts Spy, 20 November 1783: 2.

709 LAMANON 1784a; LAMANON 1784b; LAMANON 1799. Salon-de-Crau is today known as Salon-en-Provence.

710 Royal Gazette, 25 October 1783: 2: Reprint from London, 4 August 1783. It is unclear whether the newspaper article refers to 23 June or 23 July 1783.

has during the past summer, overspread that [French] kingdom – a phenomenon that has excited the attention of the greatest philosophers."[711] An almost verbatim report was printed in the *Norwich Packet* (Connecticut) on 4 December 1783, in the *Newport Mercury* (Rhode Island) on 6 December 1783, and in the *Rivington's New York Gazette* on 20 December 1783.[712]

In late December 1783, an article in the *Independent Ledger* from Boston remarked that newspapers had reported on more violent storms over the past 12 months than in any other year. The report concluded that "It may not be improper to add, that the European papers contain accounts of a variety of unusual natural phenomena that have appeared in the course of the last twelve months." By which they meant the Calabrian earthquakes, the uncommon fog, the new island Nýey (which they mistakenly placed in the Hebrides), and the "burning and eruption of vast bodies of sulphurous matter from the bowels of the earth."[713] By the end of the year 1783, the American periodicals were up to speed. Comparisons between the phenomena in Europe and North America were drawn, but for the most part, Americans remained unconcerned.

Province of Quebec

In Labrador, weather journals from three different coastal settlements called Nain, Okkak, and Hoffenthal contain strong evidence for the presence of the dry fog (Figure 46). The Moravian Church founded these settlements between 1771 and 1782 to proselytize the migratory Inuit tribes.[714] Weather observers made several daily recordings as well as qualitative remarks about the sky, sea ice in the bay, and the northern lights. The weather observations were presented at the Royal Society in London by British-American architect Benjamin LA TROBE in 1786. In the same year, the German astronomer Johann Daniel TITIUS (1729–1796) published them in his monthly extract in the *Wittenbergisches Wochenblatt*, a weekly newspaper in the German Territories.[715] TITIUS remarked that the mention of vapor at the coast of Labrador confirmed that the mist was present across the entire Northern Hemisphere.[716]

---

711 The Connecticut Courant and Weekly Intelligencer, 2 December 1783: 2.
712 Norwich Packet or The Chronicle of Freedom, 4 December 1783: 3; The Newport Mercury from Newport, Rhode Island, 6 December 1783: 3; Rivington New York Gazette and Universal Advertiser, 20 December 1783: 3.
713 Independent Ledger, 22 December 1783: 3: Report from Boston, 22 December 1783.
714 DEMARÉE, OGILVIE 2006. Hoffenthal is today's Hopedale in Canada.
715 Meteorological Observations at Nain & Okak at Labrador, Hudson's Bay Company, MA/143; and Meteorological Observations at Okkak and at Hoffenthal, Hudson's Bay Company, Labrador, Newfoundland, 1782–1786, MA/144, Archives of the Royal Society, London, UK; Wittenbergisches Wochenblatt, 19 May 1786: 153–160.
716 Wittenbergisches Wochenblatt, 19 May 1786: 156: "Es giebt uns daher diese Bemerkung auf der Küste von Labrador in den Ursachen der damaligen Dunstluft eine große Aufklärung. Denn es scheint, daß diese Dunstluft über die ganze nördliche Halbkugel, wo nicht weiter, möge gegangen sein."

Figure 46: The locations of Okkak, Nain, and Hoffenthal in Labrador.

The weather observer in Okkak was either James BRANAGIN or Johannes BECK.[717] On 20 June 1783, he first mentioned what could be presumed to be the fog: "Heazey [sic] and sun shine." On 3 July 1783, he remarked: "Heazey sun shine and rain. For several days thick smoke fog [sic] throw the Air as from a great fire so we pose the Ykas [?] let some great woods on fire, they do so sometimes." Thus, he surmised that this smoke was a result of forest fires started by local Inuit tribes. On 23 October 1783, the "heazey" weather was mentioned for the last time.[718] The diary mentions phenomena similar to those in Europe, including paraselenae and parhelia. In Nain, descriptions of the fog, written by Daniel KRÜGELSTEIN, were in a similar vein.[719] In the

---

717 DEMARÉE, OGILVIE 2006: 428. A table lists the names of the different weather observers in the Moravian settlements, if they are known.
718 Meteorological Observations at Okkak and at Hoffenthal, Hudson's Bay Company, Labrador, Newfoundland, 1782–1786, MA/144, Archives of the Royal Society, London, UK, 2–4 July 1783. The name of the tribe might also be Elaas [?].
719 Meteorological Observations at Nain & Okak at Labrador, Hudson's Bay Company, MA/143; Archives of the Royal Society, London, UK.

neighboring settlement of Hoffenthal, an anonymous weather observer witnessed "dark mist with much rain" and mentioned that the summer of 1783 was "very dark, [with] little sunshine."[720]

Richard STOTHERS briefly mentions the occurrence of the haze in Labrador in his 1996 paper "The Great Dry Fog of 1783." Gaston DEMARÉE and Astrid OGILVIE added detail about the Moravian settlement in their response to his paper in 1998. The latter publication also delivers evidence from sailors who navigated the waters around Newfoundland and reported "a thick fog."[721] These reports from Labrador are clear evidence that the dry fog also reached North America. Deliberate burning practices and forest or brush fires were (and are) common in North America, and they cause "smoky" weather and even "dark days" in the southern parts of the Province of Quebec and the United States.[722] The timing of this "smoky" weather from mid-June to October 1783 and the known dates for the Laki eruption fit too perfectly to dismiss them as anything but the Laki haze. That the weather observer in Okkak looked for explanations of this unusual "heazey" weather on a local level is fascinating. The observer never knew that this "smoke" had spread further than his locality and therefore did not look further than his locality for an explanation.

Alaska

Thorvaldur THORDARSON argues that the Arctic, including the Faroe Islands, Lapland, and Alaska, was extremely cold during the summer of 1783. As a result, the trees in these regions show narrow tree rings and a low density for the latewood, which is the growth of conifer trees that occurs later in the year. THORDARSON attributes this to the "noxious dews" from the Laki eruption.[723] An analysis of tree rings in northwestern Alaska reveals that the maximum latewood density for 1783 has the lowest value in more than four centuries, possibly even the lowest in nine centuries.[724] Temperature reconstructions for northern Alaska reveal that the summer temperature (May to August) in 1783 was around 4 °C below the mean.[725] Unfortunately, there are very few written histories about this area for this period. Therefore, Gordon JACOBY and his colleagues utilize transcriptions penned by William OQUILLUK in 1973 of oral traditions passed down by the Kauwerak

---

720 Meteorological Observations at Okkak and at Hoffenthal, Hudson's Bay Company, Labrador, Newfoundland, 1782–1786, MA/144, Archives of the Royal Society, London, UK.
721 STOTHERS 1996: 79; DEMARÉE, OGILVIE, ZHANG 1998: 727.
722 CAMPANELLA 2007; PYNE 2007.
723 THORDARSON 2005: 213–214.
724 The study was conducted in 1999, which means the 400-year-period refers to the time prior to 1999; JACOBY, WORKMAN, D'ARRIGO 1999: 1366, 1370. In a recent study using trees from Alaska, EDWARDS et al. (2021) demonstrated that conventional methods for assessing maximum latewood density overestimates cooling during the summer, whereas a detailed analysis of early- and latewood cells draws a clearer picture.
725 JACOBY, WORKMAN, D'ARRIGO 1999: 1365–1367.

people from northwest Alaska. These stories refer to four disasters, the third of which they believe happened in 1783. Although exact years are missing from the text, a rough time frame can be established by ascertaining the generation during which an event took place. The oral tradition of the third disaster describes their ancestors' survival during "the time summertime did not come."[726]

The spring had been typical: the snow had melted, and the migratory birds had arrived. But then suddenly, at the end of June 1783, the sky became overcast and the weather turned cold again. Then, in the middle of summer, the rivers froze and it snowed, making fishing and collecting plants impossible. Warmer weather did not return until April of the following year. A young woman and her son were the only survivors of their village; they traveled 300 kilometers, surviving on very little until they met survivors from elsewhere who told similar stories of starvation and death. JACOBY and his team deduced that this oral tradition probably dates back to a time before 1800, a period of population decrease in the area between 1779 and 1791, and before the arrival of Europeans. From tree ring analysis, we know that the impact of the 1815 Tambora eruption on northwestern Alaska was not that severe; therefore, it is probable that this tradition refers to the effects of the Laki eruption.[727]

The Eastern Hemisphere

The Baltic
Estonian environmental historian Priit RAUDKIVI reconstructed the weather conditions of the summer of 1783 in the Baltic region: there, it was foggy and a sulfuric smell lingered in the air, the cause of which damaged plant life. RAUDKIVI analyzed death records and concluded that mortality was high in Estonia, particularly in the south. Whereas in England and France, the mortality crisis only started in August, in Estonia, it began in June.[728] From Latvia and Estonia, one account regarding air pollution in the region was published a decade after the eruption.[729] According to this source, from 1794, June 1783 was "insufferably hot," July was characterized by frequent thunderstorms and rain, and August saw many storms.[730] Finland saw significant crop failure during and after the Laki eruption. By virtue of its location near and inside the Arctic Circle, Finland's agriculture is particularly vulnerable to unseasonal frost. In particular, between 1300 and 1930, Finland experienced 1.7 major crop failures per decade on average. During the eighteenth century, there was an average of 3.3 per decade; this meant that the grim specter of famine haunted the country three years

726 OQUILLUK 1973; JACOBY, WORKMAN, D'ARRIGO 1999: 1365–1367.
727 JACOBY, WORKMAN, D'ARRIGO 1999: 1369–1370.
728 RAUDKIVI 2014.
729 RAUDKIVI 2016: 193–194.
730 SNELL 1794: 148–149.

per decade. More than one crop failure in a row was often devastating in pre-modern societies. From 1783 to 1785, Finland experienced three consecutive years of major crop failure.[731]

Northwestern Siberia

From tree ring records at two locations in the Ural Mountains in Siberia, dendrochronologists have established that during the summer of 1783, a particularly severe temperature event occurred in this region. They analyzed Siberian junipers and larches in the Polar Ural Mountains and the Yamal Peninsula, reconstructed tree ring data for the years between 742 and 2003 CE, and found frost rings and light tree rings for 1783. Based on the growth period for the trees in this region, it is evident that there were frosts in late June and early July 1783. In fact, the whole summer season is believed to have been extremely cold. Cooling as a consequence of a strong volcanic eruption impacted the second half of the growing season for those trees. Their growth during 1783 was the lowest for the past 500 to 600 years.[732]

North Africa

Between 1783 and 1785, Constantin-François VOLNEY (1757–1820), a French philosopher, traveled through Egypt and Syria. Upon his return to France, he published a travel diary. He outlined how, in July and August 1783, he regularly observed a fog during sunrise that vanished later in the day. He also noticed that the sky was often overcast, and "the sun was often invisible the whole afternoon [. . .] I was frequently so enveloped in a white, humid, warm, and opake [sic] mist, so as not to be able to see four paces before me."[733] The foggy appearance of the atmosphere seemed to be so familiar to him that he rather noticed its absence than its presence: "On my return from Suez, [. . . .] between the 24th and 26th of July, we had no fog during the two nights we passed in the desert."[734] These descriptions of fog, mist and even vapors indicate that some type of fog was also visible in Egypt, possibly as far south as 29 degrees north of the equator.[735]

Linguist Vermondo BRUGNATELLI and environmental scientist Alessandro TIBALDI have discovered one mention of the Laki haze in North Africa in a chronicle of events on the island of Djerba, in today's southern Tunisia (33 degrees north), written by Muhammad b. Yusef AL-MUSABI in 1792/1793. The text describes a mist "similar to smoke" and a red sun that lasted for about half a month. However, the dating is not precise;

---

731 MYLLYNTAUS 2009: 80, 83.
732 HANTEMIROV, GORLANOVA, SHIYATOV 2000: 170–172; HANTEMIROV, GORLANOVA, SHIYATOV 2004.
733 VOLNEY 1788: 345, 347.
734 VOLNEY 1788: 346.
735 VOLNEY 1788: 351. Contemporary scholars were aware of the geographic reach of the Laki haze: MOURGUE DE MONTREDON (1784: 754–763) knew that the dry fog affected not only Europe but also Asia and North Africa.

the source only indicated that it took place before the beginning of the Islamic year 1198, which began on 26 November 1783.[736]

It has been well-established that distant volcanic eruptions can impact the Nile River floods and the weather in Egypt.[737] The volcanic matter that the Laki eruption ejected into the atmosphere had a detrimental effect on the African and Indian Ocean monsoon circulations. Usually, the Indian Ocean monsoon brings rain to the Ethiopian highlands in early summer. The rain then feeds the Nile River. In July, the Nile River floods reach Cairo; the peak of the floods usually takes place in August or September. ZAMBRI and colleagues argue that asymmetric cooling in the Northern Hemisphere after the Laki eruption in June 1783 shifted the Intertropical Convergence Zone southward; this interrupted the monsoon rains over the Indian Ocean and eastern Africa, which led to a low flow of the Nile River in 1783 and 1784 and triggered drought and famine in Egypt. The decrease in precipitation also triggered droughts in India.[738]

India
An unusual El Niño produced dry conditions in wide parts of India, which led to drought and scarcity between 1781 and 1784. In the north, it affected Kashmir and Punjab; in the west, Rajasthan; and in the east, Uttar Pradesh. The lack of rain often disrupted agricultural employment patterns, leaving workers without an income. In part, the scarcity was fueled by hostile invasions and pests, including swarms of rats, locusts, or ants. In South India, there was famine between 1782 and 1783. North India also experienced drought during this time: in September and October 1783, the lack of rain was particularly unusual and resulted in famine, which lasted until early 1784.

In South Asia, this period of scarcity and famine in 1783 and 1784 is remembered as the "great Chalisa famine," which is estimated to have caused as many as 11 million deaths; victims succumbed to starvation or the epidemics that followed. Some areas suffered from depopulation due to this famine and did not recover until the 1790s. In this part of the world, the unpredictable weather of the 1780s caused as many as six notable famines between 1780 and 1791. While other severe famines occurred in India before and after this decade, such as in 1770s, such a high frequency of this type of event in a single decade is unusual.[739]

---

736 BRUGNATELLI, TIBALDI 2020: 73.

737 MANNING et al. 2017.

738 ZAMBRI et al. 2019b: 6787. On the connections between ENSO and the Nile River, see BELL 1970; ORTLIEB 2004. Alan MIKHAIL details the consequences for Egypt, including diseases, theft, and violence, in the aftermath of the Laki eruption; OMAN et al. 2006a; MIKHAIL 2015; MIKHAIL 2017: 184–197; DAMO-DARAN et al. 2018: 534–538.

739 There was an El Niño-Southern Oscillation event in the Pacific between 1782 and 1784, which was unusual because it lasted longer than 24 months; D'ARRIGO et al. 2011; DAMODARAN et al. 2018: 521. The El Niño was followed by La Niña conditions that lasted from 1785 to 1790. Other scholars, such as

Japan

In Japan, from 9 May until 5 August 1783, Mount Asama erupted. Asama is Honshu's most active volcano, located 140 kilometers northwest of Tokyo. The Global Volcanism Program of the Smithsonian Institution rates this eruption as a four on the index of volcanic explosivity.[740] The climax of the eruption occurred on 4 August 1783 and lasted almost 15 hours, during which time it produced pumice falls, pyroclastic surges, and lava flows that killed approximately 1,400 people directly.[741] At the time of the eruption, Japan was suffering through the Great Tenmei Famine, one of the most severe in its history.[742] The famine began in 1782; most of the country's food reserves had already been used by the time of the eruption.[743] Japan saw very unstable and variable weather during the 1780s, particularly in 1783, 1784, and 1786, which all had unusually wet and cool summers. The weather led to terrible summer rice harvests in the north and east of the country.[744] Thus, 1783 is known as a year without a summer in Japan.[745]

Europeans only learned of Mount Asama's eruption in the 1820s from an account by Isaac TITSINGH (1745–1812), a Dutch scholar, merchant, and ambassador based in Nagasaki. His description of the eruption was based on Japanese sources and posthumously published in French in 1820 and English in 1822.[746] The eruption of Mount Asama had a minimal effect on the climate of the Northern Hemisphere, according to studies of the GISP2 ice core in Greenland.[747] The Laki eruption likely caused most of the volcanic cooling of the mid-1780s.

China

The westerlies are prevailing winds from the west to the east in the mid-latitudes between 60 and 30 degrees north. These winds carried the Laki haze to Europe, the Middle East, and even the Altai Mountains in China, where the haze was said to have arrived on 1 July 1783.[748] Richard STOTHERS discovered that the dry fog only lasted for 17 days in the Altai Mountains, whereas it lasted upward of two months in Europe.[749]

Richard GROVE (2007: 80), believed that the El Niño lasted from 1789 to 1793; DAMODARAN et al. 2018: 522–530, 540–541.
740 Global Volcanism Program: Asamayama.
741 RICHARDS 2003: 177.
742 THORDARSON 2005: 214–215.
743 DAMODARAN et al. 2018: 534.
744 DAMODARAN et al. 2018: 534–535.
745 MIKAMI, TSUKAMURA 1992; JACOBY, WORKMAN, D'ARRIGO 1999: 1368.
746 TITSINGH 1820 (French); TITSINGH 1822: 98–100 (English).
747 ZIELINSKI et al. 1994; JACOBY, WORKMAN, D'ARRIGO 1999: 1366.
748 STOTHERS 1996; JACOBY, WORKMAN, D'ARRIGO 1999: 1366.
749 STOTHERS 1996: 81.

Gaston DEMARÉE and his colleagues argue that the dry fog possibly reached further east. A Chinese chronicle for the Henan province mentions "severe dry fog," and there are accounts of dark skies during 1783. Another chronicler recounted how a re-occurring haze made the sky so dark it was impossible to see. These accounts, unfortunately, do not give exact dates or further indications as to the nature of this haze or dry fog. Perhaps these instances were local dust storms or a fog caused by the eruption of Mount Asama.[750] Based on multi-proxy records, a new study has established that China, especially the eastern parts of the country, experienced drought in the years following the Laki eruption, which was followed by locust swarms, famine, and disease.[751]

The Winter of 1783/1784: A Touch of Frost

While the summer of 1783 had been unusually warm in western Europe, the winter of 1783/1784 was extremely cold. Much of Europe and the eastern United States witnessed winters that were colder than average between 1783 and 1786. The cooling in the winter of 1783/1784 reduced temperatures by as much as 3 °C below the mean.[752] 1783 to 1786 might have even been the coldest period in the second half of the eighteenth century.[753] When the low temperatures of the winter of 1783/1784 are compared to the Central European Temperature series from 1760 to 2009, we see that the winter was 4.5 °C colder than the reference period of 1961 to 1990. Only 18 winters, including 1783/1784, between 1500 and 2009, are estimated to have had such low mean temperatures.[754]

The Winter in Europe

In Iceland, winter came early that year: in September 1783, uncommonly thick snow covered much of Iceland's lowlands, and severe sea ice was present.[755] Icelanders had to house their surviving cattle and sheep throughout the frosty period of the year, from mid-October to April. Some parts of Iceland saw continuous frost during that

750 DEMARÉE, OGILVIE, ZHANG 1998.
751 GAO, YANG, LIU 2021: 196.
752 LUDLUM 1966: 64–68.
753 THORDARSON 2005: 215.
754 BRÁZDIL et al. 2010: 182–184.
755 OGILVIE 1986: 71.

period, and by February 1784, all fjords were frozen, which had not happened in 39 years.[756] Many counties faced a cold spring and an equally chilly and wet summer in 1784.[757] Thick ice remained in vast swathes across Iceland until May 1784, and the ground remained frozen well into July.[758] During the twentieth century, the mean winter temperature was −0.9 °C in west Iceland and −1.7 °C in north Iceland. In 1783/ 1784, the temperatures stayed below −15 °C for much of the winter, leading to an average temperature of 5 °C below the 225-year average.[759] Astrid OGILVIE describes 1783/ 1784 as the most interesting period of the eighteenth century for environmental historians and, conversely, the most devastating for the Icelanders.[760]

In Europe, the cold came at the end of 1783, with piercing winds from the northeast. The extreme cold and snowy weather lasted, for the most part, until late February 1784 (Figure 47). Some places had snowfall as late as May that year.[761] Deep snow made travel across land difficult, and the frozen rivers made passage for ships impossible. In addition, many towns had shortages of food and firewood, which meant the authorities had to step up to prevent the masses from freezing and starving in their homes.[762] In Paris, the municipal authorities imposed a rationing system that privileged bakeries so that they would have enough fuel to bake bread and therefore feed the people.[763] Travel across the Channel was also fraught with difficulties; in early January 1784, John ADAMS, future president of the United States, traveled from London to the Hague and later told Benjamin FRANKLIN about the unpleasant journey:

> At Harwich we were obliged to wait Several Days for fair Weather, which when it arrived brought Us little Comfort as it was very cold And the Wind exactly against Us. [. . .] So unsteady a Course, and Such a tossing Vessell that We could not keep a fire, the Weather very cold and the Passengers all very Seasick. [. . .].[764]

FRANKLIN agreed with ADAMS: "The season has been, and continues [to be], uncommonly severe, and you must have suffered much."[765] In another letter, FRANKLIN remarks: "We

---

756 Morning Herald and Daily Advertiser, 25 August 1784: Extract of a Letter from Copenhagen, 30 July 1784; WOOD 1992: 60.
757 OGILVIE 1986: 71.
758 THORDARSON 2005: 215.
759 WOOD 1992: 64; MIKHAIL 2017: 190–192.
760 OGILVIE 1986: 72.
761 Bayerische Akademie der Wissenschaften 1784: 48.
762 GLASER 2008: 235.
763 FRANKLIN 2014: 258. In the face of firewood scarcity, Benjamin FRANKLIN received a request asking if he could help in the construction of a coal-fired oven. FRANKLIN 2017: 99 (note): "Ordonnance de police, concernant la distribution des bois à brûler aux boulangers et au public. Du 7 Février 1784." For the shortage of bread, also see KAPLAN 1996: 53–58, 76–77. In Saxony, the flooding destroyed many mills along the River Elbe, which led to a shortage of flour; POLIWODA 2007: 75.
764 FRANKLIN 2014: 498, letter from John ADAMS to FRANKLIN, The Hague, 24 January 1784.
765 FRANKLIN 2014: 533, FRANKLIN responded to ADAMS on 5 February 1784, from Passy.

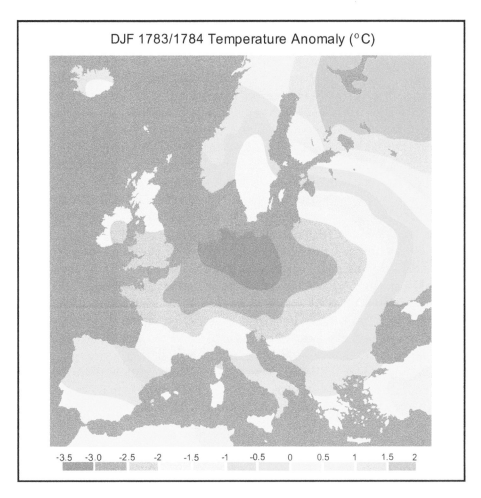

Figure 47: Temperature anomaly for December 1783, January and February 1784 based upon 31-year mean, 1770–1800.

have had a terrible Winter too, here, such as the oldest Men do not remember; and indeed, it has been very severe all over Europe."[766]

Records from the weather observatory in Mannheim reveal that it snowed 29 times between 24 December 1783 and 21 February 1784, with some snowfall continuing for several days without interruption. By 28 January 1784, the snow was roughly 154 centimeters high.[767] According to Karl Ludwig GRONAU, a keen weather observer, the winter in Mannheim that year was not as cold as 1739/1740 [−21.1 °C], but was, however, as cold as 1709 [−19.4 °C]. The coldest temperature that he measured in 1784 was on 7 January

---

766 FRANKLIN 2017: 28, letter from FRANKLIN to Charles THOMSON, from Passy, 9 March 1784.
767 GLASER 2008: 235; Societas Meteorologica Palatina 1784, 1–75.

at 8 a.m.: the Fahrenheit thermometer showed, as in 1709, 3 °F below zero [−19.4 °C], and the Réaumur thermometer showed −15.75 °Ré [−19.7 °C].[768]

In Munich, the coldest temperature of 1783 was on the morning of 31 December, −12 °Ré [−15 °C]. In December, the snow was so plentiful that trees collapsed under its sheer weight.[769] January 1784 remained extremely cold. For 28 days, not once did the thermometer rise above freezing. The total snow load reached a depth of ten shoes [ca. 292 centimeters].[770] In 1784, the coldest temperature in Munich was on 6 January 1784, −13.8 °Ré [−17.3 °C]. In February 1784, the thermometer remained below freezing for 21 days; the mountains of snow reached terrifying heights. Snow continued to fall until April, with frigid weather as late as 30 May 1784.[771]

Similarly, in Switzerland, the second half of the winter of 1783/1784 was extraordinary. In Bern, as much as 150 centimeters of snow fell from 12 to 13 March 1784; in the uplands, 270 centimeters fell. Bern was blanketed in snow for 154 days. For comparison, in the extreme winter of 1962/1963, the city saw 86 snow days. The cherry blossoms in Bern only began to bloom on 14 May 1784, three weeks later than usual. Geneva saw an estimated total snowfall of 240 centimeters. In Switzerland, the severity of the winter of 1783/1784 was only surpassed by the winter of 1739/1740.[772] The following winter, 1784/1785, was also said to be "extremely severe."[773]

The snowy conditions impacted game and birds, which could hardly find enough to eat. Wolves became more aggressive in their pursuit of food, attacking farm animals and even people.[774] In northern and southern Germany, the cold was so intense that birds froze in the skies and fell to the ground.[775] According to Ulrich BRÄKER, in Switzerland, the summer of 1784 was silent, devoid of birdsong.[776]

---

768 GRONAU 1784: 246–247, 251–252: In addition to his Fahrenheit and his Réaumur thermometer, GRONAU also had a Rosenthal thermometer that showed 861 °Rosenthal [−19.5 °C]. Furthermore, he found it odd that there had been another very severe winter exactly a century prior, in 1684. He feared that the proponents of the 100-year-calendar, which he did not regard very highly, would use this as an argument for its reliability.
769 Bayerische Akademie der Wissenschaften 1783: 48.
770 VERDENHALVEN 2011: 19–20, in Munich, one foot or shoe (*Fuß* was often used interchangeably with *Schuh*) was 29.186 centimeters, albeit since 1811. In 1783, the length might have been slightly different.
771 Bayerische Akademie der Wissenschaften 1784: 48, 69–70, 75–76.
772 PFISTER 1975: 84.
773 WOOD 1992.
774 Wiener Zeitung, 14 February 1784: 302; Provinzialnachrichten aus den Kaiserlich Königlichen Staaten und Erbländern, 18 February 1784: 211; Münchner Zeitung, 19 February 1784: 110.
775 These reports exist for northwestern Germany and also Bavaria. Bayerische Akademie der Wissenschaften 1783: 48; SANTEL 1997: 111.
776 PFISTER 1975: 84; BRÄKER 2010.

British climatologist Gordon MANLEY calculated that central England had an average temperature of 7.8 °C in 1783, which made it the coldest year of the 1780s. It remained the coldest year until 1814 when the average temperature was 7.7 °C.[777] Between 1770 and 1922, the winters of 1783/1784 (0 °C mean temperature) and 1784/1785 (0.3 °C mean temperature) were the second- and third-coldest winters on record in central England, behind 1794/1795 (−0.2 °C) and before 1788/1789 (1 °C).[778]

Sometimes the cold was so severe that taking notes with a pen was impossible. Scottish printer and publisher William STRAHAN wrote to Benjamin FRANKLIN from London on 1 February 1784: "The Frost is now so intense here this Morning, that there is no getting [obscured by inkblots: any Pen?] to write."[779] Gilbert WHITE noted on 26 December 1783 that the ground froze over. Two days later, he added that the "ground [was] so icy that people get frequent falls." On 31 December 1783, the temperature had been between −10 and −8.6 °C, resulting in "Ice [appearing] under people's beds" and "Water bottles burst[ing] in chambers."[780]

The cold at the end of 1783 in England was so intense that frost intruded into homes. In the new year, there was a thaw on 2 January 1784, which resulted in flooding in Selbourne until a hard frost returned on 6 January. In and around Selbourne, frost and snow lasted until April 1784.[781] The Reverend James WOODFORDE had similar encounters with snow and ice: the night between 28 and 29 January 1784 was, he remarks, "one of the coldest nights we have had yet." The frost continued after that, and the following days remained "bitter cold."[782] On 7 February 1784, WOODFORDE traveled to Norwich by horse-drawn carriage; on 8 February, he was due to preach at the cathedral there. Travel conditions, however, had not been ideal: it was freezing, and the snow on the road had been so deep that the "Tract of Wheels" could not be seen. "The Snow in some Places was almost up to the Horses Shoulders," which made the roads almost impassable. Fortunately, WOODFORDE eventually made it to Norwich, where the streets were "nothing but Ice, very dangerous walking about, very bad also for Carriages."[783]

---

777 MANLEY 1974: 393–398. The coldest year on average out of the series 1659–1973 was 1740 with 6.8 °C and the 1690s (1692: 7.7 °C, 1694: 7.7 °C, 1695: 7.2 °C, 1698: 7.7 °C).

778 G. T. WILLIAMS, Private Weather Diary for Greenwich and Somerset House, MET/2/1/2/3/161, National Meteorological Library and Archive, Met Office, Exeter, UK.

779 FRANKLIN 2014: 528.

780 Gilbert WHITE, "The Naturalist's Journal," 1783, Add MS 31848, British Library, London, UK: 154–155.

781 Gilbert WHITE, "The Naturalist's Journal," 1784, Add MS 31849, British Library, London, UK.

782 James WOODFORDE, Weather Diary for Weston Longville, Norfolk, vol. 10: 1782–1784, MET/2/1/2/3/467, National Meteorological Library and Archive, Met Office, Exeter, UK, 1784: 212.

783 James WOODFORDE, Weather Diary for Weston Longville, Norfolk, vol. 10: 1782–1784, MET/2/1/2/3/467, National Meteorological Library and Archive, Met Office, Exeter, UK, 1784: 212.

The winter of 1783/1784 seemed interminable: WOODFORDE noticed the frost creeping through his door as late as 1 April 1784. On 2 April, he stated, "I never knew so severe nor so long a Winter as this has turned out [. . .]. The Land has not been free from Snow since the 23rd of December last."[784] He was not the only one to make such observations, for Thomas BARKER mentions that "the Spring was very backward and frosty till mid-April, and frequent frost till the first week in May, then suddenly hot and remarkably fine, the grass and leaves remarkably green, and everything came on at a great rate."[785] Unfortunately, on 25 and 26 May 1784, a "sharp rime" came back and killed off the young shoots.[786]

Many European countries were affected by the frequent and heavy snowfall; some of the descriptions in newspapers were so over-the-top they almost seemed outlandish.[787] However, contemporary weather observers, modern climate historians, and scientists are in good agreement that the winter of 1783/1784 was indeed extreme. In the Low Countries, people could not travel by foot or horse, such was the severity of the weather.[788] In Paris, Benjamin FRANKLIN described the winter as unprecedented: "[. . .] such another [winter] in this Country is not remembered by any Man living. The Snow has lain thick upon the Ground ever since Christmas, and the Frost constant."[789] This snow lasted until mid-April 1784. The Seine froze over, which made transportation of firewood difficult, and a shortage ensued.[790] Vienna was also affected by a firewood shortage. The Danish Straits were completely frozen; it was possible to travel from Denmark to Sweden with sleds and carriages. Jutland was covered in as much as one meter of snow as late as mid-April. In early 1784, the Zuiderzee, a shallow bay northwest of the Dutch Republic, was frozen. The Dutch traveled on lakes and along the frozen coast with ice skates and sleds.[791] In Rotterdam, the ice was thick enough to support the "largest fair on ice ever known."[792] In Italy, blizzards and heavy snowfall affected Rome throughout February 1784. The frost was so bad that many lemon trees were damaged or destroyed.[793]

---

784 James WOODFORDE, Weather Diary for Weston Longville, Norfolk, vol. 10: 1782–1784, MET/2/1/2/3/467, National Meteorological Library and Archive, Met Office, Exeter, UK, 1784: 227; EDWARDS 1998, introduction.
785 Thomas BARKER, Private Weather Diary for Lyndon Hall, Leicestershire, MET/2/1/2/3/227, National Meteorological Library and Archive, Met Office, Exeter, UK: 242–243.
786 Thomas BARKER, Private Weather Diary for Lyndon Hall, Leicestershire, MET/2/1/2/3/227, National Meteorological Library and Archive, Met Office, Exeter, UK: 213.
787 Hamburger Adreß- und Comptoir-Nachrichten 1784 (quoted after GLASER 2008: 235).
788 DE POEDERLÉ 1784: 347–349.
789 FRANKLIN 2014: 558.
790 FRANKLIN 2017: liii, xlix; MIKHAIL 2017: 190–192.
791 SCHMIDT 2000: 259; THORDARSON 2005: 215. The Dutch had learned to utilize their frozen channels and lakes throughout the coldest decades of the Little Ice Age, DEGROOT 2018a: 109–151.
792 GRATTAN, BRAYSHAY, SADLER 1998: 32–33.
793 Gentleman's Magazine, March 1784: 221; GRATTAN, BRAYSHAY, SADLER 1998: 32–33; THORDARSON 2005: 215.

Ice Drift and Flooding in Europe

The extreme cold and heavy snow were not the only problems throughout the winter. Huge ice blocks piled up in the rivers, especially near bridges; in Mannheim, men and women went to the Neckar River to admire the "snow and ice colossuses." One spectator commented on their regret that the ice blocks could "not be kept as a memory aid of the greatest cold of the eighteenth century for the ensuing ages."[794] Another report from Heidelberg from 22 January 1784 describes the "monstrous icebergs" that came down the Neckar River and damaged houses, walls, and the bridge; it compares the calamitous situation to the "Calabrian convulsions [*calabrische Verwüstungen*]."[795]

In much of Europe, the winter of 1783/1784 was characterized by freezing temperatures, the accumulation of large amounts of snow, and severe flooding when a sudden thaw occurred. Rudolf Brázdil and his colleagues have identified three flooding phases in central and western Europe: late December 1783, late February/early March 1784, and late March/early April 1784. The first phase mainly affected Britain, Ireland, France, the Low Countries, and historical Hungary. The second had a more substantial impact and affected France, central Europe, and the Danube River in southeast central Europe. The third phase affected mainly historical Hungary.[796]

For the German Territories, the second phase was the most devastating. German climate scientist Rüdiger Glaser determined that a blocking high-pressure system above eastern Europe resulted in a southerly current of warm air toward central Europe. Around 23 February 1784, temperatures in western Europe suddenly increased above freezing, which led to thaw, the melting of snow, and a sudden increase in water levels in the rivers.[797] The clash of the warm Mediterranean air and the cold air from the north led to severe rain, which added to the swelling.[798] It was not only high water that affected populations near these rivers: the influx of warm air also led to

---

794 Münchner Zeitung, 23 January 1784: 49: Report from Mannheim, no date. "Täglich wird diese Schnee- und Eiskolosse von sehr vielen hiesigen Inwohnern beiderlei Geschlechts besuchet, und bewundert. Schade, daß sie der Nachwelt nicht als ein Gedächtnisstük der grösten Kälte des 18. Jahrhunderts aufbewahret werden kann."

795 Königlich Privilegirte Zeitung, 7 February 1784: 107: Report from Heidelberg, 22 January 1784. "Es ist fast nicht zu viel gesagt, wenn man unser Unglück eine calabrische Verwüstung nennt. Mauern ohne Dach, ungeheure Eisberge, halb eingerissene Häuser, Schiffe zu oberst an die abgeworfene Dachung hingezwängt, Gebäude ohne Thüren und Fenster, halb zwischen dem Eis hervorstehendes erstarrtes Vieh, die weltberühmte gedekte Brücke ohne Seitenwände, welche man selbst hat abnehmen müssen, um eine noch höhere Wasserschwellung abzuwenden."

796 Brázdil et al. 2010. Several of the accounts of the different regions that had been affected by the ice drifts and floods can be found here: Weikinn 2000.

797 Glaser 2008: 235.

798 In Vienna, for instance, the temperatures rose from freezing to 10 °C between 25 and 26 February; Brázdil et al. 2010: 177.

the ice breaking up around 27 and 28 February 1784.[799] Rivers cascaded through the landscape and carried with them huge masses of debris, including ice, uprooted trees, parts of damaged mills, and other masonry.[800] Peak flooding occurred over the three days between 28 February and 1 March 1784. In many towns, people lost their homes, and some lost their lives. Afterward, homelessness, hunger, and petty theft became serious problems in many municipalities.[801]

The 2 March 1784 issue of the *Königlich Privilegirte Zeitung* was the first newspaper to feature reports from Hamburg and Vienna about sudden thawing. Many more reports about flooding (*Sündfluth*), ice floes, and destruction were to follow in the March editions of almost all other newspapers. Slowly, different stories poured in from all over the German Territories and beyond, describing different stages of the flooding event.[802]

Cologne, located on the Rhine River, was famously affected by ice drifts and severe flooding.[803] Over the winter of 1783/1784, the ice on the Rhine River was thick enough to walk on for 47 days. On 27 February 1784, at 5 a.m., the ice broke with a rumble; by 7 a.m., half the city was flooded (Figure 48). The extent of the flooding must have been genuinely terrifying; Cologne was at the mercy of an "ocean of ice."[804] An anonymous author, who published a book about *Das Arme Köln* ("The poor Cologne") in 1784, wrote that their curiosity and fear made them climb one of Cologne's highest towers:

---

799 SCHMIDT 2000: 258–263; MUNZAR, ELLEDER, DEUTSCH 2005: 8–24; BRÁZDIL et al. 2010: 173–178; SARTOR 2010: 73–76.

800 SCHMIDT 2000: 258–263; GLASER 2008: 236.

801 Anonymous 1784: 79–84; WEICHSELGARTNER 2001: 107–109; DEGROOT et al. 2021: 546.

802 Königlich Privilegirte Zeitung, 2 March 1784: Report from Hamburg, 24 February, and letter from Vienna, 15 February 1784. The term *Sündfluth* was used by Königlich Privilegirte Zeitung, 11 March 1784: Report from Nuremberg, 20 February 1784. The Berlinische Nachrichten featured a report about "a kind of *Sündfluth*" on 11 March 1784: Report from Nuremberg, 29 February 1784. The term *Sündfluth* refers to a large and severe flood that affected large parts of a country. The first syllable either refers to the German word *Sünde* (sin), i.e., human sinfulness caused the flooding, or it derives from the word *Sund* (sound, as in Fehmarn Sound) meaning the word *Sündfluth* simply means a flood of water; ADELUNG 1793–1804, 504. All newspaper issues of the Königlich Privilegirte Zeitung mention reports of flooding across the German Territories: Cochem, Bamberg, Cologne, Koblenz, Frankfurt am Main, etc., but also mentioned reports of flooding affecting Maastricht, Paris, or Warsaw. In the Berlinische Nachrichten, the news about the "endless icefield[s]" that have "piled up like rocks" began on 9 March 1784; in the Hamburgischer Unpartheyischer Correspondent, the reports began on 6 March 1784, with news of "such a high flood" in Lüneburg from 3 March 1784.

803 THELEN 1784; SPATA 2017; DEGROOT et al. 2021.

804 Anonymous 1784: 40. "Wie saßen itzt wie mitten in dem Meere, rund um her nur Wasser und Eis."

Figure 48: Floodmarks at the portal of St. Maria in Lyskirchen, Cologne. This church shows two floodmarks, one above the door, marked by a line, and one to the right of the entrance with a line indicating the height of the flood. The water reached approximately 3.50 meters above street level.

> God! Which spectacle did I witness with my own eyes? I saw an entire ocean and mountains of ice floes between Cologne and [the nearby town of] Bensberg; there was nothing but air and water. Deutz, Mülheim, Rodenkirchen, and other towns near the river, were not visible apart from the tips of their towers and the roofs of their houses.[805]

In Cologne, the floodwaters reached 33 centimeters above the peak height of the St. Mary Magdalene's flood of 1342, which had affected half of central Europe.[806]

In Cologne and elsewhere, the aftermath of the flood was characterized by the cold and a lack of firewood and food. In the eighteenth century, insurances against

---

805 Anonymous 1784: 6. "Gott! Welch ein Schauspiel öfnete sich hier meinen Augen? Ich sah ein ganzes, ganzes Meer, und Berge von Eisschollen, zwischen Köln und Bensberg schier nichts als Luft und Wasser. Deuz, Mülheim, Rodenkirchen, und die dem Strome nächst gelegenen Oerter ließen nichts von sich sehen als die Spitzen der Thürmen und die Dächer der Häuser."
806 SPATA (2017: 10) gives 13.3 meters as the height of the 1342 flood in Cologne and 13.63 meters as the height of the water in 1784. For more information on the St. Mary Magdalene's flood, see: BAUCH 2014; BAUCH 2019; BAUCH, SCHENK 2020: 2–3.

flooding were uncommon; instead, losses were mitigated by working harder. The Church played an important role by raising money for flood victims, distributing goods, organizing processions, and bringing people together to pray.[807] Private individuals, wealthy and poor alike, made charitable efforts: flood victims were housed and sheltered, cooked for, and given clothes, shoes, and blankets. The local governing body did not actively prevent flooding; its actions were primarily reactive. The authorities handed out food and firewood and issued orders that flooded houses had to be scrubbed and disinfected; some severely damaged dwellings were demolished. They also offered direct financial aid, either as a handout or tax exemption.[808]

GLASER regards the flooding event of 1784 as one of the worst environmental catastrophes in central Europe during the early modern period. Many floodmarks from 1784 still exist in various cities along several rivers in Germany; these high-water marks help historians reconstruct the flooding event.[809] In some cities, the floodmarks indicate that this was the highest flooding ever recorded, while in others, it was second to the aforementioned St. Mary Magdalene's flood of 1342.[810] Many artists immortalized the flooding, ice drifts, and their aftermath in various European cities: below is one such example.[811]

The painting by an unknown artist shown in Figure 49 was made for a zograscope in Augsburg and would have been displayed at fairs or entertainment parlors. One could inspect it closely by looking through a "peep hole" into a "peep box," which would have been lit up from behind by a candle. This technique was very popular in Europe from the 1740s onward. Often these paintings were exaggerated for entertainment's sake.[812]

In Würzburg, the ice started to break up on 27 February 1784. Large ice floes and floating tree trunks (*Holländerbäume*) caused much damage. On 28 February, the Main River had significantly risen and continued to do so until 29 February at around 4 a.m. At this point, it reached the entrance arch of the building that today houses Würzburg's town hall, where a floodmark remains.[813]

---

807 WEICHSELGARTNER 2001: 97, 107–113.

808 Anonymous 1784: 79–83; WEICHSELGARTNER 2001: 98–99; BRÁZDIL 2010: 179–180; DEGROOT et al. 2021.

809 GLASER 2008: 233–236.

810 GLASER 2008: 238; BRÁZDIL et al. 2010: 185; BÖRNGEN 2011: 119; BRÁZDIL et al. 2012: 147–148.

811 Famous examples are a series of paintings by Ferdinand KOBELL that depicted the consequences of the ice drift and flooding along the Neckar River in Heidelberg; Friedrich RÜBNER painted the collapse of a bridge and damage to houses close to the Regnitz River in Bamberg; Franz ERBAN produced a copperplate engraving of the flood in Prague; DEURER 1784; FRICKE 1988; GLASER 2008: 237. In Prague, the flooding of the Vltava/Moldau River resulted in several casualties, buildings were dislodged, bridges were carried away, mills had become unusable, and gardens, fields, and trees were buried under mud; BRÁZDIL, VALÁŠEK, MACKOVÁ 2003: 313–314; MUNZAR, ELLEDER, DEUTSCH 2005: 8–17; GLASER 2008: 237.

812 This image is also featured in GLASER, HAGEDORN 1990: 11. The image is mirrored in their publication; more info on *laterna magica*; FAULSTICH 2006: 30.

813 GLASER 2008: 236–237.

Figure 49: Contemporary depiction of flooding in Würzburg in 1784.

Figure 49 depicts Würzburg on the Main River and shows distressed residents climbing onto roofs and boats to survive. Debris is flowing on the river. Cannons are being fired from the Marienberg Fortress, aimed at the ice, in a desperate attempt to break it up to prevent further damage.[814] Several residents can be seen on the bridge, equipped with long sticks; they are trying to push the ice and debris through the now-submerged bridge opening. Würzburg experienced no loss of life during this flooding event, but the financial losses were devastating; damage to infrastructure took years to repair.[815]

Many other riverine cities were affected by the flooding. The list is too lengthy to name them all. In Austria, Vienna was affected by floods that began on 26/27 February 1784 and lasted until 10 March.[816] Further along the Danube River, Regensburg was also affected. On 27 February 1784, the Danube River rose by three feet [94 centimeters].

---

814 DEMARÉE 2006: 895. Ice jams formed at obstacles, such as bridges, where a lot of debris gathered. This led to very high water levels that produced floods and caused dikes to burst. Gaston DEMARÉE regards ice jams as the main reason for flooding along the Rhine River and the Meuse River.
815 GLASER, HAGEDORN 1990: 1–14; SCHOTT 2004, vol. 2: 37–39.
816 STRÖMMER 2003: 209–211; ROHR 2020: 204–207.

This was followed by an ice drift and a further rise of nine feet [282 centimeters].[817] In Munich, the temperatures rose on 22 February and the thawing began.[818] The large amounts of snow melted quickly and "the saddest inundations followed."[819] France and the Low Countries were also affected by ice drifts and flooding in late February 1784. In Paris, ice on the Seine River broke up and began to melt on 26 and 27 February flooding low-lying parts of the city.[820]

Northern Germany suffered the consequences of the rapid thaw, too: heavy winter snow blanketed county Bentheim until 22 February 1784, when temperatures rose, resulting in flooding at the Vechte River. Local records state that nobody could remember such a terrible flooding event. In February and March 1784, more flooding followed at many other northern German rivers.[821] Between 7 and 10 March 1784, the Elbe River flooded the area around Dannenberg. Strong winds pushed large blocks of ice upstream, which damaged bridges.[822] By 22 March 1784, the Elbe River was navigable again, and Hamburg was able to receive ships from London and Lisbon.[823] After the flooding had occurred, the cold returned once more. It brought snow with it. Then, in May 1784, it slowly became warmer.[824]

Although the severe cold and flooding cost lives, the mortality rate of the late summer of 1783 remained larger than that of the severe winter of 1783/1784.[825] Nevertheless, the February–March 1784 flooding in much of central Europe was one of "the most disastrous events during the past millennium."[826] Like the events of the summer, some interpreted the floods as divine intervention, while others searched for rational explanations.[827]

The Winter in North America

The winter in North America had been one of the harshest on record.[828] In Okkak, one of the Moravian settlements in Labrador, it started snowing on 21 July 1783. More snow

---

817 König 1784: 52–54; Societas Meteorologica Palatina 1784: 53–54. According to Verdenhalven (2011: 19–20), in Regensburg, one foot was 31.374 centimeters before 1811.

818 Bayerische Akademie der Wissenschaften 1784: 69–70, 75–76.

819 Bayerische Akademie der Wissenschaften 1784: 48. "[. . .] denn es fieng an aufzuthauen, der Schnee schmolz, und die traurigsten Ueberschwemmungen folgten nach."

820 Franklin 2017: 294. See also McCloy 1941: 7–12; Demarée 2006; Brázdil et al. 2010.

821 Hamm 1976: 118–119; Santel 1997: 111; Poliwoda 2007: 59–84.

822 Puffahrt 2008: 20.

823 Königlich Privilegirte Zeitung, 27 March 1784: Report from Hamburg, 22 March 1784.

824 Strömmer 2003: 210–213.

825 Brázdil et al. 2010: 182.

826 Brázdil et al. 2010: 185.

827 Glaser 2008: 236.

828 Ludlum 1966.

followed in August, with "thick snow" on 21 September and throughout October 1783. On 26 November 1783, the bay near Okkak froze over. It snowed as late as 15 May 1784 and again on 8 June. On 3 July 1784, one observer wrote: "Ice on the ponds not broke[n] yet, [. . .] never so late before." Later in July 1784, the bay was "covered over with new ice, ice comes in again from [the] sea." On 2 August 1784, one weather observer summarized the preceding months thusly: "The winter was not very cold we had not much snow till new year but from new year till May I never seen so much snow and there is very much yet and very much Ice all over the sea as far as the eye can see."[829]

Further south, in the eastern United States, the winter of 1783/1784 was perhaps the coldest of the past 250 years; the temperatures were 4.8 °C below the 225-year mean.[830] Records from this winter reveal that New England saw the most prolonged period of temperatures below 0 °C in its history.[831] In early December 1783, the Schuylkill River in Philadelphia froze over.[832] The Delaware River, which also runs through Philadelphia and flows into the Atlantic, was frozen solid from December 1783. The icy conditions continued.[833] In fact, all the rivers in Pennsylvania were frozen in the winter "so as to bear wagons and sleds with immense weights." It was so cold that the thermometers "stood several times at 5 degrees below 0" [−5 °F = −20.5 °C].[834]

In Trenton, ice on the Delaware River "had frozen to an amazing thickness." However, in late January 1784, heavy rains broke up the ice, at least for a few days.[835] Richard BACHE called the winter "remarkable severe & tedious" in a letter to Benjamin FRANKLIN, his father-in-law, on 7 March 1784. He was "looking impatiently for the approach of Spring." He told of vessels in the bay off Philadelphia that had been stuck there for ten weeks, unable to travel to the city because of the ice, which had stagnated business.[836] On 18 March 1784, after three months of ice, the Delaware River opened at Philadelphia, and vessels were free to resume travel. "Such has been the severity of the weather and continued frost, that the inhabitants of Philadelphia have not experienced

---

829 Meteorological Observations at Okkak and at Hoffenthal, Hudson's Bay Company, Labrador, Newfoundland, 1782–1786, MA/144, Archives of the Royal Society, London, UK.
830 LUDLUM 1966; LUDLUM 1968; THORDARSON 2005: 215; D'ARRIGO et al. 2011: L05706; MIKHAIL 2017: 190–192. Apart from 1783/1784, the other severe winters occurred in the eastern United States in 1740/1741 and 1779/1780.
831 LUDLUM 1966: 64; WOOD 1992: 64–65.
832 Pennsylvania Packet, 18 March 1784: 2.
833 The Providence Gazette and Country Journal, 17 January 1784: 3: Report from Philadelphia. "For some Days past our Harbour has been shut up by Ice."
834 The American Museum, or Universal Magazine 1789: 253; this report had previously been published in the Columbian Magazine, November 1786.
835 United States Chronicle, 26 February 1784: 2: Report from Trenton, 27 January 1784.
836 FRANKLIN 2017: 21, letter from Richard BACHE to FRANKLIN, Philadelphia, 7 March 1784.

a winter since the year 1750–1751, that has been more intensive cold, disagreeable and distressing."[837] Around the same time, the frozen Schuylkill River suddenly broke up, and flooding followed. The river carried "large bodies of ice," flooded houses, drowned livestock, and forced people to retire to their second floor.[838] "Retiring" was a euphemism used in the United States and Europe when one had to evacuate the ground or lower floors because of flooding.[839]

In New Haven, by mid-February 1784, the weather had been so cold that the western part of the Long Island Sound was frozen "for several miles – at White-Stone, to such a degree, that people pass to and from Long Island on the Ice."[840] In Boston, the harbor froze over on the last days of 1783, a freeze that lasted for at least three weeks. Although the frozen rivers and harbors hindered the shipping industry and commerce, they did provide certain advantages.[841] Frozen waterways enabled easy transport across rivers and lakes. However, this was far from a safe endeavor: sometimes, the ice broke, and people and animals were lost to the frigid waters. Similar fates befell those trying to evacuate stranded ships.[842]

As late as March 1784, American newspapers reported that the Chesapeake Bay was covered in ice. Harbors and channels in the region were closed for boats for the longest time in the region's history.[843] In February 1784, the winter "produced ice in the harbor of Charlestown [Charleston] strong enough for skating on, which is very uncommon here."[844]

When FRANKLIN received updates about the winter in North America, particularly the status of the ratification of the Treaty of Paris, he remarked: "The Winter it seems has been as severe in America as in Europe, and has hindered the Meeting of a full Congress [. . .],"[845] and so, the necessary number of delegates to ratify the Treaty of

---

837 Norwich Packet, 1 April 1784: 3. A very similar report can be found in the Massachusetts Spy, 8 April 1784: Report from New York, 25 March 1784.

838 Pennsylvania Packet, 18 March 1784: 2.

839 GLASER 2008: 236. GLASER mentioned that Ludwig VAN BEETHOVEN, at the time 12 years old, was a famous witness to the ice floods. Together with his family, who were staying at a friend's house in Bonn, he had to "retire" to the first floor when the floods came.

840 Norwich Packet, 26 February 1784: 3–4: Report from New-Haven, 17 February 1784.

841 American Herald and the General Advertiser, 19 January 1784: 3.

842 Spooner's Vermont Journal, 25 February 1784: 3: Report from Hartford, 10 February 1784; Vermont Gazette, 14 February 1784: 2: Report from New-London, 29 January 1784; United States Chronicle, 26 February 1784: 2: Report from Philadelphia, 7 February 1784.

843 LUDLUM 1966: 32–35, 64–68; FRANKLIN 2014: 560–563, founding Father Robert MORRIS wrote FRANKLIN a letter from the Office of Finance in the US on 12 February 1784. He expressed his hope that the "intemperate" season would end soon and ships would be able to depart Chesapeake Bay by early April.

844 New-York Packet, 18 March 1784: 3: Report from Philadelphia, 10 March 1784.

845 FRANKLIN 2017: 46, letter from Benjamin FRANKLIN to David HARTLEY, written at Passy, 11 March 1784.

Paris. Ice in the harbors and adverse weather conditions further delayed the transport of the signed Treaty of Paris from the United States to Europe.[846] On 30 March 1784, FRANKLIN informed John JAY by letter that the courier, Josiah HARMER, had reached him and handed over the ratified treaty. Ice in the New York harbor had stopped HARMER's ship from setting sail for about a month, during which time, some passengers aboard had frozen to death. HARMER was the one to inform FRANKLIN about the "uncommonly severe" winter in America.[847]

Once spring came around, the ice and snow began to melt. Now, news started to pour in from further afield. The severity of the winter had been underestimated; in mid-April 1784, news reached the northeast that ice (or ice floes) had gone as far south as Jamaica.[848] Jamaica is in the tropics, so ice is rare, although snow is not unknown on the elevated parts of the island.[849]

In mid-May 1784, news also reached the United States that on 28 February 1784, "a small flight of snow fell" in Bermuda, and the temperature "was lower than ever it was known before."[850] Bermuda is north of the Caribbean and benefits from the warm waters of the Gulf Stream, which usually ensures its islands remain snow- and ice-free.[851] However, in the winter of 1783/1784, there was at least some snow in late February.

In mid-May 1784, news reached the eastern seaboard that "the Mississippi has been fast bound up by ice."[852] The river had frozen over at New Orleans between 13 and 19 February 1784. When the ice broke, ice blocks were seen about 100 kilometers south of New Orleans in the Gulf of Mexico.[853] The reports indicate that even in 1784, this was regarded as something extraordinary.[854]

---

846 SMITH 1963: 420–430; FRANKLIN 2017: xlix.
847 FRANKLIN 2017: 89, Benjamin FRANKLIN in a letter to John JAY, written at Passy, 30 March 1784. FRANKLIN 2017: 105, letter by Benjamin FRANKLIN and John JAY to David HARTLEY, 31 March 1784.
848 Massachusetts Spy, 14 April 1784: 3.
849 CHENOWETH 2003: 77.
850 Massachusetts Spy, 13 May 1784: 3: Report from Bermuda, 1 March 1784. The temperature measured outside was 44 °F, which is 6.6 °C.
851 FORBES, "Bermuda's Climate, Weather & Hurricane Conditions."
852 Massachusetts Spy, 13 May 1784: 3: Extract from a Letter from St. John's, East-Florida, 14 February 1784. Similar accounts were also published in other newspapers, such as the Salem Gazette, 18 May 1784; South-Carolina Weekly Gazette, 26 May1784; and Spooner's Vermont Journal, 2 June 1784.
853 LUDLUM 1966: 154; WOOD 1992: 66; THORDARSON 2005: 215; MIKHAIL 2017: 190–192.
854 LUDLUM 1966: 222–223. David M. LUDLUM mentions that Detroit also experienced one of the severest winters of the eighteenth century in 1784.

Searching for a Connection

The winter of 1783/1784 was indeed extreme on both sides of the Atlantic, and those who experienced it first-hand were aware of this fact. Just as in the summer, weather observers and newspaper correspondents searched for precedents. In some but not all cases, they found previous weather events that were comparable. Did contemporaries consider this most severe of winters and this most unusual of summers as part of the same phenomenon?

There was certainly some speculation about the matter. In Austria, an anonymous clerk believed "this special Earth vapor [*Erddunst*], which had caused the extraordinary heat and the many thunderstorms, also had the potential to cause the extraordinary amounts of snow and cold during this winter." And indeed, during the winter, there were further reports of haze.[855] The *Münchner Zeitung* reported that the haze had returned for just a few days, appearing during the snowfall, just as it had during the rain of the summer.[856]

Karl Ludwig GRONAU found records that indicated a fog was present in the same year as an intense cold in 1739/1740 and 1775/1776, just as in 1783/1784. According to the documents available to him, the winter of 1739/1740 was the coldest of the eighteenth century, and the temperature during the winter of 1775/1776 reached −8 °F [−22.2 °C]. Ultimately, he left it up to his readers to speculate whether the fog of 1783 had something to do with the cold of 1784.[857]

In his letter to Manchester in 1785, Benjamin FRANKLIN also postulated a connection between the dry fog and the cold of the winter. The dry fog had:

> rendered [the sun's rays] so faint in passing through [to the Earth]. [. . .] Of course, their summer effect in heating the earth was exceedingly diminished. Hence the surface was early frozen. Hence the first snows remained on it unmelted and received continual additions. Hence the air was more chilled, and the winds more severely cold. Hence perhaps the winter of 1783/1784 was more severe than any that had happened for many years.[858]

---

855 StAKI, Karton 220, Nr. 41 NR, fol. 266v (quoted after STRÖMMER 2003: 208). "Der in dem vorigen Sommer so merkwürdige Nebl, ließ sich im heurigen Winter [1784], nur wenige Tage ausgenommen, deutlich bemerken. Mitten unter den Schneeflocken erkannte man ihn meist, so wie er in dem vorigen Sommer stets mitten im Regen sich gezeigt hatte. Es ist hieraus billig zu schließen, daß dieser besondere Erddunst, so wie er im vorigen Sommer die ausserordentliche Hitze, und die verheerende Gewitter verursachet, auch diesen Winter den Stoff zu den ausserordentlichen Schnee, und Kälte verursachet haben mag."

856 Münchner Zeitung, 9 February 1784: 86–87.

857 GRONAU 1784: 252–253.

858 FRANKLIN 1785: 359–361.

FRANKLIN suggested that further research on previous harsh winters after unusually long-lasting summer fogs should be conducted.[859] Given the historical precedents of unusual hazes preceding cold winters, FRANKLIN's argument seemed very plausible.

Outlook

Cold Temperatures Continued

In Europe, another punishing winter followed in 1784/1785. The summer of 1785 was cool and dry in Europe and the United States.[860] According to current scholarship, the Laki eruption is believed to have impacted the weather over much of the Northern Hemisphere for up to three years.[861]

In England, the spring of 1784 came late, and then it was "wet & cold till Mid-July." The harvest came in late too, but it was "plenty." The winter of 1784/1785 was "severe" and brought with it heavy snow.[862] On 8 December 1784, Gilbert WHITE called the weather "Siberian." The day after, the snow on WHITE's property was 30 to 40 centimeters deep. One of his neighbors, Thomas HOAR, lost 41 sheep to the heavy snowfall. In his diary, WHITE concludes, "No such snow since January 1776."[863] On the following day, WHITE writes, "Extreme frost!!! Temperature below zero!"[864] The snow had fallen uninterrupted for 24 hours. Food items, such as bread, cheese, meat, potatoes, and apples, froze if they "were not secured in cellars under ground."[865] WHITE was curious about the true temperature outside, so he experimented with two different thermometers: "We hung out two thermometers, one made by Dollond & one by B. Martin: the latter was graduated only to 4 below ten, or 6 degrees short of zero [6 °F=−14.4 °C]: so that when the cold became intense, & our remarks interesting, the mercury went all into the ball & the instrument was of no service."[866] WHITE was frustrated with his instruments; both were inadequate and he was unable to assess the exact temperature outside.

The cold weather continued into the spring; on 2 March 1785, WHITE claimed that the ground became "as hard as iron." On 14 March, a "fierce frost" crept through doors and windows and into the bedrooms of all and sundry. On 23 March, "colds & coughs [were]

---

859 FRANKLIN 1785: 359–361.
860 KINGTON 1980: 30; WOOD 1992: 68. The period from August 1784 to July 1785 was the driest consecutive 12-month period since the rainfall series began in 1727 for England and Wales (at least until 1980 when this study was published).
861 THORDARSON, SELF 1993; THERRELL 2005: 203–207; THORDARSON 2005: 215.
862 Meteorological Observations at Dundee, Scotland, 1782–1801, "General Results from a Meteorological Journal Kept at Crescent near Dundee," MA/45, Archives of the Royal Society, London, UK.
863 Gilbert WHITE, "The Naturalist's Journal," 1784, Add MS 31849, British Library, London, UK.
864 If the temperature was −1 °F, that means it reached a low of −18.3 °C.
865 Gilbert WHITE, "The Naturalist's Journal," 1784, Add MS 31849, British Library, London, UK.
866 Gilbert WHITE, "The Naturalist's Journal," 1784, Add MS 31849, British Library, London, UK.

frequent."[867] Correlating well with these observations, J. ROTHWELL called March 1785 the coldest month on record for Greater Central England.[868] Additionally, WHITE noticed the last frost of the season as late as 22 May.[869] The spring of 1785 was the coldest in Europe between 1500 and 2004.[870]

In the German Territories, it seemed that the cold of early 1785, March in particular, exceeded that of the winter of 1783/1784. A report in the *Münchner Zeitung* was adamant that March 1785 was by all measures a more challenging month than the March of the previous year. It goes so far as to claim that it outdid the coldest years on record.[871] The *Augsburgisches Extra-Blatt* made similar claims and remarked, in a way typical of the time, that this was the worst March in living memory.[872] In Schnürpflingen, in southern Germany, the snow came in late October 1784 and lasted until spring 1785; at Easter [27 March], the town's people were still trudging through snow as swallows froze to death in the skies above them.[873] According to the *Ephemerides* of the Societas Meteorologica Palatina, in Sagan, Silesia, 1785 was the coldest year of the observation period between 1781 and 1792. The average air temperature in 1785 was 6.3 °C, compared to 7.9 °C for that 12-year period. This cold spell drew to a close after 1786.[874]

In Austria, temperatures plummeted in November 1784, marking the beginning of a bitterly cold spell which lasted at least until April 1785. The winter of 1784/1785 was even colder than the previous year and was followed by flooding and ice drift events. The winter of 1785/1786 was cold again and succeeded by yet more flooding events along the Danube River, among others. It was remarked that 1784 through 1786 had produced poor-quality wines.[875] In the Czech Lands, the slightly warmer summer of 1783 was followed by unseasonably cold temperatures until the end of 1786; the two winters following the eruption were particularly cold. Spring 1785 was the coldest on record, according to data at the Prague-Klementinum observatory, whose records date back to 1775.[876] The years that followed 1783 were much the same in the Czech Lands as they were in the rest of Europe, with significant cooling and the frequent flooding that usually follows.[877]

---

867 Gilbert WHITE, "The Naturalist's Journal," 1785, Add MS 31849, British Library, London, UK.
868 Anonymous, Private Weather Diary, "Greater Central England" Climatological series 56BC–AD2010 Archive, MET/2/1/2/3/545, National Meteorological Library and Archive, Met Office, Exeter, UK.
869 Gilbert WHITE, "The Naturalist's Journal," 1785, Add MS 31849, British Library, London, UK.
870 XOPLAKI 2005: L15713, Figure 1; PÍSEK, BRÁZDIL 2006.
871 Münchner Zeitung, 4 March 1784: 149.
872 Augsburgisches Extra-Blatt, 10 March 1785: Report from Frankfurt, 5 March 1785.
873 Gemeindechronik Schnürpflingen, Alb-Donau-Kreis, schnuerpflingen.de.
874 PRZYBYLAK et al. 2014: 2408, 2414.
875 STRÖMMER 2003: 213–223.
876 BRÁZDIL, VALÁŠEK, MACKOVÁ 2003; PÍSEK, BRÁZDIL 2006; BRÁZDIL et al. 2017: 150.
877 PÍSEK, BRÁZDIL 2006; BRÁZDIL et al. 2017: 160.

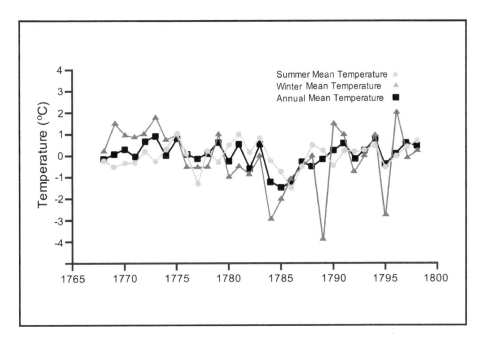

Figure 50: Mean surface temperatures for the late eighteenth century. This temperature reconstruction is based on data from Europe and the northeastern United States analyzed over a 31-year period, 1768 to 1798.

As this graph (Figure 50) reveals, the annual mean temperature dropped after the eruption. While 1783 was overall rather warm, 1784 was significantly colder. In fact, 1784 had been 1 °C colder than the 31-year mean. 1785 was the coldest year of the period, with 1786 being about as cold as 1784. Only after 1786 did the weather become warmer again. The summer of 1788 was almost the same temperature as that of 1783. The weather took another downturn between 1788 and 1790: the winter of 1788/1789 was long and severe. The Thames was covered in a sheet of ice so thick that Londoners were able to hold a frost fair on it, the first since the winter of 1739/1740.[878]

Some scholars argue that the Laki eruption "helped trigger" the French Revolution in 1789.[879] This idea, however, has been debunked by French climate historian Emmanuel LE ROY LADURIE, who argues that the harvests in France had been good up until 1787 (inclusive). In 1788, a modest yield was followed by subsistence riots and social unrest that lasted well into the summer of 1789. LE ROY LADURIE suggests any perceived connection between the Laki eruption and the French revolution is simply

---

878 KINGTON 1980: 32.
879 NEALE 2010; GARNIER 2011: 1053–1054; SPENCE 2014.

an "old Anglo-Saxon 'historiographic legend.'"[880] David MᴄCᴀʟʟᴀᴍ also called it an "overly simplistic historical assumption."[881] There is no doubt that the Laki eruption impacted various regions differently. Some were plunged into famine, some endured pollution and cold, and many were rife with fear and confusion. In most places, hardship was not evenly divided between the rungs of social standing. These regions did not exist in a vacuum; the Laki eruption, in many instances, only stirred up preexisting conditions and sentiments.[882]

Europe endured several extreme weather events in the wake of the Laki eruption. The most acute and immediate was a strange polluting fog that persisted for months. Thereafter, a stiflingly hot summer and several colder-than-usual seasons tormented the continent further. Excess mortality ensued in the late summer and autumn of 1783 and the winter and spring of 1784. The fine particulates of the haze that plagued Europe, which took their toll over months, coupled with bouts of dysentery or fever, most likely contributed to this. It seems that outside of Iceland, the Laki eruption did not trigger a societal or agricultural crisis. These findings are in good agreement with the research of Rudolf Bʀázᴅɪʟ and his team into the impact of the Laki eruption on the Czech Lands, which shows reasonably steady agricultural output and little change in grain prices. The Laki eruption seems not to have caused a socio-economic crisis. Indeed, the social impact is described as "negligible," especially in comparison to the Tambora eruption, which (as mentioned) caused a "year without a summer" in 1816. A profusion of misfortunes followed Tambora, including a poor grain harvest, an increase in grain prices, and a lack of food, followed by widespread hunger, increased crime, and an overall societal crisis.[883]

Natural scientists are still debating whether the Laki eruption triggered the cold period between 1783 and 1786 or whether it was caused by natural climatic variability.[884] The Laki eruption occurred during the Little Ice Age (1250/1300–1850) and, more specifically, the Dalton Minimum (1760–1850).[885] During this overall colder period, the weather became even more variable. Some scientists argue that El Niño-Southern Oscillation played a role: estimates of the exact dates differ, but Vinita Dᴀᴍᴏᴅᴀʀᴀɴ and her colleagues argue in *The Palgrave Handbook for Climate History* that there

---

880 Gᴀʀɴɪᴇʀ 2011, commentary by Emmanuel Lᴇ Rᴏʏ Lᴀᴅᴜʀɪᴇ, translated from French. Lᴇ Rᴏʏ Lᴀᴅᴜʀɪᴇ (2006: 77–180) documented the climate throughout the 1780s for France. Here he also discusses the seven seasons between 1787 and 1789 that he believed to have "triggered" the French Revolution; Mᴄᴄᴀʟʟᴀᴍ 2019: 218, note 77.

881 Mᴄᴄᴀʟʟᴀᴍ 2019: 199.

882 Mᴄᴄᴀʟʟᴀᴍ 2019: 218.

883 Bʀázᴅɪʟ et al. 2017: 159–160.

884 For this debate, see D'Aʀʀɪɢᴏ et al. 2011; Pᴀᴜsᴀᴛᴀ et al. 2015; Pᴀᴜsᴀᴛᴀ et al. 2016; Zᴀᴍʙʀɪ et al. 2019b: 6787; Dᴀᴡsᴏɴ, Kɪʀᴋʙʀɪᴅᴇ, Cᴏʟᴇ 2021.

885 Wᴀɢɴᴇʀ, Zᴏʀɪᴛᴀ 2005.

were prolonged El Niño and La Niña phases from 1782 to 1784 and from 1785 to 1790, respectively.[886] Climate scientists of the future are charged with the further exploration of the climate forcing associated with the Laki eruption.

Annus Arcanus

In 1784, Jean SENEBIER (1742–1809), a Swiss naturalist and librarian in Geneva, detailed how some believed the cause of the dry fog to be the earthquakes in Calabria, others, the new volcano that formed on 8 June 1783 in Iceland.[887] These two theories were deemed possible for decades. Over time, the suggestion that the vapors had come from the north – owing to the prevailing wind directions at the time – grew in popularity.

> [. . .] An eruption was taking place at the time in Iceland, and there can be no doubt of the volcanic smoke having affected the atmosphere, of the connection where the fog prevailed. The mountain of Skaptefeld vomited its columns of fire furiously during the period the fog lasted, and the wind blew chiefly from the Northwest. I traced the fall of ashes from the Orkneys through the Faroe Islands to Jutland. & some fine dust was noticed to have fallen in Germany.[888]

John Thomas STANLEY speculated that the dry fog he had witnessed in Switzerland in the summer of 1783 was caused by a volcanic eruption in Iceland, even going as far as placing the eruption in the "Skaptefeld" region. STANLEY wrote these memoirs after the French Revolution (possibly in the 1790s); therefore, it is unsurprising that he was aware that the dry fog was a Europe-wide phenomenon.[889]

On 28 March 1787, a Bavarian naturalist and member of the Electoral Academy of Science, Franz Xaver EPP (1733–1789), gave a lecture about the "so-called vapor, which did not only appear in Bavaria but in all of Europe in 1783." EPP stated that in Bavaria, the dry fog lasted from mid-June to the end of August 1783. He then detailed the various theories on the origin of the fog that were circulating, illustrating that even in 1787, there was no consensus among naturalists. EPP dismissed the idea that it might have been a sea fog, as claimed by the Swedish scholar D. GIßLER, citing the distance of Bavaria from the ocean and the fact that the fog was as intense there as anywhere else. Others suggested it was a normal "sun smoke" that occurred naturally during hot and dry days, but this, too, he dismissed: an extraordinary phenomenon such as this needed to have an extraordinary origin. Another possible explanation was that a

---

886 DAMODARAN et al. 2018.
887 SENEBIER 1784: 410. Jean SENEBIER also served as the observer for the Geneva weather station of the Societas Meteorologica Palatina; Societas Meteorologica Palatina 1783: 417–435.
888 John Thomas STANLEY, MS, JRL 722, John Rylands Library, University of Manchester, Manchester, UK: 95–96.
889 John Thomas STANLEY, MS, JRL 722, John Rylands Library, University of Manchester, Manchester, UK: 103.

large forest fire in Finland or Sweden was the source of the smoke, but Epp, once again, was not convinced.[890]

Franz Xaver Epp concluded that the most likely origin was a "fire in Iceland" in 1783. He summarized the most important information from Sæmundur Magnússon Hólm's book, including the passages on the negative impact on the animals in Iceland and the tribulations of the Icelanders.[891] Epp wondered whether this large volume of "smoke, haze, and vapors" from Iceland was enough to fill the atmosphere of an entire continent and whether it could travel such a great distance. He concluded that if the smoke of a forest fire from Finland could reach the German Territories, so too could smoke from Iceland.[892]

The Montgolfier brothers' new invention allowed Epp to clear up an old misconception: the idea that the wind was calm at great heights in the sky. The people who had risked traveling in hot-air balloons in the four years prior could attest that it could be rather windy in the upper atmosphere, far above the highest peaks. These winds "were certainly capable of spreading the Icelandic smoke far and wide."[893] Epp doubted that the smoke was detrimental to the health of those in Europe. He surmised that as the potency of chimney smoke lessens with distance, so too should the smoke from Iceland; therefore, although he believed that Icelanders smelled sulfur, he was certain the people of Bavaria had not.[894] He wrongly believed that the "ocean of fire," the lava in Iceland – which he thought to be the source of the dry fog – ceased to exist in August 1783. Epp established a link between the dry fog in Europe and an eruption in Iceland. The exact nature of the connection remained obscure. Like many other naturalists at the time, he did not dare repudiate the most popular explanation entirely; therefore, he added, "I do not want to deny that the numerous and devastating earthquakes of 1783 also added to the summer fog."[895]

A possible link between the Icelandic volcanic eruption(s) and the dry fog remained a topic of speculation; the year of awe remained a year of mystery, *annus arcanus*. In 1806, the Krünitz encyclopedia published a volume that included an entry for *Nebel* (fog), which mentioned the strange fog of 1783, but neglected to broach the topic of its cause. Even 23 years after the Laki haze, its connection to an Icelandic volcanic eruption was not widely accepted.[896] The *Physical Dictionary* from 1833 lists several possible theories regarding the origin of the dry fog: it suggests it might have been the manifestation of an overload of electricity in the air from the tail of a comet;

---

890 Epp 1787: 1–24.
891 Epp 1787: 1–24.
892 Epp 1787: 24. "Rauch, Dunst und Dampfe."
893 Epp 1787: 25. "Diese waren gewiß fähig den Isländischen Rauch weit und breit zu vertheilen."
894 Epp 1787: 25–28.
895 Epp 1787: 30. "Ich will nicht verneinen, daß die in dem Jahre 1783 so zahlreichen und verheerenden Erderschütterungen zu diesem Sommernebel auch was beygetragen haben."
896 "Nebel" in Krünitz 1806, vol. 101: 759–760.

vapors produced by earthquakes or volcanic eruptions; or smoke from peat fires. The article concludes that with so many different hypotheses, the phenomenon should not remain a mystery for much longer.[897] In the late eighteenth century, the natural sciences had not yet reached the point at which they could dissociate unrelated phenomena.[898] This would change in the late nineteenth century, as we will see in the next chapter.

---

897 BRANDES et al. 1833: 34–53.
898 DEMARÉE, OGILVIE 2001: 224.

4 The Mystery Remains

Outside of Iceland, the true cause of the dry fog of 1783 and the many other phenomena remained literally and figuratively unmapped. Months would pass before the fog dissipated, years before its consequences would be fully revealed, and an entire century before the fog of ignorance that veiled Europe lifted. Ignorance, in this case, is defined as a lack of knowledge or awareness. While the eruption raged on this island just south of the Arctic Circle, the people of mainland Europe were oblivious.[1] Europeans strove to find the truth; the results of this search offer a unique glimpse into the late-eighteenth-century train of thought. An eruption at Mount Hekla or the newly emerged island off the Icelandic coast were singled out as possible sources of the dry fog; however, severe earthquakes in southern Italy were considered a much more likely explanation.

Myth and Legend

Europeans had known about Iceland and its volcanoes since medieval times. Battered by wind and wave, fire and ice, and home to volcanoes, geysers, waterfalls, and the aurora borealis, this wild island was irresistible to the curious minds of the Enlightenment.

Beyond the famous Icelandic sagas, only a few reliable descriptions of Iceland were available to Europeans during the Middle Ages and the Renaissance. Publications, for the most part, describe Iceland as an exotic destination. Tales of mountains spitting fire on a near-constant basis and terrifying monsters inspired awe and fear in equal measure. In some cases, the natives made efforts to clean up the image of the island. For example, in the late sixteenth century, Arngrímur JÓNSSON (1568–1648), an Icelandic scholar, wrote a treatise to disabuse outsiders of the notion that Iceland was, as some seemed to think, a kind of hell on Earth. For its part, the Royal Society of London collected news from Iceland where possible, presumably to elucidate the peculiarities of the island and cast some light wherever the shadow of ignorance might fall.[2]

Maps like Figure 51, drawn in the late sixteenth century by the Barbantian cartographer Abraham ORTELIUS (1527–1598), reaffirmed the idea of "Iceland as hell," held by most Europeans. The map shows a violently erupting Hekla while terrifying-looking creatures patrol a sea laden with ice and floating debris. Swedish cartographer Olaus MAGNUS (1490–1544) had similarly mapped Iceland in 1555.[3]

---

1 GRATTAN, MICHNOWICZ, RABARTIN 2007: 154.
2 AGNARSDÓTTIR 2013: 16.
3 OSLUND 2011: 39. For a history on other historical maps of Iceland, see WALTER, BAUDACH 2019.

Figure 51: Abraham ORTELIUS' *Islandia*, ca. 1590.

In the eighteenth century, the Danish central administration began sending naturalists to Iceland, as they knew that full and accurate descriptions of the conditions would be beneficial for initiatives such as agricultural reforms.[4] The Icelandic sagas often stoked the curiosity of outsiders and an increasing number of foreigners started to visit from the early eighteenth century onward. Some visitors were probably disappointed – after a long, strenuous, and most likely costly journey – to find very little from the times of the *Landnáma*. That so little survived from that time is not surprising given Icelanders' almost endemic poverty and the lack of building materials such as stone or timber: most houses were made of turf and, from the moment of completion, had a relatively short lifespan.[5]

The exoticness of a place is a comparison to one's point of reference – usually a home country or country of residence. Most travelers introduced in this chapter came from other countries in Europe, such as Denmark, the British Isles, Germany, Sweden, or France, to which they then compared Iceland. The island was undoubtedly remote, yet it was still a European territory, part of the Danish-Norwegian kingdom, even if it was marginalized. Icelanders were white Christian Protestants; in terms of race and religion, Iceland was comparable with much of western and northern Europe. It certainly could not match the mysteriousness of the Far East or the Pacific Islands.[6]

Europeans probably perceived Iceland as unique but not "utterly foreign."[7] Icelanders were quite well-educated for their time and impressed travelers they encountered with their linguistic abilities: most often, these travelers would meet the governor or one of the four district governors, who spoke English, German, French, or Danish. As Danish was the administrative language in Iceland and all the other parts of the Danish kingdom, all officials, as well as Icelanders who had received a university education in Copenhagen, spoke Danish. Other Icelanders might only have spoken Icelandic. Visitors regarded Icelandic as a highly sophisticated Germanic language and respected the literary heritage of the sagas. If all else failed, some Icelanders could converse with visitors in Latin, a skill that garnered much admiration and was a testament to their education. For most of the eighteenth century, cities were largely non-existent in Iceland. Reykjavík could not offer the same materialistic standards as other European capitals (Figure 52).[8]

The Danish jurist and amateur naturalist Niels Horrebow (1712–1760) visited Iceland between 1749 and 1751. He sought to conduct astronomical and meteorological observations. In 1752, he published a travelogue in Danish detailing his time in Iceland. Translations in German, Dutch, English, and French followed from 1753 to 1764.[9]

---

4 Oslund 2011: 74–75.
5 Oslund 2011: 5.
6 Agnarsdóttir 2013: 11; Oslund 2011: 7–10, 19.
7 Oslund 2011: 17.
8 Oslund 2011: 24–26.
9 Horrebow 1752; Horrebow 1753; Horrebow 1758; Horrebow 1764; Agnarsdóttir 2013: 17.

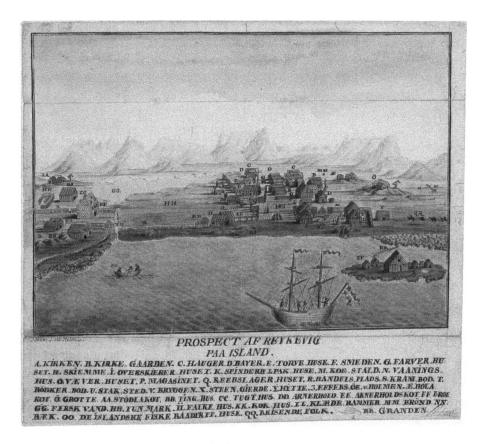

Figure 52: Sæmundur Magnússon Hólm, Reykjavík, ca. 1785.

HORREBOW's travelogue was used and cited by virtually all travelers to Iceland in the late eighteenth century.[10]

Another aim of HORREBOW's expedition was to refute the popular but sensationalistic book *Nachrichten von Island* (News from Iceland), which was published in 1746 by Johann ANDERSON, the mayor of Hamburg, and based on tales told by sailors. The outlandish claims made by ANDERSON include that Iceland saw constant volcanic eruptions and earthquakes. After his two-year-long stay in Iceland, HORREBOW was able to debunk these claims and others in his book titled *Trustworthy News from Iceland*.[11] HORREBOW was also interested in showing that there was no connection between Hekla, Vesuvius, and Etna, as Athanasius KIRCHER had suggested: HORREBOW pointed out that the Italian volcanoes had been active in the early 1750s, while Hekla had been quiet. He believed

---

10 MCCALLAM 2019: 203.
11 ANDERSON 1746; HORREBOW 1758; AGNARSDÓTTIR 2013: 17; MCCALLAM 2019: 199.

volcanism had a chemical cause, proof of which, he said, was the flammable sulfuric air released through volcanic vents. Nicolas LÉMERY and Georges-Louis LECLERC, Comte DE BUFFON, had previously propagated these explanations.[12]

In June 1750, Icelandic students Eggert ÓLAFSSON (1726–1768) and Bjarni PÁLSSON (1719–1779), undergraduates at the University of Copenhagen studying natural science, philosophy, and medicine, traveled to Iceland and made the first recorded summit of Hekla. The volcanic peak had been known to Europeans as the gate to hell since medieval times, but ÓLAFSSON and PÁLSSON did not find any evidence of this at its peak.[13]

Following their successful ascent and their graduation from the University of Copenhagen, the pair received funding from the Danish Royal Society to travel around Iceland between 1752 and 1757 and conduct a survey of the country's natural resources. Additionally, they were to investigate the culture and customs of the Icelanders. Like HORREBOW, they aimed to portray Iceland more accurately than previous reports.[14] During their exploration, they witnessed Katla's large 1755/1756 eruption. They documented a *jökulhlaup*, flooding that affected 50 farms, a layer of ash and pumice, and even volcanic bombs.[15] During their expedition, they kept diaries and notes, which they subsequently worked into a treatise. While ÓLAFSSON died early in an accident, PÁLSSON became Iceland's first general surgeon in 1760. Their detailed treatise, 1,100 pages long, was published in 1772 as *Reise igiennem Island* (Travel through Iceland) in Danish. It was translated into German in 1774, French in 1802, and English in 1805.[16]

In the spring of 1767, the French naval officer Yves-Joseph DE KERGUELEN TRÉMAREC (1734–1797) led another expedition that mainly explored Iceland from the sea and the Western Fjords. The main reason for their presence in Icelandic waters was to protect French fishing vessels. DE KERGUELEN published an account of his travels in French in 1771.[17] In his book, he paints the Icelandic population in a rather positive light but describes the Icelandic landscape as volatile: volcanoes and "ice mountains" threatened the traveler, who faced the real possibility of being engulfed by ice and lava at the same time.[18] Despite not exploring Iceland by foot or horse at great length, he

---

12 HORREBOW 1758: 9–10, 15; MCCALLAM 2019: 199–200.

13 Hekla, likely the most well-known volcano in Iceland, was said to be the mouth of hell. Several early modern maps portray Hekla as erupting, or "perpetually vomiting flames," as Karen OSLUND (2011: 39) writes when describing Abraham ORTELIUS' map from 1570. David MCCALLAM (2019: 200–201) traced the idea of Hekla as the mouth to hell back to Cistercian monks in the twelfth century, an idea that was then reinforced by German physician Kaspar PEUCER in the fifteenth century; see also CHESTER, DUNCAN 2007.

14 HJALMARSSON 1988: 88; AGNARSDÓTTIR 2013: 18.

15 MCCALLAM 2019: 201–202.

16 ÓLAFSSON, PÁLSSON 1772; ÓLAFSSON, PÁLSSON 1774; ÓLAFSSON, PÁLSSON 1802; ÓLAFSSON, PÁLSSON 1805; OGILVIE 1992; AGNARSDÓTTIR 2013: 18.

17 KERGUELEN TREMAREC 1771.

18 AGNARSDÓTTIR 2013: 20.

made claims about mineral deposits, the exploitation of which would have aligned with French colonial ambitions.[19]

The English naturalist and explorer Sir Joseph BANKS, famous for taking part in James COOK's 1768–1771 voyage, financed and organized an expedition to Iceland in 1772 with the aim of advancing botany and geology.[20] Reaching the island on 28 August 1772, at the end of the Icelandic summer, the trip was too late in the year to observe the plant life. Instead, BANKS focused on geological explorations: he was specifically interested in observing Hekla.[21] The future Archbishop of Uppsala, Uno VON TROIL (1746–1803), the Swedish botanist Daniel Carl SOLANDO, together with artists, servants, and a French cook, accompanied him on this voyage. Highlights of the six-week-long expedition included a 350-mile round trip on horseback to visit and climb the imposing Mount Hekla and visits to Þingvellir and Geysir. Despite their best efforts, they did not witness any erupting volcanoes, which must have been disappointing. To observers in the late eighteenth century, this should have proven that Iceland was not covered in countless volcanoes producing "fiery vomit" all the time. The group might even have dined with Governor THODAL and Bjarni PÁLSSON; the latter would surely have told BANKS about his summit of Hekla. Although it is not certain that BANKS and PÁLSSON met in Iceland, BANKS likely knew that he was not the first to climb the mountain. Nevertheless, upon their return to Britain in November 1772, an erroneous rumor was printed in the newspapers claiming that BANKS and his companions had been the first to make it to Hekla's summit.[22]

VON TROIL published a report about their expedition in Swedish in 1777, which was translated into several languages over the following seven years. In addition to his observations, VON TROIL corresponded with three Icelanders, making his account quite reliable.[23] The book, and many of the versions published later in other languages, also included a detailed map (Figure 53).

BANKS' expedition and volcanological fieldwork built upon previous travel reports and added to the growing body of knowledge regarding volcanoes in mid-to-late eighteenth-century Europe. These expeditions began to tip the balance in favor of reliable accounts of Iceland as opposed to exaggerated and often unreliable sailors' tales. BANKS' expedition and VON TROIL's report would later inspire others to visit, such as John Thomas STANLEY (1766–1850) in 1789 and William Jackson HOOKER (1785–1865) in 1809.[24] By the end of the eighteenth century, mainland Europeans no longer regarded Iceland as a mythical

---

19 AGNARSDÓTTIR 2013: 19–20; MCCALLAM 2019: 203–204.
20 AGNARSDÓTTIR 2013: 22–23. See AGNARSDÓTTIR (2016) for detailed insights into Sir Joseph BANKS and his Iceland expedition.
21 MCCALLAM 2019: 204–205.
22 AGNARSDÓTTIR 2013: 22–23; MCCALLAM 2019: 197–213.
23 The three learned Icelanders with whom he corresponded were Bishop Hannes FINNSSON, the scholar and Reverend Gunnar PÁLSSON, and Hálfdán EINARSSON; TROIL 1779; TROIL 1780; AGNARSDÓTTIR 2013: 22–23; MCCALLAM 2019: 197–213.
24 MCCALLAM 2019: 198, 211.

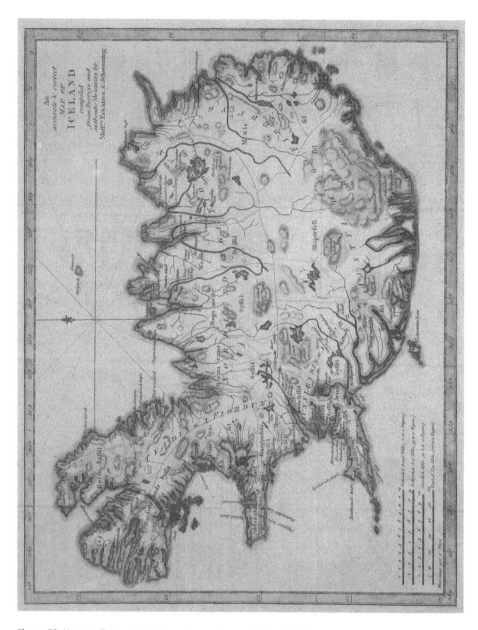

Figure 53: Uno von Troil, *An accurate and correct map of Iceland*, 1780.

land of fire; instead, a clear scientific interest in Iceland's landscape and even the customs of its people grew, which will become apparent in the following subchapters.

A Search in Vain

During the Laki eruption, the Danish central administration was far away and poorly informed; what little help they eventually sent arrived too late for many Icelanders.[25] Two reports written between 1783 and 1785 are described below. These examples illustrate that early reports about the eruption were riddled with inaccurate information.

Sæmundur Magnússon Hólm

Sæmundur Magnússon Hólm's book on the Laki eruption was the first to be written on the topic outside of Iceland.[26] Hólm was an Icelandic poet, artist, and scholar, who was born in the Meðalland region, one of the Fire Districts in southern Iceland, which explains his interest in the eruption and its aftermath. At the time of the eruption however, Hólm was in Copenhagen and relied on letters and reports sent with the merchant ships back to Denmark. His book seems to be primarily based on accounts by Jón Eiríksson (1728–1787) and Skúli Magnússon (1711–1794). He might have supplemented these accounts with additional information gleaned from letters that Icelanders in Copenhagen had received from friends and family.[27] The central administration allegedly denied him access to government documents in Copenhagen.[28] As far as we know, the book was produced by the author unprompted by any authority.

Hólm tried to convey what he had gathered about the "fires" and their extent; however, his descriptions were vague. Perhaps because of the way he gathered his information, his account contained mistakes and was, in some cases, exaggerated: for instance, on a map, he drew mountains that did not exist and depicted the lava flow incorrectly.[29]

Hólm's book titled *Jordbranden paa Island i Aaret 1783* (About the Earth fire in Iceland in the year 1783) was published in 1784 in Danish and later the same year in German.[30] It was not translated into Icelandic, perhaps because it was intended for an audience unfamiliar with the country.[31] The *Ephemerides* of the Societas Meteorologica

---

25 Herrmann 1907: 94–97; Karlsson 2000b: 180–181. The shipment arrived in April 1784, almost one year after the start of the Laki eruption.
26 He was born as Sæmundur Magnússon in Hólmaseli in Meðalland, which is one of the Fire Districts. When he went to Denmark to study philosophy and art at the Royal Academy of Art in Copenhagen, he added "Hólm" as a family name. In 1787, he became the reverend of a parish in Snæfellsnes.
27 Thordarson 2003: 7.
28 Rafnsson 1984b: 261–162.
29 Thordarson 2003: 7; Pálsson 2004: 76–77.
30 Hólm 1784a; Hólm 1784b; Thoroddsen 1925: 30; Thordarson 2003: 7.
31 Demarée, Ogilvie 2001: 223; Thordarson 2003: 7.

Palatina for 1783, published in 1785, contained an extract from Hólm's report. Johann Jakob HEMMER wrote a highly critical introduction for the report, in which he highlighted some of its inaccuracies and remarked that the translator had trouble understanding what HÓLM was referring to at times. HÓLM's publication, though inaccurate in parts, provides valuable information about how the dry fog affected Denmark.[32]

Magnús STEPHENSEN and Hans VON LEVETZOW

Magnús STEPHENSEN and Hans VON LEVETZOW's voyage, ostensibly a mission to help Iceland and investigate the eruption's effects on its society and economy, was also an opportunity for research.[33] STEPHENSEN and VON LEVETZOW wrote a detailed report on their expedition and published it in Danish in the spring of 1785 as *Kort Beskrivelse over den nye Vulcans lidsprudning i Vester-Skaptefields-Syssel paa Island i Aaret 1783*, which translates as "Short Description About the New Volcano in Vester-Skaftafellssýsla in Iceland in the Year 1783."[34] The publication includes a map that inaccurately details the location of the lava flows (Figure 54).[35] This report, based on first-hand experiences rather than hearsay, was written in the immediate aftermath of the eruption while the country was still heavily affected by famine. Just like many of the other naturalists' travelogues, such publications were quickly translated into other European languages and read by contemporaries at the time.[36]

Upon reading Magnús STEPHENSEN's travelogue carefully, it becomes apparent that he traveled to the highlands north of Klaustur; however, it seems unlikely that he made it as far north as the Laki fissure, as he would then have concluded that it was the origin of the lava flow. STEPHENSEN's travels in this area took place between 22 and 24 July 1784. He found it difficult to make progress in the highlands; the lava was still hot, smelly, and in places hard to bypass. The air was thick with smoke, drastically reducing visibility, and the lakes and rivers were still hot.[37] The tephra deposits of ten to 15 centimeters on the ground made this endeavor even more difficult.[38] STEPHENSEN was expecting to find a stereotypical volcano, that is, a conical mountain with smoke rising from its top; in this case, he would not, and indeed could not. He describes several mountains in his report, many with smoke rising from them, but the Laki fissure eluded him.[39] According to Jón STEINGRÍMSSON, locals already knew that the eruption was produced by a fissure; however, this information was not conveyed to STEPHENSEN.[40]

---

32 Societas Meteorologica Palatina 1783: 689; THORDARSON 2003: 7.
33 For more information on STEPHENSEN and VON LEVETZOW's journey to Iceland, see Chapter Two.
34 STEPHENSEN 1785; RAFNSSON 1984b: 261–262.
35 THORDARSON 2003: 7–9.
36 STEPHENSEN, EGGERS 1786; STEPHENSEN 1813; THORODDSEN 1925: 31–32; DEMARÉE, OGILVIE 2001: 223.
37 STEPHENSEN 1786: 326–330; THORODDSEN 1925: 58; THORDARSON 2003: 7–9.
38 THORDARSON 2003: 7–9.
39 STEPHENSEN 1786: 326–330; THORDARSON, HÖSKULDSSON 2014: 9.
40 ÞÓRARINSSON 1984: 35; STEINGRÍMSSON 1998: 35–36.

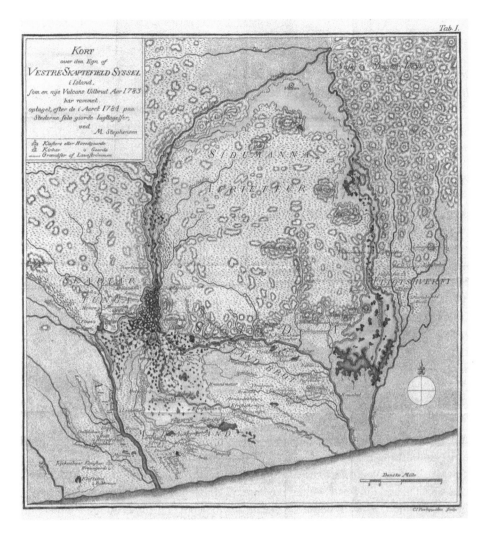

Figure 54: Map of the Laki lava flows by Magnús STEPHENSEN, 1784. The rough shape of the lava flows seems correct; however, the area from which the lava originated is lacking detail. STEPHENSEN did not travel to the location of the Laki fissure. Instead, he expected a single "Vulcan" at the most northern point of the lava (area in yellow).

In his report, STEPHENSEN explains at great length that he does not believe the source of the fire was what he calls an Earth fire mountain (*Jordbranden/Erdbrand*) but rather what he refers to as a fire-spitting mountain (*Feuerspeyen*). The fact that the lava flow was consistent and that new lava flows emerged – the first lava flow along the Skaftá River and then the second lava flow along the Hverfisfljót River – pointed to that conclusion, he argues.

> This should be enough to answer my question; I hope to have brought sufficient arguments that the subterraneous fire that erupted in Iceland now was no Earth fire but a fire-spitting [one]. The question remains where the location of this fire-spitting may have been.[41]

Naturalists and geologists at the time usually described volcanoes as *Earth fires* or *fire-spitting mountains*, the phrases seem to have been interchangeable, so it is interesting that STEPHENSEN puts great emphasis on the distinction between the terms; perhaps he did so deliberately to distract from the fact that he had not traveled very far into the highlands.

Icelandic historian Sveinbjörn RAFNSSON accuses Magnús STEPHENSEN of heavily relying upon Jón STEINGRÍMSSON's work without crediting him; Thorvaldur THORDARSON and Stephen SELF make the same allegation. STEPHENSEN did indeed use large parts of STEINGRÍMSSON's text, albeit excluding the latter's religious notions; perhaps he felt that the religious overtone of STEINGRÍMSSON's interpretation of the event was too eccentric for an enlightened European audience.[42]

The largest contrast between STEPHENSEN's and HOLM's publications is that STEPHENSEN emphasized the singularity of this "catastrophe." HÓLM, on the other hand, did not emphasize singularity at all: quite the opposite. He very much assumed that comparable events must have occurred in the past, basing his assumption solely on his awareness of the many large lava fields visible in the Fire Districts and elsewhere in Iceland.[43]

Sveinn PÁLSSON

A native Icelander, Sveinn PÁLSSON (1762–1840) (Figure 55) lived in the northern region of Iceland during the eruption – an area that was severely affected. In the spring of 1784, PÁLSSON wrote a short report about the Laki eruption, giving first-hand accounts of the ashfall and the frosty winter, the latter of which he blamed on the haze produced by the eruption. PÁLSSON penned a second-hand report on the eruption's impact on southern Iceland based on a letter he received from a friend living in the Síða region.[44]

From 1787 to 1791, PÁLSSON attended university in Copenhagen to study medicine. In the Icelandic translation of his travel journal, published in 1945, the foreword mentions how studying in Copenhagen opened a new world for PÁLSSON: he learned many new things and visited museums, art galleries, playhouses, and concerts, all of which

---

41 STEPHENSEN 1786: 325–326. "Dies mag zu Beantwortung der ersten Frage dienen, wobey ich zur Genüge gezeigt zu haben hoffe, daß das in Island jetzt ausgebrochene unterirdische Feuer kein Erdbrand, sondern ein Feuerspeyen gewesen sey. Es frägt sich nun weiter, wo der Ort des Feuerspeyens gewesen sey."
42 RAFNSSON 1984b: 261–262; THORDARSON 2003: 7–9.
43 RAFNSSON 1984b: 261–262.
44 THORDARSON 2003: 9.

Figure 55: Portrait of Sveinn PÁLSSON drawn by Sæmundur M. HÓLM in 1798 with red chalk. This connection illustrates how small the scientific community in Iceland was at the end of the eighteenth century and is an example of how most scientists knew each other personally.

he greatly enjoyed. During his time at the university, he frequently attended lectures by German naturalist and professor Christian Gottlieb KRATZENSTEIN (1723–1795), who was one of the first non-Icelanders to suspect a volcanic eruption as the cause of the strange fog of 1783.[45]

In 1789, Danish veterinarian Peter Christian ABILDGAARD (1740–1801), Danish-Norwegian zoologist Martin VAHL (1749–1804), and a host of other academics founded the Natural History Society (*naturhistorie-selskabet*), which existed until 1804. The Society gave travel research grants to four scholars.[46] In 1791, PÁLSSON received one of these grants, which was initially supposed to cover four years of field research on Iceland's natural history (1791–1795).[47] The travel research grant totaled 900 *ríkisdalir*, with an additional 58 *ríkisdalir* for the purchase of tools. He used the money for his

---

45 PÁLSSON 1945, vol. 1: xix, xx.
46 PÁLSSON 1945, vol. 1: xxi–xxii; STRØM 2006: 82–85; STRØM 2017: 62–65.
47 PÁLSSON 2004: xx–xxi.

fare to Iceland from Copenhagen, the support of two local guides, and the many horses he needed throughout his travels.[48]

PÁLSSON used the methods of Enlightenment-era science. As a first step, he went to see the natural phenomena with his own eyes; then, he concentrated on collecting data, which he would subsequently analyze and classify to create an overview of his findings.[49] At this time, between 1783 and around 1800, Icelanders were still struggling with the aftermath of the Laki eruption and faced a near-constant struggle with poverty.[50]

Initially, PÁLSSON's field research had focused on botany and biology; eventually, he became interested in Iceland's geology – primarily its glaciers but also its volcanic eruptions and hot springs. A journey through the Icelandic highlands in the eighteenth century was far from a pleasant experience: equipment and food had to be transported by horse across rough and unknown terrain, and ice-cold rivers had to be forded waist deep. The topography of these riverbeds was often unknown; getting stuck was a common occurrence. Traveling was not made easier by the frequent, almost daily, course changes of the glacial rivers, particularly in the summer – the main travel period of the year – when the sun melted the glaciers or when *jökulhlaups* (glacial outburst floods) occurred.[51] Sometimes fording rivers could be life-threatening; here, PÁLSSON depended on the expertise of his guides.[52] In the summers from 1791 to 1794, he explored Iceland and kept a travel journal of his observations (Figure 56); he was frustrated during the winters when the frigid weather compelled him to stay inside.[53]

The Natural History Society had its own journal, *Skrivter af Naturhistorie-Selskabet*, which was written mainly by the Society's directors and the recipients of the Society's travel research grants. Five volumes were published between 1790 and 1802. Excerpts from PÁLSSON's travel diary were published there too. The first excerpt was published in 1792, covering his departure from Copenhagen on 2 July 1791, his journey to Iceland, and the beginning of his travels through the country until 7 September 1791, presumably just before the last ships of the year returned to Denmark.[54] In 1793, the second excerpt was published, detailing his explorations from 7 September 1791 until April 1792.[55] Finally, the third and last excerpt from his travel diary was published in 1793, outlining his journeys from May 1792 until 20 July 1792.[56]

---

48 PÁLSSON 1945, vol. 1: xxiii. Price series of Icelandic currency only go back to 1849; therefore, an accurate estimate of the value of *ríkisdalur* from 1783 is impossible. Table 12.25 in JÓNSSON, MAGNÚSSON 1997, 637; personal correspondence with Prof. Guðmundur JÓNSSON, University of Iceland, 16 February 2020.
49 BJÖRNSSON 2017: 157.
50 PÁLSSON 1945: xxiii.
51 HERRMANN 1907: 73–74.
52 PÁLSSON 2004: 76.
53 PÁLSSON 1945, vol. 1: xxiii.
54 PÁLSSON 1792: 222–234.
55 PÁLSSON 1793a: 122–146.
56 PÁLSSON 1793b: 157–194.

A reviewer in *Laerde Efterretninger* (which translates to "learned intelligence") criticized the publications: PÁLSSON had received funds from the Natural History Society to describe the natural history of Iceland; the reviewer concluded that PÁLSSON's writings did not satisfy expectations. PÁLSSON may have been aware of this criticism, as he too believed he had not fulfilled the goals set out for him by the Society. Still, he continued to send letters and additional excerpts from his travel journal and later sent his book on Icelandic glaciers (*islensk Jisbierge*) to Copenhagen: the deliberation protocols of the Society mention the receipt of these documents in 1795 and 1796.[57]

PÁLSSON's manuscript makes it clear that he was aware of at least 70 volcanic eruptions in Iceland since 874. For most of these eruptions, very few reports existed. However, he noticed that people had been writing stories about major natural wonders, such as the Laki eruption, for many centuries to preserve them for future generations. He mentions the texts by HÓLM and STEPHENSEN, but because of their inaccuracies, he consigns them to the "solemn feast of eternal forgetfulness."[58]

On 30 July 1794, PÁLSSON arrived in the Síða region. Before continuing into the highlands and to the origin of the Skaftá Fires, he had to give his horses a few days' rest and find a new and knowledgeable guide. 11 years after the eruption, STEINGRÍMS-SON's report and map gave PÁLSSON a rough idea of where he needed to go. PÁLSSON describes the mountains in the Síða region as "some of the greenest" he had ever encountered. This, no doubt, was a result of the green moss growing slowly over the *Skaftáreldahraun*, the Laki lava fields.[59] He was aware that the eruption, including its gases, ash, and lava, had wreaked havoc in this region just a decade before.[60]

When PÁLSSON, his guide, and their horses began their trip into the highlands, they likely left early in the day. On that day, the sun rose at 2:46 a.m. They were met with adverse weather conditions almost immediately: it was windy and raining.[61] In the evening, when they set up their tent in Lauffell, the temperature was 12.5 °C. Lauffell is a mountain about 17 kilometers northwest of Klaustur; this peak roughly marks the halfway point between Klaustur and the Laki fissure. Once they had set up their tent and the wind died down, the men decided to explore further. They forded the Hellisá River and climbed up Blágil and Galti, two hills (621 meters above sea level) just south of the southwestern end of the Laki fissure. On a clear day, it is possible to see Mount Laki from this point. When darkness began to set in – sunset that day was around 8:51 p.m. – the men began their return journey.[62]

---

57 STRØM 2006: appendix, 57–58, 86, 89.
58 PÁLSSON 1945, vol. 2: 555, 556 (quote). "[. . .] af því að svo er að sjá sem sú hin sama hafi verið hátíðlega dæmd til eilífrar gleymsku."
59 PÁLSSON 1945, vol. 2: 555–557.
60 PÁLSSON 1945, vol. 1: 354.
61 PÁLSSON 1945, vol. 2: 559.
62 PÁLSSON 1945, vol. 2: 559–560.

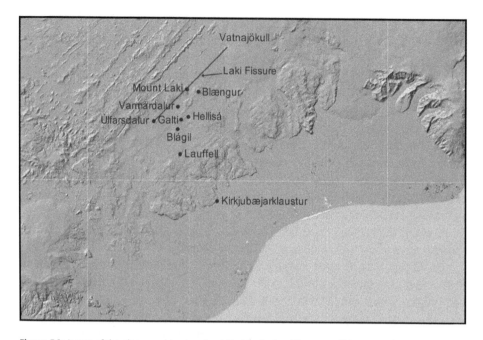

Figure 56: A map of the places and landmarks visited by Sveinn PÁLSSON and his companion.

On the second day of their journey, rain and fog greeted them as they peeked out of their tent. The wind had picked up as well. In his journal, PÁLSSON wrote of his concern over the outcome of the day. PÁLSSON and his guide decided to leave the horses behind for the day and explore the fissure on foot. One of PÁLSSON's goals was to measure the thickness of the lava, for he believed it had been partially eroded by wind and water. He collected samples of the large colorful rocks he encountered during the day, some of which he subsequently sent to the Natural History Society.[63]

Traversing the *Skaftáreldahraun* was made even more strenuous by the flimsy Icelandic shoes that PÁLSSON wore. The terrain was uneven, with many cracks and holes, and at times they could only make progress by moving along on all fours. The ground was loose in some parts and hollow in others; PÁLSSON had to pay attention to avoid falling. Some of the lava was still smoking and releasing a sulfuric smell. The craters that made up the fissure were wide and deep. PÁLSSON thought the Laki fissure was comparable to a large canyon. Interestingly, PÁLSSON was aware that they were the first humans to come across the lava field at this site and remarks upon it in his writings. When nightfall came, PÁLSSON placed a Danish coin on top of a rock on the fissure, as a sign of their arrival, before returning to his tent.[64]

---

63 PÁLSSON 1945, vol. 2: 560–562.
64 PÁLSSON 1945, vol. 2: 561–563, 588–589.

The next day, on 1 August 1794, the men decided to climb a mountain named Blængur. It is 833 meters above sea level, slightly higher than Mount Laki (812 meters). For once on this trip, the men were blessed with good weather. They set off with their horses and reached the top of the mountain sometime after noon. From their new vantage point, they could observe the "spine" of the fissure just to the northwest of Blængur. Here, PÁLSSON used the time to rest, update his journal, and draw the fissure. He also took notes on Mount Laki and the surrounding valleys, mountains, glaciers, and rivers, such as the Skaftá and Hverfisfljót. PÁLSSON took measurements with his barometer and thermometer whenever he had the opportunity. Sveinn PÁLSSON was the first person to put the Laki fissure on the map (Figure 57). He was also the first to carefully examine the location and understand the nature of this fissure eruption.[65]

After descending the mountain, the men explored the extent of the fissure further: they followed it toward where it cut through a valley called Úlfarsdalur but, once again, nightfall prevented them from traveling further.[66] PÁLSSON was "furious" that he and his companion had such limited time to explore the fissure and the extent of the lava it produced. After two full days of exploring, they returned to Klaustur.[67]

So it was that the source of the 1783 lava flow was discovered: more than ten years after the eruption. It had not come from a single cone-shaped mountain but from a row of "conical hills, stretching in nearly a direct line" within a valley called Varmárdalur.[68] The Laki fissure consists of around 140 craters, not 20, as PÁLSSON claims; that said, some craters are easier to identify than others. He correctly gives the direction as southwest to northeast. All of this gives him some leeway to criticize HÓLM and STEPHENSEN for the inaccuracies of their descriptions. Sveinn PÁLSSON believed that STEINGRÍMSSON's account of the Laki eruption was "the most probable truth."[69] HÓLM's account was understandably inaccurate, given that he had been in Copenhagen at the time. STEPHENSEN simply had not explored the highlands enough to identify the real source of the eruption.

PÁLSSON incorrectly surmised that the Laki eruption did not have a connection with a glacier. However, he knew that *jökulhlaups* had occurred between 1783 and 1785 along the coast of Vestur-Skaftafellssýsla. It was difficult for him to explain the connection between the subaerial, non-glacial Laki eruption and the *jökulhlaups*. PÁLSSON could tell from historical records that volcanic eruptions within the Vatnajökull

---

65 THORODDSEN 1925: 33; PÁLSSON 1945, vol. 2: 565–565.
66 PÁLSSON 1945, vol. 2: 568–569.
67 PÁLSSON 1945, vol. 2: 584. "Að lokum hlýtur hver maður, sem þekkir leið þá, er höfundur fór á tveim dögum, að falla alveg í stafi yfir flýtinum á honum, er hann les, hvernig hann varð dag og nótt að klängrast fram og aftur í heitu hrauninu, ganga á Miklafell með miklum erfiðismunum, klifra upp á hið mjög svo háa fjall Blæng, gera tilraun á þrem stöðum til að komast yfir hraunið og nota jarð- og steinborana á jafnmörgum stöðum, athuga og yfirlíta allar mögulegar hrauntegundir og önnur furðuverk, enda má það furðulegt kallast, að hann, fylgrarmenn og hestar skyldu endast til alls þessa á svo takmörkuðum tíma!"
68 PÁLSSON 2004: 120.
69 PÁLSSON 1945, vol. 2: 554; 571–599.

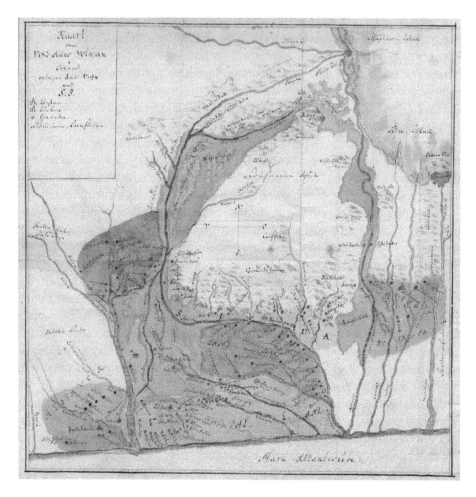

Figure 57: Map of the Laki lava flows drawn by Sveinn Pálsson, 1794. Although the extent of the lava flow is similar to the depiction in Stephensen's map, the area where the lava originated is depicted in greater detail here. There is no single "Vulcan" anymore, but instead, Pálsson depicts a row of craters and vents near the Skaftá River.

glacier often produced *jökulhlaups* along the coast on the *sandur* plains.[70] Today, these *jökulhlaups* are explained by the fact that the Grímsvötn system, of which the Laki eruption was a part, is under a glacier and was active from 1783 to 1785.

Pálsson criticizes Stephensen for believing that the Laki eruption had been a singular event in history. He refutes this notion by referring to several other devastating volcanic eruptions in Iceland's past, such as the Reykjanes Fires in the thirteenth

---

70 Pálsson 1945, vol. 2: 554, 567–576.

century, another flood basalt event, and the 1362 Öræfajökull eruption, the largest historical eruption in Iceland in terms of explosivity.[71]

Furthermore, PÁLSSON correctly assumed that the lava in the Skaftá gorge had come
from the western side of the fissure during the earlier phase of the eruption and that
the lava in the Hverfisfljót gorge had come from the northeastern part of the fissure
beginning at the end of July 1783.[72] In his writings, PÁLSSON speculates that the fissure
had only two distinct eruptive phases, a view widely accepted for almost two centuries.[73]
THORDARSON and SELF have since proven that the eruption had at least ten eruptive
episodes (14 if one includes the eruptive episodes at Grímsvötn).[74] However, it is remarkable that PÁLSSON was able to deduce that it was two different phases that flooded the
two separate rivers with lava, one after the other.

The earthquakes before the eruption were due – so PÁLSSON believed – to fire coming
into contact with flammable minerals that burnt slowly, such as peat or coal. According
to PÁLSSON, this sort of fire was inextinguishable, for it burned deep within the Earth for
as long as the mineral that fueled it lasted and could thus go on for several years.[75]

In 1794, the Natural History Society terminated PÁLSSON's travel research grant
one year earlier than they had initially negotiated.[76] What led to the discontinuation
of PÁLSSON's grant? It was most likely a combination of three things. First, the Natural
History Society was in a position of financial precarity. Initially, the Society had quite
a lot of members, each of whom paid an annual fee, including some royal members
who were in the position to pay more than the fixed amount. Over time, the number
of members dwindled, as did available funds.[77]

Second, in July 1795, a fire destroyed the publishing house of Nicolaus MØLLER
and Son, and with it the 1795 volume of *Skrifter af Naturhistorie-Selskabet*. This would
have been the fourth volume to appear in the series. The gap in publications between
1793 and 1797 was explained in the foreword of the re-written fourth volume when it
appeared.[78] A fifth and final volume was published in 1799/1802.[79] In 1804, the Society
was dissolved and, after some back and forth, its natural history collections were

---

71 PÁLSSON 1945, vol. 2: 576–577.
72 THORDARSON 2003: 9–10.
73 HELLAND 1886; THORODDSEN 1879; THORODDSEN 1894; THORODDSEN 1925; ÞÓRARINSSON 1969.
74 THORDARSON, SELF 1993; THORDARSON, SELF 2003: 9–10, 26–27.
75 PÁLSSON 1945, vol. 2: 579. PÁLSSON also attributed the strong 1784 earthquake in southern Iceland to
these flammable materials.
76 PÁLSSON 1945, vol. 1: 554.
77 PÁLSSON 1945, vol. 1: xxii; STRØM 2006: 85; Strøm 2017: 65.
78 Skrifter af naturhistorie-selskabet 4, no. 1 (1797), and no. 2 (1798). For the foreword, see no. 1.
79 Skrifter af naturhistorie-selskabet 5, no. 1 (1799), and no. 2 (1802). A sixth volume was written in
1810 but only published in 1818, it contained a register with all the contributions; STRØM 2006: appendix, 52.

taken over by the Royal Natural History Museum (*Det Kongelige Naturalienmuseum*) in Copenhagen.[80]

Third, the Society had heard very little from PÁLSSON. Allegedly, he had not responded to two letters they had sent him. PÁLSSON states in his book that he had only received one of the letters, to which he had replied.[81] Given the distance between Iceland and Copenhagen and the insufficient infrastructure within Iceland, it is possible that the root of the dissatisfaction on both sides was simply a matter of miscommunication and missed connections. Initially, some members of the Society had seemed disappointed with the direction PÁLSSON's explorations had taken and PÁLSSON's self-critical assessment of his work might have exacerbated this notion. In the long run, however, they warmed to PÁLSSON's findings. The deliberation protocols from March 1796 reveal that Christian ROTHE and Niels TØNDER LUND (1749–1809), directors of the Society, discussed the successes of PÁLSSON's trip. ROTHE reported that PÁLSSON had sent receipts for the instruments he purchased for the journey along with reasons why he believed the voyage was not entirely successful. He expressed his wish to keep the instruments and to continue making further observations. TØNDER LUND responded, saying he believed PÁLSSON had indeed fulfilled the Society's purpose, which was to extend the boundaries of science. He stated that the Society had welcomed PÁLSSON's observations and affirmed that he believed PÁLSSON should be allowed to keep his travel gear and tools.[82] In the end, self-doubt and insufficient funds proved to be the main obstacles on this expedition.

Despite the discontinuation of the grant, PÁLSSON persisted, using the long winters, in particular the winter of 1794/1795, to finish the manuscript about his Icelandic travels. In 1795, he then sent it to Copenhagen, perhaps hoping the Society's financial difficulties had been resolved. In the manuscript, he maps and describes Iceland's glaciers and their interactions with volcanic eruptions; he also offers a new system for classifying glaciers and describes their movements as well as their moraines, glacial rivers, and *jökulhlaups*.[83] The Natural History Society never published his manuscript: it simply remained in their archive.[84] The maps he drew are considered to be among the best that existed at the time. His map of Vatnajökull (then sometimes called Klofajökull) remained the most detailed of the glacier until the Danish government started mapping it in 1902 at a modern 1:50,000 scale.[85]

It was only in 1945 that PÁLSSON's glacier and volcano treatise, maps, and cross sections were finally published in their entirety, fully edited, and translated from

---

80 STRØM 2006: 85; STRØM 2017: 65.
81 PÁLSSON 1945, vol. 1: 346.
82 STRØM 2006: appendix, 89.
83 PÁLSSON 2004: xx–xxi; OGILVIE 2005: 279–280.
84 BJÖRNSSON 2017: 157–158.
85 PÁLSSON 2004: xxvii, 110.

Danish into Icelandic.[86] In 2004, the Icelandic Literary Society published an English translation.[87] The editor of PÁLSSON's work in 1945, Jón EYÞORSSON, describes PÁLSSON as "the best educated of all of his fellow Icelanders in many fields, [and] one of the most prolific writers of his age."[88] Had it been published, his treatise might have been received as a major contribution to the field of glaciology – and would perhaps have connected the Laki eruption and the phenomena in Europe much earlier. As it was, his ideas remained undiscussed in scholarly circles for some time.[89]

Severe hardship characterized the aftermath of the eruption throughout Iceland. Even still, PÁLSSON was able to use the limited means he was offered by the Natural History Society to carry out field research and finish his manuscript. Following his fellowship with the Natural History Society, PÁLSSON had several careers: he worked as a botanist, explorer, glaciologist, natural historian, meteorologist, and later a farmer and physician.[90] In 1795, PÁLSSON married Þórunn BJARNADÓTTIR (1778–1836), whom he had met during one of the winters when he was writing up his manuscript. She was the daughter of Bjarni PÁLSSON, the naturalist and explorer, who, together with Eggert ÓLAFSSON, had traveled throughout Iceland in the 1750s. In 1803, Sveinn PÁLSSON became Iceland's third general surgeon.[91]

A Clearer Picture

In the late eighteenth century, Iceland had already established itself as a magnet for scientific exploration and tourism for those interested in the sagas.[92] The eruption and its aftermath, far from discouraging exploration, motivated scientists to make the journey, particularly geologists: Iceland was an ideal site to study volcanic eruptions, their effects on the landscape, and how volcanoes might have shaped the Earth. A dichotomy existed here: the very thing that brought despair and destruction to the Icelanders now seemed so attractive to outside explorers. News of volcanic eruptions in Iceland drew naturalists and wealthy amateurs who were keen to see the changes that a volcanic eruption could produce in a landscape. These stories likely inspired the expeditions of John Thomas STANLEY, George MACKENZIE, and Henry HOLLAND, among others.[93]

---

86 BJÖRNSSON 2017: 158; PÁLSSON 1945.
87 PÁLSSON 1945; PÁLSSON 2004: xx–xxi.
88 PÁLSSON 2004: xxii.
89 PÁLSSON 2004: xxv–xxvi.
90 PÁLSSON 2004: xx–xxi.
91 PÁLSSON 1945, vol. 1: xxiii.
92 OSLUND 2011: 35.
93 OSLUND 2011: 39, 43.

Upon spotting an Icelandic volcano, one traveler described it as "ugly, barren, and desolate, although scientifically intriguing." Later, in the nineteenth century, these barren landscapes were celebrated as quintessentially Icelandic.[94]

English naturalist John Thomas STANLEY traveled to Iceland in 1789. Initial reports from the expedition were published in the *Transactions of the Royal Society of Edinburgh*.[95] The expedition's other participants kept diaries, which were posthumously published in the 1970s.[96] STANLEY remarks that the Icelanders lived in "savage" conditions but praises their moral and intellectual abilities, describing them as equal to "the most civilized communities in Europe."[97]

As was typical for affluent young men at the time, STANLEY had embarked on a grand tour of Europe during the summer of 1783; he had wished to see the Alps, but the dry fog conspired to ruin his plans.[98] Another grand tour in 1787 brought STANLEY to Italy, where he climbed both Vesuvius and Etna. By the time he traveled to Iceland in 1789, STANLEY had directly and indirectly experienced volcanoes and the effects of their eruptions. It is, therefore, unsurprising that he was fascinated by Iceland's volcanism. Perhaps it was reading Joseph BANKS' report on his ascent of Hekla almost two decades earlier that inspired STANLEY to attempt the climb himself.[99] Whatever the reason, he climbed to the summit and placed a British flag there.[100]

STANLEY also painted a watercolor of an erupting Hekla with two human climbers in the foreground, looking at the volcano. This painting depicts how he visualized the event in his mind's eye, as it had last erupted in 1766 and would not erupt again until 1845.[101] Historian Karen OSLUND suggests the painting represents the human "posture of powerlessness toward the mountain in the face of nature's might. The diabolic black and red colors of the painting are particularly evocative of Hekla's status in legend."[102] Many subsequent paintings of Hekla were based on STANLEY's sketches and descriptions; most did not allude to the hardships Icelanders experienced after significant eruptions.[103]

STANLEY wrote about his experience with the dry fog of 1783 sometime after the French Revolution, as is clear from the context of his writing.[104] Therefore, it is

---

94 OSLUND 2011: 39.
95 STANLEY 1794.
96 WEST 1970.
97 AGNARSDÓTTIR 2013: 24.
98 John Thomas STANLEY, MS, JRL 722, John Rylands Library, University of Manchester, Manchester, UK; WAWN 1981; WAWN 1989.
99 WAWN 1981; MCCALLAM 2019: 209.
100 WEST 1970, vol. 1: 208; AGNARSDÓTTIR 2013: 23–24; MCCALLAM 2019: 212.
101 Global Volcanism Program: Hekla.
102 OSLUND 2011: 41.
103 OSLUND 2011: 42.
104 John Thomas STANLEY, MS, JRL 722, John Rylands Library, University of Manchester, Manchester, UK: 103.

possible that he wrote his memoirs about his grand tour to Switzerland after his travels to Iceland. It is unsurprising that his experiences in the land of fire and ice convinced him that Iceland had something to do with the fog of 1783. He writes, "the Mountains of Skaftafell [which] vomited its columns of fire precisely during the period the fog lasted."[105]

In 1809, Joseph BANKS sent his protégé William Jackson HOOKER to Iceland as part of the PHELPS-JÖRGENSEN expedition.[106] Before this trip, BANKS shared an account of his journey to Hekla with HOOKER. Historian David McCALLAM argues that toward the late eighteenth and early nineteenth century, the perception of Hekla changed from that of a mountain synonymous with constant and violent fire-spitting to a symbol of "scientific and quasi-imperial 'conquest.'"[107]

One year later, in 1810, the Scottish geologist Sir George MACKENZIE (1780–1848) traveled to Iceland accompanied by British physicians Richard BRIGHT (1789–1858) and Henry HOLLAND (1788–1873). They sought to explore Iceland's volcanic regions. A book about their adventures and observations was published in 1811.[108]

MACKENZIE was a member of the Edinburgh Royal Society.[109] There was much debate at the time, within the Society and in wider academic circles, about the natural processes that formed the Earth. MACKENZIE wanted to find evidence to support and advance the theory championed by Scottish geologist James HUTTON, who had suggested that a great heat inside of the planet was responsible for the formation of rock: an idea called Plutonism. The counter to this was Neptunism, a theory that suggested that crystallization processes in the oceans had formed the sediments. Unlike many of his contemporaries, HUTTON imagined the Earth was in a cycle of constant formation and erosion. His theory of perpetual upheaval and renewal seemed to be confirmed by the frequency of volcanic eruptions. Furthermore, HUTTON emphasized that geological processes occurred gradually, regularly, and repeatedly over long periods, which became an accepted geological doctrine that, in the 1820s, finally ended the conflict between Neptunism and Plutonism.[110]

Between 1814 and 1815, Ebenezer HENDERSON (1784–1858) traveled around Iceland for 13 months. Unlike most of his contemporaries, he spent a winter there; he also traveled further afield than most. HENDERSON was a Scottish Calvinist minister and missionary whose trip was supported by the British and the Foreign Bible Society. The main purpose of his visit was to evangelize, and he managed to distribute at least

---

105 John Thomas STANLEY, MS, JRL 722, John Rylands Library, University of Manchester, Manchester, UK: 95–97.
106 AGNARSDÓTTIR 2013: 24; JÖRGENSEN 2016. William Jackson HOOKER was a British botanist who later in his life became the director of Kew Gardens in London.
107 HOOKER 1813, vol. 2: 105–119; McCALLAM 2019: 212.
108 MACKENZIE, HOLLAND 1811; HOLLAND 1987; OGILVIE 2005: 278–279.
109 For more information on George MACKENZIE, see WAWN 1982.
110 HUTTON 1788; HOLLAND 1987: 254–255; OSLUND 2011: 43. The geology debate is outlined in PORTER 1977.

5,000 copies of the Icelandic translation of the Bible to people whom he described as grateful.[111] HENDERSON also wrote about his adventures: he published a travelogue about his journey in 1818. His publication was special because Sveinn PÁLSSON had been able to give HENDERSON a handwritten copy of his treatise on glaciers and volcanoes. Long before PÁLSSON's manuscript would be published, HENDERSON was able to study it and share some of its revolutionary content with the world. HENDERSON built on a map drawn by PÁLSSON; both had understood that Skaftárjökull was a part of the Vatnajökull ice cap (Figure 58).[112]

> It not only appears to have been more tremendous in its phenomena than any recorded in the modern annals of Iceland, but it was followed by a train of consequences the most direful and melancholy, some of which continue to be felt to this day. Immense floods of red-hot lava were poured down from the hills with amazing velocity, and, spreading over the low country, burnt up men, cattle, churches, houses, and every thing they attacked in their progress. Not only was all vegetation in the immediate neighborhood of the volcano destroyed by the ashes, brimstone, and pumice, which it emitted; but, being borne up to an inconceivable height in the atmosphere, they were scattered over the whole island, impregnating the air with noxious vapors, intercepting the genial rays of the sun, and empoisoning whatever could satisfy the hunger or quench the thirst of man and beast.[113]

HENDERSON's vivid description gives readers a good idea of the horror of the eruption. His evocative text further details the repercussions of the *móðuharðindin*. It is intriguing to read that in 1814 and 1815, at the time of HENDERSON's journey, the echoes of the eruption's consequences remained. HENDERSON describes the Laki fissure as a stretch of conical hills running in a direct line; he also details the landmarks surrounding the fissure.[114] Unsurprisingly, he refers to the fissure as the "Skaptár volcano," as in Iceland, the eruption was known as "the Skaftá Fires." HENDERSON was the first published author to clearly identify the volcano as a fissure which was certainly a unique assertion in 1818, the publication date of his two volumes; his report, however, did not change the trajectory of the scientific debate.

From his footnotes, we can tell that HENDERSON critically engaged with the materials he used: he realized that STEPHENSEN's report included mistakes, which he "altered according to Mr. Paulson's MS. [manuscript]."[115] This suggests that he found PÁLSSON's work more credible than STEPHENSEN's. HENDERSON then linked the eruption to the dry fog in Europe: "The quantity of ashes, brimstone, &c. thrown up into the atmosphere was so great, that nearly the whole European horizon was enveloped in obscurity [. . .]."[116] Just as his peers had in Europe, HENDERSON attempted to bring together

111 AGNARSDÓTTIR 2013: 24.
112 OSLUND 2011: 43; BJÖRNSSON 2017: 157–158.
113 HENDERSON 1818: 274.
114 HENDERSON 1818: 276.
115 HENDERSON 1818: note on 287.
116 HENDERSON 1818: 287–288.

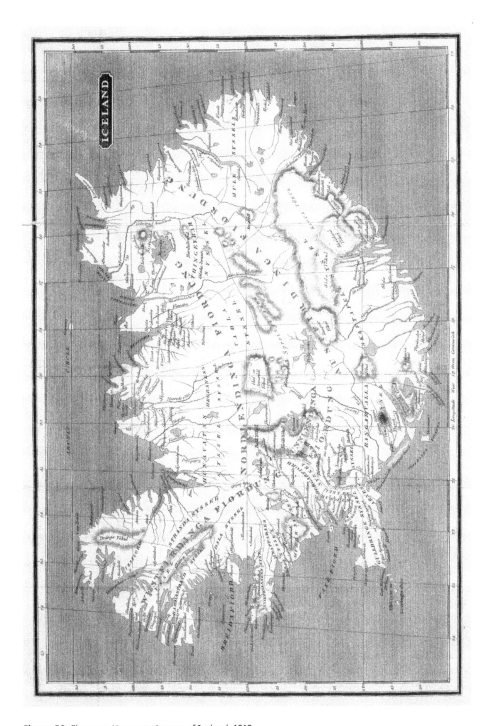

Figure 58: Ebenezer HENDERSON's map of Iceland, 1818.

all the phenomena witnessed that year rather than distinguishing one cause for one effect. Despite this, even in the 1830s, dictionary articles on the topic of the fog still repeated the plethora of theories that had been debated all the way back in 1783.[117]

Perhaps if the Tambora eruption of 1815, the largest in recorded history, had happened in full view of the outside world, there would have been more interest in HENDERSON's publication just three years later. The Tambora eruption dramatically changed the weather in large parts of Europe and North America, causing a year without summer in much of the world in 1816. Famously, this inspired novelist Mary SHELLEY, on a retreat in Switzerland, to write *Frankenstein*. Although reports about the Tambora eruption eventually circulated, its occurrence remained unconnected to the cold weather of the following year for almost one century.[118]

In 1830, British geologist Charles LYELL (1797–1875) wrote his *Principles of Geology*, which became the foremost geology textbook of the nineteenth century. The text explains that Iceland plays host to a volcanic eruption or large earthquake at least once every 20 to 40 years.[119] It reaffirms how "terrible" Hekla's eruptions are: "So intense is the energy of the volcanic action in this region, that some eruptions of Hecla have lasted six years without ceasing. Earthquakes have often shaken the whole island at once."[120] He states that volcanic and seismic activity changes the landscape: rivers change their course, new lakes appear, hills sink, and "new islands have often been thrown up near the coast, some of which still exist, while others have disappeared."[121]

The text mentions both the Nýey and "Skaptár Jokul" eruptions in Iceland in 1783. LYELL understood the direction of the spreading axis in the center of Iceland: "Many cones are often thrown up in one eruption, and in this case they take a linear direction, running generally from the northeast to the southwest, from the northeastern part of the island where the volcano Krabla [Krafla] lies, to the promontory Reykjanes."[122] He believed that

> [t]he convulsions of the year 1783 appear to have been more tremendous than any recorded in the modern annals of Iceland; and the original Danish narrative of the catastrophe, drawn up in great detail, has since been substantiated by several English travelers, particularly in regard to the prodigious extent of country laid waste, and the volume of lava produced.[123]

---

117 BRANDES et al. 1833: 34–53.
118 SYMONS 1888: 29, here, the eruption is listed as "Tomboro." HUMPHREYS 1913: 369; KRÄMER 2015: 23–26; BRÖNNIMANN, KRÄMER 2016: 11. Daniel KRÄMER has established that in 1913, William Jackson HUMPHREYS proposed a connection between the cold weather and the Tambora eruption.
119 Today we know that Iceland sees a volcanic eruption every three to five years on average.
120 LYELL 1830: 371.
121 LYELL 1830: 371.
122 LYELL 1830: 361, 372 (quote).
123 LYELL 1830: 372.

In an accompanying footnote, he mentions STEPHENSEN and HENDERSON's publications. He also gives credit to "Mr. Paulson's" manuscript and states that PÁLSSON had visited "the tract" in 1794. LYELL further details the course of the eruption, the drying-up of the rivers, and their subsequent engorgement with lava.[124]

In his conclusion, LYELL seems mesmerized by the sheer volume of "melted matter produced in this eruption." He calculated that the two rivers, which had been filled with lava during the eruption, had a length each of 40 to 50 miles and that the Skaftá River was 12 to 15 miles in breadth and the Hverfisfljót River seven. He estimated the height of the lava to be between 100 and 600 feet and claims that this volume exceeds any other. The Skaftá lava, at its peak, "rival[ed] or even surpass[ed] in height Salisbury Craigs and Arthur's Seat."[125] He was not aware of any "ancient strata" or "igneous rocks of such colossal magnitude" – but concedes that the geologists so far had "hitherto investigated but a small part of the globe."[126]

Frederick HAMILTON-TEMPLE-BLACKWOOD (1826–1902) was a British diplomat and politician better known as Lord DUFFERIN. His travels took him to Iceland, Norway, and Spitsbergen; during these voyages he wrote many letters to his mother. Upon his return, he turned these letters into a book, *Letters from High Latitudes*, which became very popular and was translated into several languages. His book was instrumental in turning Iceland into a tourist destination.[127] Travelogues like these, said to describe "the real Iceland," were much sought after by Europeans during the late Enlightenment. After the Napoleonic Wars, travelogues in many European languages became widely available.[128]

Other scholars who traveled to Iceland in the second half of the nineteenth century and produced notable publications include William PREYER (1841–1897) and Ferdinand ZIRKEL (1838–1912), who together published a travelogue in 1862. PREYER, an English-born zoologist, and ZIRKEL, a German geologist, had both traveled to Iceland and the Faroe Islands. Their publication included scientific observations and a chronology of past Icelandic eruptions, including the eruption of "Skapárjökull."[129]

---

124 LYELL 1830: 372–374.
125 LYELL 1830: 374–375.
126 LYELL 1830: 376. Today we know that there are several large igneous provinces around the planet (see also Figure 6 in this book). Their volume exceeds the Laki eruption's lava volume by orders of magnitude.
127 DUFFERIN 1856; HANSSON 2009.
128 AGNARSDÓTTIR 2013: 24.
129 PREYER, ZIRKEL 1862: 462–468. This source details the consequences of the Laki eruption for Iceland, but there is no reflection of the consequences of this eruption outside of Iceland.

Rediscovering Sveinn PÁLSSON

It took almost a century for the Laki fissure to find its way back into the scientific discourse: when it did, things moved quickly. In the 1880s and 1890s, the Laki fissure once again drew explorers to Iceland; literature about those journeys invigorated interest in the topic.

Icelandic geologist and geographer Þorvaldur THORODDSEN (1855–1921) stumbled upon Sveinn PÁLSSON's unpublished manuscript, *The Physical, Geographical, and Historical Descriptions of the Icelandic Ice Mountains*, in an archive in Copenhagen.[130] In 1879, THORODDSEN published "The Volcanic Eruption in Iceland in the Year 1783" in *Geografisk Tidsskrift*, a Danish geography journal. In his paper, THORODDSEN describes the fissure and details PÁLSSON's travels and findings alongside mentions of STEPHENSEN's journey.[131] As early as 1880, THORODDSEN had done extensive work on Icelandic volcanism and published a chronology of past volcanic eruptions since around 900 CE.[132]

Amund HELLAND (1846–1918), a Norwegian geologist, glaciologist, and later a professor, traveled extensively between 1875 and 1877, visiting Norway, Iceland, Greenland, Italy, Germany, and England.[133] HELLAND likely heard mention of Sveinn PÁLSSON's manuscript whilst researching in Copenhagen.[134] In the early summer of 1881, HELLAND returned to Iceland on a Swedish cargo ship.[135] It is safe to assume he was interested in Iceland's geology, volcanoes, and glaciers.

A footnote in THORODDSEN's posthumously published German manuscript clarifies that HELLAND and THORODDSEN knew each other. When HELLAND came to Reykjavík, he asked THORODDSEN to recommend an interesting volcano to visit; THORODDSEN suggested the Laki crater row "without reservations."[136] THORODDSEN further supplied HELLAND with a handwritten copy of Sveinn PÁLSSON's manuscript.[137] HELLAND then made the journey to the fissure, arriving on 14 August 1881 and staying in its proximity until 18 August. According to Þorvaldur THORODDSEN, only one other person had visited Laki

---

130 THORODDSEN (1879: 67) states that he found PÁLSSON's manuscript about the glaciers at the Royal Library (Kgl. Bibliothek), in the Ny Kgl. Saml. (new royal collection), nr. 1094 b-c. THORODDSEN (1879: 70) further writes that he found PÁLSSON's manuscript titled "Tillaeg Beskrivelserne over den Volcan, der braendte i Skaptafells syssel Aar 1783, sam let ved en Rejse i Egnene 1793 og 1794 med et kort" in the "isl. Lit. Selsk. Arkiv fol nr. 23," presumably also in Copenhagen.
131 THORODDSEN 1879.
132 THORODDSEN 1880.
133 KRISTJÁNSSON, KRISTJÁNSSON 1996: 28; BRYHNI 2009.
134 BJÖRNSSON 2017: 157–158; HELLAND 1886: 8.
135 KRISTJÁNSSON, KRISTJÁNSSON 1996: 29.
136 THORODDSEN 1925: note on 34. "Ich hatte 1879 über diese Ausbrüche, die mich sehr interessierten, geschrieben, und da A. Helland in Reykjavík mich frug, ob ich ihm einen besonders interessanten Vulkan nachweisen könnte, schlug ich ihm ohne Bedenken die Kraterreihe des Laki vor, worauf er dann auch hinreiste."
137 KRISTJÁNSSON, KRISTJÁNSSON 1996: 31.

since PÁLSSON's visit: an Icelandic politician, Jón GUÐMUNDSSON, who had made the trip in September 1842 while residing in Klaustur; this made HELLAND the first foreigner to visit the Laki fissure.[138]

Upon his return to Norway, HELLAND published papers on the various parts of his journey around Iceland, focusing particularly on matters of volcanology and geomorphology.[139] He also published several parts of PÁLSSON's manuscript in 1881, 1882, and 1884 in *Den Norske Turistforeningens Årbok*, the Yearbook of the Norwegian Travel Association. The title of these Danish publications translates as "Descriptions of the Icelandic volcanoes and glaciers by the Icelander Sveinn Pálsson."[140] HELLAND, the first geologist to inspect the Laki fissure, was appointed professor of geology at the University of Kristiania, today's Oslo, in 1883.[141]

In 1882, Scottish geologist Archibald GEIKIE (1835–1924) published his seminal work, *Text-Book of Geology*.[142] From the book, it is clear that the Laki eruption – at this point still referred to as "the eruption of Skaptar-Jökull" – caused great quantities of fine dust to fall on Caithness in Scotland. "In the year 1783, during an eruption of Skaptar-Jökull, so vast an amount of fine dust was ejected that the atmosphere over Iceland continued loaded with it for months afterward. It fell in such quantity over parts of Caithness – a distance of 600 miles – as to destroy the crops." This led GEIKIE to conclude that distant volcanoes can have far-reaching effects.[143] Although GEIKIE established a connection between the Laki eruption – albeit under a different name – and the ash fall observed in Scotland that year, he did not explicitly establish a relationship between the Laki eruption and the dry fog of 1783.

As late as 1887, a German encyclopedia's entry for *Herauch*, another word for *Höhenrauch*, stated that the dry fog of 1783 was caused by "great volcanic eruptions in Calabria and Iceland."[144] It was serendipitous that THORODDSEN and HELLAND had published about the Laki eruption in 1879, 1881, and 1882, since in late 1883, a colossal volcanic eruption in the tropics would alter the scientific debate and cast new light on old questions.

---

138 THORODDSEN 1925: 33–35.
139 KRISTJÁNSSON, KRISTJÁNSSON 1996: 31–33.
140 HELLAND 1881; HELLAND 1882; HELLAND 1884.
141 KRISTJÁNSSON, KRISTJÁNSSON 1996: 28, 33.
142 KRISTJÁNSSON, KRISTJÁNSSON 1996: 33.
143 GEIKIE 1882: 219.
144 "Herauch" in MEYERs Konversations-Lexikon 1887, vol. 8: 402. "Ebenso wie der H. des Jahrs 1783 aus den großartigen vulkanischen Ausbrüchen erklärt wird, die in diesem Jahr in Kalabrien und Island stattfanden [. . .]."

The Krakatau Eruption of 1883

On 27 August 1883, a volcanic eruption produced what was probably the loudest noise in recorded history. Its source was more than 12,000 kilometers away from the Laki fissure, yet its intellectual and physical reverberations would reach all the way to Iceland and beyond. The source of this loud noise was Krakatau, a volcano in the Indonesian Sunda Strait between the islands of Java and Sumatra. The island of Krakatau itself was uninhabited, but the Sunda Strait was busy with fishermen and other vessels.[145] The volcano erupted violently, reaching a six, "colossal," on the volcanic explosivity index. The noise of the eruption was audible from Perth and Alice Springs in Australia (3,000 and 3,500 kilometers away, respectively) to Rodrigues, an island near Mauritius in the Indian Ocean (4,700 kilometers away) (Figure 59).[146] The volcano is known as Krakatoa; however, Krakatau is its Indonesian name.

An earlier, smaller eruption had occurred in May 1883, news of which quickly traveled to Europe. Readers of newspapers were getting used to the name "Krakatau" with its various spellings. The eruption's final stage lasted from 26 to 27 August 1883; four final explosions almost destroyed the entire island of Krakatau and submerged most of it under the sea. The ejected ash and gases turned the area around the volcano dark. Pyroclastic flows, a tsunami, and invisible pressure waves followed. The tsunami reached Batavia, today's Jakarta, around two and a half hours later.[147] In South Africa, the resulting tsunami reached a height of half a meter. In Europe, a sea level change caused by the tsunami was registered. The eruption had a death toll of around 35,000 people, mostly caused by the tsunami.[148]

The Tambora eruption, although much larger than Krakatau, had not furthered the scientific understanding of dust transportation over long distances. Krakatau deposited a tephra layer a few centimeters thick 1,500 kilometers away and even produced a light layer of dust on ships some 6,000 kilometers away.[149] The ejecta penetrated high into the atmosphere, the dust of which later changed the apparent color of the sun and the moon. Furthermore, it gave the impression that both were encircled by halos. These optical effects lasted for as long as three years after the eruption.[150] The dust particles traveled westward around the globe; two weeks after the eruption, they formed a band around the equator. Later the particles formed belts in the Northern and Southern Hemispheres, moving gradually to higher latitudes. Starting in early November 1883,

---

145 SCHRÖDER 2003: 391.
146 SYMONS 1888: 78–88; WINCHESTER 2005: 236–237.
147 SCHRÖDER 2003: 389; WINCHESTER 2005: 146–154, 167–169, 177–178, 193–214, 228–234.
148 Global Volcanism Program: Krakatau; SCHRÖDER 2003: 392; WINCHESTER 2005: 221.
149 WILKENING 2011: 322. Precise maps of floating pumice and volcanic dust observed in the Indian Ocean until October 1883 can be found in SYMONS 1888: plate 4, 56.
150 WILKENING 2011: 322.

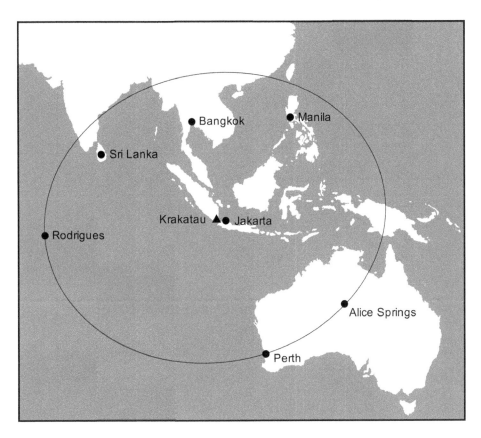

Figure 59: The explosive sound of the Krakatau eruption. The explosion was audible within the encircled area on the map.

unusual skies were reported as far away as England and Denmark.[151] The observed phenomena included amazing sunsets and a prolonged twilight period that featured a lurid afterglow. Policy analyst Ken WILKENING tells of a tale from London on 8 November 1883, when an afterglow was so intense that people alerted the fire brigade.[152] In the second half of November 1883, these phenomena also appeared in North America and Iceland (Figure 60).[153]

Something that was only apparent thanks to the scientific instruments of the time was the extent of the atmospheric pressure wave produced by the final explosive phase of the eruption. This phenomenon was recorded by observatories and amateur weather watchers alike.[154] Early in 1884, The Royal Society of London formed the so-

---

151 SCHRÖDER 2003: 393.
152 WILKENING 2011: 322.
153 KIESSLING 1885; SCHRÖDER 2003: 393.
154 SYMONS 1888: 58–88; SCHRÖDER 2003: 391–392.

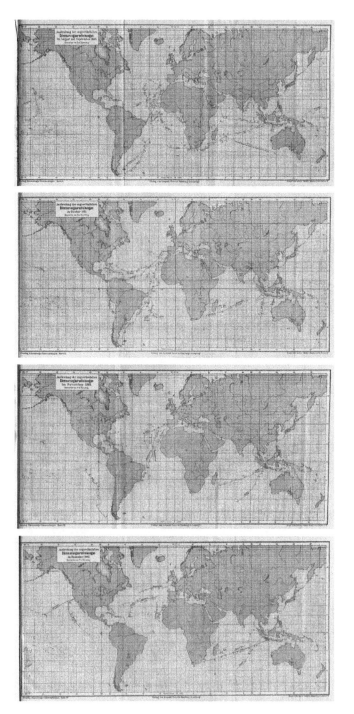

Figure 60: The extent of the prolonged twilight appearances between August and December 1883 as depicted by Johann KIESSLING, 1888.

called Krakatoa Commission, which consisted of 13 members. The committee's task was to collect all the facts they could gather about the eruption's impact, particularly concerning pumice and ashfall, irregularities in pressure, and tidal waves. They procured data from ship logs, weather stations, eyewitness statements, newspaper reports, and any other source they deemed appropriate. The committee worked for almost five years and published a 500-page report in 1888.[155] A large portion of this comprehensive text dealt with the observed optical phenomena in the eruption's aftermath. The report revealed that a single natural event could have a marked impact on the entire planet.[156]

Hypotheses recorded by the committee show that some believed the volume of volcanic ejecta produced by the eruption was insufficient to be responsible for the wide-ranging and long-lasting effects around the globe. Echoes of the previous century could be heard in their beliefs that the tail of a comet, or perhaps "needles of ice, or a cyclone in the sun's photosphere," could be to blame. The committee, however, eventually concluded that the eruption of Krakatau was responsible for the phenomena observed from 1883 to 1886. They asserted that the dust was injected into the atmosphere to a height of 30 kilometers and that this "smoke stream," consisting of dust and water vapor, traveled with the trade winds two and a half times around the equator, with some dust spreading into the mid-latitudes.[157]

The text reads:

[Judging] by the quantity of materials ejected, or by the area and duration of the darkness caused by the volcanic dust, the eruption of Krakatoa must have been on a much smaller scale than several other outbursts which have occurred in historic times. The great eruptions of Papandayang in Java, in 1772, of Skaptar Jokull (Varmárdalr [sic]) in Iceland, in 1783, and of Tomboro [sic] in Sumbawa, in 1815, were all accompanied by the extrusion of much larger quantities of material, than that thrown out of Krakatoa in 1883.[158]

The committee erroneously believed the Laki eruption, here still named after the Icelandic glacier and the valley of its location, to be larger than the Krakatau eruption.

In the committee's publication, one can find mention of the "Skaptar Jökull" alongside several remarks on the unusual atmospheric phenomena witnessed in 1783. In particular, the authors assert, "The year 1783 was remarkable for a thick dry mist or fog which spread over Europe in June, and continued, more or less, for three or four months." The authors of the report were even aware of the effects of the dry fog:

In some places, objects at 5 kilometres (3 miles) distance could not be distinguished; the sun was red and was invisible sometime after rising, and before setting. [. . .] The fog of 1783 commenced

---

155 WINCHESTER 2005: 242–252; WILKENING 2011: 323; PYLE 2017: 154–158; MORGAN, no date.
156 WILKENING 2011: 322.
157 WILKENING 2011: 323.
158 SYMONS 1888: 29. The spelling of Varmárdalr (without the u at the end) is identical to the spelling used by Þorvaldur THORODDSEN (1880: 458–468). This might have served as the basis for the committee's background information on the 1783 eruption.

about the same day (18 June) at places distant from each other, such as Paris, Avignon, Turin, and Padua. [. . .] The sun's disc was altogether obscured at rising and setting, and half-obscured during the daytime. The sky was blood-red at rising and setting of the sun. Great consternation prevailed in northern Europe.[159]

The dramatic events of 1815 are also mentioned: "Tomboro [sic], Sumbawa. 7 to 12 April 1815. This eruption was the greatest since that of Skaptar Jokull in 1783. For three days, there was darkness at a distance of 300 miles."[160] Furthermore, the report found:

> Of the period since 1750, thirteen years may be named as specially marked by numerous widespread or great eruptions, [. . .]. In 1783, 1831, and 1883, the sun was seen rayless, or like the moon, in some parts of the world, and in these years, the sunset after-glows were most conspicuous and long-enduring.[161]

The context of the report is important; in 1883, information spread much more quickly than ever before, leading some scholars to believe that a greater number of volcanic eruptions was occurring than ever before. However, this illusory notion came down to the fact that the world simply had improved methods of communication.

In 1888, Johann KIESSLING (1839–1905), a professor at the University of Hamburg, analyzed the optical phenomena witnessed in the aftermath of the Krakatau eruption, particularly the afterglow seen during dusk. He published his findings as *Untersuchungen über Dämmerungserscheinungen zur Erklärung der nach dem Krakatau-Ausbruch beobachteten atmosphärisch-optischen Störung* (Investigations into Twilight Phenomena to Explain the Atmospheric-Optical Disturbance Observed after the Krakatau Eruption).[162] He looked to the past to find historical precedents and found what he believed to be a comparable event in 1783: a fog that turned the sun red. KIESSLING had read reports of violent volcanic activity within Iceland "which had a causal relation to this fog."[163]

The main factor that led to a global recognition that the Krakatau eruption had considerable and far-reaching consequences was telegraphy. Krakatau was located within the Dutch East Indies in a colony founded in 1800 after the Dutch East India Company (*Vereenigde Oost-Indische Compagnie*, VOC) went bankrupt. The colony's

---

159 SYMONS 1888: 388–389.
160 SYMONS 1888: 393.
161 SYMONS 1888: 403.
162 KIESSLING 1888.
163 KIESSLING 1888: 26–27. "Ausser dem südlichen Italien war jedoch auch Island im Sommer 1783 der Schauplatz heftiger vulkanischer Thätigkeit gewesen, welche mit der Entstehung dieses Nebels in ursächlichem Zusammenhang zu stehen scheint." Johann KIESSLING used Karl Ernst Adolf VON HOFF's compilation of volcanic eruptions as his source for the foggy appearance just before the earthquakes in Calabria; HOFF 1841, vol. 2: 48–53.

capital, Batavia, was located around 150 kilometers from the volcano.[164] The fact that Batavia was a hub of international activity accelerated the speed at which the news of the eruption traveled to Europe and beyond. In October 1883, the Dutch also initiated a scientific commission to study the Krakatau eruption. Among the scientists in the commission was Dutch geologist Rogier Diederik Marius VERBEEK (1845–1926), who lived in Java at the time. He was an eyewitness and documented the events in a journal, which he later published as a report.[165]

Samuel MORSE (1791–1872) helped invent the electric telegraph in the first half of the nineteenth century and by the 1880s, networks were well-established.[166] By August 1883, the world had become a "global village."[167] In 1755, it took four weeks for the news about the Lisbon earthquake to be printed in the *Hamburgischer Unpartheyischer Correspondent*. In 1783, it took almost three months for the news of an Icelandic eruption to reach Denmark. By contrast, in 1883, the *Hamburgischer Unpartheyischer Correspondent* first mentioned the Krakatau eruption on 29 August, only two days after it had occurred, thanks to a telegram from the Reuters news agency. On 31 August 1883, the newspaper followed up with a detailed report about the eruption featuring a mention of the tsunami, which was compared to the tsunami that followed the 1755 Lisbon earthquake: this clearly indicates that the catastrophe in Lisbon was still present in the European consciousness.[168] As early as 6 September 1883, the academic journal *Nature* reported on the Krakatau eruption; the article was titled "The Java Upheaval." It followed up on this with another report, titled "Scientific Aspects of the Java Catastrophe," on 13 September 1883.[169] German meteorologist Wilfried SCHRÖDER opined that the Krakatau eruption and its volcanic dust gave scholars and amateur weather observers a chance to think deeply about geological processes, global wind systems, and the atmosphere.[170]

This volcanic eruption in the Dutch East Indies, separated from Iceland by a great distance and the Laki eruption by a century, significantly impacted the understanding of the 1783 eruption and volcanoes in general. A century later, Europe witnessed phenomena comparable to those of 1783; most were, at the same time, fully aware that a distant volcano on another continent had exploded into life. The fact that information about this colossal volcanic eruption in the Dutch East Indies arrived in a timely fashion, i.e., while scientists were observing those strange phenomena, made it much easier for them to connect the dots between the eruption, the blood-red sunsets, and the dust in the air.

---

164 WINCHESTER 2005: 35–41, 141.
165 VERBEEK 1886; MORGAN without year.
166 WENZLHUEMER 2013.
167 McLUHAN 1962; WINCHESTER 2005: 182.
168 WILKE 2014.
169 Anonymous 1883a; Anonymous 1883b.
170 SCHRÖDER 2003: 393–394.

The Dots Connected

Amund HELLAND and Þorvaldur THORODDSEN

In 1886, Amund HELLAND published a Danish-language book on Laki's craters and lava flows. In his 40-page publication, he describes the eruption and its consequences and outlines his experiences on his journey to the fissure.[171] He only spent a few days at the Laki fissure, and therefore he did not explore the entire extent of the lava, leading to some errors in his map illustrations (Figure 61). Nevertheless, the book and the map serve as evidence that he did, in fact, visit the fissure.

HELLAND's book included a new hypothesis:

> It is generally known that an unusual glow of the sky could be observed around sun rise and fall in the years after the Krakatoa eruption in the Strait of Sunda. This could be observed in India in September 1883 and in Norway in November 1883. As a rule, this sky glow can be explained with volcanic particles, which comes from Krakatau [. . .], these particles can stay in the atmosphere for months.[172]

As he observed the dust in Oslo, he was drawn to conclude that it must be present everywhere in the Earth's atmosphere. "Interestingly, similar phenomena could be observed in Denmark and Iceland after the eruption of 1783, and even then, the sky had changed its color and can be connected with the eruption."[173] HELLAND also included a quote from HÓLM's 1784 report, which mirrored his own description of the sky. He realized that it must have looked as spectacular in 1783 as it had in Norway the previous year – the latter, thanks to Krakatau.[174]

100 Years after PÁLSSON

In 1886, only 20 to 25 percent of Iceland's volcanoes were known. Most scholars traveled the same routes, almost exclusively within populated regions. Few ventured into the highlands: Þorvaldur THORODDSEN was an exception to this rule. Between 1882 and 1898, THORODDSEN extensively explored and mapped Iceland; for instance, he was the first geologist to explore Vatnajökull from all sides.[175] His primary focus was on volcanoes and landscape morphology.

---

171 HELLAND 1886: 8, 24–27.
172 HELLAND 1886: 33. "Som bekjendt har man i de senere aar efter det store udbrud i Krakatoa i Sundastrædet iagttaget en eiendommelig golden af himmelen, der har ledsaget solens opgang og nedgang. Denne glod blev allerede i september 1883 iagttaget i Indien og her i Norge i november 1883."
173 HELLAND 1886: 33. HELLAND refers to Oslo as Kristiania, which was Oslo's name from 1877 to 1925. "Interessant er det nu, at lignende fænomener er iagttaget i Danmark og paa Island efter udbrudet i 1783, og at man allerede da satte denne himmelens farve i forbindelse med udbrudet."
174 HELLAND 1886: 34.
175 HERRMANN 1907: 70–71; THORODDSEN 1914; THORODDSEN 1925: 9, 36.

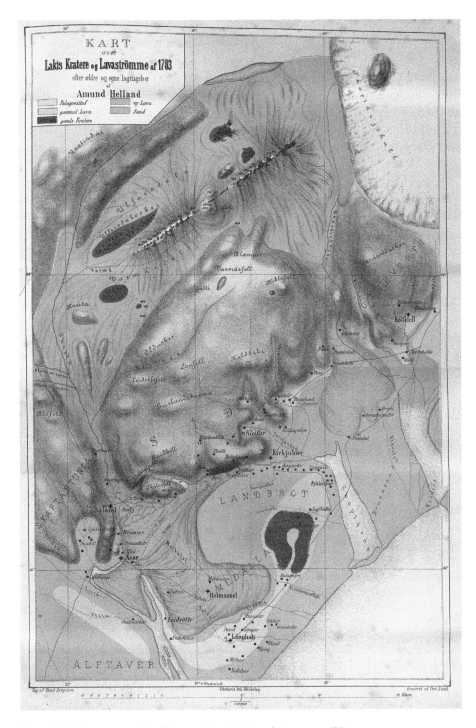

Figure 61: Laki's craters and lava flows, as drawn by Amund HELLAND, ca. 1881.

From 18 to 21 July 1893, THORODDSEN followed the Skaftá River through the Síða region to the Laki fissure. In 1794, PÁLSSON had noticed a subterraneous heat coming from the vents as well as "ugly smelling smoke" and "an eerie roaring from deep down."[176] By the time of THORODDSEN's expedition, the area had cooled down entirely. While visiting the "crater row," he was fortunate enough to enjoy clement weather and used the opportunity to take measurements. He estimated that the fissure had a length of 30 kilometers, the lava covered 565 square kilometers with a volume of 12.3 cubic kilometers, and the tephra and scoria volume totaled three cubic kilometers – all of which come very close to modern estimates.[177]

After his visit to the Laki fissure, THORODDSEN traveled further north. On 22 July 1794, he came across an even larger fissure that had produced a similar scar in the landscape. He called the fissure *Eldgjá*, the "fire fissure." It was 75 kilometers long. THORODDSEN assumed that this fissure erupted around 950, shortly after the settlement of Iceland.[178] In his subsequent publications in Danish, English, and German, he extensively discusses PÁLSSON's manuscript and findings on volcanoes and glaciers in the context of his own research. Additionally, THORODDSEN drew a geological map of Iceland. He used Björn GUNNLAUGSSON's map from 1844 but made his own adjustments, particularly to the depictions of the highlands and the regions that Icelandic cartographer GUNNLAUGSSON (1788–1876) himself had never visited. THORODDSEN's map shows the Vatnajökull ice cap as a single glacier, the first map to do so. Furthermore, his map featured the Skaftá River, Mount Laki, and Eldgja.[179]

The practice of using color schemes for geological maps was developed in the late seventeenth and early eighteenth century; THORODDSEN added color to his map, albeit in a slightly different manner than today's cartographers. Today, the color scheme used in geological maps shows lithographical units and temporal patterns of rock succession (Figure 62).[180] A significant color difference on a geological map shows a hiatus, i.e., an unconformity of the rock succession. This means that a particular stratum was unpreserved, in some cases due to erosion.[181] With their color schemes, modern geological maps inform the reader about the different units of rocks, their ages, and

---

176 THORODDSEN 1925: 59. "Als Sveinn Pálsson 1794 die Kraterreihe besuchte und die Krater bestieg, bemerkte er noch die unterirdische Wärme, denn aus den Rissen 'stieg der abscheuliche und häßlich riechende Rauch hervor und gleichzeitig hörte man tief unten ein unheimliches Sausen.' Jetzt scheinen die Lavaströme völlig abgekühlt zu sein."

177 THORODDSEN 1894: 295.

178 THORODDSEN 1893: 177–183; THORODDSEN 1925: 67: "Eine offene Spalte dieser Größe, die große Lavaströme ausgegossen hat, findet nicht ihresgleichen auf Island und vielleicht nicht auf der ganzen Erde." The Eldgjá eruption is estimated to have occurred between 934 and 939. For further information on the eruption and modern dating: THORDARSON et al. 2001; McCORMICK, DUTTON, MAYEWSKI 2007; KOSTICK, LUDLOW 2015; OPPENHEIMER et al. 2018; EBERT 2019; EBERT 2021.

179 THORODDSEN 1880; THORODDSEN 1882a; THORODDSEN 1882b; THORODDSEN 1925; BJÖRNSSON 2017: 158.

180 Commission for the Geological Map of the World, BOUYSSE, 2014.

181 FRIEDRICH 2019.

Figure 62: Þorvaldur Thoroddsen's geological map of Iceland, 1906. The white color indicates glaciers; the dotted light green marks out the *sandur* plains, dark green the lava fields, and orange the highlands. The red dots indicate the locations of volcanic fissures.

the nature of the contact zones between them. Therefore, "geological maps are generally the most important compilations of geological work."[182]

Thoroddsen remarks that although Amund Helland had been the first foreign naturalist to visit the Laki fissure, he was only there for a short time on a trip marred by bad weather and fog. Thoroddsen holds that although Helland had made several interesting observations, his map, based on speculations rather than measurements, was inadequate.[183] Although Helland's map work does indeed have shortcomings, he is responsible for bestowing upon the eruption the name that stuck.[184]

What's in a Name?

The Laki eruption has been known by many different names since 1783. Within Iceland, the eruption is known as *Skaftáreldar* ("Skaftá Fires") or *Skaftárgígar* ("Skaftá craters").[185] Older names such as *Síðueldur* ("a fire in the Síða region") are still in use. Many remember the eruption for its consequences and, therefore, refer to it as *móðuharðindin* ("the famine of the mist"). Other names reference nearby landmarks, most obviously Mount Laki. Sometimes, particularly in the European newspapers, it was called "Skaftárjökull" in its various spellings, as it was believed – not unreasonably – that the eruption had occurred underneath the glacier.

Initially, as the location of the eruption was unclear, the event was referred to as "fire(s)," a common term used for large volcanic eruptions in Iceland at the time. An observer might not easily distinguish the entire fissure as (the source of) an individual phenomenon, given that it is composed of many craters.[186] Outside of Iceland, during the late eighteenth century, the concept of a fissure volcano or flood basalt event was quite foreign. In the popular imagination, volcanoes resembled Vesuvius, Etna, or Stromboli.

Amund Helland noticed the plurality of names in circulation and then proposed a solution. He suggested naming the row of craters after the mountain in its center: Laki, a short and easy-to-pronounce name. Thus, today, in non-Icelandic literature, the eruption is called the Laki eruption or *Lakagígar*, which means the "Laki craters."[187] Mount Laki, although volcanic in origin, did not erupt in 1783: a prior eruption at the Grímsvötn system during the Holocene, probably the Botnahraun eruption of 4550 BCE ± 500 years, produced this hyaloclastite palagonite volcano.[188] This term indicates that the area was, at the time,

---

182 Friedrich et al. 2018: 174. For a history of assigning colors to rock units, see also: Schäfer-Weiss, Versemann 2005; Oldroyd 2013; Bressan 2014.
183 Thoroddsen 1894: 294–295.
184 Kleemann 2020.
185 Karlsson 2000b: 178.
186 Thoroddsen 1925: 46–47.
187 Thoroddsen 1925: 46–47.
188 Einarsson, Sveinsdóttir 1984: 48. Information on the Botnahraun eruption can be found in the "eruptive history" section in Global Volcanism Program: Grímsvötn.

covered by a glacier. In the eighteenth century, this area was remote and sparsely populated; shepherds sometimes came here to herd their sheep. The shape of the mountain reminded them of the third compartment of the stomach of a dairy cow, which looks like the fanned pages of an open book. This is called *omasum* in English and *laki* in Icelandic.[189] In Iceland, the lava field produced by the eruption is called *Skaftáreldahraun*.[190]

Field Trips to the Laki Fissure

Soon, photographs would make Iceland accessible to faraway spectators. While photography had been around since the 1820s, the cumbersome equipment necessary made it impractical for anyone on the move. During the 1890s and 1900s, photography made several advances: film replaced photographic plates and soon after, it became possible to switch film during daylight hours.[191]

When Tempest ANDERSON (1846–1913), a British surgeon and amateur photographer, traveled to the Laki fissure in the summer of 1890, he did so because he had read Lord DUFFERIN's *Letters from High Latitudes* and wanted to visit "the craters that had never been visited."[192] It was only upon his return that he learned that Amund HELLAND and Sveinn PÁLSSON had already visited the fissure and called it *Lakis Kratere*. ANDERSON himself liked the name *Skaptár Lava* better.[193] It appears that the name bestowed upon the fissure by Helland was not yet firmly established, even in the early twentieth century. ANDERSON was the first person to take photographs of the Laki fissure, which were published in 1903 (see Figure 63 for the photo from 1903 and Figure 64 for a modern color photo). He also captured several other volcanoes from around the globe on film and later published the results in his book.[194]

It remains unclear whether ANDERSON's photograph of Laki made it sufficiently obvious that the "Skaptá Lava" was, in fact, a fissure volcano. Further research is needed to establish whether this image and its distribution changed public perception of the Laki fissure or had any impact on scientific discourse at the time.

Other scholars who visited the Laki fissure in the early twentieth century and wrote about their experiences include German geologists Karl SAPPER in 1906 and Hans RECK in 1908.[195] Karl SAPPER (1866–1945) stayed at the fissure for three days.[196]

---

189 The history of the name is described and illustrated by ÁMUNDASON (2019). Additionally, the entry about Laki's name was also featured on Icelandic milk cartons. I thank Óðinn MELSTED and his father, Eyjólfur MELSTED, for this information.
190 Katla Geopark Project: 10–11.
191 BLANCHARD 1998: 508–513.
192 DUFFERIN 1873.
193 ANDERSON 1903: 122.
194 THORODDSEN 1925: 35–36.
195 SCHWARZBACH 1983: 42.
196 ÞÓRARINSSON 1969: 910–912.

PLATE LXIV.   [*To face page* 124.

Figure 63: A photograph of one crater in the Laki fissure, taken by Tempest ANDERSON, ca. 1903. ANDERSON published his book and its photographs in 1903. This one is called "Plate LXIV. Iceland. A Crater, Skaptár lava."

He was aware of Þorvaldur THORODDSEN and Amund HELLAND's publications and of the fact that the fissure had erupted in 1783. He notes that Laki was also known as "Varmárdalr" or "Skaptárjökull" (in its various spellings).[197] SAPPER also visited Eldgjá and concluded that these two "large eruptions" (*Riesenausbrüche*) "without a doubt outdid all other volcanic eruptions in historical time."[198]

In the summer of 1907, German geology student Walter VON KNEBEL and landscape painter Max RUDLOFF traveled through and studied some of the most remote areas of Iceland. Ultimately, they failed to return from their trip. This prompted their friend

---

[197] SAPPER 1917a: 65; SAPPER 1917b.

[198] SAPPER 1917a: 72. "Unter den selten tätigen Ausbruchsstellen sind aber einige, die sich durch ganz besondere Intensität des Ausbruchs ausgezeichnet haben, so vor allem Laki 1783 und Eldgjá um 930. Es waren diese Riesenausbrüche, die auf der Erde in geschichtlicher Zeit nicht ihresgleichen haben, sofern man die Lavaförderung zum Maßstab nehmen will [. . .]."

Figure 64: In contrast, one crater of the Laki fissure in August 2016. The size becomes apparent when looking at the person walking along a path at the very bottom of the image.

and fellow geology student Hans RECK (1886–1937) and VON KNEBEL's fiancée, Ina VON GRUMBKOW (1872–1942), to travel to Iceland in June 1908 to find their remains. RECK and VON GRUMBKOW traveled on horseback for 11 weeks. They did not, however, find the remains of the two Germans who had vanished.[199] The trip likely inspired RECK to write his doctoral dissertation about Icelandic "mass eruptions." It also brought RECK and VON GRUMBKOW closer, for they married in 1912.[200]

Among other sights, Hans RECK and Ina VON GRUMBKOW's journey took them to the Laki fissure. In 1909, VON GRUMBKOW published a travelogue about her journey through Iceland, which was dedicated to her late fiancé. It is likely that Ina VON GRUMBKOW was the first woman to visit the Laki fissure. One chapter of VON GRUMBKOW's book describes her travels to Laki; from her account, we learn that RECK "tried to use" HELLAND's map to find the fissure. This was difficult because sandstorms had significantly changed the landscape since HELLAND's visit in 1881.[201] RECK and VON GRUMBKOW relied on local guides, one of whom, Sigurður, was the son of the man who had guided THORODDSEN in 1894. The journey to Laki was exhausting: some rivers were impassable; a fog reduced the visibility; the horses were tired; and at times, RECK and VON GRUMBKOW had no idea where they were. Once they arrived at their campsite near the fissure, they were exhausted. The next day, RECK, VON GRUMBKOW, and Sigurður hiked four hours to reach

---

199 MAIER 2003: 86–87.

200 MAYR 2013: 232.

201 GRUMBKOW 1909, chapter VII.

and climb Mount Laki. VON GRUMBKOW writes of the "wonderful panorama up until the edge of Vatnajökull." She further describes how the palagonite mountain, Mount Laki, had been pulled apart by "a deep volcanic fissure," and estimates the fissure to be 25 kilometers long.[202] Her book includes a photograph taken from Mount Laki (Figure 65).

Abb. 19. Blick von der Höhe des Berges Laki auf die
östliche Kraterreihe.

Figure 65: A photograph of the eastern part of the Laki fissure in 1908, taken by Ina VON GRUMBKOW from Mount Laki.

VON GRUMBKOW describes the view from Mount Laki as "an extremely odd view of this grand expanse, a dead landscape that has offered an entirely unchanged image for more than 120 years."[203] Nevertheless, the forms and colors of the landscape, especially the lava fields, deeply fascinated her. Thus, she drew the fissure and explored it

---

202 GRUMBKOW 1909, chapter VII. "Wohl hatte ich eine dunkle Vorstellung von einem Ruhetag, aber der berühmte Berg Laki war doch zu verlockend nahe, – so ging ich nur zu gern mit [. . .]. Eine mühevolle Kletterei begann hügelauf, hügelab, über Geröllhalden, Schneeflecke, alte moosbewachsene, jüngere zackige Lava; nach fast vier Stunden ununterbrochenen Wanderns hatten wir die Höhe des Berges Laki erreicht, von wo wir ein wunderbares Panorama bis zum Rande des Vatna-Jökull genossen. Der 850 m hohe, durch eine tiefe vulkanische Spalte in zwei Hälften gerissene Palagonitberg liegt in der Mitte der sich von Südwest bis Nordost auf 25 km erstreckenden Kraterreihe."
203 GRUMBKOW 1909, chapter VII. "Einen überaus eigenartigen Anblick gewährte diese riesenweite, tote Landschaft, die seit mehr als 120 Jahren völlig unverändert dasselbe Bild bietet."

alone while her companions busied themselves somewhere else. From her travelogue, it becomes apparent how profoundly this landscape affected her:

> The cold dusk hours of this melancholic scene of loneliness were overwhelming. At the same time, despite the rain showers, it was wonderfully compelling, and it was difficult to turn away from this place; barely any human eyes have ever seen its peculiarities.[204]

The wealth of information that these naturalists compiled is impressive, given the scant number of days they spent at the fissure. PÁLSSON visited the Laki fissure in July 1794, HELLAND in August 1881, THORODDSEN in July 1893, ANDERSON in 1903, SAPPER in 1906, and RECK and VON GRUMBKOW in the summer of 1908.[205] As Sigurður ÞÓRARINSSON noted in 1968, if one counts the days these geologists spent at the Laki fissure and puts them together, the total is less than one month.[206]

Lifting the Fog of Ignorance

When Magnús STEPHENSEN attempted to explore the Síða region, he concluded that the "Skaftá Fires" came from some unknown cone-shaped volcano. Sveinn PÁLSSON found the source of the "fires" in 1794 and identified it as a fissure. Despite his best efforts, his findings did not impact the scientific discourse at the time. Þorvaldur THORODDSEN and Amund HELLAND revitalized interest in the Laki eruption in the 1880s. After the Krakatau eruption in 1883, the scientific community paid more attention to the far-reaching effects of volcanic eruptions and the variety of different phenomena they produced worldwide. The Laki fissure had now been put on the map. Geologists from several (mostly European) countries traveled to Iceland in ever-increasing numbers to study the exotic and dynamic landscape where the fissure sprang to life.

At the time of the Laki eruption, those concerned with the emerging discipline of geology were in the midst of a debate that centered around the conflicting ideas of Neptunism and Plutonism. Volcanoes were still considered accidents of nature, and scientific belief in the biblical Flood was only slowly going out of fashion.[207] The Laki eruption occurred during an intellectual revolution: the Enlightenment. The dry fog that followed

204 GRUMBKOW 1909, chapter VII. "Überwältigend wirkte in den kalten Dämmerstunden dieser melancholischen Szenerie die Einsamkeit, die trotz der Schauer, die sie erregte, doch wunderbar fesselte und schwer trennte ich mich von dem Ort, den fast nie Menschenaugen in seiner ganzen Eigenart sehen."
205 GUNNLAUGSSON 1984: 96, plate 7. This source includes two maps of the route that Karl SAPPER took to and along the Laki fissure. GRUMBKOW (1909) states in chapter VII that, once they arrived at the Laki fissure, the horses would get to enjoy four days of rest.
206 ÞÓRARINSSON 1969: 910–912.
207 RAPPAPORT 2007.

sparked widespread speculation. Various plausible explanations emerged; some suggested volcanic eruptions, potentially in Iceland (Nýey or Hekla) or the German Territories (Gleichberg, Cottaberg, Gottesberg, or Roßberg), were the source of this oppressive mist. The most popular theory suggested that the fog appeared as a result of the Calabrian earthquakes, themselves the result of a subsurface revolution within the Earth's bowels. Further manifestations of this revolution at surface level included earthquakes in France and the German Territories and the spontaneous emergence of a new island near Iceland.

Consulting the primary sources, it becomes clear that the European contemporaries' lack of awareness ran deep and could effectively be described as ignorance. The news about an Icelandic volcanic eruption reached Europe in September 1783. By that time, the dry fog had dissipated, as had any great interest in pursuing an understanding of its cause. If the mist was assumed to be a product of the Calabrian earthquakes, as it generally was, then news about an Icelandic volcanic eruption would not have moved naturalists to revisit the issue.

Sveinn PÁLSSON discovered the Laki fissure as early as ten years after the eruption; however, due to a series of unfortunate events, his manuscript remained unpublished until 1945. He managed to hand a copy to Ebenezer HENDERSON, who used PÁLSSON's findings to suggest that the Laki fissure had caused the fog in Europe. In 1818, the Laki eruption was already a 35-year-old event; HENDERSON's groundbreaking description of the Laki fissure did not seem to have much impact. So, from 1818 onward, a book that describes the Laki fissure accurately was essentially ignored.[208]

It took until the 1880s for interest in the eruption to be rekindled. This was as a result of Þorvaldur THORODDSEN's paper and Amund HELLAND's re-publications of parts of PÁLSSON's papers. The Krakatau eruption of 1883 and its intellectual consequences were the real breakthrough. They led the scientific community to finally connect the dry fog and the blood-red sunsets and sunrises of 1783 to the Laki eruption.

I conclude that the connection between the dry fog and the Laki eruption was only fully understood in the aftermath of the 1883 Krakatau eruption. Scientists and amateur weather observers noticed that the global impacts of the Krakatau eruption were similar to those in the wake of the Tambora and Laki eruptions. A vital contributing factor to the unfolding mystery was the almost instant communication between different parts of the globe via telegraphy. Contemporaries were aware of the volcanic eruption in the Dutch East Indies as they observed its impact on Europe. This device helped lift the fog of ignorance. Only in retrospect, it seems, could the dry fog of 1783 be attributed to the Laki eruption.

---

208 The name of the book was *Iceland or the Journal of a Residence in that Island During 1814 and 1815*; HENDERSON 1818.

5 Conclusions: A Modern Perspective on the Laki Eruption

Over the months that the Laki fissure brought misery and darkness to Iceland, much had happened elsewhere. The invention of "flying machines," balloons fueled by hot air or hydrogen, heralded a new era: that of aviation. Another volcano, Mount Asama, worsened an already dire famine for the people of Japan. In Europe, a once-in-a-lifetime celestial spectacle, a spectacular meteor that scorched a trail across the atmosphere, amazed onlookers. The United States of America became a fully recognized and independent nation.[1] As the volcano simmered down, the world outside Iceland trundled on.

Iceland's unique geology is a consequence of its location; it sits on the Mid-Atlantic Ridge and on a mantle plume. It is only when we consider deep time that we come to realize that the interminable mechanisms of geology still shape Iceland; it is a country in flux. Icelanders were challenged by their new homeland from the moment they first set foot on the island, which resulted in hard-won resilience. While the settlers were undoubtedly accustomed to ice and snow, this island presented them with the unfamiliar: expressions of volcanism. Iceland is one of the most volcanically active countries in the world. Although these "fires" do not feature heavily in the Icelandic sagas, we can assume that volcanic eruptions occurred as frequently back then, during the first few centuries after settlement, as they do today: that is to say, every three to five years. By the turn of the first millennium, the settlers understood that previous eruptions had produced the layers of lava on which they stood.[2] It is remarkable how few Icelanders lost their lives as a direct consequence of a volcanic eruption over the past 1,150 years, given that they live so close to 30 active volcanic systems.[3]

Iceland is located just below the Arctic Circle; its raw climate became all the more volatile with the onset of the Little Ice Age in the mid-thirteenth century. Settlers adapted to these new conditions by adopting a mixture of agriculture (mostly grass growing) and animal husbandry, supplemented by foraging in the highlands. They were, consequently, able to absorb some of the environmental shocks presented by this wild new landscape. However, when several unfavorable events transpired simultaneously, such as an epidemic or an epizootic, sea ice traveling far south, or a volcanic eruption, a subsistence crisis was not far off.

The eruption of the Laki fissure lasted from 8 June 1783 to 7 February 1784; it produced almost 15 cubic kilometers of lava and 122 megatons of sulfur dioxide and other volcanic gases. It was, undoubtedly, an exceptional environmental event; after eight

---

1 FRANKLIN 2011: liv.
2 LACY 1998: 15.
3 GUÐMUNDSSON et al. 2008: 263.

https://doi.org/10.1515/9783110731927-005

months of activity, its lava covered 600 square kilometers. How did the Icelanders cope with this? The lava consumed farmsteads and churches in the Síða region in southern Iceland, but no human lives were lost to it directly. However, ash and gases poisoned fields, ponds, rivers, animals, and even the desperate human population. The Icelanders lost 76 percent of their horses, 79 percent of their sheep, and half of their cattle between 1783 and 1785.[4] Starvation, malnutrition, and increased susceptibility to disease followed; around 10,000 Icelanders, one-fifth of the population, perished. The Laki eruption did not take place in a vacuum; in Iceland and beyond, it served to aggravate pre-existing social and political problems.[5]

Significant flood basalt events are rare: they occur only twice per millennium. The last large-scale flood basalt event before the Laki eruption, Eldgjá, happened in the 930s, just after the settlement of Iceland had begun. Through trial by volcanic fire, Icelanders have, throughout their history, learned how to fend for themselves and identify ways to mitigate the effects of volcanic eruptions. Isolated by hundreds of kilometers of ocean, they forged a communal resilience rather than a dependence on Copenhagen: help, in the form of resources or refuge, often came from areas within Iceland that were less affected.[6]

In 1783, Iceland was a Danish dependency. Merchants came in the spring with goods and returned to Copenhagen in the late summer or autumn, carrying news and various exports. For this reason, only in September 1783 did word of the disaster reach the Danish capital. By that time, it was too late in the season to send help, which only arrived in the spring of 1784. Icelanders were left alone to bear the brunt of the eruption. Bureaucratic hurdles put in place by the Danish central administration prevented an efficient crisis response. The hardened inhabitants of Iceland dealt with the calamity visited upon them without help or aid for almost a year. A gargantuan natural catastrophe, the *móðuharðindin*, was worsened by inaction.

The jet stream carried the ash and gases beyond Iceland: what impact did the Laki eruption have on the Northern Hemisphere? In Europe, contemporaries first observed unusual phenomena in the sky as early as mid-June 1783. Most notably, these included hazy skies, a red sun, and severe thunderstorms. Numerous earthquakes, subterraneous rumblings, sulfuric smells, heat waves, and even meteor sightings characterized the summer of 1783. The dry fog made the sun look "blood-red" at sunrise and sunset; it eclipsed the stars and reduced visibility at times to a distance of two kilometers. It affected plants, causing them to wither prematurely. Metal surfaces also fell victim to the haze, often turning green and rusty overnight. Moreover, these gases and particles affected human health. Myriad issues were documented, including respiratory problems, headaches, sore eyes, and itchy throats. That many records of

4 HENDERSON 1818: 275; FISHER, HEIKEN, HULEN 1997: 170; OPPENHEIMER 2011: 286.
5 McCALLAM 2019: 218.
6 DUGMORE, VÉSTEINSSON 2012: 76.

the summer of 1783 exist bolsters the idea that unusual weather is much more likely to be recorded in detail than typical weather.

What makes the summer of 1783 even more interesting is that this volcanic eruption transpired unbeknown to those outside of Iceland. The dry fog was observed in the Northern Hemisphere from North America to western China and from the Arctic to North Africa; it perturbed weather patterns far beyond these regions and its origin was a mystery. The thick and sulfuric dry fog that had blanketed most of Europe throughout the summer moved many to think about the weather. The concurrence of all these unusual phenomena was a point of interest for both the intellectual elite and the citizenry more broadly.

How were the dry fog and the other phenomena perceived? It is often not explicitly said that the population was frightened; however, the sheer volume of records suggests a high level of concern. Naturalists developed theories and newspaper editors, by all appearances, felt obliged to print them, at times alongside their own speculations. As well as immediate health concerns, Europeans broached the question of whether these phenomena were portents of things to come. They wondered whether a fog like this had occurred in the past. Those who cared to look found several precedents in chronicles and the recollections of the elderly.

Concerning the media landscape, newspapers allowed for a form of collected knowledge that could be shared, aided by a rise in literacy, allowing many to participate in the debate. Moreover, newspapers were often read aloud in public spaces for those who were illiterate. It was also possible for readers to send letters or comments to a newspaper in response to specific reports. In many ways, the mediascape was not dissimilar to today's, with misinformation often creating confusion. However, it is likely that most stories aimed at having a calming effect on the population. Historian Matthias GEORGI reported a similar result in his study on the impact of real and imagined earthquakes in England in 1750.[7]

The weather and unusual phenomena were omnipresent in the discourse during the summer of 1783. However, once the dry fog had passed in August and September 1783, it quickly became yesterday's news. Just as today, people were fickle; by September, the media discourse had shifted to something new. Had the search for the origin of the dry fog continued into September, the truth may have been uncovered sooner. Nevertheless, we should not be too judgmental, or future generations may look back at us with an equally deprecating eye. Indeed, engagement with the media at present has intensified: news is quickly forgotten, and people are perhaps more distracted and fickle than ever.

My contribution is a detailed account of the Laki eruption's impact on the Northern Hemisphere and the German Territories in particular. German sources reveal a multifaceted picture of the Laki eruption's effect on the continent. The dry fog was visible in the German Territories: there were regional differences and some short interruptions,

---

7 GEORGI 2009: 55. He only came across one report that clearly aimed at stoking fears.

but, in general, it remained from around 16 June to September or possibly October 1783. It is challenging to ascertain whether reports from after this period document instances of the dry fog in question or regular, wet, and seasonal fog.

What explanation strategies did contemporaries develop? In 1783, the Enlightenment inspired naturalists to search for the meaning behind observed phenomena. There was no shortage of fantastic ideas regarding the origin of the fog. These included a theory involving lightning rods, one that suggested a subterraneous revolution was the source, another that blamed volcanic eruptions within the German Territories, and even one that proposed the fog had its genesis in the tail of a meteor. The dialogue was versatile and international: discourse analysis of these phenomena shows that postulations that originated in one country often inspired those in other countries. Many explanations influenced one another and gradually became intertwined. Theories prompted rebuttals, the most famous of which were the scathing criticisms of DE LALANDE's claim that the fog had been caused by rain and flooding. Most naturalists refused to accept that such an extraordinary weather event could have an ordinary explanation. The scientific curiosity that was sparked prompted experiments and increased the will to further develop the scientific disciplines.

Various interpretations of the turbulent events of the summer co-existed: some commentators thought it to be a sign of the imminent apocalypse, although this was rare. Naturalists attempted to assuage these fears whenever they arose. Many interpretations were fluid and not mutually exclusive. One could hold that a given phenomenon had a scientific explanation and simultaneously believe that it was ultimately the manifestation of God's will. Far away from continental Europe, but not outside the Enlightenment's sphere of influence, Jón STEINGRÍMSSON – the "fire priest" – held such views: given his profession as a Lutheran pastor, naturally, he interpreted the events that unfolded in his parish religiously. God disapproved of the inclinations of his parishioners, and so, with ample warning, He sent lava to their doorstep. That God had given enough notice was proof of His mercy, as was the lava's slow flow. Such religious interpretations did not stop STEINGRÍMSSON, a student of the Enlightenment, from observing nature and conducting experiments. He threw stones into the lava to see whether they would melt; they did not, and with this he reassured his parishioners of the robustness of the land they lived on.

The coexistence of fluid interpretations was common during the Enlightenment. Previously, historians believed that the eighteenth century in Europe was a period of transition, a shift away from religious interpretations of events in favor of rational and scientifically sound explanations. This is not strictly true; competing ideas continued to exist.[8] Religious and scientific explanations at points complemented each other.[9] Just as earthquakes shook the lands of the European continent, so too did the

---

8 WEBER 2015: 14–16.
9 JAKUBOWSKI-TIESSEN 1992: 79.

Enlightenment and the transformative processes of the natural sciences shake the intellectual world.

Several modern historians have studied the existence of contrary ideas during the Enlightenment period in Europe. Christian PFISTER has demonstrated that it is misguided to assume one pattern of interpretation instantly replaced another.[10] Dieter GROH, Michael KEMPE, and Franz MAUELSHAGEN agree with this notion and argue that the eighteenth century was characterized more by the plurality of ideas and interpretations than the displacement of previously long-held beliefs by new thinking.[11] Manfred JAKUBOWSKI-TIESSEN and François WALTER demonstrate that seemingly contrary views were routinely held by one person, just as contradictory discourses coexisted throughout the eighteenth century and beyond.[12] My findings, based on discourse analysis of the debate about the unusual weather witnessed in the summer of 1783, are congruent with these studies. They also demonstrate that many attempts were made to find an overarching theory that would explain as many of the strange phenomena as possible. Religious doctrine and scientific reasoning were often used in tandem to explain whatever mystery presented itself.[13] This practice is called *physicotheology*.[14]

The Enlightenment was not one continuous episode but rather a multifaceted movement with regional and temporal differences. Although many different explanations circulated simultaneously, naturalists failed to find the inalienable truth about the origin of the dry fog, which increased the likelihood that religious interpretations would coexist with scientific ones.[15]

Applying the concept of societal teleconnections allows for the analysis of the far-flung effects of the Laki eruption, including those that only occurred some time after. Indeed, this concept also allows me to cast a wider net and shine a light on the real and imagined consequences of the eruption and the physical, emotional, and intellectual responses; some of which were intertwined with one another. The eruption was indeed responsible for the Laki haze, bizarre blood-red sunsets, and sulfuric smells. What is still contested is whether it precipitated the numerous thunderstorms of the summer of 1783 and the colder-than-average seasons that followed. Today we know that the heat of that summer occurred coincidentally and independently from the Laki eruption, and it would likely have been warmer still had the eruption not occurred.[16] Given the temporal proximity of the heat wave and the dry fog, it is unsurprising that naturalists in 1783 saw a connection between the two.

---

10 PFISTER 2002: 215.
11 GROH, KEMPE, MAUELSHAGEN 2003: 26; MISSFELDER 2009: 93–94.
12 JAKUBOWKSI-TIESSEN 1992: 107; WALTER 2010: 72.
13 WEBER 2015: 359.
14 ALT 2007: 34; REITH 2011: 92; BLAIR, GREYERZ 2020.
15 WEBER 2015: 359.
16 ZAMBRI et al. 2019b.

The echoes of Iceland's formation, etched in its landscape, reveal a process that usually remains hidden under the ocean. In Iceland, naturalists could witness the "cycles of collapse and renewal" that shape our planet.[17] Since the late eighteenth century, scientists have used the very landscapes around them to develop more sophisticated theories about the Earth's formation. The study of Icelandic geology was a ponderous and perilous undertaking. Given the deadly and tumultuous aftermath of the Laki eruption, it is unsurprising that it was more than a decade before the naturalist Sveinn PÁLSSON explored the highlands in an attempt to document Iceland's natural history. As we now know, PÁLSSON was the first person to set foot on the Laki fissure in the vast and rugged Icelandic highlands. He wrote a detailed report about this trip and the rest of his expedition. This should have settled the debate over the dry fog's origin. Circumstances, however, conspired against PÁLSSON, and so the fog of ignorance lingered.

The Tambora eruption of 1815 triggered the following year's cold summer; this was only brought to light in 1913. As the link between the Tambora eruption and the cold summer of 1816 remained obscure for some time, this event did nothing to elucidate the matter at hand: that being, the mystery of the fog of 1783.[18] It would take a century of progress and an eruption at Krakatau, 100 years hence, to finally reveal the origin of this mist. Had the Laki eruption and the fog been connected before Tambora erupted, it would have undoubtedly shed light on the events of 1816. As environmental historian Liza PIPER puts it, it is often "only with the benefit of historical hindsight that we can see the manifold ways" that a volcanic eruption like Tambora can affect the environment, economy, and society.[19]

When was a connection between the Laki eruption and the dry fog of 1783 firmly established? An Icelandic geologist, Þorvaldur THORODDSEN, discovered PÁLSSON's written work in an archive in Copenhagen in the 1870s. In 1883, Krakatau erupted with a blast that was audible thousands of kilometers away; telegraphy spread the news of the eruption far and wide. Scientists and amateur weather observers took note of strange atmospheric phenomena and brilliant sunsets around the globe. The idea that a distant volcanic eruption could affect regions far away from the actual volcano led some geologists, such as Amund HELLAND and Þorvaldur THORODDSEN, to think about the Laki eruption and the dry fog. Both visited the Laki fissure in the late nineteenth century. Between 1783 and 1883, knowledge about volcanic eruptions and geology increased significantly. From 1883 to 1888, the British Krakatoa Commission set out to compile as much information as possible about a colossal eruption on the other side of the world that they believed influenced the very skies above them. The nature of their investigation led them to revisit the events of 1783.

---

17 OSLUND 2011: 44–45.

18 The connection was made by American physicist William Jackson HUMPHREYS in 1913; KRÄMER 2015: 26.

19 PIPER 2009: 117.

Laki's legacy is the story of a century-long mystery solved. Its legacy is also a contradiction: In Iceland, it was the worst disaster in history, whilst in mainland Europe, it was all but forgotten by September of that year. The eruption is better remembered in mainland Europe today, 250 years later, than in the years immediately following it. Its cultural legacy was all the more subtle: the tumultuous summer of 1783 found its way into William COWPER's poem *The Task*, in which he commented on the Calabrian earthquakes and the dry fog. He writes:

> [. . .] Fires from beneath, and meteors from above, portentous, unexampled, unexplained, have kindled beacons in the skies, and th' old and crazy earth has had her shaking fits more frequent, and foregone her usual rest. Is it a time to wrangle, when the props and pillars of our planet seem to fail, and Nature with a dim and sickly eye to wait the close of all? [. . .][20]

Another example of a poem that mentions the strange weather of that summer is Peter Gottlieb LINDEMAYR's poem *Hänts Leutel sagts mä do*. However, apart from a few mentions in poetry, the fog does not seem to have inspired other art, such as paintings – at least none that have survived to the present day; the strange weather of 1783 did not leave much by way of cultural memory. Physical reminders exist of the upheavals after the eruption: numerous floodmarks scattered here and there along several European rivers indicate the great heights the water reached on those fateful days in 1784. The records and chronicles remain; the mist was the backdrop to many an important story and the catalyst to countless more.

Continental Drift, Plate Tectonics, and Mantle Plumes

In 1783, natural scientists struggled to find scientific principles that shed light on what they were witnessing. Volcanism was scarcely understood; theories were cobbled together by scientists with what little information they could glean from observations. Over the succeeding century, the understanding of geological mechanisms grew and the enigma of the volcano unraveled. Undoubtedly, two of the greatest leaps forward came in the twentieth century: the idea that the continents are in flux and the discovery of mantle plumes. Both of these concepts are crucial to the understanding of Icelandic volcanism.

With the help of a map, it is easy to imagine that Africa and South America were once bound together before they were torn apart by seafloor spreading, a process that continually creates new seafloor on both sides of the Mid-Atlantic Ridge. Furthermore, fossil evidence and occurrences of similar rock formations on lands separated by a vast ocean point to a prehistoric supercontinent. In light of these findings, in 1912, the German polar explorer, astronomer, and meteorologist Alfred WEGENER (1880–1930) formulated the theory of continental drift. In 1915, he published his book

20 COWPER 1785, book 2.

Die Entstehung der Kontinente und Ozeane (The Origin of Continents and Oceans).[21] WEGENER's ideas were treated with scorn; he died on an expedition to Greenland in 1930, many years before the significance of his theory was appreciated.[22]

In the 1950s, new technologies enabled scientists to work out the age of the sea-floor based on the effect of the periodic reversal of the magnetic poles. They proved that seafloor closer to the divergent boundary of the mid-oceanic ridges is younger than that which is further away. Subsequently, in the 1960s, the theory of *plate tecton-ics*, a refinement of the theory of continental drift, was established. Bathymetric data collected from ships also enabled scientists, such as Marie THARP (1920–2006), to draw maps of the ocean floors; this revealed the presence of mid-oceanic ridges in oceans around the world, amounting to a total length of 65,000 kilometers.[23] Plate tectonics became a major focus of subsequent scholarship; with this theory, scientists made sig-nificant progress.[24]

Canadian geologist J. Tuzo WILSON, who brought about this change with his semi-nal 1965 paper on plate tectonics, had published another paper on mantle plumes two years prior.[25] American geophysicist W. Jason MORGAN built upon WILSON's work on mantle plumes and published two papers on the topic in 1971 and 1972.[26] Understand-ably, the study of plate tectonics took precedence over other avenues of research. Only in the 1990s did the theory of mantle plumes really gain momentum.[27] Although some geoscientists still question the existence of the Iceland mantle plume, its pres-ence is generally accepted.[28]

American geophysicist David T. SANDWELL, in an attempt to explain why the theory of plate tectonics took so long to be accepted, reasoned thusly: throughout early modern his-tory and most of modern history, naturalists and geologists explored the Earth backward. Scholars speculated using land-based observations; with this method, they developed ideas on the formation of the Earth.[29] Ideally, one should begin with planetary-scale

---

21 WEGENER 1915.

22 KEHRT 2013, Biography of Alfred WEGENER.

23 SEARLE 2013: 1.

24 There are several publications that detail the history spanning from the theory of continental drift to plate tectonics; ORESKES 2003; FRANKEL 2017.

25 WILSON 1965; WINCHESTER 2005: 99–112.

26 MORGAN 1971; MORGAN 1972.

27 CONDIE 2001: xi, 1.

28 See DAVIES 1999. Those who question the existence of the mantle plume include, for instance, Don ANDERSON (2007) and Gillian FOULGER (2010). For more debate on the topic of the existence of mantle plumes, see: PARK 2010: 53–54; GROTZINGER, JORDAN 2017: 325–327; FRIEDRICH et al. 2018: 156–188. E. R. LUNDIN and A. G. DORÉ argue that the large igneous province might have been caused by the plate break-up rather than a mantle plume; LUNDIN, DORÉ 2005. Many geoscientists, however, have accepted that mantle plumes play a substantial role in the formation of flood basalt events: MORGAN 1971; VINK 1984; COURTILLOT et al. 1988; WHITE, McKENZIE 1989; COFFIN, ELDHOLM 1994; JULL, McKENZIE 1996; SAUN-DERS 1997: 46.

29 SANDWELL 2003: 334.

observations and then test these theories on smaller-scale environments. In early modern times, naturalists extrapolated theories of the Earth's formation from studies of the continental crust. They were unaware that continental crust was much older than oceanic crust. Some continental crust is up to four billion years old. In contrast, oceanic crust is rarely older than 200 million years and is much lighter. When push comes to shove, oceanic crust gets subducted deep into the Earth's mantle. One of the first scholars to develop theories on a planetary scale was the Austrian geologist Eduard SUESS (1831–1914), who formulated ideas about the orogeny of the Alps and the existence of a supercontinent called Gondwana.[30] It was only in the second half of the twentieth century that substantial technological advances allowed scientists to read the geology of planet Earth more effectively: satellites in space measured the motion of the tectonic plates, and seismometers reliably identified earthquakes at plate boundaries, even at depths of almost 700 kilometers.[31]

History and Geology

Geology, like history, is inescapable. Even those who live in aseismic regions of the world cannot escape its influence. The eyes that witnessed the fog were stung by its elements, and the tongues that uttered prayers and contrived explanations, tasted its sulfur. Those in the German Territories and beyond were not mere witnesses to the happenings of the summer of 1783: they also took part.

Why is an environmental history of a volcanic eruption important? A study of a volcanic eruption is an ideal platform for combining different disciplines, such as environmental history, climate history, history of science, and the natural sciences. The processes that produce a volcanic eruption are geological; these processes impact the surrounding landscape, the chemical makeup of the atmosphere and, therefore, the weather and climate. All this can have repercussions on societies near and far.

It is fascinating to study eighteenth-century naturalists' comprehension of volcanic eruptions and how their knowledge changed over time. In the future, our current understanding of science will provide historians with a snapshot of our time as well. Modern science is not a "repository of unalterable truth." As historian of geology Rhoda RAPPAPORT puts it, "[o]ur task as historians is to study when and why men thought as they did."[32] An understanding of modern geology and climate allows one to appreciate the evolution of science over the past 250 years.

Geology should play a more significant role in environmental history; it shaped the land that shaped history. Interdisciplinarity, however, is a two-way street. As geology is

---

30 SUESS 1883.
31 SANDWELL 2003: 345.
32 RAPPAPORT 2007: 131.

important to history, so too is history important to geology. History reveals the consequences of geological phenomena on society. Without an interdisciplinary approach, many connections within this story would have remained obscure. As historians, we often view past events through the eyes of those who experienced them. An interdisciplinary approach offers fresh perspectives on established topics and opens new doors, particularly to abstract issues and those at the crossroads between the environment and society. Should we not cast an eye over the episodes of history from a different, current perspective? Exploring the history of volcanoes through the lens of a geologist can reveal eruption styles or patterns of activity that took place thousands of years ago and have gone unrecorded. Moreover, if an active volcano has a lengthy recurrence period, generations of people may believe it to be extinct, with grave implications.[33] A greater understanding of natural phenomena can shed new light on old questions, just as it did a century after Laki. This same logic applies to climate history: studies like this one may reveal extreme weather events and climatic patterns that have not yet occurred in historical times. Interdisciplinarity does not dilute the potency of the disciplines involved; instead, it elevates them. It affords the scholar the chance to discover answers just outside the realm of their individual pursuit and offers the historian, specifically, a higher-resolution view of the past.

Science helps establish not only why something happens but also where it can happen and whether it will happen again. It tells us that earthquakes can occur around Aachen and that reports of an earthquake there in 1783 are likely true. Geologists are able to produce this knowledge, which helps historians critically assess contemporaries' statements rather than taking them at face value. Science provides incontestable evidence that the Gleichberg did not erupt in 1783 and explains why *jökulhlaups* take place after sub-glacial eruptions. Through geology, we can appreciate how large igneous provinces might have formed in many parts of the globe; indeed, this is an area of research also relevant to the understanding of extraterrestrial geology. Large igneous provinces, and perhaps mantle plumes, play significant roles in the evolution of Mars and Venus. Unlike on Earth, plate tectonics does not play a role on these planets.[34]

This environmental history of the Laki eruption can help historians of other periods identify signs of volcanic eruptions, possibly far from their actual source. I list various observable indicators of volcanic activity in Chapters Two and Three. In recent decades, natural scientists have established a good chronology of historical volcanic eruptions using ice cores, tree rings, and other proxies.[35] Historians and volcanologists

---

33 Dugmore, Vésteinsson 2012: 73, 79.
34 Condie 2001: xi–xiii, 88–95.
35 Sigl et al. 2015.

endeavor to determine the exact place and time of the eruptions they study.[36] Proxies can indicate when an eruption transpired, with an uncertainty of plus or minus five years. Historical sources, however, are high-resolution sources that can pinpoint events down to an exact day.

Relative to those volcanic events that occurred earlier, we are reasonably well-versed in the chronology of the Laki eruption. From the perspective of climate history, it is a luxury to know an eruption's exact start and end dates. The sources show that one event can be interpreted differently in different regions or even by different authors within the same region. Consequences of volcanic eruptions include withered vegetation, blood-red sunsets, respiratory problems, sore throats, and stinging eyes. Historians can look out for mentions of one or more of these in their sources as they scour the past. Volcanic cooling is another consequence; however, this is often harder to pinpoint, particularly within the Little Ice Age.

Through close collaboration with geologists and volcanologists, historians can reconstruct a more reliable chronology of past volcanic eruptions. Proxies such as ice cores, tree rings, sediment layers, and historical sources can aid in the search for as-yet-undiscovered volcanic eruptions. Thanks to ice core records, we have a good idea of when significant volcanic eruptions happened in the last two millennia; however, scientists have yet to identify the locations of all these eruptions. In 1808/1809, a tropical eruption occurred that was comparable to Tambora in scale, yet we are in the dark as to its location.[37] Knowledge of previously unknown eruptions at particular volcanoes can help us improve our understanding of recurrence periods. It could lead us to conclude that the next eruption at a particular volcano might occur much sooner than previously estimated; this would give local populations time to prepare for this event.

In 2010, Iceland's volcanism made global headlines when Eyjafjallajökull erupted (VEI 4).[38] The eruption's ash plume filled Europe's skies for several days; one consequence was the grounding of all transatlantic air traffic, stranding seven million passengers. The estimated cost of this for related industries was 4.7 billion USD.[39] If a long-lasting volcanic haze, similar to the dry fog of 1783, were to occur today and linger at altitudes

---

36 One example of such a collaboration is the group Volcanic Impacts on Climate and Society (VICS) from Past Global Changes (PAGES). Positive recent examples are the identification of the location of the 1258 Samalas eruption (LAVIGNE et al. 2013) and the 1458 Kuwae eruption in Vanuatu (GAO et al. 2006, although these findings have recently been disputed by HARTMAN et al. 2019). Medieval sources also give indications of perturbed weather patterns and unusual natural phenomena: BAUCH 2017.
37 Alvaro GUEVARA-MURUA et al. (2014) found descriptions of a haze and an afterglow of the sun between December 1808 and February 1809 in primary sources from Colombia and Peru, indicating that the eruption probably took place in late November or early December 1808.
38 Global Volcanism Program: Eyjafjallajökull.
39 DAVIES et al. 2010; ELLERTSDÓTTIR 2014: 129–137.

of eight to 12 kilometers, the situation for the aviation industry would be appreciably worse. Ash particles can cause abrasions on airplanes, damage navigational instruments and engines, and even induce engine failure.[40] Recently, airplane manufacturers have started to develop sophisticated infrared and ultraviolet cameras, aptly named "Airborne Volcanic Object Imaging Detectors (AVOID)," that can help pilots avoid encounters with ash.[41]

This is just one obvious potential consequence of a future eruption; undoubtedly, there are many more. Indeed, as elements of our global economy are so intertwined and interdependent, one economic woe will probably beget another. Living in a significantly more interconnected and globalized world has many advantages; however, a globalized economy is quite vulnerable to the vicissitudes of nature, such as volcanic eruptions. The wind direction and atmospheric conditions will determine the regions affected by the next large Icelandic eruption. If this theoretical eruption occurred when circulation patterns were different, for instance, in another season, there could be unexpected outcomes. The next flood basalt event in Iceland has the potential to be much bigger. A larger eruption could produce more lava and gases, have a prolonged eruption period, and have an even more significant impact on health and the environment.

While it is likely we will know the location of a future volcanic eruption almost instantly, it is still essential to consider the history of knowledge of past volcanic eruptions in future studies. Technology such as telegraphy enabled the relatively fast spread of news about the Krakatau eruption in 1883; however, this was not the case in the pre-modern period. Future studies should investigate when links were established between other volcanic eruptions and their corresponding visible phenomena.

Future Research

This book elucidates the complex mindset of contemporaries in the late Enlightenment and how they perceived and interpreted the phenomena they witnessed; contemporary accounts also reveal how the authors' own health, or that of the people around them, was affected. The fears they expressed, directly or indirectly, reflect common concerns of the time. A detailed analysis of the descriptions of the phenomena written by contemporaries reveals complex and, at times, contradicting perceptions of reality. Over time and across different regions the dry fog had different intensities, which means that not all phenomena were experienced everywhere. In

---

40 GUÐMUNDSSON et al. 2008: 252; USGS, "Volcanic Ashfall Impact Working group: Aviation"; GUFFANTI, CASADEVALL, BUDDING 2010. In 2014, no guidelines defined when a volcanic plume of sulfur dioxide is hazardous for air traffic: SCHMIDT et al. 2014a.
41 DAVIES 2014.

the future, a qualitative study of how countries other than Germany, Britain, or France perceived the dry fog would be very desirable.

John GRATTAN, Sabina MICHNOWICZ, and Roland RABARTIN suggest that the Laki eruption was responsible for the deaths of millions of people in India and Egypt: they argue that it affected monsoonal rains, which, in turn, resulted in famine and drought. Consequently, they call the Laki eruption "one of the greatest natural disasters in human history."[42] Burial records for France and Britain reveal an elevated mortality rate in the late summer and autumn of 1783 and in the winter and spring of 1784. They suggest that an estimated 36,000 extra deaths occurred in France and England after the eruption.[43] A careful analysis of burial records from parishes across different regions in the German Territories and other parts of Europe in 1783 and 1784 would establish a clearer picture of the mortality rate during the Laki eruption and its aftermath.[44] This mortality spike had regionally different expressions; therefore, it would be interesting to determine whether it struck elsewhere in Europe and, if so, where. It would be particularly interesting to examine whether an excess mortality rate occurred soon after a thick, dry fog with a sulfuric smell was reported. In many regions, there was a time lag of a few months between the peak of the dry fog and the onset of the mortality crisis. Burial records from this time vary drastically; these records only sometimes include information on a cause of death, information which could be helpful to establish whether the Laki haze was responsible. Such a study would be useful for today's health and civil protection ministries to plan for mitigation in the face of the next large Icelandic volcanic eruption.

Further research into the perceptions of the cold winters and seasons that followed 1783 would also be advantageous.

The Bigger Picture: Lessons for the Present and Future

In modern Iceland, Jón STEINGRÍMSSON's knowledge was put to good use. In 1973, an eruption took place on the populated island of Heimaey. The island is part of the Vestmannaeyjar volcanic system at Iceland's southernmost tip.[45] The volcano *Eldfell* (literally the "fire mountain") started erupting and a fissure opened up a mere 200 meters from the closest house. Within the first six hours, almost all 5,300 residents had been evacuated to the Icelandic mainland; the Icelandic State Civil Defense Organization had prepared for a scenario like this.

---

42 GRATTAN, MICHNOWICZ, RABARTIN 2007: 156–157, 157 (quote).
43 WITHAM, OPPENHEIMER 2004; COURTILLOT 2005: 636. During the 2003 heat wave, an extra 16,000 people died; the Laki haze probably caused excess mortality of a similar, if not greater, magnitude.
44 WITHAM, OPPENHEIMER 2004; GRATTAN et al. 2005; MICHNOWICZ 2011.
45 Catalogue of Icelandic Volcanoes: Vestmannaeyjar; ÞÓRARINSSON et al. 1964.

The eruption lasted for about five months, emitted toxic gases, and boiled the sea-water wherever the lava entered the ocean.[46] At that time, STEINGRÍMSSON's descriptions of the water from the rivers cooling down the lava flow in Kirkjubæjarklaustur inspired Icelandic geologist Thorbjörn SIGURGEIRSSON to suggest that seawater be pumped onto the lava, which cooled the top layer and slowed its flow. The eruption ended in early July 1973, having destroyed approximately 40 percent of the town.[47] The lava flow that entered the sea formed an ideal natural barrier for Heimaey's harbor, protecting it from the North Atlantic waves.[48] The response to the eruption has inspired other cities prone to these problems to follow the Icelanders' proactive example.[49]

The dry fog and the many unusual phenomena of 1783 presented contemporary naturalists with a significant, abstract problem of unknown origin. People are afraid of the unknown; this fear can be debilitating. Perhaps this explains the fixation on local explanations. It took no effort to see something in the vicinity and point a finger. If no local explanation sufficed any theory that provided an answer, even from further afield, would do. That the origin of the fog could be established was the critical factor. If the fog over Europe could be put down to the simple process of precipitation and evaporation, then the fear, too, would evaporate. If the fog was explained away as the mists from the tail of a comet then perhaps it, like the fireball that streaked across the sky, would be finite and end rather soon. And if it was the Calabrian earthquakes that had caused the fog, then at least the source of the problem could be defined.

Today, we live in a world of uncertainty. This uncertainty leads us to grasp for the familiar and the safe. It is easy to understand why someone would want to willfully and knowingly tread water if the future seems threatening. A familiar short-term solution can look decidedly more attractive than a long road of tough choices. Concerning efforts to tackle climate change, it is true that perhaps some may be afraid not of inaction but of action, especially if this action is sure to threaten their livelihood. All of us must, however, face the unfamiliar. To reduce anthropogenic climate change, humankind must make certain sacrifices that we have, until now, been unable to.

When the fog of 1783 dispersed, so did the search for its origin. If we are not constantly confronted with climate change as a problem, perhaps we will neglect the quest for a solution. The unfortunate fact is that if we are not all looking for an answer then, collectively, we have forgotten the question and an answer may elude us. During the Laki eruption, small pockets of intellectuals were left unsatisfied by most

---

46 SIGURGEIRSSON 1973; ÞÓRARINSSON et al. 1973; WILLIAMS, MOORE 1988: 9; FURMAN, FREY, PARK 1991; LACY 1998: 13; MATTSON, HÖSKULDSSON 2003; Catalogue of Icelandic Volcanoes: Vestmannaeyjar.
47 LACY 1998: 13.
48 THORDARSON 2010: 285.
49 WILLIAMS, MOORE 1988: 18.

theories and sought answers in the unfamiliar; they were not unsettled by carrying with them the burden of a question for so long. The contemporaries of 1783 could not have prevented the fog but perhaps today we can proactively tackle the threats of anthropogenic climate change.

Many accepted truths coexisted in 1783, but accepted truths can sometimes lead us astray. Even those who proposed that an Icelandic volcanic eruption had caused the unusual weather often could not dismiss the accepted truth at the time: that the Calabrian earthquakes were to blame. The distinction between the accepted and actual truth is crucial today. In 1783, a far-flung volcanic eruption had a seemingly improbable impact on Europe, but this *actual* truth remained unknown. In many ways, this can be related to anthropogenic climate change. While the scientific consensus is that fossil fuel emissions have changed and are still changing the planet's climate, some still dismiss this because it does not agree with their accepted truth.

The challenges of the past can seem remote in the present. Modern amenities and technologies can give us a false sense of security. The basic nature of the human experience has not changed since 1783: we still breathe air, and we still rely on agriculture to produce our food. Both fresh air and vegetation could be compromised by the pollution caused by a future Icelandic eruption. This prompts the question: have we learned from the Laki eruption of 1783? Today, generally speaking, Europeans are better nourished and healthier than the Europeans of 1783; our atmosphere, however, is much more polluted. Many more people have asthma today, which creates a larger group that is potentially at risk during a volcanic pollution event. Demographic changes have led to much larger percentages of elderly people in the population, who are also at risk in the event of an eruption.[50]

Fortunately, some countries are already planning for this kind of scenario: for instance, the United Kingdom has listed a Laki-style Icelandic eruption on the National Risk Register for Civil Emergencies.[51] The German Federal Office of Civil Protection and Disaster Assistance has a similar risk analysis list; the specifics are not public, but it includes volcanic eruptions. In the event of a future volcanic eruption that affects Germany, the different state governments would respond individually. The federal government can, however, offer assistance. In the past, Germany's civil protection system has successfully tackled major events, such as flooding in 2002 and 2013.[52] Proactive measures could

---

50 GUÐMUNDSSON et al. 2008: 251, 264.

51 Cabinet Office (UK), 2017: 25–26.

52 Personal correspondence with Alexander ESSER, Federal Office of Civil Protection and Disaster Assistance | Bundesamt für Bevölkerungsschutz und Katastrophenhilfe, Referat II.1 – Grundsatzangelegenheiten des Bevölkerungsschutzes, Ehrenamt, Risikoanalyse, Abteilung II – Risikomanagement, internationale Angelegenheiten. More information on the air quality stations can be found in Umweltbundesamt 2013; the case of the Eyjafjallajökull eruption is discussed on pages 78–79.

include the development of tephra-resistant airplane engines, suitable face masks, or methods of protecting crops from acidic pollution.[53]

It is important to draw hope from the dramatic stories of past challenges.[54] Trials of the past can serve to fortify us, embolden us, and provide us with the courage to overcome our present challenges. Icelanders and many other inhabitants of the Northern Hemisphere in 1783 had fewer means and lesser technology than we do today, but they endured. The thirst for knowledge was very much present in the late eighteenth century. Many cherished the discoveries made in the different fields of science: 400,000 people braved the cold in Paris in December 1783 to witness the flight of the hydrogen-filled *La Charlière*; many enthusiastically embraced the idea of the lightning rods; and Sveinn PÁLSSON wandered through the challenging terrain of the Icelandic highlands to further human understanding of Iceland's natural history. The thrill of discovery still exists today and is something inherent in humanity, which inspires optimism.

The events of that extraordinary year – 1783 – encourage us not to underestimate human kindness: some Icelanders gave shelter to their neighbors as ash fell from the sky; many Danes participated in fundraising efforts to help the Icelanders in their hour of need; the fishermen who discovered the Nýey eruption were genuinely worried about the fate of the Icelanders and made sure to check on them; newspapers were committed to informing their readers, believing that adequate information would stop them from panicking; and in the winter of 1783/1784, churches collected money for people affected by the severe flooding.

Outlook: The Present and the Future

The 200-Year Anniversary of the Laki Eruption

How did the memory of the Laki eruption evolve over two centuries? Much had changed over that period: most notably, on 17 June 1944, Iceland had become an independent republic. In the early twentieth century, the Icelandic economy broke out of the cycle of poverty and stagnation due to the mechanization of its fishing industry.[55] After the Second World War, Iceland was on the fast lane to becoming a truly modern nation. In 1949, it became a founding member of NATO. Two years later, it agreed to enter into a defense treaty with the US, which led to huge investment in the country's

---

53 Some developments are already underway. In the aftermath of the 2010 Eyjafjallajökull eruption, special jet engine coatings were tested to resist volcanic ash damage: DREXLER et al. 2011.
54 MAUCH 2019.
55 KARLSSON 2000b: 239–242, 287–291; MAGNÚSSON 2010: 30–31.

infrastructure.[56] Iceland also utilized its natural resources by using sand and gravel for construction and by harnessing geothermal energy.[57] In 2018, Iceland had around 350,000 inhabitants.[58] The number of tourists coming to Iceland rises almost every year: 303,000 tourists visited in 2000, 489,000 in 2010, 807,000 in 2013, and 2,224,603 in 2017.[59] Tourism has become an integral part of the economy. Today, Iceland is part of Europe and "a thoroughly modern country."[60]

Figure 66: A stamp commemorating the Laki eruption, released in 1970, featuring a painting by Sveinn Ólafsson.

The many reports and descriptions of this eruption, and the fissure itself, torn into the landscape, immortalize the event which caused so much hardship and pain. In 1970, a painting of the Laki fissure by Sveinn Ólafsson featured in a stamp that was created as part of a nature conservation series, *náttúruvernd* in Icelandic. As part of this series, iconic images of several other Icelandic volcanoes had previously been turned into stamps to celebrate the country, such as Hekla, Surtsey, and Eldfell. The stamp (pictured here as Figure 66), according to historian Karen Oslund, presents the volcano as an unthreatening feature of the landscape, a mere spectacle that complements the bright blue sky on the horizon. The fissure is still visible and reminds Icelanders and foreigners alike what nature, specifically Icelandic nature, is capable of. In the eighteenth and nineteenth centuries, volcanic eruptions were often depicted as awe-inspiring, destructive forces of nature; Ólafsson's painting puts the unique landscape at center stage. Volcanoes are no

---

56 Lacy 1998: 233–249; Karlsson 2000b: 313–318, 336–341; Magnússon 2010: 232–235; Oslund 2011: 26. On the strategic importance of Iceland during the Cold War, see: Whitehead 1998; Ingimundarson 2011.
57 Lacy 1998: 18–19.
58 Lacy 1998: 3–8.
59 Icelandic Tourist Board, Tourism in Iceland in Figures, 2018.
60 Oslund 2011: 27.

longer a symbol of destruction or despair. "Icelandic nature is extreme, unpredictable, and even wild, but people live within this wilderness, and their character has been formed by the struggle with this nature."[61]

It is not the nature of the Icelandic landscape that has changed since the eighteenth century but, rather, how the Icelanders see it. They no longer consider volcanic eruptions as catastrophes; instead, they view them with pride and awe. Now, with a curious eye, they can study and understand their country and its moods and, in doing so, protect their people and property.[62] In 1975, the Laki fissure was deemed a natural monument. It is now part of the Vatnajökull National Park, which was established in 2008 and is Europe's largest. It covers 13,200 square kilometers and includes the entirety of Vatnajökull as well as some surrounding land, including Eldgjá.[63]

Figure 67: "Lakagígar" by Finnur JÓNSSON, 1940. The painting was turned into a stamp in 1983.

---

61 OSLUND 2011: 53–54.
62 OSLUND 2011: 58–59.
63 Vatnajökull National Park: Lakagígar, Langisjór and Eldgjá brochure and map, 2011; BALDURSSON et al. 2018.

In 1983, Finnur JÓNSSON's painting of the Laki fissure was also turned into a stamp (Figure 67) to commemorate the 200-year anniversary of the eruption. OSLUND further observes that during the first half of the twentieth century, the portrayal of Icelandic nature changed – instead of focusing on the human perspective, depictions now focus on nature itself. The landscape is celebrated in its own right, as is the triumphant struggle to overcome the challenges presented by this wilderness. The ever-present threat of volcanic eruptions is part of this rugged island's unique character.[64]

Skafáreldahraun – The Laki Lava Field

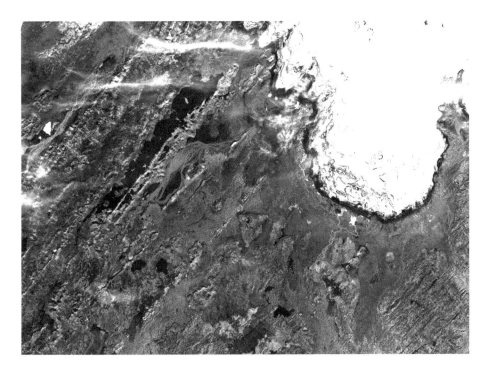

Figure 68: A satellite image of the Laki fissure and Vatnajökull.

The Laki fissure is one of several in southeastern Iceland. Satellite images and drones change our understanding of remote landscapes and make it easier to grasp the geological forces that have shaped our planet (Figure 68). Eldgjá, the Great Þjórsá Lava, and the Laki eruption are the three largest terrestrial basalt eruptions of the last 10,000 years.[65] Eldgjá and Laki combined produced more than half of all the lava

---

64 OSLUND 2011: 52–53, 60.
65 THORDARSON, HÖSKULDSSON 2008: 198.

ejected from Icelandic volcanoes since settlement in the ninth century.[66] The lava produced by these eruptions precipitated some of the longest postglacial lava flows on the planet.[67] The fissures produced by the Eldgjá and Laki events are parallel and quite close to one another, marking the rifting area of the Eastern Volcanic Zone. The remnants of these eruptions are still visible in the landscape of the highlands between Vatnajökull and Mýrdalsjökull. Laki alone left 600 square kilometers of lava fields which, over time, have been blanketed in moss (Figure 69).

Figure 69: The Laki fissure in 2016. The southwestern part of the fissure, as seen from Mount Laki.

The Laki lava field is known as *Skafáreldahraun* (Figure 70). The lava morphology of its surface is unique; the lava field does not solely consist of *pahoehoe lava*, which is smooth, fluid basaltic lava, nor *'A'ā lava*, which is rough, rubbly basaltic lava. Therefore, the term *rubbly pahoehoe* was introduced to describe and differentiate it from other types of lava fields.[68] Its green moss grows very slowly in the harsh climate of the Icelandic highlands and is always under threat from a constant stream of plodding hikers.[69]

Today, lava fields are a tourist destination and have, in recent years, been marketed as such. One can even buy lava-field-shaped candy called *hraun* (lava field) in the Icelandic supermarket: it is a puffed rice cake covered in chocolate.[70]

---

66 THORDARSON, HÖSKULDSSON 2014: 131.
67 THORDARSON, HÖSKULDSSON 2008: 211.
68 GUILBAUD 2005.
69 Katla Geopark Project: 10.
70 OSLUND 2011: note 51, 186.

Figure 70: *Skaftáreldahraun*, the Laki lava, 2016.

Once, it took a few days on horseback to get from Kirkjubæjarklaustur to the Laki fissure. Today, you can make that journey within three hours. The distance of 50 kilometers is exceptionally rough, with dirt roads that are often flooded. "The public is encouraged to visit [. . .], under certain conditions," states the brochure for Vatnajökull National Park, where the Laki fissure is located. The brochure further clarifies that one condition is adherence to a protection order put in place to help visitors "enjoy the area without damaging the volcanic features or vegetation."[71] Action was necessary as the ever-increasing number of visitors was having a detrimental effect on the environment. Visitors to the National Park are asked to stay on the wooden walkways built by volunteers.[72]

Accompanying signs warn visitors of the consequences of stepping on the moss. Nevertheless, a lot of plants near the walkway have turned brown or are dead because tourists have stepped on them too often (Figure 71).

From the moment of settlement, Icelanders began to clear land for grazing; happily, the felled trees also met their fuel needs. As a consequence, much of Iceland's forest cover was lost. Without trees, the soil dried up and was subsequently eroded by the elements. Soil erosion, aridification, and desertification are huge problems in today's Iceland. Presently, 40 percent of Iceland is considered "severely eroded."[73] In 1989, the Ministry for the Environment and Natural Resources was founded. It had two goals: to utilize natural resources and preserve the environment.[74] Naturally, future volcanic eruptions, severe storms, avalanches, or floods might hinder some of these efforts.[75]

---

71 Vatnajökull National Park, Lakagígar, Langisjór and Eldgjá brochure and map, 2011.
72 LACY 1998: 47–48. Volunteers work toward preservation by building paths to prevent tourists from stamping on the vegetation; KLEEMANN 2016.
73 STREETER, DUGMORE, VÉSTEINSSON 2012: 3665.
74 LACY 1998: 47–48.
75 HJÁLMARSSON 2009: 214–216.

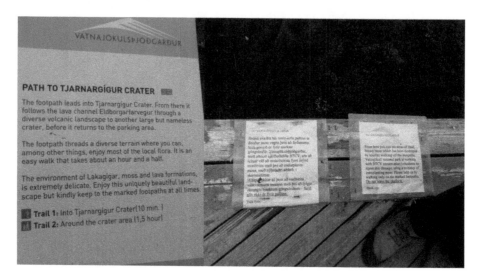

Figure 71: An information sign and a footpath near one of the Laki craters.

Mount Laki was formed by volcanic activity under a thick layer of ice, possibly in 4550 BCE (± 500 years).[76] In 1783, a fissure formed in this exact spot. In the future, another fissure and a crater row might form somewhere parallel to the Laki and Eldgjá fissures.

Just as Iceland is being wrenched apart by geological forces, Europe has, throughout its history, been wrenched apart by different rulers, economic interests, religious beliefs, or ideological frameworks. Iceland's remoteness shielded it – for the most part – from these often bloody conflicts. Day by day, centimeter by centimeter, Iceland grows; it is a great stage where a drama featuring earthquakes, eruptions, and volcano-tectonic rifting episodes that Icelanders simply refer to as "fires" is unfolding.[77]

Over the past 1,150 years, Icelanders have learned to understand the quirks of their homeland. They have seen many mortality crises but have always picked themselves up and started anew. Icelandic society has never collapsed despite interminable volcanism, a volatile climate, centuries of foreign rule, poverty, demographic crises, and economic setbacks. Today, the country is well-prepared to deal with the moods of its fire-spitting landscape. Icelanders are still only tenants of the land. Volcanoes remain the natural kings and queens of Iceland; they can be violent, eccentric, and despotic, but they can also be benevolent, supplying Iceland with hot pools and almost unlimited geothermal energy.

---

76 EINARSSON, SVEINSDÓTTIR 1984: 48. Information on the Botnahraun eruption can be found in the "eruptive history" section in Global Volcanism Program: Grímsvötn.
77 STEINGRÍMSSON 2002: 3.

Iceland and Its Volcanoes in the Future

Iceland is one of the most active volcanic regions on our planet.[78] The different volcanic systems are monitored continuously. Since the Gjálp eruption at Grímsvötn in 1996, every eruption has been predicted by seismic activity, which gives civil defense authorities some time to evacuate people from the threatened parts of the country. Technological and economic improvements in Iceland have undoubtedly saved many lives; only two fatalities directly linked to volcanism occurred in Iceland in the twentieth century. Increasing tourism places not only the fragile moss at risk but also the tourists themselves; some volcanoes, such as Hekla, a popular hiking spot, often give very little warning before erupting.[79]

By monitoring and analyzing Iceland's volcanoes and geology, experts have learned a lot. We know, for instance, that the Grímsvötn volcanic system is in a high-frequency eruption period, which will last, it is estimated, for another 50 years. It is predicted to peak between 2030 and 2040. Bárðarbunga follows a similar pattern of high- and low-frequency periods. Both systems are above the center of the Iceland mantle plume. Major rifting episodes occurred at Bárðarbunga in 1477 and at Grímsvötn between 1783 and 1785, both of which coincided with high-activity periods.[80]

Recent fissure eruptions in Iceland include the Holuhraun eruption, which began in 2014 and lasted until 2015 and was part of the Bárðarbunga volcanic system, and the 2021 Geldingadalir and 2022 Fagradalsfjall eruptions on the Reykjanes Peninsula.[81] These eruptions were small compared to Laki.[82] Webcams were quickly installed nearby to broadcast the eruption worldwide. Many scientists flew to Iceland to witness this impressive event first hand. Smaller flood basalt events, such as the Holuhraun eruption, happen roughly every 40 to 50 years. Fortunately, in Iceland, large flood basalt events, such as Laki or Eldgjá, take place only every 200 to 500 years.[83] They produce extraordinarily large amounts of lava, disgorged from magma chambers and deep-seated magma reservoirs. Based on observations made during historical times, it is estimated that within one Icelandic volcanic system the recurrence period of flood basalt events is likely to be hundreds or even thousands of years.[84]

From historical descriptions, researchers know that a future Laki-style eruption may be preceded by increased earthquake activity days or even weeks before it begins. Due to the lack of instrumental records from 1783, we do not know whether the

---

78 BYOCK 2011: 27.
79 GUÐMUNDSSON et al. 2008: 251, 255, 263.
80 THORDARSON, LARSEN 2007: 138, 143.
81 Veðurstofa Íslands: "Fagradalsfjall Eruption," 2021; Global Volcanism Program: Krýsuvík-Trölladyngja.
82 Holuhraun produced about 1.5 cubic kilometers of lava and occurred in a relatively remote area.
83 SCHMIDT et al. 2015b.
84 THORDARSON, LARSEN 2007: 138.

system would likely show inflation or dilation. Still, it appears likely it would show one of these characteristics prior to an eruption.[85]

In Iceland, the most severe volcanic events we can expect are of three different sorts. The first is a major flood basalt event similar to the Laki eruption. The second is a VEI 6 Plinian-style eruption, such as the one that occurred in 1362 at Öræfajökull. Luckily, eruptions like these only occur once or twice per millennium. The third is volcanic activity at Katla, which is almost guaranteed to be followed by *jökulhlaups*, potentially threatening the residential and agricultural areas to the west.[86]

Flood basalt eruptions, like the Laki eruption, are low-probability but high-impact events. Depending on meteorological conditions, they have the potential to affect regions as far away as southern and eastern Europe. Outside Iceland, the Netherlands, Belgium, and the United Kingdom may be among the worst affected areas.[87] Economic losses from volcanic eruptions and their consequences are to be expected in the near future.

Today, famine is not a great danger for Icelandic society. Agriculture is more stable, and the nation's wealth allows it to import goods from elsewhere. However, infrastructure such as roads (particularly Highway 1, which surrounds the island and is a key transportation route), power lines, telecommunications equipment, and industrial complexes could be affected by a large eruption. Iceland is sparsely populated; around 70 percent of all Icelanders live in or near Reykjavík. Fortunately, the Reykjavík metropolitan area is safe from lava flows, except for the suburbs in the very south and east.[88] Volcanic pollution, however, and the discomfort it brings upon those with respiratory problems may be more difficult to avoid.[89]

At the end of the last ice age, the vast ice sheet that covered Iceland began to retreat; an immense weight, and thus pressure, had been removed from the land and mantle below. This change led to a rebound effect, which may have destabilized the magma chambers.[90] The ice melt increased the magma production in the magma chambers below by a factor of 30, compared to the production rate during glaciation.[91] During the early Holocene, there was an increase in volcanic activity in Iceland, which geophysicist Magnús GUÐMUNDSSON and his colleagues describe as "a major peak in volcanism and lava production."[92] This prompts the question of whether another increase in volcanic activity can be expected now that Iceland's glaciers are retreating due to anthropogenic climate change.

---

85 Catalogue of Icelandic Volcanoes: Grímsvötn.
86 GUÐMUNDSSON et al. 2008: 251.
87 SCHMIDT et al. 2011: 15710–15714; ÁGÚSTDÓTTIR 2015: 1674; HEAVISIDE, WITHAM, VARDOULAKIS 2021.
88 GUÐMUNDSSON et al. 2008: 251, 263.
89 CARLSEN et al. 2021.
90 GUÐMUNDSSON et al. 2008: 265.
91 PAGLI, SIGMUNDSSON 2008.
92 GUÐMUNDSSON et al. 2008: 265.

During the Little Ice Age, a larger area of Iceland was covered in ice as the glaciers surged during the coldest periods. Glaciers are permanent ice caps consisting of dense ice and snow; they exist even during the summer.[93] Today, only four glaciers are left in Iceland, of which Vatnajökull is the largest, covering an area of 8,000 square kilometers. Iceland's glaciers have been retreating for over 100 years; over the next one or two centuries, it is possible they will disappear entirely. Since 1890, Vatnajökull has been continuously retreating and has lost around ten percent of its ice mass.[94] In 2019, for the first time, Iceland officially mourned the loss of a glacier, Okjökull, to anthropogenic climate change.[95]

Geoscientists Carolina PAGLI and Freysteinn SIGMUNDSSON estimate that an additional 0.014 cubic kilometers of magma are produced under Vatnajökull every year due to the thinning of the ice shield.[96] The changes caused by melting glaciers will primarily affect the Eastern Volcanic Zone: Grímsvötn, Bárðarbunga, and Katla. This volcanic zone is already Iceland's most active. Vatnajökull's ice shield today measures 50 kilometers in radius; it was 180 kilometers after the last glaciation. As ice shields melt, their influence on magma production within magma chambers is lessened.[97] In the long run, when Iceland loses most of its ice, there will be a substantially decreased chance of *jökulhlaups* threatening human and non-human lives and infrastructure.[98]

In 1783, the people of Iceland could only react to the eruption and mitigate its ill effects; this approach to the disaster has been the only option for most of human history, and it is a habit that is hard to break. Today, we have the opportunity to be more proactive in our approach to natural calamities; perhaps we should embrace this opportunity. If we only react to environmental disasters and mitigate their deleterious consequences, we may feel like we are in a duel with nature, bogged down in a constant struggle. A proactive approach would feel more like a cooperation with nature. The danger posed by Icelandic volcanic eruptions is ever-evolving, but so too is our understanding of them.

---

93 SIGURÐSSON 2005: 242–248.
94 GUÐMUNDSSON et al. 2008: 265; PAGLI, SIGMUNDSSON 2008.
95 JAKOBSDÓTTIR 2019.
96 PAGLI, SIGMUNDSSON 2008.
97 MACLENNAN et al. 2002; GUÐMUNDSSON et al. 2008: 265; ÁGÚSTDÓTTIR 2015: 1674.
98 Similar effects can be expected from presently ice-capped stratovolcanoes that will lose their ice cover over the next decades and centuries, such as Mount Erebus in Antarctica or the Aleutian volcanoes in Alaska: GUÐMUNDSSON et al. 2008: 265; PAGLI, SIGMUNDSSON 2008.

Illustrations

Figure 1 Volcanic outputs injected into the atmosphere. Graphic created by Jack WALSH. Used with permission. It is adapted from COOPER et al. 2018: 240; ROBOCK 2000 —— **11**

Figure 2 Northern Hemisphere temperature variations and global volcanic aerosol forcing, 500 BCE to 2000 CE. SIGL et al. 2015: 545. It was modified by Michael SIGL. Used with permission —— **14**

Figure 3 The different layers of the atmosphere. Graphic created by Jack WALSH. Used with permission. It is adapted from FLANNERY 2007: 21 —— **15**

Figure 4 A model of Earth's internal density structure. This graphic was created by the United States Geological Service (USGS) and is in the public domain. Edited by Mikhail RYAZANOV. https://upload.wikimedia.org/wikipedia/commons/archive/e/e9/20190801114540%21Earth_cutaway_schematic-en.svg (29 March 2020) —— **36**

Figure 5 Dynamics of the possible effects of a dense layer in the lower mantle. Graphic created by Jack WALSH. Used with permission. It is adapted from KELLOGG, HAGER, HILST 1999: 1882, Figure 1 —— **37**

Figure 6 Large igneous provinces. Map created by Jack WALSH. Used with permission. The data are from CONDIE 2001, 55; COFFIN, ELDHOLM 1994 —— **38**

Figure 7 Proposed trajectories of the Iceland mantle plume over the past 100 million years. MARTOS, Yasmina M.; JORDAN, Tom A.; CATALÁN, Manuel; JORDAN, Thomas M.; BAMBER, Jonathan L.; VAUGHAN, David G.: Geothermal Heat Flux Reveals the Iceland Hotspot Track Underneath Greenland. In: Geophysical Research Letters 45 (2018), 8214–8222, here 8216, Figure 3. Used with permission —— **39**

Figure 8 The North Atlantic's basalt structures. Map created by Jack WALSH. Used with permission. It is adapted from THORDARSON, LARSEN 2007: 120, Figure 1 —— **40**

Figure 9 Iceland's deformation zones today. Map created by Jack WALSH. Used with permission. It is adapted from THORDARSON, LARSEN 2007: 121, Figure 2 —— **42**

Figure 10 Iceland's 30 volcanic systems. Map created by Jack WALSH. Used with permission. It is adapted from THORDARSON, HÖSKULDSSON 2014: 11 —— **43**

Figure 11 The Grímsvötn volcanic system. PAGNEUX, Emmanuel, in: GUÐMUNDSSON, M.T., LARSEN, G. (15 November 2019). Grímsvötn: Figure 1 of 16. The map is based on previous works by JÓHANNESSON et al. 1990; JÓHANNESSON and SÆMUNDSSON 1998; GUÐMUNDSSON et al. 2013. The base data is based on Iceland GeoSurvey, IMO, NLSI. The base map is by IMO. Retrieved from http://icelandicvolcanos.is/?volcano=GRV# (10 April 2021). Used with permission —— **44**

Figure 12 Map of eruptions in historical times featured in this book. Map created by Jack WALSH. Used with permission. It is adapted from RABARTIN, ROCHER 1993: 26, which in turn was inspired by Sigurður ÞÓRARINSSON —— **48**

Figure 13 A map of Iceland, ca. 1700. "Novissima Islandiæ Tabula" was created by Peter SCHENK and Gerard VALK around 1700. This map is in the public domain. https://islandskort.is/en/map/show/91 (9 March 2020) —— **56**

Figure 14 A map of Iceland's counties. Map created by Jack WALSH. Used with permission —— **57**

Figure 15 The location of the glaciers, the spreading axis, and the location of the Laki lava. Map created by Jack WALSH. Used with permission —— **61**

Figure 16 The different segments of the Laki fissure. Map created by Jack WALSH. Used with permission. It is based on data from THORDARSON et al. 2003b: 15, Figure 3 —— **66**

Figure 17 Tephra fall during the Laki eruption. Map created by Jack WALSH. Used with permission. It is adapted from THORDARSON, SELF 2003: 3, Figure 1 —— **75**

Figure 18 Danish trading posts in Iceland during the time of the Danish trade monopoly, 1602 to 1787. Map created by Jack WALSH. Used with permission —— **77**

Figure 19 Athanasius KIRCHER, *Subterraneus Pyrophylaciorum*, 1665. This copperplate print is part of KIRCHER's *Mundus Subterraneus*; it was reproduced from the 1678 edition, vol. 1: 194. RBC Q155 .K6 1678 F. Courtesy of the Department of Special Collections, Stanford University Libraries. This image is in the public domain —— **91**

Figure 20 North America in 1784. Map created by Jack WALSH. Used with permission. This map was adapted from a map created by Esemono, which is in the public domain. https://en.wikipedia.org/wiki/File:Non-Native_Nations_Claim_over_NAFTA_countries_ 1784 (9 March 2020) —— **95**

Figure 21 Europe ca. 1783. Map created by Jack WALSH. Used with permission. This map is adapted from a map created by Bryan RUTHERFORD, which is licensed under a CC BY-SA 4.0 international license. https://en.wikipedia.org/wiki/File:Europe_1783-1792_en.png (9 March 2020) —— **96**

Figure 22 The view from Benjamin FRANKLIN's terrace in Passy on 21 November 1783. Anonymous artist, *Vue de la terrasse de Mr. Franklin a Passi*. Paris: Le Vachez, 1783 [?]. Bibliothèque nationale de France, département Estampes et photographie: FOL-IB-1. This image is in the public domain. https://gallica.bnf.fr/ark:/12148/btv1b550014819 (9 March 2020) —— **98**

Figure 23 The ascent of *La Charlière* in the Jardin des Tuileries in Paris on 1 December 1783. Anonymous engraver, *Jacques Charles: Charlière*, 1783. Antoine Louis François Sergent dit SERGENT-MARCEAU, United States Library of Congress. This image is in the public domain. https://commons.wikimedia.org/wiki/File:Jacques_Charles_Luftschiff.jpg#/ media/File:Jacques_Charles_Luftschiff.jpg (9 March 2020) —— **99**

Figure 24 The locations of the five main earthquakes in Calabria in February and March 1783. Map created by Jack WALSH. Used with permission. It is adapted from GRAZIANI, MARAMAI, TINTI 2006: 1055 —— **103**

Figure 25 The Strait of Messina as seen from the north when the earthquake struck. Unknown artist, *Le celebre pour les Vaisseaux autre fois si dangereux detroit de Faro di Messina*, hand-colored copper engraving, no date. Photograph by The Film Museum. This image is in the public domain. https://www.flickr.com/photos/36461985@N08/ 25229827725/in/album-72157664333261669/ (9 April 2021) —— **104**

Figure 26 The earthquake of 5 February 1783. Hand-colored copper engraving, ca. 1790. Unknown artist, *Vue de la Ville de Regio dil Messinae et ces alentour detruite par le terrible tremblement de Terre arrivée le Cinq Fevrier de l'année 1783*, ca. 1790. Published by Jacques-Simon Chereau in Paris. Bibliothèque nationale de France, département Estampes et photographie: LI-72 (7)-FOL. This image is in the public domain. https://gallica.bnf.fr/ark:/12148/btv1b69494071# (22 March 2021) —— **105**

Figure 27 Map of weather stations in 1783 that featured in the research for this book. Additional weather stations that are not pictured in this map for reasons of space are: Peißenberg (Bavaria); Somerset House (England); Cambridge (Massachusetts); Edinburgh (Scotland); Moscow (Russia); St. Petersburg (Russia). Map created by Jack WALSH. Used with permission. The inspiration for this map was one that featured the weather stations in the 1780s; KINGTON 1988: 23, Figure 4 —— **110**

Figure 28 The beginning of the haze in Europe in June 1783. The idea for this map is based on GRATTAN et al. 2005: 647. Map created by Jack WALSH. Used with permission. This map displays the dates when the Laki haze first appeared in different cities and regions, based on the following data: 14 June: Dijon, Societas Meteorologica Palatina 1783: 447; STOTHERS 1996: 80. 15 June: Edinburgh, George and James WARROCH in Prestonpans,

1783–1807, CS96/4401, National Records of Scotland, Edinburgh, UK. 16 June: Mannheim, Societas Meteorologica Palatina 1783: 10; Würzburg, Societas Meteorologica Palatina 1783: 10; Bündten, SALIS-MARSCHLINS 1783: 393; Rome, Societas Meteorologica Palatina 1783: 520; Prague, Societas Meteorologica Palatina 1783: 380. 17 June: Scotland, Scottish Meteorological Society, MET1, National Records of Scotland, Edinburgh, UK, June 1783; Middelburg, Societas Meteorologica Palatina 1783: 637; Oberrhein/Obernai, Bas-Rhin, Société Royale de Médecine, June 1783, http://meteo.aca demie-medecine.fr/_app/visualisation.php?id=5354 (9 March 2020); Erfurt, Societas Meteorologica Palatina 1783: 238; Munich, Societas Meteorologica Palatina 1783: 267; Saxony, the Ore Mountains, KIESSLING 1888, vol. 1: 27–28; Sagan, Societas Meteorologica Palatina 1783: 339; LE VIVIER, Pyrenees-Orientales, Société Royale de Médecine, June 1783, http://meteo.academie-medecine.fr/_app/visualisation.php?id= 1628 (9 March 2020). 18 June: Göttingen, Societas Meteorologica Palatina 1783: 665; Frankfurt am Main, Société Royale de Médecine, June 1783, http://meteo.academie-medecine.fr/_app/visualisation.php?id=10503 (9 March 2020); Mark Brandenburg, KIESSLING 1888, vol. 1: 27–28; Regensburg, König 1783: 288; Saint Gotthard, Societas Meteorologica Palatina 1783: 175; Andechs, Societas Meteorologica Palatina 1783: 88; Tegernsee, Societas Meteorologica Palatina 1783: 288; Bern, PFISTER 1975: 87. 19 June: Berlin, Societas Meteorologica Palatina 1783: 108–110; Franeker, VAN SWINDEN 2001: 73; Peißenberg, Societas Meteorologica Palatina 1783: 310; Gdansk, KIESSLING 1888. 20 June: Perthshire, James Stuart MACKENZIE, Weather Reports Journal of James Stuart Mackenzie at Belmont Castle, Perthshire, 1771–1799, RH4/100, National Records of Scotland, Edinburgh, UK; Lyon, Rhône, Société Royale de Médecine, June 1783, http://meteo.academie-medecine.fr/_app/visualisation.php?id=8607 (9 March 2020). 20 June: Ofen, Wiener Zeitung, 9 July 1783: Report from Ofen, 2 July 1783. 21 June: Metz, Moselle, Société Royale de Médecine, June 1783, http://meteo.academie-medecine.fr/_app/visualisation.php?id=5578 (9 March 2020); Arras, Pas-de-Calais, Société Royale de Médecine, June 1783, http://meteo.academie-medecine.fr/_app/ visualisation.php?id=2414 (9 March 2020); Suffolk, J. FENTON, The Weather, etc. from June 27, 1779 to December 30, 1786 (at Nacton, Suffolk), Bib. No. 178296, National Meteorological Library and Archive, Met Office, Exeter, UK. 22 June: Oslo, STOTHERS 1996: 80–81; Spydeberg, Norway, KÖNIG 1783, Observationes Spydbergenses; Leicestershire, Thomas BARKER, Private Weather Diary for Lyndon Hall, Leicestershire, MET/2/1/2/3/227, National Meteorological Library and Archive, Met Office, Exeter, UK; Lausanne, VERDEIL 1783: 11. 23 June: Buda, Societas Meteorologica Palatina 1783: 129–130; Selbourne, Gilbert WHITE, "The Naturalist's Journal," 1783, Add MS 31848, British Library, London, UK: 31. 24 June: Grampians, T. HOY, Private Weather Diary for Gordon Castle, Grampian, Scotland, MET/2/1/2/3/486, National Meteorological Library and Archive, Met Office, Exeter, UK. 25 June: London, Unknown, Private Weather Diary for London (and additional locations), Gentleman's Magazine, MET/2/1/2/3/255, National Meteorological Library and Archive, Met Office, Exeter, UK; James WOODFORDE, Weather Diary for Weston Longville, Norfolk, vol. 10: 1782–1784, MET/2/1/2/3/467, National Meteorological Library and Archive, Met Office, Exeter, UK; Moscow, Societas Meteorologica Palatina 1783: 618–619, The dry fog might have reached St. Petersburg as early as June 15 [N.S.]; Moscow, Societas Meteorologica Palatina 1783 [unclear whether 25 June is O.S. or N.S.]; North Sea, VAN SWINDEN 2001: 76. 26 June: Lisbon, STOTHERS 1996: 80–81. [DEMARÉE, OGILVIE 2001: 231–232, state the dry fog reached Lisbon on 22 June 1783.] 27 June: Padua, Societas Meteorologica Palatina 1783: 555 ━━ **116**

Figure 29   The Laki eruption's emissions of sulfur dioxide. Graphic created by Jack WALSH. Used with permission. It is based on data from ZAMBRI et al. 2019a: 4, Figure 1 —— **119**

Figure 30   The hazards of sulfur dioxide to health and vegetation. Graphic created by Jack WALSH. Used with permission. It is based on data from MEYER 1977 —— **124**

Figure 31   The sulfuric smell and its impact on health and vegetation, late June 1783. Map created by Jack WALSH. Used with permission. The map is based on the following data: 20 June 1783 (circa): Groningen, the Netherlands: BRUGMANS wrote of a fog that had appeared in Holland and Utrecht before 20 June with no apparent sulfuric odor, and in Gelderland and Overijssel, where "a sulfuric smell was admixed to the fog." BRUGMANS 1783: preface. 23 June 1783: Arras, France, Monsieur BUISSART for the Société Royale de Médecine: "Frost," based on hearsay; Société Royale de Médecine, Arras, Pas-de-Calais, June 1783. 23 June 1783: Franeker, the Netherlands: "Across the haze, the sun was perceived deep red, with brilliance at the edge; even at midday, we were able to gaze at the sun with our naked eyes without injury. Objects scattered further were scarcely and only unintelligibly perceived. These were the usual effects of this haze, but on the 24th day it brought with it as a companion a sulfurous odor very readily perceived by the senses, crawling through everything, even closed houses. Men with delicate lungs experienced that same sensation, as if they were turned towards a place in the neighborhood of burning sulfur. They were unable to contain a cough, as soon as they were exposed to air. I myself experienced this, and many others, first in the city, then in the country. The heat was great enough. The sulfurous haze of this day brought very great loss to the vegetable realm, [. . .]. In the afternoon of the 24th many experienced very troublesome headaches and respiratory difficulties, similar to that which they experienced while the atmosphere around us was filled with the vapor of burned sulfur. Asthmatics experience a return of asthma. [. . .] In general it is possible to establish that it was not the origin of fructification that was injured, but only the leaves, which immediately began after midday on the 24th, but variously in various plants. Certain ones were covered with spots, which increased gradually and soon caused drooping of leaves. Some leaves were not entirely spoilt; they continued to quicken, but the places, in which they had been affected, were soon made into little holes. Others faster than a minute turned from green to brown, black, gray, or white. Others kept their color, but began to droop, so, that they were reduced to powder at the touch of one's fingers. A very great abundance of leaves fell. [. . .] Moreover the injury, and fallings of leaves, lasted for some time." VAN SWINDEN 2001: 73–74. 23 June 1783: Hardwick, England: "All these vegetables appeared exactly as if a fire had been lighted near them, that had shriveled and discoloured their leaves." CULLUM 1784. 23 June 1783: Königlich Privilegirte Zeitung, 19 July 1783: Report from Thuringia from 4 July 1783. In Thuringia, from 23 June onwards, many trees were covered by a "corrosive moist." which caused the leaves to look brown or even black; they then shrunk and withered. Other plants withered as well. 23 June 1783: Selbourne, England: "The blades of wheat in several fields are turned yellow & look as if scorched by frost." This was also the first day the dry fog appeared; Gilbert WHITE, "The Naturalist's Journal," 1783, Add MS 31848, British Library, London, UK: 128. 23–24 June 1783 (overnight): eastern parts of England, *Sherborne Mercury*, 14 July 1783 (quoted after GRATTAN, PYATT 1994): "Throughout most of the eastern counties there was a most severe frost in the night between 23rd and 24th of June. It turned most of the barley and oats yellow, to their very great damage; the walnut trees lost their leaves, and the larch and firs in plantation suffered severely." 23–24 June 1783 (overnight): Schleswig-Holstein: "The second half of June had warm days but cool nights. There was rime and frost during the night from 23 to 24 June; the buckwheat was touched by the great frost." KUSS 1826: 170. 23–24 June 1783 (overnight): Dannenberg/Elbe: "During the night from the 23rd to the 24th, we had a sharp night frost, which damaged fruit and other crops, the foliage of all kinds, as well as grass, and crops, turned black from this night frost, some fell off, others turned yellow [. . .], after this night's frost, a new, strong and stinking dry fog, or so-called vapor, came and lasted every day until 18 July." Elias Diderich THÖRL, no title, Archiv Dannenberg: 95–96. A transcript of this record was given to me by Wolfgang JÜRRIES. 24 June 1783, Brussels: first description of sulfuric smell; DE POEDERLE 1784: 336. 24 June 1783, Groningen: on 24 June, the fog was so intense that one could taste it with each breath. The smell began to dissipate over the following days, although the dry fog remained: on the morning of 28 June, the sulfuric smell had vanished entirely; Sebald Justinus BRUGMANS, as above. 24 June 1783, Groß Hesepe, Emsland: a few days before Saint John's Eve (24 June), a smell that a local chronicler compared to "heated hay" had already manifested itself. On 24 June, the smell of sulfur became apparent. Locals compared it to the smell of burnt gunpowder and decay. The comparison to burnt gunpowder strongly suggests what they smelt was really sulfur dioxide, which is known to have a sharp smell similar to that of burnt matches. Locals "felt a sulfuric or saltpeter-like taste in their mouth on 24 June and the days after, some even complained about an itchy throat or breathing difficulties. The water in some rain tanks also started to have a saltpeter-like taste." SANTEL 1997: 108–121. 24 June 1783, Königlich Privilegirte Zeitung, 22 July 1783: 714: Report from Hildburghausen from 24 June 1783. "All the forests in the area are white instead of green; the whole sky looks like chalk; the fog is true natural sulfur, which spoils everything that it touches; the sun and moon always set in a blood-red color. For eight days now, inside the mountain there has been a horrendous and frightening bashing, as if cannons were being fired; then, finally, the whole mountain opened up under the plumes of thick sulfuric smoke; and in the entire area you can hear a constant terrible roaring and rushing (*Sausen und Brausen*) from the opening." 24 June 1783, Nacton, England: a particular sale at a market "began notwithstanding the close heat in the day, it was commonly reported that there was a sharp freezing rhyme this morning, which on the succeeding day caused the leaves to drop from the Trees." J. FENTON, The Weather, etc. from June 27, 1779 to December 30, 1786 (at Nacton, Suffolk), Bib. No. 178296, National Meteorological Library and Archive, Met Office, Exeter, UK. 24 June 1783, Selbourne: "Sun, sultry, misty & hot [. . .]. This is the weather that men think injurious to hops." Gilbert WHITE, as above. 24 June 1783, Weston Longville, England: "a smart frost this evening," James WOODFORDE, Weather Diary for Weston Longville, Norfolk, vol. 10: 1782–1784, MET/2/1/2/3/467, National Meteorological Library and Archive, Met Office, Exeter, UK. 24 June 1783, Weston Longville: "A smart Frost again this Night," James WOODFORDE as above. 24–25 June 1783 (overnight), Groß Hesepe: another chronicle for Groß Hesepe dramatically describes this "poisoning thaw," which withered everything it touched and covered all the land, water, and the ocean; SANTEL, SANTEL 1992: 71–94. 24–25 June 1783 (overnight), Plön, Schleswig-Holstein: a strong dew damaged the grain crop. The grain turned yellow [. . .] but later recovered again; based on the notes of Heinrich Christian STRUCK, a mason from Plön, in: KINDER 1904 (1976): 304–307. 25 June 1783, Franeker: "In the morning of the 25th day the fields showed a very sad appearance. The green color of the trees and plants had disappeared and the earth was covered in dropping leaves. He would easily have believed that it was October or November. But happily, it befell that not all plants were equally affected; certain uninjured ones remained standing." VAN SWINDEN 2001: 75. 27 June 1783 (before this date), Heilbronn, Königlich Privilegirte Zeitung, 12 July 1783: Report from Lautern from 29 June 1783. Sometime before 27 June, "the air smelled like sulfur and the sun also seemed very red and fiery in the evening." 30 June 1783, Leicestershire, England: it was an evening with "thick smoaky [sic] air & smell of fens" with winds from the North to the East. The "smell of fens" is probably in reference to hydrogen sulfide, a colorless gas that is characterized by the foul odor of rotten eggs; Thomas BARKER, Private Weather Diary for Lyndon Hall, Leicestershire, MET/2/1/2/3/227, National Meteorological Library and Archive, Met Office, Exeter, UK. Late June 1783, Koblenzer Intelligenzblatt, 18 July 1783: people in the Lower Rhine region had to cover their faces with sheets because "the heaviest smell of sulfur" was almost unbearable when leaving their houses. Late June 1783, the Netherlands: "The dry fog was very strong in the Netherlands, for half a week it did not become day[light], when the people wanted to go outside, they had to put on sheets because of the strong sulfuric smell." Bayerische Akademie der Wissenschaften 1783: 45 —— **126**

Figure 32 Synoptic weather maps, 23 to 28 June 1783. Graphic created by Jack WALSH. The maps are based on data from John KINGTON's The Weather of the 1780s Over Europe. Reproduced with permission of Cambridge University Press through PLSclear —— **127**

Figure 33 Dispersal of the Laki haze. Graphic created by Jack WALSH. Used with permission. It is adapted from THORDARSON, SELF 2001: 70 —— **128**

Figure 34 Peak temperatures of the August 1783 heat wave. Graphic created by Jack WALSH. Used with permission. It is based on the following data: the temperatures given in the Ephemerides are usually in Réaumur; for the purpose of easier comparability, they have all been converted into Celsius here: Rome, Societas Meteorologica Palatina 1783: 523; Copenhagen, Societas Meteorologica Palatina 1783: 603; Andechs, Societas Meteorologica Palatina 1783: 90; Geneva, Societas Meteorologica Palatina 1783: 424; Marseille, Societas Meteorologica Palatina 1783: 506; Padua, Societas Meteorologica Palatina 1783: 220; London, Unknown, Private Weather Diary for London (and additional locations), Gentleman's Magazine, MET/2/1/2/3/255, National Meteorological Library and Archive, Met Office, Exeter, UK; Franeker, VAN SWINDEN 2001: 80; Munich, Bayerische Akademie der Wissenschaften 1783: 48; this value is corroborated by the Societas Meteorologica Palatina 1783: 270; Düsseldorf, Societas Meteorologica Palatina 1783: 157; Mannheim, Societas Meteorologica Palatina 1783: 13; Göttingen, Societas Meteorologica Palatina 1783: 668; Dijon, Societas Meteorologica Palatina 1783: 450, Societas Meteorologica Palatina 1783: 111; Würzburg, Societas Meteorologica Palatina 1783: 383, Hamburg, KUSS 1826: 172 —— **133**

Figure 35 Temperature anomaly for June, July, and August 1783 based upon 31-year mean, 1770–1800. Graphic created by Jack WALSH. Used with permission. It is adapted from Alan ROBOCK's graphic, kiss.caltech.edu/workshops/geoengineering/presentations1/robock_sc.pdf (9 March 2020). The reconstruction is based upon data from LUTERBACHER et al. 2004 —— **135**

Figure 36 Earthquakes in Europe in 1783. Map created by Jack WALSH. Used with permission. It is based on the following data: 2 February 1783, Vallespir, France, 4.99 moment magnitude (Mw), https://emidius.eu/SHEEC/maps/query_eq/external_call.php?eq_id=6122. 5 February 1783, 03:00, Scuol, Switzerland, 3.79 Mw, https://emidius.eu/SHEEC/maps/query_eq/external_call.php?eq_id=8476. 5 February 1783, 12:00, Calabria, 7.0 Mw, https://emidius.eu/SHEEC/maps/query_eq/external_call.php?eq_id=1384.

6 February, 00:20, Calabria/Messina, 6.20 Mw, https://emidius.eu/SHEEC/maps/query_eq/external_call.php?eq_id=1387. 7 February 1783, 13:10, Calabria, 6.61 Mw, https://emidius.eu/SHEEC/maps/query_eq/external_call.php?eq_id=1390. 18 February 1783, Central Slovenia, 4.30 Mw, https://emidius.eu/SHEEC/maps_nomdp/query_eq/external_call.php?eq_id=96051. 1 March 1783, 01:40, Calabria, 5.94 Mw, https://emidius.eu/SHEEC/maps/query_eq/external_call.php?eq_id=1393. 9 March 1783, Stemnytsa, Greece, 5.94 Mw, https://emidius.eu/SHEEC/maps/query_eq/external_call.php?eq_id=8093. 23 March 1783, 05:00, Athani, Greece, 6.64 Mw, https://emidius.eu/SHEEC/maps/query_eq/external_call.php?eq_id=8094. 25 March 1783, 03:00, Basse-Durance, France, 4.44 Mw, https://emidius.eu/SHEEC/maps/query_eq/external_call.php?eq_id=6125. 28 March 1783, 18:55, Calabria, 6.98 Mw, https://emidius.eu/SHEEC/maps/query_eq/external_call.php?eq_id=1396. 29 March 1783, 18:30, Lorca, Spain, 4.52 Mw, https://emidius.eu/SHEEC/maps_nomdp/query_eq/external_call.php?eq_id=200820. 14 April 1783, 23:00, Porto, Portugal, 5.00 Mw, https://emidius.eu/SHEEC/maps_nomdp/query_eq/external_call.php?eq_id=18224. 22 April 1783, 02:45, Komárno, Hungary, 5.35 Mw, https://emidius.eu/SHEEC/maps/query_eq/external_call.php?eq_id=8347. 22 April 1783, 11:00, Komárno, Hungary, 3.80 Mw, https://emidius.eu/SHEEC/maps_nomdp/query_eq/external_call.php?eq_id=100367. 26 April 1783, 02:10, Milazzo, Sicily, 4.72 Mw, https://emidius.eu/SHEEC/maps/query_eq/external_call.php?eq_id=1398. 31 May 1783, 12:00, Komárno, Hungary, 4.46 Mw, https://emidius.eu/SHEEC/maps_nomdp/query_eq/external_call.php?eq_id=100369. 1 June 1783, Istanbul, Turkey, 5.43 Mw, https://emidius.eu/SHEEC/maps_nomdp/query_eq/external_call.php?eq_id=26831. 1 June 1783, Síða region, Iceland, earthquake swarm, mid-May to June 8, 1783, more than 4.0 Mw, ÞÓRARINSSON 1984: 35; THORDARSON et al. 2003b: 19. 8 to 11 June 1783, Síða region, Iceland, strong earthquakes continued, damaging farmsteads, STEINGRÍMSSON 2002: 181. 15 June 1783, Síða region, Iceland, earthquakes continued, STEINGRÍMSSON 1998: 30. 21 June 1783, 18:00, Belledonne, near Lyon, France, 3.95 Mw, https://www.emidius.eu/SHEEC/maps/query_eq/external_call.php?eq_id=6127; QUENET 2005: 541. 22 June 1783, Schweidnitz, Silesia, Severe thunderstorm, probably not an earthquake, Berlinische Nachrichten, 1 July 1783: 605: Report, 25 June 1783. Königlich Privilegirte Zeitung, 1 July 1783, Hamburgischer Unpartheyischer Correspondent, 4 July 1783. 29 June 1783, 04:30, Florence, Italy, "another weak earthquake," Hamburgischer Unpartheyischer Correspondent, 19 July 1783: Report from Florence, 30 June 1783; Königlich Privilegirte Zeitung, 24 July 1783: 723: Report from Florence, 30 June 1783. July 1783, Austria, Königlich Privilegirte Zeitung, 31 July 1783: 742: Report from Austria, 19 July 1783. 6 July 1783, 09:56, Vallée de l'Ouche, near Dijon, France, 5.14 Mw; VI (Mercalli), https://www.emidius.eu/SHEEC/maps/query_eq/external_call.php?eq_id=6131; QUENET 2005: 541; Berlinische Nachrichten, 26 July 1783: 693: Report from Paris 14 July 1783. 20 July 1783 [?], Gorgona, Italy, lightning struck powder magazine on Gorgona; people in Livorno thought it was an earthquake, Hamburgischer Unpartheyischer Correspondent, 5 August 1783: Report from Italy, 20 July 1783. 20 July 1783, Tripoli, Lebanon, Journal historique et littéraire, 1 November 1783: 363; AMBRASEYS 2009: 613. 28 July 1783, Val di Ledro, Italy, 4.80 Mw, https://emidius.eu/SHEEC/maps/query_eq/external_call.php?eq_id=1400. 8 August 1783, 03:00, between Aachen, Germany, and Maastricht, the Netherlands, QUENET 2005: 541; DEMARÉE, OGILVIE 2001: 226; Königlich Privilegirte Zeitung, 21 August 1783: 802: Report from Cologne, 10 August 1783. 10 August 1783, Devon, England, mild tremors, Exeter Flying Post, 14 August 1783 (quoted after GRATTAN, BRAYSHAY 1995: 6). 11 August 1783, 04:00,

Lucerne, Switzerland, 3.9 Mw, https://www.emidius.eu/SHEEC/maps_nomdp/query_eq/external_call.php?eq_id=200976. 15 November 1783, 09:15, S. Severo, Italy, 5.17 Mw, https://emidius.eu/SHEEC/maps_nomdp/query_eq/external_call.php?eq_id=6258. 20 November 1783, 12:58, S. Gregorio, Calabria, 5.17 M, https://emidius.eu/SHEEC/maps_nomdp/query_eq/external_call.php?eq_id=6259. 9 November 1783, 04:00, Cambrai, France, 4.59 Mw, VI Mercalli, QUENET 2005: 541; https://emidius.eu/SHEEC/maps/query_eq/external_call.php?eq_id=6135. 10 December 1783, 05:00, Komárno, Hungary, 3.48 Mw, https://emidius.eu/SHEEC/maps_nomdp/query_eq/external_call.php?eq_id=100371 (all 9 March 2020) —— **145**

Figure 37 Earthquake history in Europe. This map shows the distribution of earthquakes between the years 1000 and 2006, as compiled by the SHARE European Earthquake Catalog (SHEEC). WOESSNER et al. 2015: 3358, Figure 2. This map is licensed under a CC BY 4.0 license —— **147**

Figure 38 Earthquake risk in today's Germany, Austria, and Switzerland. GRÜNTHAL, Gottfried; MAYER-ROSA, Dieter; LENHARDT, Wolfgang A.: Abschätzung der Erdbebengefährdung für die D-A-CH-Staaten – Deutschland, Österreich, Schweiz. In: Bautechnik 75, no. 10 (1998), 753–767, here 764, Figure 6. I thank Professor GRÜNTHAL for providing me with an extracted version of this map featuring a map legend —— **148**

Figure 39 The total lunar eclipse of 10 September 1783. Graphic created by Jack WALSH. Used with permission. It is adapted from a graphic provided by NASA, which is in the public domain. https://eclipse.gsfc.nasa.gov/5MCLEmap/1701-1800/LE1783-09-10T.gif (9 March 2020) —— **154**

Figure 40 The last observed occurrence of the Laki haze, based on data compiled for this book. Map created by Jack WALSH. Used with permission. Sometimes it is hard to distinguish when the dry fog ends and normal humid fogs commences. The dates are based on the following data: Regensburg, 19 August 1783, "thick fog," KÖNIG 1783, here: Regensburg. Edinburgh, Scotland, 21 August 1783, "foggs [sic] which are continuing w[ith] the wind at the North East," Scottish Meteorological Society, MET1, National Records of Scotland, Edinburgh, UK. Buda, 27 August 1783, Societas Meteorologica Palatina 1783: 134. Berlin, 29 August 1783 (?), "Aer nebulosis post mer," Societas Meteorologica Palatina 1783: 112. Peißenberg, 31 August 1783, Societas Meteorologica Palatina 1783: 329. Belmont Castle, Scotland, 31 August 1783, "a very uncommon fog," James Stuart MACKENZIE, Weather Reports Journal of James Stuart Mackenzie at Belmont Castle, Perthshire, 1771–1799, RH4/100, National Records of Scotland, Edinburgh, UK. Erfurt, 1 September 1783 (?), Societas Meteorologica Palatina 1783: 242. Grampians 1 September 1783, "fogg" [sic], T. HOY, Private Weather Diary for Gordon Castle, Grampian, Scotland, MET/2/1/2/3/486, National Meteorological Library and Archive, Met Office, Exeter, UK. Weston Longville, England, 2 September 1783, "Morn' hazy and warm, afternoon ditto," James WOODFORDE, Weather Diary for Weston Longville, Norfolk, vol. 10: 1782–1784, MET/2/1/2/3/467, National Meteorological Library and Archive, Met Office, Exeter, UK. Selbourne, England, 3 September 1783, "red sunshine," 14 October "haze," Gilbert WHITE, "The Naturalist's Journal," 1783, Add MS 31848, British Library, London, UK: 139–146. Middelburg, 3 September 1783, Societas Meteorologica Palatina 1783: 642. Copenhagen (according to HÓLM), 4 September 1783, Societas Meteorologica Palatina 1783: 605. Minehead, Somersetshire, England, 10 September 1783, "hazy," John ATKINS, "The Meteorological Journal for the Year 1783. Kept at Minehead in Somersetshire by Mr. John Atkins. Presented at the Royal Society in London on 19 January 1786," MA/166, Archives of the Royal Society, London, UK. Munich, 28 September 1783 (?), Societas Meteorologica

Palatina 1783: 271. Lyndon Hall, Leicestershire, England, late September 1783, "The latter part of the month [September] very fine, warm and pleasant, often thick air." Thomas BARKER, Private Weather Diary for Lyndon Hall, Leicestershire, MET/2/1/2/3/227, National Meteorological Library and Archive, Met Office, Exeter, UK. Nain, Labrador, 30 September 1783, "smoky sky," Meteorological Observations at Nain & Okak at Labrador, Hudson's Bay Company, MA/143, Archives of the Royal Society, London, UK, here: 10 September 1783. Prestonpans, Scotland, 1 October 1783, "hazy," George and James WARROCH in Prestonpans, 1783–1807, CS96/4401, National Records of Scotland, Edinburgh, UK. Mannheim, 5 October 1783, "vapors tenuis [fine haze]," Societas Meteorologica Palatina 1783: 16. Sagan, 11 October 1783, "Fog, like smoke," Societas Meteorologica Palatina 1783: 346. Okkak, Labrador, 23 October 1783, "heazey and sun shine," Meteorological Observations at Okkak and at Hoffenthal, Hudson's Bay Company, Labrador, Newfoundland, 1782–1786, MA/144, Archives of the Royal Society, London, UK. Hoffenthal, Labrador, 16 November 1783, "[. . .] although we have had much snow this year yet it most all melted, and blown away and the land most over all haze." Meteorological Observations at Okkak and at Hoffenthal, Hudson's Bay Company, Labrador, Newfoundland, 1782–1786, MA/144, Archives of the Royal Society, London, UK —— **156**

Figure 41 The locations of the different German "volcanic eruptions" in 1783. Map created by Jack WALSH. Used with permission —— **179**

Figure 42 A contemporary depiction of the Great Meteor. Henry ROBINSON, "An accurate representation of the meteor," as seen at Winthorpe, Nottinghamshire, England, on 18 August 1783. Astronomy: a meteor shower in the night sky. Mezzotint after H. Robinson, 1783. © The Trustees of the British Museum. Shared under a Creative Commons Attribution-NonCommercial-ShareAlike 4.0 International (CC BY-NC-SA 4.0) license. Used with permission —— **192**

Figure 43 The approximate path of the meteor on 18 August 1783. Map created by Jack WALSH. Used with permission. The idea for this map comes from PAYNE 2011: 21; the data are based on my sources —— **193**

Figure 44 Paul SANDBY's watercolor, "The Meteor of 1783, as seen from the East Angle of the North Terrace, Windsor Castle." This image is in the public domain. Yale Center for British Art, Paul Mellon Collection, B1993.30.115. https://collections.britishart.yale.edu/catalog/tms:4107 (9 April 2021) —— **194**

Figure 45 First appearance of the dry fog across the Northern Hemisphere. Map created by Jack WALSH. Used with permission. The data are based on sources mentioned throughout this book —— **205**

Figure 46 The locations of Okkak, Nain, and Hoffenthal in Labrador. Map created by Jack WALSH. Used with permission —— **209**

Figure 47 Temperature anomaly for December 1783, January and February 1784 based upon 31-year mean, 1770–1800. Graphic created by Jack WALSH. Used with permission. It is adapted from Alan ROBOCK's graphic. kiss.caltech.edu/workshops/geoengineering/presentations1/robock_sc.pdf (9 March 2020). The reconstruction is based upon data from LUTERBACHER et al. 2004 —— **217**

Figure 48 Floodmarks at the portal of St. Maria in Lyskirchen, Cologne. Photograph by Katrin KLEEMANN, May 2019 —— **223**

Figure 49 Contemporary depiction of flooding in Würzburg in 1784. Unknown artist, *La grande desolation arrivée a Wurzbourg par le débordement du Mayn*, copper engraving, Augsburg, ca. 1785. The image is in the public domain and is located at the Stadtarchiv Würzburg, Karten und Pläne B 88 —— **225**

Figure 50 Mean surface temperatures for the late eighteenth century. This temperature reconstruction is based on stations in Europe and the northeastern United States analyzed over a 31-year period, 1768 to 1798. Graphic created by Jack WALSH. Used with permission. It is based on data from THORDARSON, SELF 2003: 17, Figure 9 —— **233**

Figure 51 Abraham ORTELIUS' *Islandia*, ca. 1590. This image is in the public domain. https://myndir.islandskort.is/map/Kortgerd_Abrahams_Orteliusar_10/Islandia_2/136/ 2012-09-04-11-05-33.jpg (9 April 2021) —— **239**

Figure 52 Sæmundur Magnússon HÓLM, Reykjavík, ca. 1785. Hólm 1784a. This image is in the public domain. https://islandskort.is/en/map/show/1130 (9 March 2020) —— **241**

Figure 53 Uno VON TROIL, *An accurate and correct map of Iceland*, 1780. The map featured in Uno VON TROIL's publication and was drawn by Jón EIRÍKSSON and Gerhard SCHØNING, who had used Eggert ÓLAFSSON and Bjarni PÁLSSON's book as their source of information. This map is in the public domain. https://islandskort.is/en/map/show/27 (9 March 2020) —— **244**

Figure 54 Map of the Laki lava flows by Magnús STEPHENSEN, 1784. Photograph by Helgi BRAGASON. Landsbókasafn Íslands Háskólabókasafn | National and University Library of Iceland. https://baekur.is/skra/JPG/2821948 (20 April 2021). This image is in the public domain —— **247**

Figure 55 Portrait of Sveinn PÁLSSON drawn by Sæmundur M. HOLM in 1798 with red chalk. This image is in the public domain. https://commons.wikimedia.org/wiki/File:Sveinn_P% C3%A1lsson.jpg (9 April 2021) —— **249**

Figure 56 A map of the landmarks visited by Sveinn PÁLSSON and his companion. Map created by Jack WALSH. Used with permission. The basemap is called ÍslandsDEM útgáfa 0 and it was created by Landmælingar Íslands (National Land Survey of Iceland), it is licensed under a CC BY 4.0 international license. https://gatt.lmi.is/geonetwork/srv/eng/cata log.search#/metadata/e6712430-a63c-4ae5-9158-c89d16da6361 (15 March 2021) —— **252**

Figure 57 Map of the Laki lava flows drawn by Sveinn PÁLSSON, 1794. Sveinn PÁLSSON, Tillæg til Beskrivelse over den Volcan der brændte i Skaptafells Syssel Aar 1783, 1839–1846. Photograph by Helgi BRAGASON, Landsbókasafn Íslands Háskólabókasafn | National and University Library of Iceland, ÍB 23 fol. This image is in the public domain —— **254**

Figure 58 Ebenezer HENDERSON's map of Iceland, 1818. The map was engraved by Daniel and William Home LIZARS under the direction of HENDERSON. This map is in the public domain. https://islandskort.is/en/map/show/52 (9 March 2020) —— **261**

Figure 59 The explosive sound of the Krakatau eruption. The explosion was audible within the encircled area on the map. Map created by Jack WALSH. Used with permission —— **267**

Figure 60 The extent of the prolonged twilight appearances between August and December 1883 as depicted by Johann KIESSLING. KIESSLING 1888, vol. 1: 203ff. Bayerische Staatsbibliothek München, 4 Phys.sp. 128 m, 203ff. Used with permission —— **268**

Figure 61 Laki's craters and lava flows, as drawn by Amund HELLAND, ca. 1881. HELLAND 1886: 41. Bayerische Staatsbibliothek München, 4 Phys.sp. 131 o, 41. Used with permission —— **273**

Figure 62 Þorvaldur THORODDSEN's geological map of Iceland, 1906. Scale: 1: 750,000. This map is in the public domain. https://islandskort.is/en/map/show/608 (9 March 2020) —— **275**

Figure 63 A photograph of one crater in the Laki fissure, taken by Tempest ANDERSON, ca. 1903. ANDERSON published his book and photographs in 1903. This one is called "Plate LXIV. Iceland. A Crater, Skaptár lava." ANDERSON 1903: 124, plate LXIV. Bayerische Staatsbibliothek München, 4 Phys.sp. 7 is-1, urn:nbn:de:bvb:12-bsb00067739-3, http://daten.digitale-sammlungen.de/bsb00067739/image_279. This digitized version of this photograph is licensed under a CC BY-NC-SA 4.0 license. Used with permission —— **278**

Figure 64 In contrast, one crater of the Laki fissure in August 2016. Photograph by Katrin KLEEMANN, August 2016 —— **279**

Figure 65 A photograph of the eastern part of the Laki fissure in 1908, taken by Ina VON GRUMBKOW from Mount Laki. Ina VON GRUMBKOW, "Abbildung 19. Blick von der Höhe des Berges Laki auf die östliche Kraterreihe," 1909. GRUMBKOW 1909, chapter VII. Bayerische Staatsbibliothek München, It.sing. 1309 p. Used with permission —— **280**

Figure 66 A stamp commemorating the Laki eruption, released in 1970, featuring a painting by Sveinn ÓLAFSSON. This stamp was part of the *náttúruvernd* (nature conservation) series. Photograph by Katrin KLEEMANN. Used with permission from Myndstef —— **299**

Figure 67 "Lakagígar" by Finnur JÓNSSON, 1940. This painting depicted on the stamp is owned by the National Gallery of Iceland in Reykjavík. Finnur JÓNSSON's "Lakagígar" was turned into a stamp in 1983. Photograph by Katrin KLEEMANN. Used with permission from Myndstef —— **300**

Figure 68 A satellite image of the Laki fissure and Vatnajökull. European Space Agency – ESA, Copernicus Sentinel 2B data, 16 August 2020. Used with permission. I thank Harald ZANDLER for his help extracting the image from the Copernicus Open Access Hub —— **301**

Figure 69 The Laki fissure in 2016. The southwestern part of the fissure, as seen from Mount Laki. Photograph by Katrin KLEEMANN, August 2016 —— **302**

Figure 70 *Skaftáreldahraun*, the Laki lava, 2016. Photograph by Katrin KLEEMANN, August 2016 —— **303**

Figure 71 An information sign and a footpath near one of the Laki craters. Photograph by Katrin KLEEMANN, August 2016 —— **304**

Bibliography

Archived Primary Sources

American Antiquarian Society, Worcester, MA, USA

Ezra WHEELER Diary, 1783, Almanacs Mass. B490 1783a, American Antiquarian Society, Worcester MA, USA.

LEE Family Diaries [manuscript], 1783–1807, vol. 1 (1783), 271762, American Antiquarian Society, Worcester MA, USA.

American Philosophical Society, Philadelphia, PA, USA

Jacob HILTZHEIMER Diaries, vol. 13 (March 1783 to February 1784), Mss.B.H56d, American Philosophical Society, Philadelphia, PA, USA.

William ADAIR Meteorological Notebook, Meteorological Observations Taken at Lewes, Delaware, 1776–1788, Mss.551.5.Ad1, American Philosophical Society, Philadelphia PA, USA.

Archives of the Royal Society, London, UK

John ATKINS, "The Meteorological Journal for the Year 1783. Kept at Minehead in Somersetshire by Mr. John Atkins. Presented at the Royal Society in London on 19 January 1786," MA/166, Archives of the Royal Society, London, UK.

Meteorological Observations at Dundee, Scotland, 1782–1801, "General Results From a Meteorological Journal Kept at Crescent near Dundee," MA/45, Archives of the Royal Society, London, UK.

Meteorological Observations at Nain & Okak at Labrador, Hudson's Bay Company, MA/143, Archives of the Royal Society, London, UK.

Meteorological Observations at Okkak and at Hoffenthal, Hudson's Bay Company, Labrador, Newfoundland, 1782–1786, MA/144, Archives of the Royal Society, London, UK.

British Library, London, UK

Gilbert WHITE, "The Naturalist's Journal": The Printed Tabulated Annual Diaries, Edited by the Hon. Daines Barrington; With Records of the Barometer, Thermometer, Wind, and Weather, and Remarks Relating to Gardening, Agriculture, and [. . .], 1783, Add MS 31848, British Library, London, UK.

Gilbert WHITE, "The Naturalist's Journal": The Printed Tabulated Annual Diaries, Edited by the Hon. Daines Barrington; With Records of the Barometer, Thermometer, Wind, and Weather, and Remarks Relating to Gardening, Agriculture, and [. . .], 1784, Add MS 31849, British Library, London, UK.

Gilbert WHITE, "The Naturalist's Journal": The Printed Tabulated Annual Diaries, Edited by the Hon. Daines Barrington; With Records of the Barometer, Thermometer, Wind, and Weather, and Remarks Relating to Gardening, Agriculture, and [. . .], 1785, Add MS 31849, British Library, London, UK.

Henry WHITE, Diaries of the Rev. Henry WHITE, Rector of Fyfield, County Southampton, Brother of Gilbert WHITE, 1783, Add MS 43816, British Library, London, UK.

Houghton Library, Harvard, Cambridge, MA, USA

Caleb GANNETT, A Meteorological Register Kept at Cambridge 1783, MS Am 1360, bMS Am 1360 (7),
    Houghton Library, Harvard, Cambridge MA, USA.
Edward WIGGLESWORTH, Meteorological Journal. A.MS.s. Cambridge, 1780–1789, 1793, MS Am 1361 vol. 33,
    Houghton Library, Harvard, Cambridge MA, USA.

John Rylands Library, University of Manchester, Manchester, UK

John Thomas STANLEY, MS, JRL 722, John Rylands Library, University of Manchester, Manchester, UK.

National Meteorological Library and Archive, Met Office, Exeter, UK

Anonymous, Private Weather Diary, "Greater Central England" Climatological series 56BC–AD2010 Archive,
    MET/2/1/2/3/545, National Meteorological Library and Archive, Met Office, Exeter, UK.
G. T. WILLIAMS, Private Weather Diary for Greenwich and Somerset House, MET/2/1/2/3/161, National
    Meteorological Library and Archive, Met Office, Exeter, UK.
J. FENTON, The Weather, etc. from June 27, 1779 to December 30, 1786 (at Nacton, Suffolk), Bib. No. 178296,
    National Meteorological Library and Archive, Met Office, Exeter, UK.
James WOODFORDE, Weather Diary for Weston Longville, Norfolk, vol. 10: 1782–1784, MET/2/1/2/3/467,
    National Meteorological Library and Archive, Met Office, Exeter, UK.
T. HOY, Private Weather Diary for Gordon Castle, Grampian, Scotland, MET/2/1/2/3/486, National
    Meteorological Library and Archive, Met Office, Exeter, UK.
Thomas BARKER, Private Weather Diary for Lyndon Hall, Leicestershire, MET/2/1/2/3/227, National
    Meteorological Library and Archive, Met Office, Exeter, UK.
Thomas HUGHES, Private Weather Diary for Stroud, Gloucestershire, MET/2/1/2/3/410, National
    Meteorological Library and Archive, Met Office, Exeter, UK.
William HUTCHINSON, Private Weather Diary for Liverpool Dock, MET/2/1/2/3/230, National Meteorological
    Library and Archive, Met Office, Exeter, UK.
Unknown, Private Weather Diary for London (and additional locations), Gentleman's Magazine, MET/2/1/2/
    3/255, National Meteorological Library and Archive, Met Office, Exeter, UK.

Massachusetts Historical Society, Boston, MA, USA

Cotton TUFTS Diaries, 1748–1794, Ms. N-1686, Massachusetts Historical Society, Boston MA, USA.
Edward HOLYOKE, Medical Records, 1782–1788, Folder 2 (1783), Ms. N-1427, Massachusetts Historical
    Society, Boston MA, USA.
Mary Robie SEWALL Diary, May–October 1783, call no. Ms. N-804, Massachusetts Historical Society, Boston
    MA, USA.

National Records of Scotland, Edinburgh, UK

George and James WARROCH in Prestonpans, 1783–1807, CS96/4401, National Records of Scotland, Edinburgh, UK.

James Stuart MACKENZIE, Weather Reports Journal of James Stuart Mackenzie at Belmont Castle, Perthshire, 1771–1799, RH4/100, National Records of Scotland, Edinburgh, UK.

John ALVUS, Weather Diary for Dalkeith Kept by John Alvus, MET1/4/37, National Records of Scotland, Edinburgh, UK.

Scottish Meteorological Society, MET1, National Records of Scotland, Edinburgh, UK.

Published Primary Sources

Publications

ADELUNG, Johann Christoph (ed.): Grammatisch-kritisches Wörterbuch der Hochdeutschen Mundart mit beständiger Vergleichung der übrigen Mundarten, besonders aber der oberdeutschen. Vol. 4. Leipzig ²1793–1801.

AHLWARDT, Peter: Bronto-Theologie, oder Vernünftige und theologische Betrachtungen über den Blitz und Donner, wodurch der Mensch zur wahren Erkenntniß Gottes und seiner Vollkommenheiten, wie auch zu tugendhaften Leben und Wandel geführt werden kann. Greifswald, Leipzig 1745.

ANDERSON, Johann: Nachrichten von Island, Grönland und der Strasse Davis, zum wahren Nutzen der Wissenschaften und der Handlung. Hamburg 1746.

ANDERSON, Richard: Lightning Conductors. Their History, Nature and Mode of Application. London 1880.

ANDERSON, Tempest: Volcanic Studies in Many Lands: Being Reproductions of Photographs by the Author of Above One Hundred Actual Objects, With Explanatory Notices. London 1903.

Anonymous ("E. R."): Report from Zurich. In: Johann Ernst Fabri's Geographisches Magazin 2, no. 7 (1783), 405–406.

Anonymous: The Java Upheaval. In: Nature 28, no. 723 (6 September 1883a), 443–444, DOI: 10.1038/028443a0.

Anonymous: Scientific Aspects of the Java Catastrophe. In: Nature 28, no. 724 (13 September 1883b), 457–458, DOI: 10.1038/028457a0.

Anonymous: Wahrhafte und umständliche Nachrichten in Briefen von dem Schaden und der Verheerung, den das Erdbeben in Calabrien und Meßina im Hornung dieses Jahrs angerichtet hat. (Meist aus dem Italiänischen.) N. p., 1783. https://opacplus.bsb-muenchen.de/search?oclcno=165796332&db=100&View=default.

Anonymous: Das arme Köln bey der Ueberschwemmung im Jahr 1784, 27. Hornung. Cologne 1784.

AUBERT, Alexander: An Account of the Meteors of the 18th of August and 4th of October, 1783. By Alex. Aubert, Esq. F. R. S. and S. A. In: Philosophical Transactions 74 (1784), 112–115.

BÁRÐARSON, Guðmundur G.: Geologisk Kort over Reykjanes-Halvøen. In: Beretning om det 18. Skandinaviske Naturforskermøde i København 26.–31. August 1929, vol. 18. Copenhagen 1929, 182–190.

BARDILI, Christoph Gottfried: Ueber die Entstehung und Beschaffenheit des ausserordentlichen Nebels in unsern Gegenden. Frankfurt, Leipzig 1783.

Bayerische Akademie der Wissenschaften: Der Baierischen Akademie der Wissenschaften in München meteorologische Ephemeriden, vol. 3. Munich 1783 (1785).

Bayerische Akademie der Wissenschaften: Der Baierischen Akademie der Wissenschaften in München meteorologische Ephemeriden, vol. 4. Munich 1784 (1786).

BEROLDINGEN, Franz Cölestin von: Gedanken über den so lange angehaltenen ungewöhnlichen Nebel. Braunschweig 1783.

BLAGDEN, Charles: An Account of Some Late Fiery Meteors; With Observations. In a Letter from Charles Blagden, M. D. Physician to the Army, Sec. R. S. to Sir Joseph Banks, Bart. P. R. S. In: Philosophical Transactions 74 (1784), 201–232.

BODE, Johann Elert (ed.): Astronomisches Jahrbuch für das Jahr 1787: Nebst einer Sammlung der neuesten in die astronomischen Wissenschaften einschlagenden Abhandlungen, Beobachtungen und Nachrichten. Berlin 1784.

BORNEMANN, Karl: Procop Diwisch. Ein Beitrag zur Geschichte des Blitzableiters. In: Die Gartenlaube 38 (1878), 624–627.

BRÄKER, Ulrich: Sämtliche Schriften (ed. HOLLINGER, Christian). Munich 2010.

BRANDES, Heinrich Wilhelm; GMELIN, Leopold; HORNER, Johann Caspar; MUNCKE, George Wilhelm; PFAFF, C. H. (eds.): Nichtfeuchte, trockne Nebel. In: Johann Samuel Traugott GEHLER's Physikalisches Wörterbuch, vol. 7, part 1: N-Pn. Leipzig 1833, 34–53.

BRUGMANS, Sebald Justinus: Natuurkundige verhandeling over sen zwavelagtigten nevel den 24 Juni 1783, in de provintie van Stad en Lande en naburige landen waargenomen. Groningen 1783.

CAVALLO, Tiberius: Description of a Meteor, Observed Aug. 18, 1783. By Mr. Tiberius Cavallo, F. R. S. In: Philosophical Transaction 74 (1784), 108–111.

CHRIST, Johann Ludwig: Von der außerordentlichen Witterung des Jahres 1783, in Ansehung des anhaltenden und heftigen Höherauchs: Vom Thermometer und Barometer, von dem natürlichen Barometer unserer Gegend, dem Feldberg oder der Höhe, und von der Beschaffenheit und Entstehung unserer gewöhnlichen Lufterscheinungen, wie auch etwas von den Erdbeben. Frankfurt am Main 1783.

CHRISTMANN, Ernst; KRÄMER, Julius (eds.): Pfälzisches Wörterbuch. Vol. 3. Wiesbaden 1980.

COOPER, William: Observation on a Remarkable Meteor Seen on the 18th of August, 1783, Communicated in a Letter to Sir Joseph Banks, Bart. P. R. S. By William Cooper, D. D. F. R. S. Archdeacon of York. In: Philosophical Transactions 74 (1784), 116–117.

COTTE, Louis: Mémoire sur les Brouillards Extraordinaires des Mois de Juin et Juillet 1783. In: Journal de Physique 23 (1783), 201–206.

COWPER, William: The Task: A Poem, in Six Books. By William Cowper, . . . To which are added, by the same author, An epistle to Joseph Hill, Esq. . . . To which are added, . . . an epistle . . . and the history of John Gilpin. London 1785, 42–88.

CULLUM, John: An Account of a Remarkable Frost on the 23d of June, 1783. In a Letter from the Rev. Sir John Cullum, Bart. F. R. S. and S. A. to Sir Joseph Banks, Bart. P. R. S. In: Philosophical Transactions 74 (1784), 416–418.

DE POEDERLÉ, Baron: Précis des observations météorologiques faites à Bruxelles pendant l'année 1783. In: L'Esprit des Journaux (May 1784), 326–349.

DELUC, Jean-André: Lettres physiques et morales sur les montagnes, et sur l'histoire de la terre et de l'homme. 6 vols. Den Haag 1778–1780.

DEURER, E. F.: Umständliche Beschreibung der im Jänner und Hornung 1784 die Städte Heidelberg, Mannheim und andere Gegenden der Pfalz durch die Eisgänge und Überschwemmungen betroffenen großen Noth; nebst einigen vorausangeführten Natur-Denkwürdigkeiten des vorhergehenden Jahres. Mannheim 1784.

DOLOMIEU, Déodat Gratet de: Mémoire sur les Tremblements de Terre de la Calabre Pendant l'Année 1783. Rome 1784.

Dominicus de Gravina: Chronicon de Rebus in Apulia Gestis. Florence 2011.

DUFFERIN, Frederick Temple, Marquis of Blackwood: Letters from High Latitudes: Being Some Account of a Voyage, in 1856, in the Schooner Yacht "Foam," 85 O. M., to Iceland, Jan Mayen, & Spitsbergen. London 1856.

EDGEWORTH, Richard Lovell: An Account of the Meteor of the 18th of August, 1783. In a Letter from Richard Lovell Edgeworth, Esq. F. R. S. to Sir Joseph Banks, Bart. P. R. S. In: Philosophical Transactions 74 (1784), 118.

EHRMANN, Friedrich Ludwig: Montgolfier'sche Luftkörper oder Aerostatische Maschinen. Eine Abhandlung, worinn die Kunst sie zu verfertigen und die Geschichte der bisher damit angestellten Versuche beschrieben werden. Straßburg 1784.

ESCHELS, Jens Jacob: Lebensbeschreibung eines alten Seemanns. Hamburg 1928.

EPP, Franz Xaver: Rede über den so genannten Hehrrauch, welcher im Jahre 1783 nicht nur in Baiern sondern ganz Europa erschienen [. . .]. Munich 1787.

FISCHER, Johann Nepomuk: Beweiß, daß das Glockenläuten bey Gewittern mehr schädlich als nützlich sey. Nebst einer allgemeinen Untersuchung ächter und unächter Verwahrungsmittel gegen die Gewitter. Munich 1784.

FRANKLIN, Benjamin: Experiments and Observations on Electricity: Made at Philadelphia in America, by Mr. Benjamin Franklin, and Communicated in Several Letters to Mr. P. Collinson, of London, F.R.S. London 1751.

FRANKLIN, Benjamin: Meteorological Imaginations and Conjectures. In: Memoirs of the Literary and Philosophical Society of Manchester 2 (1785), 357–361.

FRANKLIN, Benjamin: The Papers of Benjamin Franklin. Vol. 40: May 16 through September 15, 1783 (ed. COHN, Ellen R.). New Haven CT 2011.

FRANKLIN, Benjamin: The Papers of Benjamin Franklin. Vol. 41: September 16, 1783, through February 29, 1784 (ed. COHN, Ellen R.). New Haven CT 2014.

FRANKLIN, Benjamin: The Papers of Benjamin Franklin. Vol. 42: March 1 through August 15, 1784 (ed. COHN, Ellen R.). New Haven CT 2017.

GEIKIE, Archibald: Text-Book of Geology. London 1882.

GIESSBERGER, H.: Die Erdbeben Bayerns, I. Teil. Munich 1922.

GIESSBERGER, H.: Die Erdbeben Bayerns, II. Teil. Munich 1924.

GRIMM, Jacob; GRIMM, Wilhelm (ed.): Deutsches Wörterbuch. Vol. 10. Stuttgart 1877 [Munich 1984].

GRONAU, Karl Ludwig: Einige Bemerkungen der diesjährigen Winterkälte. In: Schriften der Berlinischen Gesellschaft naturforschende Freunde 5 (1784), 246–253.

GRONAU, Karl Ludwig: Einige Bemerkungen über Nebel und Nordschein. In: Schriften der Berlinischen Gesellschaft naturforschende Freunde 6 (1785), 92–104.

GRUMBKOW, Ina von: Ísafold. Reisebilder aus Island. Berlin 1909. http://www.isafold.de/klassiker/grumb kow/kap_07.htm.

HAMILTON, William: An Account of the Earthquakes Which Happened in Italy, from February to May 1783. By Sir William HAMILTON, Knight of the Bath, F. R. S.; in a Letter to Sir Joseph BANKS, Bart. P. R. S. In: Philosophical Transactions 73 (1783), 169–208.

HARREAUX, M.: Du Brouillard Sec. In: Memoirs, Société Archéologique d'Eure-et-Loir 4 (1858), 30–31.

HELLAND, Amund: Islaendingen Sveinn Pálssons beskrivelse af islandske vulkaner og braer. In: Den Norske Turistforenings Arbog for 1881, 22–74.

HELLAND, Amund: Islaendingen Sveinn Pálssons beskrivelse af islandske vulkaner og braer. In: Den Norske Turistforenings Arbog for 1882, 19–79.

HELLAND, Amund: Islaendingen Sveinn Pálssons beskrivelse af islandske vulkaner og braer. In: Den Norske Turistforenings Arbog for 1884, 27–56.

HELLAND, Amund: Lakis kratere og lavastrømme. Kristiania [Oslo] 1886.

HENDERSON, Ebenezer: Iceland or the Journal of a Residence in that Island During 1814 and 1815. 2 vols. Edinburgh 1818.

HERRMANN, Paul: Island in Vergangenheit und Gegenwart: Reise-Erinnerungen. Leipzig 1907.

HICKMANN, Dom Robert: Observations addressées aux auteurs de ce Journal, sur la cause du brouillard extraordinaire qui a regné cette année. In: Journal encyclopédique ou universel 6, no. 3 (September 1783), 505–512.

HILLIGER, Pastor: Von der brandstig riechenden Dunstluft, Nachts des 31. Juli dieses Jahres. In: Wittenberger Wochenblatt zum Aufnehmen der Naturkunde und des ökonomischen Gewerbes 16, no. 35 (5 September 1783), 277–278.

HOFF, Karl Ernst Adolf von: Geschichte der durch Überlieferung nachgewiesenen natürlichen Veränderungen der Erdoberfläche. Chronik der Erdbeben und Vulcan-Ausbrüche. 2 vols. Gotha 1840–1841.

HOLLAND, Henry: Iceland Journal of Henry Holland, 1810 (ed. WAWN, Andrew). London 1987.

HÓLM, Sæmundur Magnússon: Om Jordbranden paa Island i Aaret 1783. Copenhagen 1784a.

HÓLM, Sæmundur Magnússon: Vom Erdbrande auf Island im Jahr 1783 (translated from Danish by HÓLM, Sæmundur Magnússon). Copenhagen 1784b.

HOOKER, William Jackson: Journal of a Tour in Iceland, in the Summer of 1809. 2 vols. London [2]1813.

HORREBOW, Niels: Tilforladelige efterretninger om Island med et nyt landkort og 2 aars meteorologiske observationer. Copenhagen 1752.

HORREBOW, Niels: Zuverlässige Nachrichten von Island nebst einer neuen Landkarte und 2 jährliche meteorologische Anmerkungen. Copenhagen and Leipzig 1753.

HORREBOW, Niels: The Natural history of Iceland: Containing a particular and accurate account of the different soils, burning mountains, minerals, vegetables, metal, stones, beasts, birds and fishes, together with the disposition, customs, and manner of living of the inhabitants. London 1758.

HORREBOW, Niels: Nouvelle Description Physique-Historique, Civile et Politique de l'Islande. Paris 1764.

HUMPHREYS, William Jackson: Volcanic Dust as a Factor in the Production of Climatic Changes. In: Journal of the Washington Academy of Sciences 3, no. 13 (1913), 365–371.

HUTTON, James: Theory of the Earth; or an Investigation of the Laws Observable in the Composition, Dissolution, and Restoration of Land upon the Globe. In: Transactions of the Royal Society of Edinburgh 1/2 (1788), 209–304.

IMBÓ, Guiseppe J.: I Terremoti Etnei. Florence 1935.

JÖRGENSEN, Jörgen: Historical Account of a Revolution in the Island of Iceland in the Year 1809 (eds. MELSTED, Óðinn; AGNARSDÓTTIR, Anna). Reykjavík 2016 (The University of Iceland Historical Document Series III).

KANT, Immanuel: Von den Ursachen der Erderschütterungen bei Gelegenheit des Unglücks, welches die westlichen Länder von Europa gegen das Ende des vorigen Jahres betroffen hat (1755). In: Preussische Akademie der Wissenschaften (ed.): Kant's Werke. Vol. I. Vorkritische Schriften I: 1747–1756. Berlin 1910, 259–260.

KANT, Immanuel: Geschichte und Naturbeschreibung der merkwürdigsten Vorfälle des Erdbebens, welches an dem Ende des 1755sten Jahres einen grossen Teil der Erde erschüttert hat (1756a). In: Preussische Akademie der Wissenschaften (ed.): Kant's Werke. Vol. I. Vorkritische Schriften I: 1747–1756. Berlin 1910, 429–462.

KANT, Immanuel: Fortgesetzte Betrachtung der seit einiger Zeit wahrgenommenen Erderschütterungen (1756b). In: Preussische Akademie der Wissenschaften (ed.): Kant's Werke. Vol. I. Vorkritische Schriften I: 1747–1756. Berlin 1910, 463–472.

KERGUELEN TRÉMAREC, Yves-Joseph de: Relation d'un voyage dans la mer du Nord, aux côtes d'Islande, du Groenland, de Ferro, de Schettland, des Orcades et de Norwége, fait en 1767 et 1768. Paris 1771.

KIESSLING, Johann: Die Dämmerungserscheinungen im Jahre 1883 und ihre physikale Erklärung. Hamburg, Leipzig 1885.

KIESSLING, Johann: Untersuchung über Dämmerungserscheinungen zur Erklärung der nach dem Krakatau-Ausbruch beobachteten atmosphärisch-optischen Störungen. Hamburg 1888.

KINDER, Johannes Christian: Witterungsberichte aus dem Ende des 18. Jahrhunderts. In: KINDER, Johannes Christian (ed.): Plön. Beiträge zur Stadtgeschichte. Plön 1904 [Reprint Kiel 1976], 304–307.

KÖNIG, Karl (ed.): Observationes meteorologicae Ratisbonae. Regensburg 1783.

KÖNIG, Karl (ed.): Observationes meteorologicae Ratisbonae. Regensburg 1784.

Konrad von Megenberg: Das "Buch der Natur." Band II: Kritischer Text nach den Handschriften (eds. LUFF, Robert; STEER, Georg). Tübingen 2003.

KRÜNITZ, Johann Georg (ed.): Oeconomische Encyclopädie, oder allgemeines System der Staats-, Stadt-, Haus- und Landwirthschaft in alphabetischer Ordnung. Vol. 9. Berlin 1776/1785.

KRÜNITZ, Johann Georg (ed.): Oeconomische Encyclopädie, oder allgemeines System der Staats-, Stadt-, Haus- und Landwirthschaft in alphabetischer Ordnung. Vol. 11. Berlin 1777.

KRÜNITZ, Johann Georg (ed.): Oeconomische Encyclopädie, oder allgemeines System der Staats-, Stadt-, Haus- und Landwirthschaft in alphabetischer Ordnung. Vol. 101. Berlin 1806.

KRÜNITZ, Johann Georg (ed.): Oeconomische Encyclopädie, oder allgemeines System der Staats-, Stadt-, Haus- und Landwirthschaft in alphabetischer Ordnung. Vol. 123. Berlin 1813.

KUSS, Christian: Jahrbuch denkwürdiger Naturereignisse in den Herzogthümern Schleswig und Holstein vom elften bis zum neunzehnten Jahrhundert. Vol. 2. Altona 1826.

LALANDE, Jérôme J. Le Français de: Ueber den Nebel im Sommer 1783. In: Magazin für das Neueste aus der Physik und Naturgeschichte 2, no. 2 (1783), 95–99.

LAMANON, Robert P. C. de: Vues sur la nature et l'origine du brouillard qui a eu lieu cette année. In: Observationes sur la physique sur l'histoire naturelle et sur les arts 24 (January 1784a), 8–18.

LAMANON, Robert P. C. de: Nieuwe gedagten over den aart en oorsprong der zeldzaamen nevels, in den jaare MDCCLXXXIII. In: Algemeene vaderlandsche letter-oefeningen 6, no. 2 (1784b), 296–308.

LAMANON, Robert P. C. de: Observations on the Nature of the Fog of 1783. In: The Philosophical Magazine 5, no. 17 (1799), 80–89.

LAPI, Giovanni: Sulla Caligine del Corrente Anno 1783. E sulla vigorosa vegetazione e fertilita' delle piante del suddetto anno congetture. Florence 1783.

LICHTENBERG, Georg Christoph: Briefwechsel, vol. II: 1780–1784 (eds. JOOST, Ulrich; SCHÖNE, Albrecht). Munich 1985.

LIER, J. van; TONKENS, J.: Hedendaagsche Historie van het Landschap Drenthe. Amsterdam 1792.

LINDEMAYR, Peter Gottlieb [LINDEMAYR, P. Maurus]: Hänts Leutel sagts mä do. In: NEUHUBER, Christian (ed.): Lieder in oberösterreichischer Mundart. Kritische Ausgabe. Vienna 2010, 77–82, 130–139.

LUDWIG, Christian to Carl Friedrich HINDENBURG: Bemerkungen über die in den nächstverflossenen Monaten allgemein sich verbreitete dicke Luft; in einem Sendschreiben von Herrn D. Christian Ludwig an Prof. Hindenburg. In: LESKE, Nathanael Gottfried; HINDENBURG, Carl Friedrich; FUNK, Christlieb Benedict (eds.): Leipziger Magazin zur Naturkunde, Mathematik und Oekonomie. Leipzig 1783, 211–221.

LÜTGENDORF, Joseph Maximilian von: Kunst den Luftball nach jenen des Herrn von Montgolfier zu verfertigen von Herrn Pingeron, verschiedene Akademien Mitglieder. S. l. 1784.

LYELL, Charles: Principles of Geology: Being an Attempt to Explain the Former Changes of the Earth's Surface, By Reference to Causes Now in Operation. Vol. 1. London 1830.

MacKAY, Andrew: The Theory and Practice of Finding the Longitude at Sea or Land: To Which Are Added, Various Methods of Determining the Latitude of a Place, and Variation of the Compass. London 1793.

MACKENZIE, George Stuart; HOLLAND, Henry: Travels in the Island of Iceland, During the Summer of the Year MDCCCX. Edinburgh 1811.

MERCIER, Louis-Sébastien: Mon bonnet de nuit. Vol. 2. Neuchâtel 1784.

MEYERS Konversations-Lexikon: Eine Encyklopädie des allgemeinen Wissens (ed. MEYER, Herrmann Julius). Vol. 8. Leipzig ⁴1887.

MONGE, Gaspard; CASSINI, Jean-Dominique; BERTHOLON, Pierre et al.: Encyclopédie méthodique, Dictionnaire de Physique. Paris 1793.

Mourgue de Montredon, Jacques Antoine: Recherches sur l'origine et sur la nature des vapeurs qui ont régné dans l'Atmosphère pendant l'été de 1783. In: Histoire de l'Académie Royale des Sciences avec les Mémoires de Mathématique et de Physique pour l'Année (1781 [1784]), 754–773.

Murr, Christoph Gottlieb von: Die Herren Stephan und Joseph von Montgolfier: Versuche mit der von ihnen erfundenen aerostatischen Maschine. Ein Auszug aus der französischen Beschreibung des Herrn Faujas de Saint-Fond. Nuremberg 1784.

Nordenskiöld, Adolf Erik: Distant Transport of Volcanic Dust. In: Geological Magazine 3 (1876).

Ólafsson, Eggert; Pálsson, Bjarni: Vice-Lavmand Eggert Olafsens og Land-Physici Biarne Povelsens Reise igiennem Island, foranstaltet af Videnskabernes Sælskab i København [. . .]. Sorø 1772.

Ólafsson, Eggert; Pálsson, Bjarni: Des Vice-Lavmands Eggert Olafsens und des Landphysici Biarne Povelsens Reise durch Island veranstaltet von der Königlichen Societät der Wissenschaften in Kopenhagen Reise durch Island. Kopenhagen 1774.

Ólafsson, Eggert; Pálsson, Bjarni: Travels in Iceland: Performed by Order of His Danish Majesty; Containing Observations on the Manners and Customs of the Inhabitants, a Description of the Lakes, Rivers, Glaciers, Hot-Springs, and Volcanoes; of the Various Kinds of Earths, Stones, Fossils, and Petrifactions; as well as of the Animals, Insects, Fishes, &c. London 1805.

Pálsson, Bjarni: Voyage en Islande, fait par ordre de S. M. danoise: contenant des observations sur les mœurs et les usages des Habitans [. . .]. Paris 1802.

Pálsson, Sveinn: Udtog af Paulsons Dagbog til og i Island fra 2 Julii til 7 Sep 1791. In: Skrifter af naturhistorie-selskabet 2, no. 1 (1792), 222–234.

Pálsson, Sveinn: Videre Udtog af Hr. Paulsons Dagbog paa hans Reise i Island fra 7 Sept. 1791 til Udgangen af April 1792. In: Skrifter af naturhistorie-selskabet 2, no. 2 (1793a), 122–146.

Pálsson, Sveinn: Udtog af Hr. Paulsons Dagbog paa hans Reise i Island fra Begyndelsen af Mai Maaned 1792. In: Skrifter af naturhistorie-selskabet 3, no. 1 (1793b), 157–194.

Pálsson, Sveinn: Ferðabók Sveins Pálssonar: dagbækur og ritgerðir; 1791–1797. 2 vols. (eds. Eyþórsson, Jón; Hannesson, Pálm; Steindórsson, Steindór). Reykjavík 1945.

Pálsson, Sveinn: Draft of a Physical, Geographical, and Historical Description of Icelandic Ice Mountains on the Basis of a Journey to the Most Prominent of Them in 1792–1794, With Four Maps and Eight Perspective Drawings (ed. and translated by Williams, Richard S. Jr.; Sigurðsson, Oddur). Reykjavík 2004.

Pigott, Edward: The Latitude and Longitude of York Determined from a Variety of Astronomical Observations; Together With a Recommendation of the Method of Determining the Longitude of Places by Observations of the Moon's Transit Over the Meridian. Contained in a Letter from Edward Pigott, Esq. to Nevil Maskelyne, D. D. F. R. S. and Astronomer Royal. In: Philosophical Transactions 76 (1786), 409–425.

Plinius der Ältere: Plinius Naturkunde. Buch II: Kosmologie. Berlin 22013 (Sammlung Tusculum).

Preyer, William T.; Zirkel, Ferdinand: Reise nach Island im Sommer 1860, mit wissenschaftlichen Anhängen. Leipzig 1862.

Renovanz, Hans Michael: Mineralogisch-geographische und andere vermischte Nachrichten von den Altaischen Gebürgen, Russisch Kayserlichen Antheils. Reval 1788.

Rittenhouse, David: A Letter to David Rittenhouse, Esq. from John Page, Esq. From David Rittenhouse, Esq. to John Page, Esq. Concerning a Remarkable Meteor Seen in Virginia and Pennsylvania. In: Transactions of the American Philosophical Society 2 (1786), 173–176.

Roberjot, l'abbé: Lettre aux Auteurs du Journal de Physique, sur un Phénomène Singulier du Brouillard de 1783. In: Observations et Mémoires sur la Physique, sur l'Histoire Naturelle et sur les Arts et Métiers 24 (1784), 399–400.

Salis-Marschlins, Johann Rudolf von: Einige Bemerkungen über den allgemeinen Dampf oder Heerrauch, der im Junius und Julius dieses Jahrs sich auch in unserer Gegend verbreitet hat. In: Der Sammler. Eine gemeinnützige Wochenschrift für Bündten 5, no. 47 (1783), 393–398.

SAPPER, Karl: Katalog der geschichtlichen Vulkanausbrüche. Strasbourg 1917a.

SAPPER, Karl: Beiträge zur Geographie der tätigen Vulkane. In: Zeitschrift für Vulkanologie 3 (1917b), 65–197.

SCHELHORN, Johann Georg: Unterhaltungen beym Donnerwetter, seinen werthesten Mitbürgern, und dem lieben Landvolk seiner Vaterstadt besonders gewiedmet. Memmingen 1783.

SENEBIER, Jean: Observation sur la Vapeur qui a régné pendant l'été de 1783, faite à Genève, par Jean Senebier, Bibliothécaire de la République. In: Observations et Mémoires sur la Physique, sur l'Histoire Naturelle et sur les Arts et Métiers 24 (1784), 404–411.

SILBERSCHLAG, Johann Esaias: Beobachtung der in der Nacht vom 10. bis 11ten September 1783 gesehenen Mondfinsterniß mit einigen physikalischen Anmerkungen begleitet. In: Schriften der Berlinischen Gesellschaft naturforschende Freunde 5 (1784), 134–147.

Skrifter af naturhistorie-selskabet 4, no. 1 (1797) and no. 2 (1798).

Skrifter af naturhistorie-selskabet 5, no. 1 (1799) and no. 2 (1802).

SNELL, Karl Philipp Michael: Beschreibung der russischen Provinzen an der Ostsee oder, Zuverlässige Nachrichten sowohl von Russland überhaupt, als auch insonderheit von der natürlichen und politischen Verfassung, dem Handel, der Schiffahrt, der Lebensart, den Sitten und Gebräuchen, den Künsten und der Litteratur, dem Civil- und Militairwesen, und andern Merkwürdigkeiten von Livland, Esthland und Ingermannland. Jena 1794.

Societas Meteorologica Palatina: Ephemerides Societatis Meteorologicae Palatinae. Observationes Anni 1783. Mannheim 1783 (1785).

Societas Meteorologica Palatina: Ephemerides Societatis Meteorologicae Palatinae. Observationes Anni 1784. Mannheim 1784 (1786).

Société Royale de Médecine: Histoire de la Société Royale de Médecine 1782–1783 (1787).

STANLEY, John Thomas: An Account of the Hot Springs near Rykum in Iceland. In: Transactions of the Royal Society of Edinburgh 3, no. 2 (1794), 127–137.

STEINGRÍMSSON, Jón: Fires of the Earth: The Laki Eruption, 1783–1784 (translated by KUNZ, Keneva). Reykjavík 1998.

STEINGRÍMSSON, Jón: A Very Present Help in Trouble: The Autobiography of the Fire-Priest (ed. and translated by FELL, Michael). New York NY 2002.

STEPHENSEN, Magnús: Kort Beskrivelse over den nye Vulcans Ildsprudning i Vester-Skaptefields-Syssel paa Island i Aaret 1783. Copenhagen 1785.

STEPHENSEN, Magnús; EGGERS, Christian Ulrich Detlev Freiherr von: Philosophische Schilderungen der gegenwärtigen Verfassung von Island, nebst Stephensens zuverlässiger Beschreibung des Erdbrandes im Jahre 1783, und anderen Beylagen. Altona 1786.

STEPHENSEN, Magnús: Account of the Volcanic Eruption in Skaptefield's Syssel. In: HOOKER, William Jackson (ed.): Journal of a Tour in Iceland in the Summer of 1809. Vol. 2. London 1813, 121–261.

SUESS, Eduard: Das Antlitz der Erde. 3 vols. Vienna 1883.

SYMONS, George James (ed.): The Eruption of Krakatoa and Subsequent Phenomena: Report of the Krakatoa Committee of the Royal Society. London 1888.

THELEN, Johann Leonhard: Ausführliche Nachricht von dem erschrecklichen Eisgange, und den Ueberschwemmungen des Rheines, welche im Jahre 1784 die Stadt Köln, und die umliegende Gegend getroffen. Cologne 1784.

THORODDSEN, Þorvaldur: De vulkanske Udbrud paa Island i Aaret 1783. In: Geografisk Tidsskrift 3 (1879), 67–80.

THORODDSEN, Þorvaldur: An Account of Volcanic Eruptions and Earthquakes Which Have Taken Place in Iceland Within Historical Times. In: The Geological Magazine (new series) 7 (1880), 458–468.

THORODDSEN, Þorvaldur: Oversigt over de islandske Vulkaners Historie. Copenhagen 1882a.

THORODDSEN, Þorvaldur: Aperçu des éruption de volcans et des tremblements de terre qui ont eu lieu en Islande pendant les temps historiques. Copenhagen 1882b.

THORODDSEN, Þorvaldur: Rejse i Vester-Skaptafells Syssel paa Island i Sommeren 1893. In: Geografisk
     Tidsskrift 12 (1893), 177–183.
THORODDSEN, Þorvaldur: Von Herrn Th. Thoroddsen über seine Forschungsreise in Island im Jahr 1893. d. d.
     Reykjavík, 14. April 1894 (translated from Danish by Ms LEHMANN-FILHÉS). In: Verhandlungen der
     Gesellschaft für Erdkunde zu Berlin 21 (1894), 289–296.
THORODDSEN, Þorvaldur: Eldreykjarmódan 1783 (The Volcanic Haze in 1783). In: KAALUND, Kristian (ed.):
     Afmælisrit til dr. phil. Kr. Kålunds bókavarðar við safn Árna Magnússonar 19. ágúst 1914. Copenhagen
     1914.
THORODDSEN, Þorvaldur: Geschichte der isländischen Vulkane (nach einem hinterlassenen Manuskript).
     Copenhagen 1925.
TITIUS, Johann Daniel: Von der anhaltenden dunstigen Luft im Juni, Juli 1783. In: Wittenberger Wochenblatt
     zum Aufnehmen der Naturkunde und des ökonomischen Gewerbes 16, no. 26 (4 July 1783), 206–208.
TITSINGH, Isaac: Mémoires et anecdotes sur la dynastie régnante des Djogouns, souverains du Japon: Avec
     la description des fetes et cérémonies observées aux differentes epoques de l'année a la cour de ces
     princes, et un appendice contenant des détails sur la poesie de Japonais, leur manière de diviser
     l'année [. . .]. Paris 1820.
TITSINGH, Isaac: Illustrations of Japan: Consisting of Private Memoirs and Anecdotes of the Reigning
     Dynasty of the Djogouns, or Sovereigns of Japan; [. . .]. London 1822.
TOALDO, Guiseppe: Observations météorologiques faites à Padoue au mois de Juin, 1783 avec une
     dissertation sur le Brouillard extraordinaire qui a régné durant ce rempu-là. In: Observations et
     Mémoires sur la Physique, sur l'Histoire Naturelle et sur les Arts et Métiers 24 (1784), 3–18.
TOALDO, Giuseppe: Meteorological Observations Made at Padua in the Month of June 1783, With a
     Dissertation on the Extraordinary Fog Which Prevailed About That Time. In: The Philosophical
     Magazine 4, no. 16 (1799), 417–422, DOI: 10.1080/14786449908677100.
TORCIA, Michael: Kurze Beschreibung des Erdbebens, welches den 5ten Februar 1783 Meßina und einen
     Theil Calabriens betroffen, translated from Italian. Nuremberg 1783a.
TORCIA, Michele: Lettre sur le tremblement de terre de Calabre. In: Le journal des sçavans (December
     1783b), 839–841.
TROIL, Uno von: Briefe, welche eine von Herrn Troil im Jahr 1772 nach Island angestellte Reise betreffen
     (ed. and translated from Swedish by MÖLLER, Johann Georg Peter). Uppsala 1779.
TROIL, Uno von: Letters on Iceland. Containing Observations on the Civil, Literary, Ecclesiastical, and
     Natural History; Antiquities, Volcanos, Basaltes, Hot Springs; Customs, Dress, Manners of the
     Inhabitants, &c. &c. Made During a Voyage undertaken in the Year 1772 by Joseph Banks [. . .].
     London 1780.
VAN SWINDEN, Jan Hendrik: Observations on the Cloud (Dry Fog) Which Appeared in June 1783 (translated
     from Latin by LINTLEMAN, Susan). In: Jökull 50 (2001), 73–80.
VERBEEK, Rogier Diederik Marius: Krakatau. Batavia 1886.
VERDEIL, François: Mémoire sur les brouillards électriques vus en Juin & Juillet 1783, et sur le tremblement
     de terre arrive à Lausanne le 6 Juillet de la même année. In: Mémoires de la Société Physiques de
     Lausanne 1 (19 July 1783), 110–137.
VOLNEY, Constantin François: Travels through Syria and Egypt, in the Years 1783, 1784, and 1785, [. . .],
     translated from French. Vol. 1. London ²1788.
WEGENER, Alfred: Die Entstehung der Kontinente und Ozeane. Braunschweig 1915.
WEIGEL, Johann Adam Valentin: Geographische, naturhistorische und technologische Beschreibungen des
     souverainen Herzogthums Schlesien. Berlin 1803.
WIEDEBURG, Johann Ernst Basilius: Über die Erdbeben und den allgemeinen Nebel 1783, [. . .]. Jena 1784.
ZEDLER, Johann Heinrich (ed.): Grosses vollständiges Universal-Lexicon aller Wissenschafften und Künste.
     Vol. 9. Halle, Leipzig 1734.

Zedler, Johann Heinrich (ed.): Grosses vollständiges Universal-Lexicon aller Wissenschafften und Künste. Vol. 23. Halle, Leipzig 1740.

Zedler, Johann Heinrich (ed.): Grosses vollständiges Universal-Lexicon aller Wissenschafften und Künste. Vol. 51. Halle, Leipzig 1747.

Ziehen, Conrad Siegmund: Vom Mehlthau und Honigthau. In: Ziehen, Conrad Siegmund (ed.): Nachricht von einer bevorstehenden großen Revolution der Erde, die insonderheit das südliche Europa und einen Theil Deutschlands treffen [. . .] 2 vols. Frankfurt am Main, Leipzig 1783, 54–56.

Newspapers

Allerneueste Mannigfaltigkeiten. Eine gemeinnützige Wochenschrift, Berlin 1783.

American Herald and the General Advertiser, Boston MA 1784.

Augsburgische Postzeitung [Augsburgische Ordinari Postzeitung von Staats, gelehrten, historischen und oeconomischen Neuigkeiten and Augspurgische Extra-Zeitung], Augsburg 1783.

Augsburgisches Extra-Blatt, Augsburg 1785.

Berlinische Nachrichten [Berlinische Nachrichten von Staats- und gelehrten Sachen], Berlin 1783–1785.

Columbian Magazine, Philadelphia PA 1786.

Das Wienerblättchen, Vienna 1783.

Dessauische Zeitung für die Jugend und ihre Freunde, Dessau 1783.

Edinburgh Advertiser, Edinburgh 1783.

Exeter Flying Post, Exeter 1783.

Felix Farley's Bristol Journal, Bristol 1783.

Frankfurter Staats-Ristretto, Frankfurt am Main 1783.

Gazette de France, Paris 1783.

Gazette de Leyde, Leyden 1783.

Gazette van Gent, Gent 1783.

Gentleman's Magazine, London 1783.

Göttingische Anzeigen von gelehrten Sachen [Göttingische Anzeigen von gelehrten Sachen unter der Aufsicht der königlichen Gesellschaft der Wissenschaften Anzeigen], Göttingen 1783–1784.

Göttingische Gelehrte Anzeigen, Göttingen 1784.

Hamburger Adreß- und Comptoir-Nachrichten, Hamburg 1784.

Hamburgischer Unpartheyischer Correspondent [Staats- und Gelehrte Zeitung des Hamburgischen unpartheyischen Correspondenten], Hamburg 1783–1785.

Hanauisches Magazin, Hanau 1783.

Hannoverisches Magazin, Hannover 1783.

Hochfürstlich-Bambergische wöchentliche Frag- und Anzeigenachrichten, Bamberg 1783.

Hochfürstlich-Bambergisches Wochenblatt, Bamberg 1783.

Independent Ledger, Boston MA 1783.

Journal de Paris, Paris 1783–1784.

Journal historique et littéraire, Luxembourg 1783.

Kjøbenhavns Adresse-Contoirs Efterretninger, Copenhagen 1783.

Koblenzer Intelligenzblatt, Koblenz 1783.

Königlich Privilegirte Zeitung [Königlich privilegirte Berlinische Staats- und gelehrte Zeitung], Berlin 1783–1785.

Massachusetts Spy, Worcester MA 1783–1784.

Meiningische Wöchentliche Nachrichten, Meiningen 1783.

Mercure de France, Paris 1782.

Morning Herald and Daily Advertiser, London 1783.

Münchner Gelehrte Zeitung, Munich 1783.

Münchner Zeitung [Münchner staats-, gelehrte, und vermischte Nachrichten aus Journalen, Zeitungen, und Correspondenzen übersetzt, und gesammelt], Munich 1783–1785.

Neue Zürcher Zeitung, Zurich 1783.

New-York Packet, New York NY 1784.

Norwich Packet or The Chronicle of Freedom, Norwich CT 1783–1784.

Parker's General Advertiser and Morning Intelligencer, London 1783.

Preßburger Zeitung, Bratislava 1783.

Provinzialnachrichten aus den Kaiserlich Königlichen Staaten und Erbländern, Vienna 1784.

Rivington New York Gazette and Universal Advertiser, New York NY 1783.

Royal Gazette, New York NY 1783.

Salem Gazette, Salem MA 1784.

Schlesische Privilegirte Zeitung, Breslau 1783.

Sherborne Mercury, Sherborne 1783.

Spooner's Vermont Journal, Windsor VT 1784.

The American Museum, or Universal Magazine: Or Repository of Ancient and Modern Fugitive Pieces etc. Prose and Poetical 4 (1789). Vol. VI, Philadelphia PA 1789.

The Connecticut Courant and Weekly Intelligencer, Hartford CT 1783.

The Connecticut Journal, New Haven CT 1783.

The Continental Journal and Weekly Advertiser, Boston MA 1783.

The Newport Mercury from Newport, Rhode Island, Newport RI 1783.

The Pennsylvania Packet and General Advertiser, Philadelphia PA 1783–1784.

The Providence Gazette and Country Journal, Providence RI 1784.

The South-Carolina Weekly Gazette, Charleston SC 1784.

United States Chronicle, Providence RI 1784.

Vermont Gazette, Bennington VT 1784.

Whitehall Evening Post, London 1783.

Wiener Zeitung, Vienna, 1783–1784.

Wittenbergisches Wochenblatt [Wittenbergisches Wochenblatt zum Aufnehmen der Naturkunde und des ökonomischen Gewerbes], Wittenberg 1786.

Zürcherische Freitagszeitung, Zurich 1783.

Published Secondary Sources

ADGER, W. Neil; EAKIN, Hallie; WINKELS, Alexandra: Nested and Teleconnected Vulnerabilities to Environmental Change. In: Frontiers in Ecology 7, no. 3 (2009), 150–157, DOI: 10.1890/070148.

AGNARSDÓTTIR, Anna: The Urbanization of Iceland in the 18th and Early 19th Centuries. In: RIIS, Thomas (ed.): Urbanization in the Oldenburg Monarchy 1500–1800. Kiel 2012, 115–140.

AGNARSDÓTTIR, Anna: Iceland in the Eighteenth Century: An Island Outpost of Europe? In: Sjuttonhundratal. Nordic Yearbook for Eighteenth-Century Studies 10 (2013), 11–38.

AGNARSDÓTTIR, Anna: Introduction: Banks and Iceland. In: AGNARSDÓTTIR, Anna (ed.): Sir Joseph Banks, Iceland and the North Atlantic 1772–1820: Journals, Letters and Documents. London 2016, 1–34.

ÁGÚSTDÓTTIR, Anna María: Ecosystem Approach For Natural Hazard Mitigation of Volcanic Tephra in Iceland: Building Resilience and Sustainability. In: Natural Hazards 78 (2015), 1669–1691, DOI: 10.1007/s11069-015-1795-6.

AHRENS, Donald C: Meteorology Today. An Introduction to Weather, Climate, and the Environment. Belmont CA 2009.

ALEXANDER, John T.: Aeromania, "Fire-Balloons" and Catherine the Great's Ban of 1784. In: The Historian 58 (1996), 497–516.

ALFARO, Raimon; BRANDSDÓTTIR, Bryndís; ROWLANDS, Daniel P.; WHITE, Robert S.; GUÐMUNDSSON, Magnús T.: Structure of the Grímsvötn Central Volcano Under the Vatnajökull Icecap, Iceland. In: Geophysical Journal International 168, no. 2 (February 2007), 863–876, DOI: 10.1111/j.1365-246X.2006.03238.x.

ALT, Peter-André: Aufklärung. Lehrbuch Germanistik. Stuttgart 2007.

AMBRASEYS, Nicholas: Earthquakes in the Eastern Mediterranean and the Middle East. A Multidisciplinary Study of Seismicity up to 1900. Cambridge UK 2009.

AMBROSE, Stanley H.: Late Pleistocene Human Population Bottlenecks. Volcanic Winter, and Differentiation of Modern Humans. In: Journal of Human Evolution 34 (1998), 623–651.

ANCHUKAITIS, Kevin J.; WILSON, Rob; BRIFFA, Keith R.; BÜNTGEN, Ulf; COOK, E. R.; D'ARRIGO, Rosanne; et al.: Last Millennium Northern Hemisphere Summer Temperatures from Tree Rings: Part II, Spatially Resolved Reconstructions. In: Quaternary Science Reviews 163 (2017), 1–22, DOI: 10.1016/j.quascirev.2017.02.020.

ANDERSON, Don Lynn: New Theory of the Earth. New York NY 2007.

ANDRÉSSON, Sigfús Haukur: Adstod einokunarverslunarinnar við Íslendinga í Móduhardindum. (Aid Provided by the Royal Monopoly Company in the Years 1783–1785.) In: GUNNLAUGSSON, Gísli Ágúst; GUÐBERGSSON, Gylfi Már; ÞÓRARINSSON, Sigurður; RAFNSSON, Sveinbjörn; EINARSSON, Þorleifur (eds.): Skaftáreldar 1783–1784: Ritgerðir og heimildir. Reykjavík 1984, 215–234.

ANDREWS, John T.; DARBY, Dennis; EBERLE, Dennis; JENNINGS, Anne E.; MOROS, Matthias; E. J. OGILVIE, Astrid: A Robust, Multisite Holocene History of Drift Ice Off Northern Iceland: Implications for North Atlantic Climate. In: The Holocene 19, no. 1 (2009), 71–77, DOI: 10.1177/0959683608098953.

ANGELL, James K.; KORSHOVER, J.: Surface Temperature Changes Following the Six Major Volcanic Episodes Between 1780 and 1980. In: Journal of Climate and Applied Meteorology 24 (1985), 937–951, DOI: 10.1175/1520-0450(1985)024<0937:STCFTS>2.0.CO;2.

ANGSTRÖM, Anders: Teleconnections of Climate Changes in Present Time. In: Geografiska Annaler 17 (1935), 242–258.

ARNDT, Johannes; KÖRBER, Esther-Beate: Einleitung. In: ARNDT, Johannes; KÖRBER, Esther-Beate (eds.): Das Mediensystem im Alten Reich der Frühen Neuzeit (1600–1750). Göttingen 2010, 1–26.

BAILLIE, Mike: Hekla 3: How Big Was It? In: Endeavour 13, no. 2 (1989), 78–81, DOI: 10.1016/0160-9327(89)90006-9.

BALDURSSON, Snorri; HANNESDÓTTIR, Hrafnhildur; GUÐNASON, Jónas; THÓRÐARSON, Thor: Nomination of Vatnajökull National Park for Inclusion in the World Heritage List. Reykjavík 2018.

BALKANSKI, Yves; MENUT, Laurent; GARNIER, Emmanuel; WANG, R.; EVANGELIOU, N.; JOURDAIN, S.; ESCHSTRUTH, C.; VRAC, M.; YIOU, P.: Mortality Induced by PM 2.5 Exposure Following the 1783 Laki Eruption Using Reconstructed Meteorological Fields. In: Scientific Reports 8, no. 1 (2018), 15896, DOI: 10.1038/s41598-018-34228-7.

BARLOW, Lisa K.; SADLER, Jon; OGILVIE, Astrid E. J.; BUCKLAND, P. C.; AMOROSI, T.; INGIMUNDARSON, J. H.; SKIDMORE, P.; DUGMORE, A. J.; McGOVERN, T. H.: Interdisciplinary Investigations of the End of the Norse Western Settlement in Greenland. In: The Holocene 7 (1997), 489–499, DOI: 10.1177/095968369700700411.

BARNETT-MOORE, Nicholas; HASSAN, Rakib; FLAMENT, Nicolas; MÜLLER, Dietmar: The Deep Earth Origin of the Iceland Plume and its Effects on Regional Surface Uplift and Subsidence. In: Solid Earth 8 (2017), 235–254, DOI: 10.5194/se-8-235-2017.

BARRETT, James H. (ed.): Contact, Continuity, and Collapse: The Norse Colonization of the North Atlantic. Turnhout 2003.

Barriendos, Mariano; Llasat, M. Carmen: The Case of the Malda Anomaly. In: Climatic Change 61, no. 1/2 (2003), 191–216, DOI: 10.1023/A:1026327613698.

Bauch, Martin: The Day the Sun Turned Blue. A Volcanic Eruption in the Early 1460s and its Possible Climatic Impact. A Natural Disaster Perceived Globally in the Late Middle Ages? In: Schenk, Gerrit J. (ed.): Historical Disaster Experiences. A Comparative and Transcultural Survey Between Asia and Europe. Heidelberg 2017, 107–138, DOI: 10.1007/978-3-319-49163-9_6.

Bauch, Martin: Die Magdalenenflut 1342 am Schnittpunkt von Umwelt- und Infrastrukturgeschichte: Ein compound event als Taktgeber für mittelalterliche Infrastrukturentwicklung und Daseinsvorsorge? In: NTM. Zeitschrift für Geschichte der Wissenschaften, Technik und Medizin 27, no. 3 (2019), 273–309, DOI: 10.1007/s00048-019-00221-y.

Bauch, Martin; Schenk, Gerrit Jasper: Teleconnection, Correlations, Causalities Between Nature and Society? An Introductory Comment on the "Crisis of the Fourteenth Century." In: Bauch, Martin; Schenk, Gerrit Jasper (eds.): Crisis of the Fourteenth Century. Berlin, Boston 2020, 1–23, DOI: 10.1515/9783110660784.

Bauer, Gerhard; Budde, Kai; Kreutz, Wilhelm; Schäfer, Patrick: "Die fernunft siget." Der kurpfälzische Universalgelehrte Johann Jakob Hemmer (1733–1790) und sein Werk. Bern 2010 (Jahrbuch für internationale Germanistik 103).

Baxter, Peter J.: Gases. In: Bayter, Peter J.; Aw, Tar-Ching; Cockcroft, Anne; Harrington, J. Malcolm (eds.): Hunter's Disease of Occupations. London 2000, 123–178.

Beech, Martin: The Great Meteor of 18th August 1783. In: Journal of the British Astronomical Association 99, no. 3 (1989), 130–134.

Begemann, Christian: Furcht und Angst im Prozeß der Aufklärung. Zu Literatur und Bewußtseinsgeschichte des 18. Jahrhunderts. Frankfurt am Main 1987.

Behringer, Wolfgang: Im Zeichen des Merkur. Reichspost und Kommunikationsrevolution in der Frühen Neuzeit. Göttingen 2003.

Behringer, Wolfgang: "Von der Gutenberg-Galaxis zur Taxis-Galaxis." Die Kommunikationsrevolution – ein Konzept zum besseren Verständnis der Frühen Neuzeit. In: Burkhardt, Johannes; Werkstetter, Christine (eds.): Kommunikation und Medien in der Frühen Neuzeit. Munich 2005, 39–56 (Historische Zeitschrift Beiheft 41).

Behringer, Wolfgang: Das Netzwerk der Netzwerke. Raumportionierung und Medienrevolution in der Frühen Neuzeit. In: Arndt, Johannes; Körber, Esther-Beate (eds.): Das Mediensystem im Alten Reich der Frühen Neuzeit (1600–1750). Göttingen 2010, 39–58.

Behringer, Wolfgang: Kulturgeschichte des Klimas. Von der Eiszeit bis zur globalen Erwärmung. Munich 2011.

Behringer, Wolfgang: Der Planet atmet. Überlegungen zu einer Globalgeschichte der Frühen Neuzeit. In: Frühneuzeit-Info 28 (2017), 25–55.

Behringer, Wolfgang: Tambora and the Year Without a Summer: How a Volcano Plunged the World into Crisis. Cambridge UK 2019.

Behringer, Wolfgang; Lehmann, Hartmut; Pfister, Christian (eds.): Kulturelle Konsequenzen der "Kleinen Eiszeit." Cultural Consequences of the "Little Ice Age." Göttingen 2005 (Veröffentlichungen des Max-Planck-Instituts für Geschichte 212).

Bell, Barbara: The Oldest Records of the Nile Floods. In: Geographical Journal 136 (1970), 569–573.

Bergþórsson, Páll: An Estimate of Drift-Ice and Temperature in Iceland in 1000 Years. In: Jökull 19 (1969), 94–101.

Bergþórsson, Páll: Sensitivity of Icelandic Agriculture to Climatic Variations. In: Climatic Change 7 (1985), 111–127.

Bergþórsson, Páll: The Effects of Climatic Variations on Agriculture in Iceland. In: Parry, M. L.; Carter, T. R.; Konijn, N. T. (eds.): The Impact of Climatic Variations on Agriculture. Dordrecht 1987, 389–414.

Bittner, Donald F.: The Lion and the White Falcon: Britain and Iceland in the World War II Era. Hamden CT 1983.

BJARNAR, Vilhjálmur: The Laki Eruption and the Famine of the Mist. In: BAYERSCHMIDT, Carl; FRIIS, Erik J. (eds.): Scandinavian Studies, The American–Scandinavian Foundation. Seattle WA 1965, 410–421.

BJÖRNSSON, Helgi; BJÖRNSSON, Sveinbjörn; SIGURGEIRSSON, Th.: Penetration of Water into Hot Rock Boundaries of Magma at Grímsvötn. In: Nature 295 (1982), 580–581.

BJÖRNSSON, Helgi: Explanation of Jökulhlaups from Grímsvötn, Vatnajökull, Iceland. In: Jökull 24 (1974), 1–26.

BJÖRNSSON, Helgi: Marginal and Supraglacial Lakes in Iceland. In: Jökull 26 (1976), 40–51.

BJÖRNSSON, Helgi: Jökulhlaups in Iceland: Prediction, Characteristics, and Simulation. In: Annals of Glaciology 16 (1992), 95–106.

BJÖRNSSON, Helgi: Subglacial Lakes and Jökulhlaups in Iceland. In: Global and Planetary Change 35 (2002), 255–271, DOI: 10.1016/S0921-8181(02)00130-3.

BJÖRNSSON, Helgi: The Glaciers of Iceland. A Historical, Cultural and Scientific Overview. Paris 2017.

BLACK, Jeremy: War for America. The Fight for Independence, 1775–1783. Stroud 2001.

BLAIR, Ann; GREYERZ, Kaspar von (eds.): Physico-Theology: Religion and Science in Europe, 1650–1750. Baltimore MD 2020 (Medicine, Science, and Religion in Historical Context).

BLANCHARD, Margaret A.: History of the Mass Media in the United States. An Encyclopedia. New York NY 1998.

BLOME, Astrid: Historica et Venditio. Zeitungen als "Bildungsmittel" im 17. und 18. Jahrhundert. In: ARNDT, Johannes; KÖRBER, Esther-Beate (eds.): Das Mediensystem im Alten Reich der Frühen Neuzeit (1600–1750). Göttingen 2010, 207–226.

BÖNING, Holger: Ulrich Bräker: Der arme Mann aus dem Toggenburg. Eine Biographie. Zürich 1998.

BÖNING, Holger: Der "gemeine Mann" als Zeitungs- und Medienkonsument. In: ARNDT, Johannes; KÖRBER, Esther-Beate (eds.): Das Mediensystem im Alten Reich der Frühen Neuzeit (1600–1750). Göttingen 2010, 227–238.

BOREHAM, Frances; CASHMAN, Kathy; RUST, Alison: Lava-River Interactions and Secondary Hazards During the 1783-84 Laki Fissure Eruption. Presented at Cities on Volcanoes 10 (2018), Naples (2 September 2018-7 September 2018), DOI: 10.13140/RG.2.2.15716.42884.

BÖRNGEN, Michael: Wasserkatastrophen in historischer Sicht. In: LOZÁN, José L.; GRAßL, Hartmut; HUPFER, Peter; KARBE, Ludwig; SCHÖNWIESE, Christian-Dietrich (eds.): Warnsignal Klima: Genug Wasser für alle? Hamburg 2011, 118–127.

BOSCANI LEONI, Simona: Wissenschaft – Berge – Ideologien: Johann Jakob Scheuchzer (1672–1733) und die frühneuzeitliche Naturforschung. Basel 2010.

BRAGADÓTTIR, Valgerður: Pons Kompaktwörterbuch, Deutsch-Isländisch. Stuttgart 2008.

BRAMHAM-LAW, Cassian W. F.; THEUERKAUF, Martin; LANE, Christine S.; MANGERUD, Jan: New Findings Regarding the Saksunarvatn Ash in Germany. In: Journal of Quaternary Science 28, no. 3 (2013), 248–257, DOI: 10.1002/jqs.2615.

BRAYSHAY, Mark; GRATTAN, John: Environmental and Social Responses in Europe to the 1783 Eruption of the Laki Fissure Volcano in Iceland: A Consideration of Contemporary Documentary Evidence. In: FIRTH, Callum R.; McGUIRE, William J. (eds.): Volcanoes in the Quaternary. Geological Society of London Special Publication 161 (1999), 173–187.

BRÁZDIL, Rudolf; DEMARÉE, Gaston R.; DEUTSCH, Mathias; GARNIER, Emmanuel; KISS, Andrea; LUTERBACHER, Jürg; MACDONALD, Neil; ROHR, Christian; DOBROVOLNÝ, Petr; KOLÁŘ, Petr; CHROMÁ, Kateřina: European Floods During the Winter 1783/1784: Scenarios of an Extreme Event During the "Little Ice Age." In: Theoretical and Applied Climatology 100 (2010), 163–189, DOI: 10.1007/s00704-009-0170-5.

BRÁZDIL, Rudolf; KUNDZEWICZ, Zbigniew W.; BENITO, Gerardo; DEMARÉE, Gaston; MACDONALD, Neil; ROALD, Lars A.: Historical Floods in Europe in the Past Millennium. In: KUNDZEWICZ, Zbigniew W. (ed.): Changes in Flood Risk in Europe. Boca Raton FL 2012, 121–166.

BRÁZDIL, Rudolf; PFISTER, Christian; WANNER, Heinz; STORCH, Hans von; LUTERBACHER, Jürg: Historical Climatology In Europe – The State Of The Art. In: Climatic Change 70, no. 3 (2005), 363–430, DOI: 10.1007/s10584-005-5924-1.

Brázdil, Rudolf; Řezníčková, Ladislava; Valášek, Hubert; Dolák, Lukáš; Kotyza, Oldřich: Climatic and Other Responses to the Lakagígar 1783 and Tambora 1815 Volcanic Eruptions in the Czech Lands. In: Geografie 122, no. 2 (2017), 147–168, DOI: 10.37040/geografie2017122020147.

Brázdil, Rudolf; Valášek, Hubert; Macková, Jarmila: Climate in the Czech Lands During the 1780s in Light of the Daily Weather Records of Parson Karel Bernard Hein of Hodonice (Southwestern Moravia): Comparison of Documentary and Instrumental Data. In: Climatic Change 60 (2003), 297–327, DOI: 10.1023/A:1026045902062.

Bridgman, Howard A.; Oliver, John E.: The Global Climate System. Patterns, Processes, and Teleconnections. Cambridge UK 2006.

Briese, Olaf: Die Macht der Metaphern. Blitz, Erdbeben und Kometen im Gefüge der Aufklärung. Stuttgart 1998.

Briffa, Keith R.; Schweingruber, Fritz Hans; Jones, Phil D.; Osborn, Timothy J.; Shiyatov, Stepan G.; Vaganov, Eugene A.: Reduced Sensitivity of Recent Tree-Growth to Temperature at High Northern Latitudes. In: Nature 391, no. 6668 (1998), 678–82, DOI: 10.1038/35596.

Brönnimann, Stefan; Allan, Rob; Ashcroft, Linden; Baer, Saba; Barriendos, Mariano; Brázdil, Rudolf; Brugnara, Yuri; et al.: Unlocking Pre-1850 Instrumental Meteorological Records: A Global Inventory. In: Bulletin of the American Meteorological Society 100, no. 12 (2019), DOI: 10.1175/bams-d-19-0040.1.

Brönnimann, Stefan; Krämer, Daniel: Tambora and the "Year Without a Summer" of 1816. A Perspective on Earth and Human Systems Science. In: Geographica Bernensia G90 (2016), DOI: 10.4480/GB2016.G90.01.

Brönnimann, Stefan; Pfister, Christian; White, Sam: Archives of Nature and Archives of Societies. In: White, Sam; Pfister, Christian; Mauelshagen, Franz (eds.): The Palgrave Handbook of Climate History. London 2018, 27–36.

Brown, Robert H.: Nature's Hidden Terror: Violent Nature Imagery in Eighteenth-Century Germany. Columbia SC 1991.

Brugnara, Yuri; Pfister, Lucas; Villiger, Leonie; Rohr, Christian; Isotta, Francesco Alessandro; Brönnimann, Stefan: Early Instrumental Meteorological Observations in Switzerland: 1708–1873. In: Earth System Science Data 12 (2020), 1179–1190, DOI: 10.5194/essd-12-1179-2020.

Brugnatelli, Vermondo; Tibaldi, Alessandro: Effects in North Africa of the 934–940 CE Eldgjá and 1783–1784 CE Laki Eruptions (Iceland) Revealed by Previously Unrecognized Written Sources. In: Bulletin of Volcanology 82 (2020), 73, DOI: 10.1007/s00445-020-01409-0.

Burke, John G.: Cosmic Debris. Meteorites in History. Berkeley CA 1986.

Byock, Jesse: Viking Age Iceland. London 2001.

Büntgen, Ulf; Hellmann, Lena: The Little Ice Age in Scientific Perspective: Cold Spells and Caveats. In: Journal of Interdisciplinary History 44, no. 3 (2014), 353–368.

Callow, Chris; Evans, Charles Morris: The Mystery of Plague in Medieval Iceland. In: Journal of Medieval History 42, no. 2 (March 2016), 254–284, DOI: 10.1080/03044181.2016.1149503.

Campanella, Thomas J.: "Mark Well the Gloom": Shedding Light on the Great Dark Day of 1780. In: Environmental History 12, no. 1 (January 2007), 35–58, DOI: 10.1093/envhis/12.1.35.

Campbell, Bruce M. S.: The Great Transition. Climate, Disease, and Society in the Late-Medieval World. Cambridge UK 2016.

Camuffo, Dario: History of the Long Series of Daily Air Temperatures in Padova (1725–1998). In: Climatic Change 53 (2002), 7–75, DOI: 10.1023/A:1014958506923.

Camuffo, Dario: Evidence from the Archives of Societies: Early Instrumental Observations. In: White, Sam; Pfister, Christian; Mauelshagen, Franz (eds.): The Palgrave Handbook of Climate History. London 2018, 83–92.

Camuffo, Dario; Enzi, Silvia: Chronology of "Dry Fogs" in Italy, 1374–1891. In: Theoretical and Applied Climatology 50 (1994), 31–33.

CAMUFFO, Dario; ENZI, Silvia: Impact of the Clouds of Volcanic Aerosols in Italy During the Last 7 Centuries. In: Natural Hazards 11 (1995), 135–161.

CAPPEL, Albert: Das Wetter und seine Aufklärer. Johann Jakob Hemmer in Mannheim. In: Photorin. Mitteilungen der Lichtenberg-Gesellschaft 10 (1986), 14–26.

CARCIONE, José M.; KOZÁK, Jan T.: The Messina-Reggio Earthquake of December 28, 1908. In: Studia Geophysica et Geodaetica 52 (2008), 661–672, DOI: 10.1007/s11200-008-0043-x.

CARLSEN, Hanne Krage; ILYINSKAYA, Evgenia; BAXTER Peter J.; SCHMIDT, Anja, THORSTEINSSON, Throstur; PFEFFER, Melissa Anne; BARSOTTI, Sara; DOMINICI, Francesca; FINNBJORNSDOTTIR, Ragnhildur Gudrun; JÓHANNSSON, Thorsteinn; ASPELUND, Thor; GISLASON, Thorarinn; VALDIMARSDÓTTIR, Unnur; BRIEM, Haraldur; GUDNASON, Thorolfur: Increased Respiratory Morbidity Associated with Exposure to a Mature Volcanic Plume from a Large Icelandic Fissure Eruption. In: Nature Communications 12, no. 1 (2021), 2161, DOI: 10.1038/s41467-021-22432-5.

CARTWRIGHT, David E.: Robert Paul de Lamanon: An Unlucky Naturalist. In: Annals of Science 54, no. 6 (1997), 585–596.

CASEY, Joan A.; GEMMILL, Alison; ELSER, Holly; KARASEK, Deborah; CATALANO, Ralph: Sun Smoke in Sweden. Perinatal Implications of the Laki Volcanic Eruptions, 1783–1784. In: Epidemiology 30, no. 3 (May 2019), 330–333, DOI: 10.1097/EDE.0000000000000977.

CHARPENTIER, Arthur: On the Return Period of the 2003 Heat Wave. In: Climatic Change 109 (2011), 245–260, DOI: 10.1007/s10584-010-9944-0.

CHENET, Anne-Lise; FLUTEAU, Frédéric; COURTILLOT, Vincent: Modelling Massive Sulphate Aerosol Pollution, Following the Large 1783 Laki Basaltic Eruption. In: Earth and Planetary Science Letters 236, no. 3–4 (2005), 721–731, DOI: 10.1016/j.epsl.2005.04.046.

CHENOWETH, Michael: The 18th Century Climate of Jamaica Derived from the Journals of Thomas Thistlewood, 1750–1786. In: Transactions of the American Philosophical Society 93, no. 2 (2003), i–153, DOI: 10.2307/20020339.

CHESTER, David K.; DUNCAN, Angus M.: Geomythology, Theodicy, and the Continuing Relevance of Religious Worldviews on Responses to Volcanic Eruptions. In: GRATTAN, John; TORRENCE, Robin (eds.): Living Under the Shadow: The Cultural Impacts of Volcanic Eruptions. New York NY 2007, 203–224.

CLAUSEN, Henrik B.; HAMMER, Claus U.: The Laki and Tambora Eruptions as Revealed in Greenland Ice Cores from 11 Locations. In: Annals of Glaciology 10 (1988), 16–22.

CLIFTON, Amy E.; KATTENHORN, Simon: Structural Architecture of a Highly Oblique Divergent Plate Boundary Segment. In: Tectonophysics 419 (2006), 27–40, DOI: 10.1016/j.tecto.2006.03.016.

COCCO, Sean: Watching Vesuvius. A History of Science and Culture in Early Modern Italy. Chicago IL 2012.

COEN, Deborah R.: The Earthquake Observers. Disaster Science from Lisbon to Richter. Chicago IL 2014.

COFFIN, Millard F.; ELDHOLM, Olav: Large Igneous Provinces: Crustal Structure, Dimensions, and External Consequences. In: Reviews of Geophysics 32, no. 1 (1994), 1–36.

COHEN, I. Bernard: Benjamin Franklin's Science. Cambridge MA 1990.

COLE-DAI, Jihong: Volcanoes and Climate. In: Wiley Interdisciplinary Reviews: Climate Change 1, no. 6 (2010), 824–839, DOI: 10.1002/wcc.76.

COLE-DAI, Jihong; SAVARINO, Joël; THIEMENS, Mark H.; LANCIKI, Alyson: Comments on "Climatic Impact of the Long-Lasting Laki Eruption: Inapplicability of Mass-Independent Sulfur Isotope Composition Measurements" by SCHMIDT et al. [2012]. In: Journal of Geophysical Research: Atmospheres 119, no. 11 (2014), 6629–6635, DOI: 10.1002/2013JD019869.

Commission for the Geological Map of the World; BOUYSSE, P.: Geological Map of the World at 1: 35 000 000. Paris 2014, DOI: 10.14682/2020GEOWORLD.

CONDIE, Kent: Mantle Plumes and Their Record in History. Cambridge UK 2001.

CONDIE, Kent: Earth as an Evolving Planetary System. Amsterdam ³2016.

COOPER, Claire L.; SWINDLES, Graeme T.; SAVOV, Ivan P.; SCHMIDT, Anja; BACON, Karen L.: Evaluating the Relationship Between Climate Change and Volcanism. In: Earth-Science Reviews 177 (2018), 238–247, DOI: 10.1016/j.earscirev.2017.11.009.

COURTILLOT, Vincent: New Evidence for Massive Pollution and Mortality in Europe in 1783–1784 May Have Bearing on Global Change and Mass Extinctions. In: Comptes Rendus Geoscience 337, no. 7 (2005), 635–637, DOI: 10.1016/j.crte.2005.03.001.

CRAIG, Martin: Renaissance Meteorology. Pomponazzi to Descartes. Baltimore MD 2011.

D'ANGELO, Michela; SAJIA, Marcello: A City and Two Earthquakes: Messina 1783–1908. In: MASSARD-GUILBAUD, Genevieve; PLATT, Harold L.; SCHOTT, Dieter: Cities and Catastrophes. Frankfurt am Main 2002, 123–140.

D'APRILE, Iwan-Michelangelo; SIEBERS, Winfried: Das 18. Jahrhundert. Zeitalter der Aufklärung. Berlin 2008.

D'ARRIGO, Rosanne; SEAGER, Richard; SMERDON, Jason E.; LEGRANDE, Allegra N.; COOK, Edward R.: The Anomalous Winter of 1783–1784: Was the Laki Eruption an Analog of the 2009–2010 Winter to Blame? In: Geophysical Research Letters 38, no. 5 (2011), 1–4, DOI: 10.1029/2011gl046696.

DAMODARAN, Vinita; ALLAN, Rob; OGILVIE, Astrid E. J.; DEMARÉE, Gaston R.; GERGIS, Joëlle; MIKAMI, Takehiko; MIKHAIL, Alan; NICHOLSON, Sharon E.; NORRGÅRD, Stefan; HAMILTON, James: The 1780s: Global Climate Anomalies, Floods, Droughts, and Famines. In: WHITE, Sam; PFISTER, Christian; MAUELSHAGEN, Franz (eds.): The Palgrave Handbook of Climate History. London 2018, 517–550.

DASTON, Lorraine: The History of Science and the History of Knowledge. In: KNOW: A Journal on the Formation of Knowledge 1 (2017): 131–154. Verburgt, Lukas M.; Burke, Peter: Introduction: Histories of Ignorance. In: Journal for the History of Knowledge 2, no. 1 (2021): 1–9. DOI: 10.5334/jhk.45.

DAVIES, Geoffrey F.: Dynamic Earth: Plates, Plumes and Mantle Convection. Cambridge UK 1999, DOI: 10.1017/CBO9780511605802.

DAVIES, Siwan M.; LARSEN, Guðrún; WASTEGÅRD, Stefan; TURNEY, Chris S. M.; HALL, Valerie A.; COYLE, Lisa; THORDARSON, Thor: Widespread Dispersal of Icelandic Tephra: How Does the Eyjafjöll Eruption of 2010 Compare to Past Icelandic Events? In: Journal of Quaternary Science 25, no. 5 (2010), 605–611, DOI: 10.1002/jqs.1421.

DAWSON, Alastair G.; KIRKBRIDE, Martin P.; COLE, Harriet: Atmospheric Effects in Scotland of the AD 1783–84 Laki Eruption in Iceland. In: The Holocene 31, no. 5 (2021), 1–14, DOI: 10.1177/0959683620988052.

DE ANGELIS, M.; LEGRAND, M.: Origins and Variations of Fluoride in Greenland Precipitation. In: Journal of Geophysical Research: Atmospheres 99, no. D1 (1994), 1157–1172.

DE BOER, Jelle Zeilinga; SANDERS, Donald Theodore: Volcanoes in Human History. Princeton NJ 2002.

DE BOER, Jelle Zeilinga; SANDERS, Donald Theodore: Earthquakes in Human History: The Far-Reaching Effects of Seismic Disruptions. Princeton NJ 2005.

DEGROOT, Dagomar: The Frigid Golden Age: Climate Change, the Little Ice Age, and the Dutch Republic, 1560–1720. Cambridge UK 2018a (Studies in Environment and Society).

DEGROOT, Dagomar: Climate Change and Society in the 15th to 18th Centuries. In: Wiley Interdisciplinary Reviews: Climate Change 9, no. 3 (2018b), e518, DOI: 10.1002/wcc.518.

DEGROOT, Dagomar; ANCHUKAITIS, Kevin; BAUCH, Martin; BURNHAM, Jakob; CARNEGY, Fred; CUI, Jianxin; DE LUNA, Kathryn; GUZOWSKI, Piotr; HAMBRECHT, George; HUHTAMAA, Heli; IZDEBSKI, Adam; KLEEMANN, Katrin; MOESSWILDE, Emma; NEUPANE, Naresh; NEWFIELD, Timothy; PEI, Qing; XOPLAKI, Elena; ZAPPIA, Natale: Towards a Rigorous Understanding of Societal Responses to Climate Change. In: Nature 591 (2021), 539–550, DOI: 10.1038/s41586-021-03190-2.

DEGROOT, Dagomar; ANCHUKAITIS, Kevin J.; TIERNEY, Jessica E.; RIEDE, Felix; MANICA, Andrea; MOESSWILDE, Emma; GAUTHIER, Nicolas: The History of Climate and Society: A Review of the Influence of Climate Change on the Human Past. In: Environmental Research Letters 17, no. 10 (2022), 103001, DOI: 10.1088/1748-9326/ac8faa.

DEMARÉE, Gaston R.: "de grote droge nevel" van 1783 in de Zuidelijke Nederlanden: een historisch-klimatologische studie. In: Tijdschrift voor Ecologische Geschiedenis 1 (1997), 27–35.

DEMARÉE, Gaston R.: The Catastrophic Floods of February 1784 in and around Belgium – a Little Ice Age Event of Frost, Snow, River Ice . . . and Floods. In: Hydrological Sciences – Journal – des Sciences Hydrologiques 51, no. 5 (October 2006), 878–898, DOI: 10.1623/hysj.51.5.878.

DEMARÉE, Gaston R.: Haarrauch, un trouble atmosphérique ou un trouble environnemental et médical au XIX siecle. In: BECKER, Karin; LEPLATRE, Olivier (eds.): La brume et le brouillard dans la science, la littérature et les arts. Paris 2014, 129–143.

DEMARÉE, Gaston; NORDLI, Øyvind: The Lisbon Earthquake of 1755 vs. Volcano Eruptions and Dry Fogs – Are Its 'Meteoric' Descriptions Related to the Katla Eruption of Mid October 1755? In: ARAÚJO, Ana Cristina (ed.): O terramoto de 1755: impactos históricos. Lisbon 2007, 117–130.

DEMARÉE, Gaston R.; NORDLI, Øyvind; MALAQUIAS, Isabel; LOPO, Domingo Gonzalez: Volcano Eruptions, Earth- & Seaquakes, Dry Fogs vs. Aristotle's Meteorologica and the Bible in the Framework of the Eighteenth Century Science History. In: Bulletin des Séances Académie Royale des Sciences d'Outre-Mer 53, no. 3 (2007), 337–359.

DEMARÉE, Gaston R.; OGILVIE, Astrid E. J.: Bons Baisers d'Islande: Climatological, Environmental, and Human Dimensions Impacts of the Lakagígar Eruption (1783–1784) in Iceland. In: JONES, Phil D.; OGILVIE, Astrid E. J.; DAVIES, T. D.; BRIFFA, Keith R. (eds.): History and Climate: Memories of the Future? New York NY 2001, 219–246.

DEMARÉE, Gaston R.; OGILVIE, Astrid E. J.: The Moravian Missionaries at the Labrador Coast and Their Centuries-Long Contribution to Instrumental Meteorological Observations. In: Climatic Change 91, no. 3–4 (2008), 423–450, DOI: 10.1007/s10584-008-9420-2.

DEMARÉE, Gaston R.; OGILVIE, Astrid E. J.: L'éruption du Lakagígar en Islande ou "Annus mirabilis" 1783. Chronique d'une année extraordinaire en Belgique et ailleurs. In: PARMENTIER, Isabelle (ed.): Études et bibliographies d'histoire environnementale. Belgique – Nord de la France – Afrique centrale. Namur 2016, 117–157.

DEMARÉE, Gaston R.; OGILVIE, Astrid E. J.: L'éruption du Lakagígar en Islande ou "Annus mirabilis" 1783. Chronique d'une année extraordinaire en Belgique et ailleurs. In: SÉMATA, Ciencias Sociais e Humanidades 29 (2017), 239–260, DOI: 10.15304/s.29.4208.

DEMARÉE, Gaston R.; OGILVIE, Astrid E. J.; ZHANG, Deer. Further Documentary Evidence of Northern Hemispheric Coverage of the Great Dry Fog of 1783. In: Climatic Change 39, no. 4 (1998), 727–730, DOI: 10.1023/a:1005319607233.

DE SYON, Guillaume: Zeppelin! Germany and the Airship, 1900–1939. Baltimore MD 2002.

DOBSON, Mary J.: "Marsh Fever" – The Geography of Malaria in England. In: Journal of Historical Geography 6, no. 4 (1980), 357–389.

DOBSON, Mary J.: Contours of Death and Disease in Early Modern England. Cambridge UK 1997.

DOMÍNGUEZ-CASTRO, F.; RIBERA, P.; GARCÍA-HERRERA, R.; VAQUERO, J. M.; BARRIENDOS, M.; CUADRAT, J. M.; MORENO, J. M.: Assessing Extreme Droughts in Spain during 1750–1850 from Rogation Ceremonies. In: Climate of the Past 8, no. 2 (February 2012), 705–722, DOI: 10.5194/cp-8-705-2012.

DREXLER, Julie M.; GLEDHILL, Andrew D.; SHINODA, Kentaro; VASILIEV, Alexander L.; REDDY, Kongara M.; SAMPATH, Sanjay; PADTURE, Nitin P.: Jet Engine Coatings for Resisting Volcanic Ash Damage. In: Advanced Materials 23, no. 21 (June 2011), 2419–2124, DOI: 10.1002/adma.201004783.

DROSS, Fritz: Gottes elektrischer Wille? Zum Düsseldorfer "Blitzableiter-Aufruhr" 1782/83. In: ENGELBRECHT, Jörg; LAUX, Stephan (eds.): Landes- und Reichsgeschichte. Festschrift für Hansgeorg Molitor zum 65. Geburtstag. Bielefeld 2004, 281–302.

DUGMORE, Andrew J.; GISLADÓTTIR, Gudrún; SIMPSON, Ian A.; NEWTON, Anthony: Conceptual Models of 1200 Years of Icelandic Soil Erosion Reconstructed Using Tephrochronology. In: Journal of the North Atlantic 2 (2009a), 1–18, DOI: 10.3721/037.002.0103.

DUGMORE, Andrew J.; KELLER, Christian; McGOVERN, T. H.; CASELY, Andrew F.; SMIAROWSKI, Konrad: Norse Greenland Settlement and Limits to Adaptation. In: ADGER, W. Neil; LORENZI, Irene; O'BRIEN, Karen (eds.): Adapting to Climate Change: Thresholds, Values, and Governance. Cambridge UK 2009b, 96–113.

DUGMORE, Andrew; VÉSTEINSSON, Orri: Black Sun, High Flame, and Flood. Volcanic Hazards in Iceland. In: COOPER, Jago; SHEETS, Payson (eds.): Surviving Sudden Environmental Change: Answers from Archaeology. Boulder CO 2012, 67–89.

DURAND, Michael; GRATTAN, John P.: Extensive Respiratory Health Effects of Volcanogenic Dry Fog in 1783 Inferred from European Documentary Sources. In: Environmental Geochemistry and Health 21 (1999), 371–376, DOI: 10.1023/A:1006700921208.

DURAND, Michael; GRATTAN, John P.: Effects of Volcanic Air Pollution on Health. In: The Lancet 257 (2001), 164, DOI: 10.1016/S0140-6736(00)03586-8.

DÜRR, Renate: Threatened Knowledge. Practices of Knowing and Ignoring from the Middle Ages to the Twentieth Century. London 2021.

DYNES, Russell Rowe: The Dialogue Between Voltaire and Rousseau on the Lisbon Earthquake: The Emergence of a Social Science View. Newark DE 1999.

EBEL, John E.: New England Earthquakes. The Surprising History of Seismic Activity in the Northeast. Lanham MD 2019.

EBERT, Stephan: Methodological Benefits of a GIS Map: The Example of the Eldgjá Eruption of the Late 930s CE and the Reliability of Historical Documents. In: Environment & Society Portal, Arcadia (Spring 2019), no. 7. Rachel Carson Center for Environment and Society, DOI: 10.5282/rcc/8560.

EBERT, Stephan: Der Umwelt begegnen: Extremereignisse und die Verflechtung von Natur und Kultur im Frankenreich vom 8. bis 10. Jahrhundert. Stuttgart 2021 (Vierteljahrschrift für Sozial- und Wirtschaftsgeschichte – Beihefte 254).

EDDY, John A.: The Maunder Minimum. In: Science 192, no. 4245 (1976), 1189–1202.

EDWARDS, Heather: Introduction. In: WINSTANLEY, R. I.; JAMESON, Peter; EDWARDS, Heather (eds.): The Diary of James Woodforde. Vol. 10: Norfolk, 1782–1784. Norfolk 1998.

EDWARDS, Julie; ANCHUKAITIS, Kevin J.; GUNNARSON, Björn E.; PEARSON, Charlotte; SEFTIGEN, Kristina; ARX, Georg von; LINDERHOLM, Hans W.: The Origin of Tree-Ring Reconstructed Summer Cooling in Northern Europe During the 18th Century Eruption of Laki. In: Paleoceanography and Paleoclimatology 37, no. 2 (2022), e2021PA004386, DOI: 10.1029/2021PA004386.

EDWARDS, Julie; ANCHUKAITIS, Kevin J.; ZAMBRI, Brian; ANDREU-HAYLES, Laia; OELKERS, Rose; D'ARRIGO, Rosanne; ARX, Georg von: Intra-Annual Climate Anomalies in Northwestern North America Following the 1783–1784 CE Laki Eruption. In: Journal of Geophysical Research: Atmospheres 126 (2021), e2020JD033544, DOI: 10.1029/2020JD033544.

EINARSSON, Þorleifur; SVEINSDÓTTIR, Edda Lilja: Nýtt kort af Skaftáreldahrauni og Lakagígum. (A New Map of the Lakagígar Crater Row and the Lava Flows of 1783–1784). In: GUNNLAUGSSON, Gísli Ágúst; GUÐBERGSSON, Gylfi Már; ÞÓRARINSSON, Sigurður; RAFNSSON, Sveinbjörn; EINARSSON, Þorleifur (eds.): Skaftáreldar 1783–1784: Ritgerðir og heimildir. Reykjavík 1984, 37–48.

ENGELSING, Rolf: Analphabetentum und Lektüre. Zur Sozialgeschichte des Lesens in Deutschland zwischen feudaler und industrieller Gesellschaft. Stuttgart 1973.

FAGAN, Brian: The Little Ice Age: How Climate Made History, 1300–1850. New York NY 2000.

FALK, Oren: The Vanishing Volcanoes: Fragments of Fourteenth-Century Icelandic Folklore. In: Folklore 118, no. 1 (April 2007), 1–22, DOI: 10.1080/00155870601096257.

FAULSTICH, Werner: Mediengeschichte von 1700 bis ins 3. Jahrtausend. Göttingen 2006.

FEDOROVA, Tanya; JACOBY, Wolfgang R.; WALLNER, Herbert: Crust-Mantle Transition and Moho Model For Iceland and Surroundings from Seismic Topography, and Gravity Data. In: Tectonophysics 396 (2005), 119–140, DOI: 10.1016/j.tecto.2004.11.004.

FERLING, John: Whirlwind: The American Revolution and the War That Won It. New York NY 2015.

FIACCO, R. Joseph; THORDARSON, Thorvaldur; GERMANI, Mark S.; SELF, Stephen; PALAIS, Julie M.; WHITLOW, Sallie; GROOTES, Peter M.: Atmospheric Aerosol Loading and Transport Due to the 1783–84 Laki Eruption in Iceland, Interpreted from Ash Particles and Acidity in the GISP2 Ice Core. In: Quaternary Research 42, no. 3 (1994), 231–240, DOI: 10.1006/qres.1994.1074.

FISHER, David A.; KOERNER, Roy M.: Signal and Noise in Four Ice-Core Records from the Agassiz Ice Cap, Ellesmere Island, Canada: Details of the Last Millennium for Stable Isotopes, Melt and Solid Conductivity. In: Holocene 4 (1994), 113–120, DOI: 10.1177/095968369400400201.

FISHER, Richard V.; HEIKEN, Grant; HULEN, Jeffrey B.: Volcanoes: Crucibles of Change. Princeton NJ 1997.

FITZHUGH, William W.; WARD, Elisabeth I. (eds.): Vikings: The North Atlantic Saga. Washington DC 2000.

FLANNERY, Tim: The Weather Makers. Our Changing Climate and What It Means for Life on Earth. London 2007.

FLEMING, James Rodger: Historical Perspectives on Climate Change. New York NY 1998.

FONSECA, Joao F. B. D.: A Reassessment of the Magnitude of the 1755 Lisbon Earthquake. In: Bulletin of the Seismological Society of America 110, no. 1 (July 2020), 1–17, DOI: 10.1785/0120190198.

FORSYTH, David A.; MOREL-A-L'HUISSIER, Patrick; ASUDEH, Isa; GREEN, Alan G.: Alpha Ridge and Iceland-Products of the Same Plume? In: Journal of Geodynamics 6, no. 1–4 (1986), 197–214, DOI: 10.1016/0264-3707(86)90039-6.

FORTEY, Richard: The Earth. An Intimate History. London 2005.

FOULGER, Gillian R.: Plates vs Plumes: A Geological Controversy. Hoboken NJ 2010.

FRANCIS, Peter; OPPENHEIMER, Clive: Volcanoes. Oxford 2004.

FRANÇOIS, Étienne: Alphabetisierung und Lesefähigkeit in Frankreich und Deutschland um 1800. In: BERDING, Helmut; FRANÇOIS, Étienne; ULLMANN, Hans-Peter (eds.): Deutschland und Frankreich im Zeitalter der Französischen Revolution. Frankfurt am Main 1989, 407–426.

FRANKE, Jörg; BRÖNNIMANN, Stefan; BHEND, Jonas; BRUGNARA, Yuri: A Monthly Global Paleo-Reanalysis of the Atmosphere from 1600 to 2005 For Studying Past Climatic Variations. In: Scientific Data 4 (2017), 170076, DOI: 10.1038/sdata.2017.76.

FRANKEL, Henry R.: The Continental Drift Controversy. Cambridge UK 2017.

FREMONT-BARNES, Gregory; ARNOLD, James R.: The Encyclopedia of the American Revolutionary War: A Political, Social, and Military History. 5 vols. Santa Barbara CA 2006.

FRICKE, Werner: Der Bericht von E. F. Deurer über das Eishochwasser von 1783. Eine Erklärung der Naturkatastrophe. In: PRÜCKNER, Helmut (ed.): Die alte Brücke in Heidelberg 1788–1988. Heidelberg 1988, 41–61.

FRIEDRICH, Anke M.: Palaeogeological Hiatus Surface Mapping: A Tool to Visualize Vertical Motion of the Continents. In: MEINHOLD, G. (ed.): Advances in Palaeogeography. Geological Magazine 156, no. 2. Cambridge UK 2019, 308–319, DOI: 10.1017/S0016756818000560.

FRIEDRICH, Anke M.; BUNGE, Hans-Peter; RIEGER, Stefanie M.; COLLI, Lorenzo; GHELICHKHAN, Siavash; NERLICH, N. S. Rainer: Stratigraphic Framework for the Plume Mode of Mantle Convection and the Analysis of Interregional Unconformities on Geological Maps. In: Gondwana Research 53 (January 2018), 159–188, DOI: 10.1016/j.gr.2017.06.003.

FRODEMAN, Robert: Geological Reasoning: Geology as an Interpretive and Historical Science. In: GSA Bulletin 107, no. 8 (August 1995), 960–968.

FRÖMMING, Urte Undine: Naturkatastrophen. Kulturelle Deutung und Verarbeitung. Frankfurt am Main 2005.

FUJII, Yoshiyuki; KAMIYAMA, K.; KAWAMURA, T.; KAMEDA, Takao; IZUMI, K.; SATOW, K.; ENOMOTO, Hiroyuki; et al.: 6000-Year Climate Records in an Ice Core from the Høghetta Ice Dome in Northern Spitsbergen. In: Annals of Glaciology 14 (1990), 85–89, DOI: 10.1017/s0260305500008314.

FURMAN, Tanya; FREY, Fred A.; PARK, Kye-Hun: Chemical Constraints of the Petrogenesis of Mildly Alkaline Lavas from Vestmannaeyjar, Iceland: The Eldfell (1973) and Surtsey (1963–1967) Eruptions. In: Contributions to Mineralogy and Petrology 109 (1991), 19–37.

GAO, Chaochao; OMAN, Luke; ROBOCK, Alan; STENCHIKOV, Georgiy L.: Atmospheric Volcanic Loading Derived from Bipolar Ice Cores: Accounting for the Spatial Distribution of Volcanic Deposition. In: Journal of Geophysical Research 112, no. D9 (August 2007), D09109, DOI: 10.1029/2006jd007461.

GAO, Chaochao; ROBOCK, Alan; SELF, Stephen; WITTER, Jeffrey B.; STEFFENSON, J. P.; CLAUSEN, Henrik Brink; SIGGAARD-ANDERSEN, Marie-Louise; JOHNSEN, Sigfus; MAYEWSKI, Paul A.; AMMANN, Caspar: The 1452 or

1453 A.D. Kuwae Eruption Signal Derived from Multiple Ice Core Records: Greatest Volcanic Sulfate Event of the Past 700 Years. In: Journal of Geophysical Research 111, no. D12 (2006), D12107, DOI: 10.1029/2005jd006710.

GAO, Chao-Chao; YANG, Lin-Shan; LIU, Fei: Hydroclimatic Anomalies in China During the Post-Laki Years and the Role of Concurring El Niño. In: Advances in Climate Change Research 12, no. 2 (2021), 187–198, DOI: 10.1016/j.accre.2021.03.006.

GARNIER, Emmanuel: Laki: une catastrophe européenne. In: L'Histoire 343 (2009), 72–77.

GARNIER, Emmanuel: Les brouillards du Laki en 1783. Volcanisme et crise sanitaire en Europe. In: Bulletin de l'Académie nationale de médecine 195, no. 4–5 (April 2011), 1043–1055, DOI: 10.1016/S0001-4079 (19)32018-7.

GEORGI, Matthias: Heuschrecken, Erdbeben und Kometen. Naturkatastrophen und Naturwissenschaft in der englischen Öffentlichkeit des 18. Jahrhunderts. Munich 2009.

GILLESPIE, Richard: Ballooning in France and Britain, 1783–1786: Aerostationa and Adventurism. In: ISIS 75, no. 2 (June 1984), 248–268.

GLACKEN, Clarence J.: Traces on the Rhodian Shore: Nature and Culture in Western Thought from Ancient Times to the End of the Eighteenth Century. Berkeley CA 1990.

GLANTZ, Michael H.: Currents of Change: El Niño's Impact on Climate and Society. Cambridge UK 1996.

GLASER, Rüdiger: Klimageschichte Mitteleuropas. 1000 Jahre Wetter, Klima, Katastrophen. Darmstadt 2001.

GLASER, Rüdiger: Klima. In: JAEGER, Friedrich (ed.): Enzyklopädie der Neuzeit. Vol. 6: Jenseits-Konvikt. Stuttgart 2007, 786–808.

GLASER, Rüdiger: Klimageschichte Mitteleuropas. 1200 Jahre Wetter, Klima und Katastrophen. Darmstadt ²2008.

GLASER, Rüdiger; HAGEDORN, Horst: Die Überschwemmungskatastrophe von 1784 im Maintal. Eine Chronologie ihrer witterungsklimatischen Voraussetzungen und Auswirkungen. In: Die Erde 121 (1990), 1–14.

GOLINSKI, Jan: Barometers of Change: Meteorological Instruments as Machines of Enlightenment. In: CLARK, William; GOLINSKI, Jan; SCHAFFER, Simon (eds.): The Sciences in Enlightened Europe. Chicago IL 1999, 69–93.

GOLINSKI, Jan: British Weather and the Climate of Enlightenment. Chicago IL 2007.

GOULD, Stephen Jay: Time's Arrow, Time's Cycle. Cambridge MA 1987.

GRAF, Friedrich Wilhelm: Entzauberung der Welt. In: JAEGER, Friedrich (ed.): Enzyklopädie der Neuzeit. Vol. 3: Dynastie-Freundschaftslinien. Stuttgart 2006, 342–344.

GRATTAN, John P.: The Distal Impact of Icelandic Volcanic Gases and Aerosols in Europe: A Review of the 1783 Laki Fissure Eruption and Environmental Vulnerability in the Late 20th Century. In: MAUND, Julian G.; EDDLESTON, Michael (eds.): Geohazards in Engineering Geology. Geological Society of London. Engineering Geology Special Publication 15 (1998), 97–103.

GRATTAN, John P.; BRAYSHAY, Mark: An Amazing and Portentous Summer: Environmental and Social Responses in Britain to the 1783 Eruption of an Iceland Volcano. In: The Geographical Journal 161, no. 2 (July 1995), 125–134.

GRATTAN, John P.; BRAYSHAY, Mark; SADLER, Jon: Modelling the Distal Impacts of Past Volcanic Gas Emissions. Evidence of Europe-Wide Environmental Impacts from Gases Emitted During the Eruption of Italian and Icelandic Volcanoes in 1783. In: Quaternaire 9, no. 1 (1998), 25–35.

GRATTAN, John P.; BRAYSHAY, Mark; SCHÜTTENHELM, Ruud T. E.: "The End is Nigh?" Social and Environmental Responses to Volcanic Gas Pollution. In: TORRENCE, Robin; GRATTAN, John (eds.): Natural Disasters and Cultural Change. London 2002, 87–106.

GRATTAN, John P.; CHARMAN, Daniel J.: Non-Climatic Factors and the Environmental Impact of Volcanic Volatiles: Implications of the Laki Fissure Eruption of AD 1783. In: The Holocene 4, no. 1 (1994), 101–106, DOI: 10.1177/095968369400400113.

GRATTAN, John P.; DURAND, Michael: Distant Volcanic Eruptions, Human Health and Mortality. In: RAYNAL, Jean-Paul; LIVADIE, Claude Albore; PIPERNO, Marcello (eds.): Hommes Et Volcans: De l'éruption à l'objet: Actes Du Symposium 15.2. Goudet 2002, 15–22.

GRATTAN, John P.; DURAND, Michael; GILBERTSON, David; PYATT, F. Brian: Human Sickness and Mortality Rates in Relation to the Distant Eruption of Volcanic Gases. Rural England and the 1783 Eruption of the Laki Fissure, Iceland. In: SKINNER, Catherine W.; BERGER, Antony R. (eds.): Geology and Health: Closing the Gap. Oxford 2003, 19–24.

GRATTAN, John P.; DURAND, Michael; SCHÜTTENHELM, Ruud: Human Illness and Vegetation Damage Induced by the Volcanogenic Dry Fog in 1783: Further Studies of the Distal Environmental Impacts of the Laki Fissure Eruption. In: JUVIGNÉ, Etienne; RAYNAL, Jean-Paul (eds.): Tephras: Chronology and Archaeology. Goudet 2001, 145–152.

GRATTAN, John P.; DURAND, Michael; TAYLOR, S.: Illness and Elevated Human Mortality in Europe Coincident with the Laki Fissure Eruption. In: OPPENHEIMER, Clive; PYLE, David M.; BARCLAY, Jennie (eds.): Volcanic Degassing. Geological Society of London Special Publication 213 (2003), 404–414.

GRATTAN, John; GILBERTSON, David D.; DILL, A: "A Fire Spitting Volcano in Our Dear Germany": Documentary Evidence for a Low-Intensity Volcanic Eruption of the Gleichberg in 1783? In: McGUIRE, William J.; GRIFFITHS, Dafydd R.; HANCOCK, P. L.; STEWART, Iain S. (eds.): The Archaeology of Geological Catastrophes. Geological Society, London, Special Publications 171 (2000), 307–315.

GRATTAN, John P.; MICHNOWICZ, Sabina; RABARTIN, Roland: The Long Shadow. Understanding the Influence of the Laki Fissure Eruption on Human Mortality in Europe. In: GRATTAN, John; TORRENCE, Robin (eds.): Living Under the Shadow: The Cultural Impacts of Volcanic Eruptions. New York NY 2007, 153–174.

GRATTAN, John P.; PYATT, F. Brian: Acid Damage to Vegetation Following the Laki Fissure Eruption in 1783 – an Historical Review. In: Science of the Total Environment 151, no. 3 (1994), 241–247, DOI: 10.1016/0048-9697(94)90473-1.

GRATTAN, John P.; PYATT, F. B.: Volcanic Eruptions Dry Fogs and the European Palaeoenvironmental Record: Localised Phenomena or Hemispheric Impacts? In: Global and Planetary Change 21, no. 1–3 (1999), 173–179, DOI: 10.1016/s0921-8181(99)00013-2.

GRATTAN, John; RABARTIN, Roland; SELF, Stephen; THORDARSON, Thorvaldur: Volcanic Air Pollution and Mortality in France 1783–1784. In: Comptes Rendus Geoscience 337, no. 7 (2005), 641–651, DOI: 10.1016/j.crte.2005.01.013.

GRATTAN, John P.; SADLER, Jon: Regional Warming of the Lower Atmosphere in the Wake of Volcanic Eruptions: The Role of the Laki Fissure Eruption in the Hot Summer of 1783. In: FIRTH, Callum R.; McGUIRE, William J. (eds.): Volcanoes in the Quaternary. Geological Society of London Special Publication 161 (1999), 161–171.

GRATTAN, John P.; SADLER, Jon: An Exploration of the Contribution of the Laki Fissure Gases to the High Summer Air Temperatures in Western Europe in 1783. In: JUVIGNÉ, Etienne; RAYNAL, Jean-Paul (eds.): Tephras: Chronology and Archaeology. Goudet 2001, 137–172.

GRAZIANI, Laura; MARAMAI, Alessandra; TINTI, Stefano: A Revision of the 1783&1784 Calabrian (Southern Italy) Tsunamis. In: Natural Hazards and Earth System Sciences 6, no. 6 (2006), 1053–1060, DOI: 10.5194/nhess-6-1053-2006.

GREENE, Jack P; POLE, J. R.: A Companion to the American Revolution. Malden MA 2000.

GREENWOOD, Norman N.; EARNSHAW, Alan: Chemistry of the Elements. Oxford 2008.

GREYERZ, Kaspar von: Religion und Natur in der Frühen Neuzeit. Aspekte einer vielschichtigen Beziehung. In: RUPPEL, Sophie; STEINBRECHER, Aline (eds.): "Die Natur ist überall bey uns." Mensch und Natur in der Frühen Neuzeit. Zurich 2009, 41–58.

GROH, Dieter; KEMPE, Michael; MAUELSHAGEN, Franz: Einleitung. Naturkatastrophen – wahrgenommen, gedeutet, dargestellt. In: GROH, Dieter; KEMPE, Michael; MAUELSHAGEN, Franz (eds.): Naturkatastrophen. Beiträge zu ihrer Deutung, Wahrnehmung und Darstellung in Text und Bild von der Antike bis ins 20. Jahrhundert. Tübingen 2003, 11–33.

GRÖNVOLD, Karl: Bergfraedi Skaftáreldahrauns. (Chemical Analysis of the Lakagígar Lava). In: GUNNLAUGSSON, Gísli Ágúst; GUÐBERGSSON, Gylfi Már; ÞÓRARINSSON, Sigurður; RAFNSSON, Sveinbjörn; EINARSSON, Þorleifur (eds.): Skaftáreldar 1783-1784: Ritgerðir og heimildir. Reykjavík 1984, 49–58.

GRÖNVOLD, Karl; ÓSKARSSON, Níels; JOHNSEN, Sigfús. J.; CLAUSEN, Henrik B.; HAMMER, Claus U.; BOND, Gerard; BARD, Edouard: Ash Layers from Iceland in the Greenland GRIP Ice Core Correlated With Oceanic and Land Sediments. In: Earth and Planetary Science Letters 135 (1995), 149–155.

GROTEFEND, Hermann: Taschenbuch der Zeitrechnung des deutschen Mittelalters und der Neuzeit. Hannover [14]2007.

GROTZINGER, John; JORDAN, Thomas: Press Sievers. Allgemeine Geologie. Berlin [7]2017.

GROVE, Jean: The Initiation of the "Little Ice Age" in Regions Round the North Atlantic. In: OGILVIE, Astrid E. J.; JÓNSSON, Trausti (eds.): The Iceberg in the Mist: Northern Research in Pursuit of a "Little Ice Age." Dordrecht 2000, 53–82, DOI: 10.1007/978-94-017-3352-6_2.

GROVE, Jean: The Initiation of the "Little Ice Age" in Regions Round the North Atlantic. Climatic Change 48 (2001), 53–82, DOI: 10.1023/A:1005662822136.

GROVE, Richard H.: The East India Company, the Australians and the El Niño: Colonial Scientists and Ideas About Global Climatic Change and Teleconnections Between 1770 and 1930. In: GROVE, Richard H. (ed.): Ecology, Climate, and Empire. Colonialism and Global Environmental History, 1400–1940. Cambridge UK 1997, 124–146.

GROVE, Richard: The Great El Niño of 1789–93 and its Global Consequences: Reconstructing an Extreme Climate Event in World Environmental History. In: The Medieval History Journal 10, no. 1–2 (October 2007), 75–98, DOI: 10.1177/097194580701000203.

GRÜNTHAL, Gottfried (ed.): European Macroseismic Scale 1998 (EMS-98). Helfent-Bertrange 1998 (Cahiers du Centre Européen de Géodynamique et de Séismologie 15).

GRÜNTHAL, Gottfried; MAYER-ROSA, Dieter; LENHARDT, Wolfgang A.: Abschätzung der Erdbebengefährdung für die D-A-CH-Staaten – Deutschland, Österreich, Schweiz. In: Bautechnik 75, no. 10 (1998), 753–767.

GUÐBERGSSON, Gylfi Már; THEODÓRSSON, Theodór: Áhrif Skaftárelda á byggd og mannfjölda í Leiðvallarhreppi og Leifahreppi. (Effects of the Lakagígar Eruption on Population and Settlement in Leiðvallar and Kleifa Communes.) In: GUNNLAUGSSON, Gísli Ágúst; GUÐBERGSSON, Gylfi Már; ÞÓRARINSSON, Sigurður; RAFNSSON, Sveinbjörn; EINARSSON, Þorleifur (eds.): Skaftáreldar 1783-1784: Ritgerðir og heimildir. Reykjavík 1984, 99–118.

GUÐMUNDSSON, Águst: Lateral Magma Flow, Caldera Collapse, and a Mechanism of Large Eruptions. In: Journal of Volcanology and Geothermal Research 34 (1987), 65–78.

GUÐMUNDSSON, Magnús T.: The Grímsvötn Caldera, Vatnajökull: Subglacial Topography and Structure of Caldera Infill. In: Jökull 39 (1989), 1–19.

GUÐMUNDSSON, Magnús T.: Subglacial Volcanic Activity in Iceland. In: CASELDINE, Christopher; RUSSELL, Andrew; HARDARDÓTTIR, Jórunn; KNUDSEN, Óskar (eds.): Iceland – Modern Processes and Past Environments. Vol. 5. Amsterdam 2005, 127–152.

GUÐMUNDSSON, Magnús T.; LARSEN, Guðrún; HÖSKULDSSON, Ármann; GYLFASON, Agust Gunnar: Volcanic Hazards in Iceland. In: Jökull 58 (2008), 251–268.

GUERRA, Corinna: If You Don't Have a Good Laboratory, Find a Good Volcano: Mount Vesuvius as a Natural Chemical Laboratory in Eighteenth-Century Italy. In: Ambix 62, no. 3 (2015), 245–265, DOI: 10.1179/1745823415Y.0000000005.

GUEVARA-MURUA, Alvaro; WILLIAMS, Caroline A.; HENDY, Erica J.; RUST, Alison C.; CASHMAN, Katharine V.: Observations of a Stratospheric Aerosol Veil from a Tropical Eruption in December 1808: Is this the Unknown ~1809 eruption? In: Climate of the Past 10 (2014), 1707–1722, DOI: 10.5194/cp-10-1707-2014.

GUILBAUD, Marie-Noëlle; SELF, Stephen; THORDARSON, Thorvaldur; BLAKE, Stephen: Morphology, Surface Structures, and Emplacement of Lavas Produced by Laki, A.D. 1783–1784. In: Kinematics and Dynamics of Lava Flows, 2005, DOI: 10.1130/0-8137-2396-5.81.

GUNN, Joel D.: A.D. 536 and its 300-Year Aftermath. In: GUNN, Joel D. (ed.): The Years Without Summer: Tracing AD 536 and its Aftermath. Oxford 2000, 5–20.

GUNNARSDÓTTIR, Margrét: Facing Natural Extremes: The Catastrophe of the Laki Eruption in Iceland in 1783–84. In: 1700-tal: Nordic Journal for Eighteenth-Century Studies 19 (2022), 72–93, DOI: 10.7557/4.6611.

GUNNARSSON, Gísli: A Study of Causal Relations in Climate and History. With an Emphasis on the Icelandic Experience. Lund 1980 (Meddelande Fran Ekonomisk-Historiska Institutionen Lunds Universitet 17).

GUNNARSSON, Gísli: Monopoly Trade and Economic Stagnation: Studies in the Foreign Trade of Iceland 1602–1787. Lund 1983.

GUNNARSSON, Gísli: Voru Móðuharðindin af manna völdum? (The Famine of the Mist and the Human Factor). In: GUNNLAUGSSON, Gísli Ágúst; GUÐBERGSSON, Gylfi Már; ÞÓRARINSSON, Sigurður; RAFNSSON, Sveinbjörn; EINARSSON, Þorleifur (eds.): Skaftáreldar 1783–1784: Ritgerðir og heimildir. Reykjavík 1984, 235–242.

GUNNLAUGSSON, Gísli Ágúst; GUÐBERGSSON, Gylfi Már; ÞÓRARINSSON, Sigurður; RAFNSSON, Sveinbjörn; EINARSSON, Þorleifur (eds.): Skaftáreldar 1783–1784: Ritgerðir og heimildir. Reykjavík 1984.

GUNNLAUGSSON, Gísli Ágúst: Vidbrögd stjórnvalda í Kaupmannahöfn við Skaftáreldum. (The Reactions of the Central Administration in Copenhagen to the Laki-Eruption and its Consequences 1783–1785). In: GUNNLAUGSSON, Gísli Ágúst; GUÐBERGSSON, Gylfi Már; ÞÓRARINSSON, Sigurður; RAFNSSON, Sveinbjörn; EINARSSON, Þorleifur (eds.): Skaftáreldar 1783–1784: Ritgerðir og heimildir. Reykjavík 1984, 187–214.

GUNNLAUGSSON, Gísli Ágúst: Family and Household in Iceland 1801–1930: Studies in the Relationship Between Demographic and Socio-Economic Development, Social Legislation, and Family and Household Structures. Uppsala 1988.

GUSTAFSSON, Harald: Political Interaction in the Old Regime. Central Power and Local Society in the Eighteenth-Century Nordic States. Lund 1994.

GUTSCHER, Marc-André.; BAPTISTA, Maria Ana; MIRANDA, J. M.: The Gibraltar Arc Seismogenic Zone (part 2): Constraints on a Shallow East Dipping Fault Plane Source for the 1755 Lisbon Earthquake Provided by Tsunami Modeling and Seismic Intensity. In: Tectonophysics 426, 1–2 (2006), 153–166, DOI: 10.1016/j.tecto.2006.02.025.

HAARLÄNDER, Stephanie. Rabanus Maurus zum Kennenlernen. Ein Lesebuch mit einer Einführung in sein Leben und Werk. Mainz 2006.

HALDON, John; MORDECHAI, Lee; NEWFIELD, Timothy P.; CHASE, Arlen F.; IZDEBSKI, Adam; GUZOWSKI, Piotr; LABUHN, Inga; ROBERTS, Neil: History Meets Palaeoscience: Consilience and Collaboration in Studying Past Societal Responses to Environmental Change. In: Proceedings of the National Academy of Sciences 115, no. 13 (December 2018), 3210–3218, DOI: 10.1073/pnas.1716912115.

HÁLFDANARSSON, Guðmundur: Mannfall í Móðuhardindum. (Loss of Human Lives Following the Laki Eruption.) In: GUNNLAUGSSON, Gísli Ágúst; GUÐBERGSSON, Gylfi Már; ÞÓRARINSSON, Sigurður; RAFNSSON, Sveinbjörn; EINARSSON, Þorleifur (eds.): Skaftáreldar 1783–1784: Ritgerðir og heimildir. Reykjavík 1984, 139–162.

HÁLFDANARSSON, Guðmundur: Iceland Perceived: Nordic, European, or a Colonial Other? In: KÖRBER, Lill-Ann; VOLQUARDSEN, Ebbe (eds.): The Post-Colonial North Atlantic: Iceland, Greenland, and the Faroe Islands. Berlin 2014, 39–66.

HAMILTON, Christopher W.; FAGENTS, Sarah A.; THORDARSON, Thorvaldur: Explosive Lava-Water Interactions I: Architecture and Emplacement Chronology of Volcanic Rootless Cone Groups in the 1783–1784 Laki Lava Flow, Iceland. In: Bulletin of Volcanology 72, no. 4 (2010), 449–467, DOI: 10.1007/s00445-009-0330-6.

HAMM, F.: Naturkundliche Chronik Nordwestdeutschlands. Hannover 1976.

HAMMER, Claus U.: Past Volcanism Revealed by Greenland Ice Sheet Impurities. In: Nature 270 (1977), 482–486.

HAMMER, Claus U.: Traces of Icelandic Eruptions in the Greenland Ice Sheet. In: Jökull 34 (1984), 51–65.

HAMMER, Claus U.; CLAUSEN, Henrik B.; DANSGAARD, Willi: Past Volcanism and Climate Revealed by Greenland Ice-Cores. In: Journal of Volcanology and Geothermal Research 11 (1981), 3–10.

HANSELL, Anna; OPPENHEIMER, Clive: Health Hazards from Volcanic Gases: A Systematic Literature Review. In: Archives of Environmental Health: An International Journal 59, no. 12 (2004), 628–639. DOI: 10.1080/00039890409602947.

HANSEN, James E.; WANG, Wei-Chyung; LACIS, Andrew A.: Mount Agung Eruption Provides Test of a Global Climatic Perturbation. In: Science 199 (1978), 1065–1068, DOI: 10.1126/science.199.4333.1065.

HANSSON, Heidi: The Gentleman's North: Lord Dufferin and the Beginnings of Arctic Tourism. In: Studies in Travel Writing 13, no. 1 (2009), 61–73, DOI: 10.1080/13645140802611358.

HANTEMIROV, Rashit M.; GORLANOVA, Ludmila A.; SHIYATOV, Stepan G.: Pathological Tree-Ring Structures in Siberian Juniper (*Juniperus sibirica* Burgsd.) and Their Use for Reconstructing Extreme Climate Events. In: Russian Journal of Ecology 31, No. 3 (2000), 167–173, DOI: 10.1007/BF02762816.

HANTEMIROV, Rashit M.; GORLANOVA, Ludmila A.; SHIYATOV, Stepan G.: Extreme Temperature Events in Summer in Northwest Siberia Since AD 742 Inferred from Tree Rings. In: Palaeogeography, Palaeoclimatology, Palaeoecology 209 (2004), 155–164, DOI: 10.1016/j.palaeo.2003.12.023.

HARTMAN, Laura H.; KURBATOV, Andrei V.; WINSKI, Dominic A.; CRUZ-URIBE, Alicia M.; DAVIES, Siwan M.; DUNBAR, Nelia W.; IVERSON, Nels A.; et al.: Volcanic Glass Properties from 1459 C.E. Volcanic Event in South Pole Ice Core Dismiss Kuwae Caldera as a Potential Source. In: Scientific Reports 9, no. 1 (August 2019), 14437, DOI: 10.1038/s41598-019-50939-x.

HARTMANN, Claus Peter: Bevölkerungszahlen und Konfessionsverhältnisse des Heiligen Römisches Reiches Deutscher Nation und der Reichskreise am Ende des 18. Jahrhunderts. In: Zeitschrift für Historische Forschung 22, no. 3 (1995), 345–369.

HASTRUP, Kirsten: Nature and Policy in Iceland, 1400–1800. An Anthropological Analysis of History and Mentality. Oxford 1990.

HEAVISIDE, Clare; WITHAM, Claire; VARDOULAKIS, Sotiris: Potential Health Impacts from Sulphur Dioxide and Sulphate Exposure in the UK Resulting from an Icelandic Effusive Volcanic Eruption. In: Science of the Total Environment 774 (2021), 145549, DOI: 10.1016/j.scitotenv.2021.145549.

HELGASON, Agnar; SIGURDARDÓTTIR, Sigrún; GULCHER, Jeffrey R.; WARD, Ryk; STEFÁNSSON, Kári: mtDNA and the Origin of the Icelanders: Deciphering Signals of Recent Population History. In: American Journal of Human Genetics 66 (2000), 999–1016, DOI: 10.1086/302816.

HELLDÉN, Ulf; BROGAARD, Rannveig: Desertification and Global Climate Change – Little Ice Age Desertification in Iceland? In: Proceedings of the 2nd International Conference on Land Degradation, International Union of Soil Sciences (IUSS), Lund University (1999), 1–29.

HIGGINS, David: British Romanticism and the Global Climate. In: JOHNS-PUTRA, Adeline (ed.): Climate and Literature. Cambridge UK 2019, 128–143.

HIGHWOOD, Ellie J.; STEVENSON, David S.: Atmospheric Impact of the 1783–1784 Laki Eruption: Part II – Climatic Effect of Sulphate Aerosol. In: Atmospheric Chemistry and Physics 3 (2003), 1177–1189, DOI: 10.5194/acp-3-1177-2003.

HILLER, J.: The Moravians in Labrador 1771–1805. In: The Polar Record 15, no. 99 (1971), 839–854.

HINZEN, Klaus-G.; REAMER, Sharon K.: Seismicity, Seismotectonics, and Seismic Hazard in the Northern Rhine Area. In: Special Paper of the Geological Society of America 425 (2007), 225–242, DOI: 10.1130/2007.2425(15).

HJÁLMARSSON, Jón R.: History of Iceland: From the Settlement to the Present Day. Reykjavík 1988.

HJÁLMARSSON, Jón R.: Die Geschichte Islands. Von der Besiedlung zur Gegenwart (translated by KLOES, Gudrun M. H.). Reykjavík ²2009.

HOCHADEL, Oliver: "Hier haben die Wetterableiter unter den Augsburger Gelehrten eine kleine Revolution gemacht." Die Debatte um die Einführung der Blitzableiter in Augsburg (1783–1791). In: Zeitschrift des Historischen Vereins für Schwaben 92 (1999), 139–164.

HOCHADEL, Oliver: Öffentliche Wissenschaft. Elektrizität in der deutschen Aufklärung. Göttingen 2003.

HOCHADEL, Oliver: Blitzableiter. In: JAEGER, Friedrich (ed.): Enzyklopädie der Neuzeit. Vol. 2: Beobachtung–
    Dürre. Stuttgart 2005, 301–304.
HOCHADEL, Oliver: "In Nebula Nebulorum": The Dry Fog of the Summer of 1783 and the Introduction of
    Lightning Rods in the German Empire. In: Transactions of the American Philosophical Society,
    New Series 99, no. 5 (2009), 45–70.
HOFBAUER, Gottfried: Der Vulkan von Oberleinleiter: Spuren eines Maars in der Nördlichen Frankenalb.
    In: Natur und Mensch, Jahresmitteilungen der Naturhistorischen Gesellschaft Nürnberg e.V. (2008),
    69–87.
HOFFMAN, Ronald: Peace and the Peacemakers: The Treaty of 1783. Charlottesville VA 1986.
HOFFMANN, Richard C.: Thoughts on a Connected Fourteenth Century. In: BAUCH, Martin; SCHENK, Gerrit
    Jasper (eds.): Crisis of the Fourteenth Century. Berlin, Boston 2020, 280–288, DOI: 10.1515/
    9783110660784.
HOLMES, Richard: The Age of Wonder: How the Romantic Generation Discovered the Beauty and Terror of
    Science. Paperback ed. London 2009.
HUFTHAMMER, Anne Karin; WALLØE, Lars: Rats Cannot Have Been Intermediate Hosts for Yersinia Pestis
    During Medieval Plague Epidemics in Northern Europe. In: Journal of Archaeological Science
    40 (2013), 1752–1759, DOI: 10.1016/j.jas.2012.12.007.
HUGHES, J. Donald: What is Environmental History? Cambridge UK 2006.
INGIMUNDARSON, Valur: The Rebellious Ally: Iceland, the United States, and the Politics of Empire 1945–2006.
    Dordrecht 2011.
IRWIN, Julie; SMITH, Jenny Leigh: Focus: Disasters, Science, and History. Introduction: On Disaster. In: Isis
    111, no. 1 (2020), 98–103, DOI: 10.1086/707818.
JACOBY, Gordon C.; WORKMAN, Karen W.; D'ARRIGO, Rosanne D.: Laki Eruption of 1783, Tree Rings, and
    Disaster for Northwest Alaska Inuit. In: Quaternary Science Reviews 18, no. 12 (1999), 1365–1371,
    DOI: 10.1016/s0277-3791(98)00112-7.
JACQUES, Eric; MONACO, Carmelo; TAPPONNIER, Paul; TORTORICI, Luigi; WINTER, T.: Faulting and Earthquake
    Triggering during the 1783 Calabria Seismic Sequence. In: Geophysical Journal International 147,
    no. 3 (2001), 499–516, DOI: 10.1046/j.0956-540x.2001.01518.x.
JAKOBSSON, Sveinn P.: Petrology of Recent Basalts of the Eastern Volcanic Zone, Iceland. In: Acta Naturalia
    Islandica 26 (1979), 1–103.
JAKUBOWSKI-TIESSEN, Manfred: Sturmflut 1717. Die Bewältigung einer Naturkatastrophe in der Frühen
    Neuzeit. Munich 1992.
JOHNSTON, Stephen T.; THORKELSON, Derek J.: Continental Flood Basalts: Episodic Magmatism Above Long-
    Lived Hotspots. In: Earth and Planetary Science Letters 175 (2000), 247–256, DOI: 10.1016/S0012-821X
    (99)00293-9.
JOLLEY, David W.; BELL, Brian R.: The North Atlantic Igneous Province: Stratigraphy, Tectonic, Volcanic and
    Magmatic Processes. Geological Society of London Special Publication 197 (2002).
JONES, E.: Climate, Archaeology, History, and the Arthurian Tradition: A Multiple-Source Study of Two Dark-
    Age Puzzles. In: GUNN, Joel D. (ed.): The Years Without Summer: Tracing AD 536 and its Aftermath.
    Oxford 2000, 5–20.
JONES, Philip; BRIFFA, Keith: The "Little Ice Age": Local and Global Perspectives. In: Climatic Change
    48 (2001), 5–8, DOI: 10.1023/A:1005670904293.
JÓNSSON, Guðmundur; MAGNÚSSON, Magnús S. (eds.): Hagskinna. Icelandic Historical Statistics. Reykjavík
    1997.
JÓNSSON, Steingrímur; VALDIMARSSON, Héðinn: Recent Developments in Oceanographic Research in Icelandic
    Waters. In: CASELDINE, Christopher; RUSSELL, Andrew; HARDARDÓTTIR, Jórunn; KNUDSEN, Óskar (eds.):
    Iceland-Modern Processes and Past Environments. Vol. 5. Amsterdam 2005, 79–92.
JULL, Matthew; MCKENZIE, Dan: The Effect of Deglaciation on Mantle Melting beneath Iceland. In: Journal of
    Geophysical Research: Solid Earth 101, no. B10 (October 1996), 21815–21828, DOI: 10.1029/96jb01308.

JUNEJA, Monica; MAUELSHAGEN, Franz: Disasters and Pre-Industrial Societies: Historiographic Trends and Comparative Perspectives. In: The Medieval History Journal 10, no. 1&2 (2007), 1–31, DOI: 10.1177/097194580701000201.

KAPLAN, Steven L.: The Bakers of Paris and the Bread Question, 1700–1775. Durham NC 1996.

KARLSSON, Gunnar: Plague Without Rats: The Case of Fifteenth-Century Iceland. In: Journal of Medieval History 22, no. 3 (1996), 263–284, DOI: 10.1016/S0304-4181(96)00017-6.

KARLSSON, Gunnar: Iceland's 1100 Years. History of a Marginal Society. London 2000a.

KARLSSON, Gunnar: The History of Iceland. Minneapolis MN 2000b.

KARPENKO, Vladimír; NORRIS, John A.: Vitriol in the History of Chemistry. In: Chemické Listy 96 (2002), 997–1005.

KEHRT, Christian: The Wegener Diaries: Scientific Expeditions into the Eternal Ice. In: Virtual exhibition, Environment & Society Portal (Rachel Carson Center for Environment and Society, 2013), DOI: 10.5282/rcc/3877.

KEIDING, Marie; ÁRNADÓTTIR, Thóra; STURKELL, Erik; GEIRSSON, Halldor; LUND, Birthe: Strain Accumulation along an Oblique Plate Boundary: the Reykjanes Peninsula, Southwest Iceland. In: Geophysical Journal International 172, no. 2 (2008), 861–872, DOI: 10.1111/j.1365-246x.2007.03655.x.

KELLOGG, Louise M.; HAGER, Bradford H.; HILST, Rob D. van der: Compositional Stratification in the Deep Mantle. In: Science 283, no. 5409 (1999), 1881–1884.

KELLY, Morgan; Ó GRÁDA, Cormac: The Waning of the Little Ice Age: Climate Change in Early Modern Europe. In: Journal of Interdisciplinary History 44, no. 3 (2014), 301–325.

KELLY, P. M.; SEAR, C. B.: The Formulation of Lamb's Dust Veil Index. In: DEEPAK, Adarsh (ed.): Atmospheric Effects and Potential Climatic Impact of the 1980 Eruptions of Mount St. Helens. Washington DC 1982 (NASA Conference Publication 2240), 293–298.

KEMPE, Michael; ROHR, Christian: Natural Disasters and Their Perception. In: Environment and History 9, no. 2, special issue (May 2003), 123–126.

KEMPE, Michael: Wissenschaft, Theologie, Aufklärung. Johann Jakob Scheuchzer (1672–1733) und die Sintfluttheorie. Epfendorf 2003a.

KEMPE, Michael: Noah's Flood: The Genesis Story and Natural Disasters in Early Modern Times. In: Environment and History 9, no. 2 (May 2003b), 151–171, DOI: 10.3197/096734003129342809.

KEPLINGER, M. L.; SUISSA, L. W.: Toxicity of Fluorine Short-Term Inhalation. In: American Industrial Hygiene Association Journal 29, no. 1 (1968), 10–18, DOI: 10.1080/00028896809342975.

KINDER, Hermann Hilgermann, Werner(eds.): DTV-Atlas Weltgeschichte. Vol. 1: Von den Anfängen bis zur Französischen Revolution. Munich [37]2004.

KINGTON, John A.: Daily Weather Mapping from 1781: A Detailed Synoptic Examination of Weather and Climate During the Decade Leading Up to the French Revolution. In: Climatic Change 3 (1980), 7–36.

KINGTON, John A.: The Weather of the 1780s Over Europe. Cambridge UK 1988.

KITTSTEINER, Heinz-Dieter: Das Gewissen im Gewitter. In: Jahrbuch für Volkskunde 10 (1987), 7–26.

KJAERGAARD, Thorkild: The Danish Revolution 1500–1800: An Ecohistorical Interpretation (translated by HOHNEN, David). Cambridge UK 1994.

KLEEMANN, Katrin: Living in the Time of a Subsurface Revolution: The 1783 Calabrian Earthquake Sequence. In: Environment & Society Portal, Arcadia (Summer 2019a), no. 30. Rachel Carson Center for Environment and Society, DOI: 10.5282/rcc/8767.

KLEEMANN, Katrin: Telling Stories of a Changed Climate: The Laki Fissure Eruption and the Interdisciplinarity of Climate History. In: KLEEMANN, Katrin; OOMEN, Jeroen (eds.): Communicating the Climate: From Knowing Change to Changing Knowledge. RCC Perspectives: Transformations in Environment and Society 4 (2019b), 33–42, DOI: 10.5282/rcc/8822.

KLEEMANN, Katrin: Active Volcanoes, Active Imaginations: Fire-Spitting Mountains and Subterraneous Roars in the German Territories in the Summer of 1783. In: Global Environment 15, no. 3 (2022a), 456–489, DOI: 10.3197/ge.2022.150302.

KLEEMANN, Katrin: Maximum Latewood Density Analysis Solves Long-Standing Mystery between Temperature Reconstructions and Historical Records. In: Paleoceanography and Paleoclimatology 37, no. 4 (2022b), DOI: 10.1029/2022PA004444.

KÖBLER, Gerhard: Historisches Lexikon der deutschen Länder. Die deutschen Territorien vom Mittelalter bis zur Gegenwart. Munich ⁷2007.

KOERNER, Roy M.; FISHER, David: Acid Snow in the Canadian High Arctic. In: Nature 295, no. 5845 (1982), 137–140, DOI: 1038/295137a0.

KOSTICK, Conor; LUDLOW, Francis: The Dating of Volcanic Events and Their Impact upon European Society, 400–800 CE. In: PCA | European Journal of Post Classical Archaeologies 5 (2015), 7–30.

KOZÁK, Jan; ČERMÁK, Vladimir: The Illustrated History of Natural Disasters. New York NY 2010.

KRÄMER, Daniel: "Menschen grasten mit dem Vieh." Die letzte grosse Hungerkrise der Schweiz 1816/17. Basel 2015.

KRAVITZ, Ben; ROBOCK, Alan: Climate Effects of High-Latitude Volcanic Eruptions: Role of the Time of Year. In: Journal of Geophysical Research 116 (2011), D01105, DOI: 10.1029/2010JD014448.

KRIDER, E. Philip: Benjamin Franklin and the First Lightning Conductors. In: Proceedings of the International Commission on History of Meteorology 1, no. 1. (2004), 1–13.

KRIDER, E. Philip: Benjamin Franklin and Lightning Rods. In: Physics Today 59, no. 1 (2006), 42–48, DOI: 10.1063/1.2180176.

KRIEGER, Martin: Der kartographische Blick auf Island zwischen Mittelalter und Neuzeit. In: WALTER, Axel E.; BAUDACH, Frank (eds.): "Das Tor zur Hölle." Island-Karten aus fünf Jahrhunderten. Eutin 2019, 13–26.

KRISTINSDÓTTIR, Kristjana: Afleiðingar Skaftárelda í Suður-Múlasýslu. (The Laki Eruption and Its Consequences in Suður-Múlasýsla). In: GUNNLAUGSSON, Gísli Ágúst; GUÐBERGSSON, Gylfi Már; ÞÓRARINSSON, Sigurður; RAFNSSON, Sveinbjörn; EINARSSON, Þorleifur (eds.): Skaftáreldar 1783–1784: Ritgerðir og heimildir. Reykjavík 1984, 179–186.

KRISTJÁNSSON, Jónas: Eddas and Sagas. Iceland's Medieval Literature (translated by FOOTE, Peter). Reykjavík ⁴2007

KRISTJÁNSSON, Leó; KRISTJÁNSSON, Kristján J.: Amund Helland og ferd hans til Íslands 1881. In: Náttúrufræðingurinn 66, no. 1 (1996), 27–33.

KRÜGER, Tobias: Die Entdeckung der Eiszeiten. Internationale Rezeption und Konsequenzen für das Verständnis der Klimageschichte. Basel 2008.

LACY, Terry G.: Ring of Seasons. Iceland, Its Culture and History. Ann Arbor 1998.

LAMB, Hubert Horace: Volcanic Dust in the Atmosphere; With a Chronology and Assessment of Its Meteorological Significance. In: Philosophical Transactions of the Royal Society of London 266, no. 1178 (1970), 425–533.

LAMB, Hubert Horace: Supplementary Volcanic Dust Veil Assessments. In: Climate Monitor 6 (1977), 57–67.

LAMB, Hubert Horace: Update of the Chronology of Assessments of the Volcanic Dust Veil Index. In: Climate Monitor 12 (1983), 79–90.

LAMB, Hubert Horace: Volcanic Loading: The Dust Veil Index. In: Carbon Dioxide Information Analysis Center (CDIAC) Datasets, 1985, DOI: 10.3334/cdiac/atg.ndp013.

LAMB, Hubert Horace: Climate, History and the Modern World. London ²1995.

LAMB, Hubert Horace: Climate: Present, Past, and Future. London 1972.

LARSEN, Guðrún: Recent Volcanic History of the Veiðivötn Fissure Swarm, Southern Iceland, an Approach to Volcanic Risk Assessment. In: Journal of Volcanology and Geothermal Research 22 (1984), 33–58.

LARSEN, Guðrún: Holocene Eruptions Within the Katla Volcanic System, South Iceland: Characteristics and Environmental Impact. In: Jökull 49 (2000), 1–28.

LARSEN, Guðrún: Katla: Tephrochronology and Eruption History. In: SCHOMACKER, Anders; KRÜGER, Johannes; KJÆR, Kurt (eds.): The Mýrdalsjökull Ice Cap, Iceland: Glacial Processes, Sediments and Landforms on an Active Volcano. Amsterdam 2010, 23–49 (Development in Quaternary Science 13).

LARSEN, Gudrún; THÓRDARSON, Thorvaldur: Gjóskan frá Skaftáreldum 1783. (Tephra Produced by the Laki Eruption). In: GUNNLAUGSSON, Gísli Ágúst; GUÐBERGSSON, Gylfi Már; ÞÓRARINSSON, Sigurður; RAFNSSON, Sveinbjörn; EINARSSON, Þorleifur (eds.): Skaftáreldar 1783–1784: Ritgerðir og heimildir. Reykjavík 1984, 59–66.

LASHER, G. Everett; AXFORD, Yarrow: Medieval Warmth Confirmed at the Norse Eastern Settlement in Greenland. In: Geology 47, no. 3 (June 2019), 267–270, DOI: 10.1130/g45833.1.

LAURING, Palle: A History of the Kingdom of Denmark (translated from Danish by HOHNEN, David). Copenhagen 1960.

LAVIGNE, Franck; DEGEAI, Jean-Philippe; KOMOROWSKI, Jean-Christophe; GUILLET, Sébastien; ROBERT, Vincent; LAHITTE, Pierre; OPPENHEIMER, Clive; et al.: Source of the Great A. D. 1257 Mystery Eruption Unveiled, Samalas Volcano, Rinjani Volcanic Complex, Indonesia. In: Proceedings of the National Academy of Sciences 110, no. 42 (2013), 16742–16747, DOI: 10.1073/pnas.1307520110.

LE PICHON, Xavier: Introduction to the Publication of the Extended Outline of Jason Morgan's April 17, 1967, American Geophysical Union Paper on "Rises, Trenches, Great Faults and Crustal Blocks." In: Tectonophysics 187 (1991), 1–22, DOI: 10.1016/0040-1951(91)90407-J.

LE ROY LADURIE, Emmanuel: Histoire humaine et comparée du Climat. II. Disettes et révolutions (1740–1860). Paris 2006.

LEHMANN, Hartmut: The Interplay of Disenchantment and Re-Enchantment in Modern European History; or, the Origin and the Meaning of Max Weber's Phrase "Die Entzauberung der Welt." In: LEHMANN, Hartmut (ed.): Die Entzauberung der Welt. Studien zu Themen von Max Weber. Göttingen 2009.

LEYDECKER, Günter: Erdbebenkatalog für Deutschland mit Randgebieten 800 bis 2008. Stuttgart 2011.

LIDE, David R.: Handbook of Chemistry and Physics. Boca Raton FL [84]2004.

LIEBERMAN, Victor: Strange Parallels. Southeast Asia in Global Context, c. 800–1830. 2 vols. Cambridge UK 2003–2009.

LINDEMANN, Margot: Deutsche Presse bis 1815. Geschichte der Deutschen Presse. Part I. Berlin 1969.

LÖFFLER, Ulrich: Lissabons Fall – Europas Schrecken. Die Deutung des Erdbebens von Lissabon im deutschsprachigen Protestantismus des 18. Jahrhunderts. Berlin 1999.

LONGO, Bernadette M.; YANG, Wei; GREEN, Joshua B.; CROSBY, Frederick L.; CROSBY, Vickie L.: Acute Health Effects Associated with Exposure to Volcanic Air Pollution (Vog) from Increased Activity at Kīlauea Volcano in 2008. In: Journal of Toxicology and Environmental Health, Part A 73, no. 20 (2010), 1370–1381, DOI: 10.1080/15287394.2010.497440.

LOUGHLIN, Sue C.; ASPINALL, Willy P.; VYE-BROWN, Charlotte; BAXTER, Peter J.; BRABAN, Christine F.; HORT, Matthias; SCHMIDT, Anja; THORDARSON, Thorvaldur; WITHAM, Claire: Large-Magnitude Fissure Eruptions in Iceland: Source Characterisation. In: British Geological Survey Open File Report OR/12/098 (2012).

LÜBKEN, Uwe: Zwischen Alltag und Ausnahmezustand. Ein Überblick über die historiographische Auseinandersetzung mit Naturkatastrophen. In: WerkstattGeschichte 38 (2004), 91–100.

LUCAS, Gavin: The Tensions of Modernity: Skálholt during the 17th and 18th Centuries. In: Journal of the North Atlantic, Special Issue 1 (2009), 75–88.

LUDLUM, David M.: Early American Winters. Vol. 1: Early American Winters 1604–1820. Boston MA 1966.

LUDLUM, David M.: Early American Winters. Vol. 2: Early American Winters 1821–1870. Boston MA 1968.

LUNDIN, Erik R.; DORÉ, Anthony G.: NE Atlantic Break-Up: A Re-Examination of the Iceland Mantle Plume Model and the Atlantic-Arctic Linkage. In: DORÉ, Anthony G.; VINING, B. A. (eds.): Petroleum Geology: North-West Europe and Global Perspectives – Proceedings of the 6th Petroleum Geology Conference. London 2005, 739–754.

LUTERBACHER, Jürg: The Late Maunder Minimum (1675–1715). Climax of the 'Little Ice Age' in Europe. In: JONES, Phil D.; OGILVIE, Astrid E. J.; DAVIES, T. D.; BRIFFA, Keith R. (eds.): History and Climate: Memories of the Future? New York NY 2001, 29–54.

LUTERBACHER, Jürg; DIETRICH, D.; XOPLAKI, Elena; GROSJEAN, Martin; WANNER, Heinz: European Seasonal and Annual Temperature Variability, Trends, and Extremes Since 1500. In: Science 303, no. 5663 (2004), 1499–1503, DOI: 10.1126/ science.1093877.

LUTERBACHER, Jürg; WERNER, Johannes P.; SMERDON, Jason E.; FERNÁNDEZ-DONADO, Laura; GONZÁLEZ-ROUCO, F. J.; BARRIOPEDRO, David; et al.: European Summer Temperatures since Roman Times. In: Environmental Research Letters 11, no. 2 (2016), DOI: 10.1088/1748-9326/11/2/024001.

LUTZ, Herbert; LORENZ, Volker: Early Volcanological Research in the Vulkaneifel, Germany, the Classic Region of Maar-Diatreme Volcanoes: The Years 1774–1865. In: Bulletin of Volcanology 75, no. 8 (2013), 743, DOI: 10.1007/s00445-013-0743-0.

LYNN, Michael R.: The Sublime Invention: Ballooning in Europe, 1783–1820. London 2010.

MABEY, Richard: Gilbert White. A Biography of the Author of the Natural History of Selbourne. Charlottesville VA 2006.

MACLENNAN, John; JULL, Matthew; MCKENZIE, Dan; SLATER, L.; GRÖNVOLD, Karl: The Link Between Volcanism and Deglaciation in Iceland. In: Geochemistry, Geophysics, Geosystems 3, no. 11 (2002), 1–25, DOI: 10.1029/2001gc000282.

MAGNÚSSON, Sigurður Gylfi: Wasteland with Words. A Social History of Iceland. London 2010.

MAIER, Gerhard: African Dinosaurs Unearthed. The Tendaguru Expeditions. Bloomington, Indianapolis IN 2003.

MALBERG, Horst: Meteorologie und Klimatologie. Eine Einführung. Berlin [5]2007.

MALILA, Jussi: On the Early Studies Recognizing the Role of Sulphuric Acid in Atmospheric Haze and New Particle Formation. In: Tellus B: Chemical and Physical Meteorology 70, no. 1 (2018), 1–11, DOI: 10.1080/16000889.2018.1471913.

MANLEY, Gordon: Central England Temperatures: Monthly Means 1659 to 1973. In: Quarterly Journal of the Royal Meteorological Society 100 (1974), 389–405.

MANN, Michael: The Little Ice Age. In: MACCRACKEN, Michel C.; PERRY, John S. (eds.): Encyclopedia of Global Environmental Change. Hoboken NJ 2002, 504–509.

MANNING, Joseph G.; LUDLOW, Francis; STINE, Alexander R.; BOOS, William R.; SIGL, Michael; MARLON, Jennifer R.: Volcanic Suppression of Nile Summer Flooding Triggers Revolt and Constrains Interstate Conflict in Ancient Egypt. In: Nature Communications, 8, no. 1 (2017), DOI: 10.1038/s41467-017-00957-y.

MARSHALL, Lauren R.; MATERS, Elena C; SCHMIDT, Anja; TIMMRECK, Claudia; ROBOCK, Alan; TOOHEY, Matthew: Volcanic Effects on Climate: Recent Advances and Future Avenues. In: Bulletin of Volcanology 84 (2022), 54, DOI: 10.1007/s00445-022-01559-3.

MARTOS, Yasmina M.; JORDAN, Tom A.; CATALÁN, Manuel; JORDAN, Thomas M.; BAMBER, Jonathan L.; VAUGHAN, David G.: Geothermal Heat Flux Reveals the Iceland Hotspot Track Underneath Greenland. In: Geophysical Research Letters 45 (2018), 8214–8222, DOI: 10.1029/2018GL078289.

MATHER, Tamsin A.; SCHMIDT, Anja: Environmental Effects of Volcanic Volatile Fluxes From Subaerial Large Igneous Provinces. In: ERNST, Richard E.; DICKSON, Alex J.; BEKKER, Andrey (eds.): Large Igneous Provinces. Hoboken NJ 2021, 103–116, DOI: 10.1002/9781119507444.ch4.

MATTHES, François E.: Glaciers. In: MEINZER, Oscar E. (ed.): Physics of the Earth: IX. Hydrology. New York NY 1949.

MATTHEWS, John A.; BRIFFA, Keith R.: The Little Ice Age: Re-Evaluation of an Evolving Cencept. In: Geografiska Annaler: Series A, Physical Geography 87, no. 1 (2005), 17–36, DOI: 10.1111/j.0435-3676.2005.00242.x.

MATTSON, Hannes; HÖSKULDSSON, Ármann: Geology of the Heimaey Volcanic Center, South Iceland: Early Evolution of a Central Volcano in a Propagating Rift? In: Journal of Volcanology and Geothermal Research 127 (2003), 55–71, DOI: 10.1016/S0377-0273(03)00178-1.

MAUCH, Christof: Introduction. In: MAUCH, Christof; PFISTER, Christian (eds.): Natural Disasters, Cultural Responses. Case Studies Toward a Global Environmental History. Lanham MD 2009, 1–16.

MAUCH, Christof: Slow Hope. Rethinking Ecologies of Crisis and Fear. In: RCC Perspectives: Transformations in Environment and Society 1 (2019), 1–43, DOI: 10.5282/rcc/8556.

MAUELSHAGEN, Franz: Netzwerke des Nachrichtenaustauschs. Für einen Paradigmenwechsel in der Erforschung der "neuen Zeitung." In: BURKHARDT, Johannes; WERKSTETTER, Christine (eds.): Kommunikation und Medien in der Frühen Neuzeit. Munich 2005, 409–432. (Historische Zeitschrift Beiheft 41).

MAUELSHAGEN, Franz: Klimageschichte der Neuzeit. Darmstadt 2010.

MAUELSHAGEN, Franz: Historische Klimaforschung: Ursprünge, Trends und Zukunftsperspektiven eines interdisziplinären Forschungsfeldes. In: Frühneuzeit-Info 28 (2017), 56–75.

MÄUSSNEST, Otto: Die isländische Vulkaneruption 1783 und das Geheimnis des Hahlrauches. In: Photorin. Mitteilungen der Lichtenberg-Gesellschaft 9 (1983), 53–59.

MAYEWSKI, Paul A.; LYONS, W. Berry; SPENCER, Mary Jo; TWICKLER, Mark S.; BUCK, Christopher F.; WHITLOW, Sallie: An Ice-Core Record of Atmospheric Response to Anthropogenic Sulphate and Nitrate. In: Nature 346 (1990), 554–556, DOI: 10.1038/346554a0.

McCALLAM, David: Un météore inédit: les brouillards secs de 1783. In: BELLEGUIC, Thierry (ed.): Ordre et désordre du Monde. Enquête sur les météores, de la Renaissance à l'âge moderne. Paris 2013, 369–388.

McCALLAM, David: Volcanoes in Eighteenth-Century Europe: An Essay in Environmental Humanities. Liverpool 2019 (Oxford University Studies in the Enlightenment 19).

McCLOY, Shelby T.: Flood Relief and Control in 18th Century France. In: Journal of Modern History 13 (1941), 7–12.

McCONNELL, Joseph R.; SIGL, Michael; PLUNKETT, Gill; BURKE, Andrea; KIM, Woon Mi; RAIBLE, Christoph C.; WILSON, Andrew I.; MANNING, Joseph G.; LUDLOW, Francis; CHELLMAN, Nathan J.; INNES, Helen M.; YANG, Zhen; LARSEN, Jessica F.; SCHAEFER, Janet R.; KIPFSTUHL, Sepp; MOJTABAVI, Seyedhamidreza; WILHELMS, Frank; OPEL, Thomas; MEYER, Hanno; STEFFENSEN, Jørgen Peder: Extreme Climate after Massive Eruption of Alaska's Okmok Volcano in 43 BCE and Effects on the Late Roman Republic and Ptolemaic Kingdom. In: Proceedings of the National Academy of Sciences 117, no. 27 (2020), 15443–15449, DOI: 10.1073/pnas.2002722117.

McCORMICK, M. Patrick; THOMASON, Larry W.; TREPTE, Charles R.: Atmospheric Effects of the Mt Pinatubo Eruption. In: Nature 373, no. 6513 (1995), 399–404, DOI: 10.1038/373399a0.

McCORMICK, Michael: Climates of History, Histories of Climate: From History to Archaeoscience. In: The Journal of Interdisciplinary History 50, no. 1 (2019), 3–30, DOI: 10.1162/jinh_a_01374.

McCORMICK, Michael; DUTTON, Paul Edward; MAYEWSKI, Paul A.: Volcanoes and the Climate Forcing of Carolingian Europe, A.D. 750–950. In: Speculum 82 (2007), 865–895, DOI: 10.1017/S0038713400011325.

MCKENZIE, Dan P.; PARKER, R. L.: The North Pacific: an Example of Tectonics on a Sphere. In: Nature 216, no. 5122 (1967), 1276–1280, DOI: 10.1038/2161276a0.

McLUHAN, Marshall: The Gutenberg Galaxy: The Making of Typographic Man. Toronto 1962.

McNEILL, John R.: The State of the Field of Environmental History. In: Annual Review of Environment and Resources 35, no. 1 (2010), 345–374, DOI: 10.1146/annurev-environ-040609-105431.

MEIDOW, Hein: Comparison of the Macroseismic Field of the 1992 Roermond Earthquake, the Netherlands, With Those of Large Historical Earthquakes in the Lower Rhine Embayment and its Vicinity. In: Geologie en Mijnbouw | Netherlands Journal of Geosciences 73, no. 2–4 (1994), 282–289.

MENELY, Tobias: "The Present Obfuscation": Cowper's Task and the Time of Climate Change. In: The Modern Language Association of America 127, no. 3 (2012), 477–492, DOI: 10.2307/41616841.

METZKE, Hermann: Lexikon der historischen Krankheitsbezeichnungen. Neustadt an der Aisch 2005.

MEYER, Beat: Medical Use and Health Effects. In: RICHTER, E. (ed.): Sulfur, Energy, and Environment. Amsterdam 1977, 242–258, DOI: 10.1016/B978-0-444-41595-0.50013-9.

MIKAMI, Takehiko; TSUKAMURA, Yasufumi: The Climate of Japan in 1816 as Compared With an Extremely Cool Summer Climate in 1783. In: HARINGTON, C. Richard (ed.): The Year Without a Summer? World Climate in 1816. Ottawa 1992, 462–476.

MIKHAIL, Alan: Ottoman Iceland: A Climate History. In: Environmental History 20, no. 2 (2015), 262–284, DOI: 10.1093/envhis/emv006.

MIKHAIL, Alan: Under Osman's Tree. The Ottoman Empire, Egypt, and Environmental History. Chicago IL 2017.

MILES, G. M.; GRAINGER, Roy G.; HIGHWOOD, Ellie J.: Volcanic Aerosols. The Significance of Volcanic Eruption Strength and Frequence For Climate. In: Quarterly Journal of the Royal Meteorological Society 128 (2003), 1–16, DOI: 10.1256/qj.03.60.

MILLER, Gifford H.; GEIRSDÓTTIR, Áslaug; ZHONG, Yafang; LARSEN, Darren J.; OTTO-BLIESNER, Bette L.; HOLLAND, Marika M.; BAILEY, David A. et al.: Abrupt Onset of the Little Ice Age Triggered by Volcanism and Sustained by Sea-Ice/Ocean Feedbacks. In: Geophysical Research Letters 39, no. 2 (2012), DOI: 10.1029/2011gl050168.

MISSFELDER, Jan-Friedrich: Donner und Donnerwort. Zur akustischen Wahrnehmung der Natur im 18. Jahrhundert. In: RUPPEL, Sophie; STEINBRECHER, Aline (eds.): "Die Natur ist überall bey uns." Mensch und Natur in der Frühen Neuzeit. Zurich 2009, 81–94.

MIX, York-Gothart: Schreiben, lesen und gelesen werden. Zur Kulturökonomie des literarischen Feldes (1700–1800). In: WOLFGANG, Adam; FAUSER, Markus (eds.): Geselligkeit und Bibliothek. Lesekultur im 18. Jahrhundert. Göttingen 2005, 283–310 (Schriften des Gleimhauses Halberstadt 4).

MORGAN, W. Jason: Rises, Trenches, Great Faults, and Crustal Blocks. In: Journal of Geophysical Research 73, no. 6 (1968), 1959–1982, DOI: 10.1029/jb073i006p01959.

MORGAN, W. Jason: Convection Plumes in the Lower Mantle. In: Nature 230 (1971), 42–43.

MORGAN, W. Jason: Plate Motions and Deep Mantle Convection. In: The American Association of Petroleum Geologists Bulletin 56, no. 2 (1972), 203–213.

MOSER, Susanne C.; FINZI HART, Juliette A.: The Long Arm of Climate Change: Societal Teleconnections and the Future of Climate Change Impact Studies. In: Climatic Change 129 (2015), 13–26, DOI: 10.1007/s10584-015-1328-z.

MÜNCH, Paul: Lebensformen in der Frühen Neuzeit 1500 bis 1800. Frankfurt am Main 1992.

MUNZAR, Jan; ELLEDER, Libor; DEUTSCH, Mathias: The Catastrophic Flood in February/March 1784 –A Natural Disaster of European Scope. In: Moravian Geographical Reports 13, no. 1 (2005), 8–24.

MYLLYNTAUS, Timo: Summer Frost. A Natural Hazard with Fatal Consequences in Pre-Industrial Finland. In: MAUCH, Christof; PFISTER, Christian (eds.): Natural Disasters, Cultural Responses. Case Studies Toward A Global Environmental History. Lanham MD 2009, 77–102.

NEUKOM, Raphael; STEIGER, Nathan; GÓMEZ-NAVARRO, Juan José: No Evidence for Globally Coherent Warm and Cold Periods Over the Preindustrial Common Era. In: Nature 571 (2019), 550–554, DOI: 10.1038/s41586-019-1401-2.

NEWFIELD, Timothy: The Climate Downturn of 536–550. In: WHITE, Sam; PFISTER, Christian; MAUELSHAGEN, Franz (eds.): The Palgrave Handbook of Climate History. London 2018, 447–493.

NEWHALL, Christopher G.; SELF, Stephen: The Volcanic Explosivity Index (VEI): An Estimate of Explosive Magnitude for Historical Volcanism. In: Journal of Geophysical Research: Oceans 87, no. C2 (1982), 1231–1238.

OESER, Erhard: Historische Erdbebentheorien von der Antike bis zum Ende des 19. Jahrhunderts. Vienna 2003.

OGILVIE, Astrid E. J.: The Past Climate and Sea-Ice Record from Iceland. Part 1: Data to A.D. 1780. In: Climatic Change 6 (1984), 131–152.

OGILVIE, Astrid E. J.: The Climate of Iceland 1701–1784. In: Jökull 36 (1986), 57–73.

OGILVIE, Astrid E. J.: Climatic Changes in Iceland A.D. c. 1500 to 1598. In: Acta Archaeologica 61 (1991), 233–251.

OGILVIE, Astrid E. J.: Documentary Evidence for Changes in the Climate of Iceland, A.D. 1500 to 1800. In: BRADLEY, Raymond S.; JONES, P. D. (eds.): Climate Since A.D. 1500. London, New York NY 1992, 92–117.

OGILVIE, Astrid E. J.: Climate and Farming in Northern Iceland, 1700–1850. In: SIGURÐSSON, I.; SKAPTASON, J. (eds.): Aspects of Arctic and Sub-Arctic History. Reykjavík 2001, 289–299.

OGILVIE, Astrid E. J.: Local Knowledge and Travellers' Tales: A Selection of Climatic Observations in Iceland. In: CASELDINE, Christopher; RUSSELL, Andrew; HARDARDÓTTIR, Jórunn; KNUDSEN, Óskar (eds.): Iceland-Modern Processes and Past Environments. Vol. 5. Amsterdam 2005, 257–288.

OGILVIE, Astrid E. J.; BARLOW, Lisa Katharine; JENNINGS, Anne E.: North Atlantic Climate c. AD 1000: Millennial Reflections on the Viking Discoveries of Iceland, Greenland, and North America. In: Weather 55 (2000), 34–45, DOI: 10.1002/j.1477-8696.2000.tb04028.x.

OGILVIE, Astrid E. J.; JÓNSDÓTTIR, Ingibjörg: Sea Ice, Climate, and Icelandic Fisheries in Historical Times. In: Arctic 52 (2000), 383–394, DOI: 10.14430/arctic869.

OGILVIE, Astrid E. J.; JÓNSSON, Trausti: "Little Ice Age" Research: A Perspective from Iceland. In: OGILVIE, Astrid E. J.; JÓNSSON, Trausti (eds.): The Iceberg in the Mist: Northern Research in Pursuit of a "Little Ice Age." Dordrecht 2000, 9–52.

OGILVIE, Astrid E. J.; WOOLLETT, James M.; SMIAROWSKI, Konrad; ARNEBORG, Jette; TROELSTRA, Simon; KUIJPERS, Antoon; PÁLSDÓTTIR, Albina; MCGOVERN, Thomas H.: Seals and Sea Ice in Medieval Greenland. In: Journal of the North Atlantic 2 (2009), 60–80, DOI: 10.3721/037.002.0107.

ÓLADÓTTIR, Bergrún Arna; LARSEN, Guðrún; SIGMARSSON, Olgeir; THORDARSON, Thorvaldur: The Katla Volcano S-Iceland: Holocene Tephra Stratigraphy and Eruption Frequency. In: Jökull 55 (2005), 53–74.

ÓLADÓTTIR, Bergrún Arna; SIGMARSSON, Olgeir; LARSEN, Guðrún; THORDARSON, Thorvaldur: Katla Volcano, Iceland: Magma Composition, Dynamics and Eruption Frequency as Recorded by Tephra Layers. In: Bulletin of Volcanology 70 (2008), 475–493, DOI: 10.1007/s00445-007-0150-5.

OLDROYD, David: Maps as Pictures or Diagrams: The Early Development of Geological Maps. In: BAKER, Victor R. (ed.): Rethinking the Fabric of Geology. Geological Society of America Special Paper 502 (2013), 41–101.

OLIVER-SMITH, Anthony: Theorizing Disaster, Nature, Power, and Culture. In: HOFFMAN, Susanna M.; OLIVER-SMITH, Anthony (eds.): Catastrophe & Culture. The Anthropology of Disaster. Santa Fe NM 2002, 23–47.

OLWIG, Kenneth Robert: Nature's Ideological Landscape. London 1984.

OMAN, Luke; ROBOCK, Alan; STENCHIKOV, Georgiy L.; SCHMIDT, Gavin A.; RUEDY, R.: Climatic Response to High-Latitude Volcanic Eruptions. In: Journal of Geophysical Research 110 (2005), D13103, DOI: 10.1029/2004JD005487.

OMAN, Luke; ROBOCK, Alan; STENCHIKOV, Georgiy L.; THORDARSON, Thorvaldur: High-Latitude Eruptions Cast Shadow over the African Monsoon and the Flow of the Nile. In: Geophysical Research Letters 33, no. 18 (2006a), DOI: 10.1029/2006gl027665.

OMAN, Luke; ROBOCK, Alan; STENCHIKOV, Georgiy L.; THORDARSON, Thorvaldur; KOCH, Dorothy; SHINDELL, Drew T.; GAO, Chaochao: Modeling the Distribution of the Volcanic Aerosol Cloud from the 1783–1784 Laki Eruption. In: Journal of Geophysical Research 111, no. D12 (2006b), DOI: 10.1029/2005jd006899.

OPPENHEIMER, Clive: Eruptions That Shook the World. Cambridge UK 2011.

OPPENHEIMER, Clive; ORCHARD, Andy; STOFFEL, Markus; NEWFIELD, Timothy P.; GUILLET, Sébastien; CORONA, Christophe; SIGL, Michael; DI COSMO, Nicola; BÜNTGEN, Ulf: The Eldgjá Eruption: Timing, Long-Range Impacts and Influence on the Christianisation of Iceland. In: Climatic Change 147, no. 3–4 (2018), 369–381, DOI: 10.1007/s10584-018-2171-9.

OPPENHEIMER, Clive; PYLE, David M.; BARCLAY, Jennie (eds.): Volcanic Degassing. Geological Society of London Special Publication 213 (2003).

OQUILLUK, William A.: People of Kauwerak: Legends of the Northern Eskimo. Anchorage AK 1973.

ORESKES, Naomi (ed.): Plate Tectonics. An Insider's History of the Modern Theory of the Earth. Boulder CO 2003.

ORTLIEB, Luc: Historical Chronology of ENSO and the Nile Flood Record. In: BATTARBEE, Richard W.; GASSE, Françoise; STICKLEY, Catherine E. (eds.): Past Climate Variability Through Europe and Africa. Dordrecht 2004, 257–278.

OSLUND, Karen: Iceland Imagined: Nature, Culture, and Storytelling in the North Atlantic. Seattle 2011.

PAGLI, Carolina; SIGMUNDSSON, Freysteinn: Will Present Day Glacier Retreat Increase Volcanic Activity? Stress Induced by Recent Glacier Retreat and Its Effect on Magmatism at the Vatnajökull Ice Cap, Iceland. In: Geophysical Research Letters 35, no. 9 (July 2008), L09304, DOI: 10.1029/2008gl033510.

PAPPERT, Duncan; BRUGNARA, Yuri; JOURDAIN, Sylvie; POSPIESZYŃSKA, Aleksandra; PRZYBYLAK, Rajmund; ROHR, Christian; BRÖNNIMANN, Stefan: Unlocking Weather Observations from the Societas Meteorologica Palatina (1781–1792). In: Climate of the Past 17 (2021), 2361–2379, DOI: 10.5194/cp-17-2361-2021.

PARK, Graham: Introduction Geology. A Guide to the World of Rocks. Edinburgh ^{2}2010.

PARKER, David E.; LEGG, T. P.; FOLLAND, Chris K.: A New Daily Central England Temperature Series, 1772–1991. In: International Journal of Climatology 12, no. 4 (1992), 317–342, DOI: 10.1002/joc.3370120402.

PARRINELLO, Giacomo: Fault Lines: Earthquakes and Urbanism in Modern Italy. New York NY 2015 (The Environment in History – International Perspectives 6).

PASSMORE, Emma; MACLENNAN, John; FITTON, Godfrey; THORDARSON, Thor: Mush Disaggregation in Basaltic Magma Chambers:Evidence from the AD 1783 Laki Eruption. In: Journal of Petrology 53, no. 12 (2012), 2593–2623, DOI: 10.1093/petrology/egs061.

PAULING, Andreas; LUTERBACHER, Jürg; WANNER, Heinz: Evaluation of Proxies for European and North Atlantic Temperature Field Reconstructions. In: Geophysical Research Letters 30, no. 15 (2003), DOI: 10.1029/2003gl017589.

PAUSATA, Francesco S. R.; CHAFIK, Leon; CABALLERO, Rodrigo; BATTISTI, David S.: Impacts of High-Latitude Volcanic Eruptions on ENSO and AMOC. In: Proceedings of the National Academy of Sciences 112, no. 45 (2015), 13784–13788, DOI: 10.1073/pnas.1509153112.

PAUSATA, Francesco S. R.; KARAMPERIDOU, Christina; CABALLERO, Rodrigo; BATTISTI, David S.: ENSO Response to High-Latitude Volcanic Eruptions in the Northern Hemisphere: The Role of the Initial Conditions. In: Geophysical Research Letters 43, no. 16 (2016), 8694–8702, DOI: 10.1002/2016gl069575.

PAYNE, Richard J.: The "Meteorological Imaginations and Conjectures" of Benjamin Franklin. In: North West Geography 10, no. 2 (2010), 1–7.

PAYNE, Richard J.: Meteors and Perceptions of Environmental Change in the Annus Mirabilis AD 1783–4. In: North West Geography 11, no. 1 (2011), 19–28.

PELLY, David F.: Scared Hunt. A Portrait of the Relationship Between Seals and Inuit. Vancouver 2001.

PETERSEN, Janet; FISHER, G. W.; TIMPANY, G.: Survey of Background Hydrogen Sulphide in Rotorua – 1996. Prepared for the Ministry of the Environment, Bay of Plenty. NIWA Report AK96058.

PÉTURSSON, Guðmundur; PÁLSSON, Páll A.; GEORGSSON, Guðmundur: Um eituráhrif af völdum Skaftárelda. (Contamination and Diseases Following Volcanic Eruptions in Iceland). In: GUNNLAUGSSON, Gísli Ágúst; GUÐBERGSSON, Gylfi Már; ÞÓRARINSSON, Sigurður; RAFNSSON, Sveinbjörn; EINARSSON, Þorleifur (eds.): Skaftáreldar 1783–1784: Ritgerðir og heimildir. Reykjavík 1984, 81–98.

PFISTER, Christian: Die Lufttrübungserscheinung des Sommers 1783 in der Sicht schweizerischer Beobachter. In: Informationen und Beiträge zur Klimaforschung / Contributions à la recherche climatologique 7 (1972), 23–29.

PFISTER, Christian: Agrarkonjunktur und Witterungsverlauf im westlichen Schweizer Mittelland 1755–1797. Bern 1975 (Beiheft 2 zum Jahrbuch der Geographischen Gesellschaft von Bern).

PFISTER, Christian: Das Klima der Schweiz von 1525–1860 und seine Bedeutung in der Geschichte der Bevölkerung und Landwirtschaft. Bern 1984.

PFISTER, Christian: Wetternachhersage. 500 Jahre Klimavariationen und Naturkatastrophen. Bern 1999.

PFISTER, Christian: Strategien zur Bewältigung von Naturkatastrophen seit 1500. In: PFISTER, Christian (ed.): Am Tag danach. Zur Bewältigung von Naturkatastrophen in der Schweiz 1500–2000. Bern 2002, 209–254.

PFISTER, Christian: Evidence from the Archives of Societies: Documentary Evidence – Overview. In: WHITE, Sam; PFISTER, Christian; MAUELSHAGEN, Franz (eds.): The Palgrave Handbook of Climate History. London 2018a, 37–47.

PFISTER, Christian: Evidence from the Archives of Societies: Institutional Sources. In: WHITE, Sam; PFISTER, Christian; MAUELSHAGEN, Franz (eds.): The Palgrave Handbook of Climate History. London 2018b, 67–81.

PFISTER, Christian; BRÁZDIL, Rudolf: Social Vulnerability to Climate in the "Little Ice Age": An Example from Central Europe in the Early 1770s. In: Climate of the Past 2 (2006), 115–129, DOI: 10.5194/cp-2-115-2006.

PFISTER, Christian; GARNIER, Emmanuel; ALCOFORADO, Marie-João; WHEELER, Dennis; LUTERBACHER, Jürg; NUNES, Maria; FATIMA, Taborda; PAULI, João: The Meteorological Framework and the Cultural Memory of Three Severe Winter-Storms in Early Eighteenth-Century Europe. In: Climatic Chance 101 (2010), 304–305, DOI: 10.1007/s10584-009-9784-y.

PFISTER, Christian; BRÁZDIL, Rudolf; LUTERBACHER, Jürg; OGILVIE, Astrid E. J.; WHITE, Sam: Early Modern Europe. In: WHITE, Sam; PFISTER, Christian; MAUELSHAGEN, Franz (eds.): The Palgrave Handbook of Climate History. London 2018, 265–295.

PFISTER, Christian; WHITE, Sam: A Year Without a Summer, 1816. In: WHITE, Sam; PFISTER, Christian; MAUELSHAGEN, Franz (eds.): The Palgrave Handbook of Climate History. London 2018a, 551–561.

PFISTER, Christian; WHITE, Sam: Evidence from the Archives of Society: Personal Documentary Sources. In: WHITE, Sam; PFISTER, Christian; MAUELSHAGEN, Franz (eds.): The Palgrave Handbook of Climate History. London 2018b, 49–65.

PFISTER, Christian; WHITE, Sam; MAUELSHAGEN, Franz: General Introduction: Weather, Climate, and Human History. In: WHITE, Sam; PFISTER, Christian; MAUELSHAGEN, Franz (eds.): The Palgrave Handbook of Climate History. London 2018, 1–17.

PFISTER, Lucas; HUPFER, Franziska; BRUGNARA, Yuri; MUNZ, Lukas; VILLIGER, Leonie; MEYER, Lukas; SCHWANDER, Mikhaël; ISOTTA, Francesco Alessandro; ROHR, Christian; BRÖNNIMANN, Stefan: Early Instrumental Meteorological Measurements in Switzerland. In: Climate of the Past 15, no. 4 (2019), 1345–1361, DOI: 10.5194/cp-15-1345-2019.

PIPER, Liza: Colloquial Meteorology. In: MACEACHERN, Alan; TURKEL, Willian J. (eds.): Method and Meaning in Canadian Environmental History. Toronto 2009, 102–123.

PÍSEK, Jan; BRÁZDIL, Rudolf: Responses of Large Volcanic Eruptions in the Instrumental and Documentary Climatic Data Over Central Europe. In: International Journal of Climatology 26, no. 4 (2006), 439–459, DOI: 10.1002/joc.1249.

PLACANICA, Augusto: Il filosofo e la catastrofe. Un terremoto del Settecento. Torino 1985.

POLIWODA, Guido: Aus Katastrophen lernen. Sachsen im Kampf gegen die Fluten der Elbe 1784 bis 1856. Cologne, Weimar, Vienna 2007.

POPE, C. Arden; DOCKERY, Douglas W.; SCHWARTZ, Joel: Review of Epidemiological Evidence of Health Effects of Particulate Air Pollution. In: Inhalation Toxicology 7, no. 1 (1995), 1–18, DOI: 10.3109/08958379509014267.

PORTER, Roy: The Making of Geology: Earth Science in Britain 1660–1815. Cambridge UK 1977.

PRINCE, Cathryn J.: A Professor, a President, and a Meteor: The Birth of American Science. Amherst NY 2010.

PROSS, Wolfgang; PRIESNER, Claus: Lichtenberg, Georg Christoph. In: Neue Deutsche Biographie (NDB). Vol. 14, 449–464. Berlin 1985.

PRZYBYLAK, Rajmund; POSPIESZYŃSKA, A.; WYSZYŃSKI, P.; NOWAKOWSKI, M.: Air Temperature Changes in Żagań (Poland) in the Period from 1781 to 1792. In: International Journal of Climatology 34, no. 7 (2014), 2408–2426, DOI: 10.1002/joc.3847.

PUFFAHRT, Otto: Historische und neuzeitliche Hochwassergeschehnisse im Raum Hitzacker, Band 2 zum Hochwasserschutz für Hitzacker und die Jeetzelniederung. Dannenberg 2008.

PUSCHNER, Uwe: Lesegesellschaften. In: SÖSEMANN, Bernd (ed.): Kommunikation und Medien in Preußen vom 16. bis zum 19. Jahrhundert. Stuttgart 2002, 193–206.

PYLE, David M.: Volcanoes. Encounters Through the Ages. Oxford 2017.

PYNE, Stephen J.: Awful Splendour. A Fire History of Canada. Vancouver 2007.

QUENET, Grégory: Les Tremblements de Terre aux XVIIe et XVIIIe siècles. La Naissance d'un Risque. Seyssel 2005.

RABARTIN, Roland; ROCHER, Philippe: Les Volcans, le climat et la Révolution Française. In: Mémoire de l'Association Volcanologique Européenne (L.A.V.E.) 1 (1993).

RAFNSSON, Sveinbjörn: Búfé og byggd við lok Skaftárelda og Móðuharðinda. (Livestock and Settlement in Iceland at the End of the Laki Eruption). In: GUNNLAUGSSON, Gísli Ágúst; GUÐBERGSSON, Gylfi Már; ÞÓRARINSSON, Sigurður; RAFNSSON, Sveinbjörn; EINARSSON, Þorleifur (eds.): Skaftáreldar 1783–1784: Ritgerðir og heimildir. Reykjavík 1984a, 163–178.

RAFNSSON, Sveinbjörn: Um eldritin 1783–1788. (Works on the Laki Eruption 1783–1784). In: GUNNLAUGSSON, Gísli Ágúst; GUÐBERGSSON, Gylfi Már; ÞÓRARINSSON, Sigurður; RAFNSSON, Sveinbjörn; EINARSSON, Þorleifur (eds.): Skaftáreldar 1783–1784: Ritgerðir og heimildir. Reykjavík 1984b, 243–264.

RAPPAPORT, Rhoda: Borrowed Words: Problems of Vocabulary in Eighteenth-Century Geology. In: The British Journal for the History of Science 15, no. 1 (1982), 27–44, DOI: 10.1017/s0007087400018926.

RAPPAPORT, Rhoda: Dangerous Words: Diluvialism, Neptunism, Catastrophism. In: HEILBRON, John L. (ed.): Advancements of Learning: Essays in Honour of Paolo Rossi. Florence 2007, 101–131.

RAUDKIVI, Priit: Islandi 1783. Aasta Vulkaanipurske Võimalikust möjust eestis. Keskkonnaajalooline arutlus. In: Acta Historica Tallinnensia 20 (2014), 51–73, DOI: 10.3176/hist.2014.1.02.

RAUDKIVI, Priit: Die Erde unter einer Aschewolke. Der Vulkanismus als historischer Faktor. In: LAUR, Mati; BRÜGGEMANN, Karsten (eds.): Forschungen zur Baltischen Geschichte II. Tartu 2016, 181–196.

REINHARDT, Olaf; OLDROYD, David Roger: Kant's Theory of Earthquakes and Volcanic Action. In: Annals of Science 40, no. 3 (1983), 247–272.

REITH, Reinhold: Umweltgeschichte der Frühen Neuzeit. Munich 2011.

RICHARDS, John F.: The Unending Frontier. An Environmental History of the Early Modern World. Berkeley CA 2003.

RICKERS, Florian; FICHTNER, Andreas; TRAMPERT, Jeannot: The Iceland–Jan Mayen Plume System and Its Impact on Mantle Dynamics in the North Atlantic Region: Evidence from Full-Waveform Inversion. In: Earth and Planetary Science Letters 367 (2013), 39–51, DOI: 10.1016/j.epsl.2013.02.022.

RIEDE, Felix: Splendid Isolation: The Eruption of the Laacher See Volcano & Southern Scandinavian Late Glacial Hunter-Gatherers. Aarhus 2017.

RIGBY, Kate: Dancing with Disaster: Environmental Histories, Narratives, and Ethics for Perilous Times. Charlottesville, VA 2015.

RISKIN, Jessica: The Lawyer and the Lightning Rod. In: Science in Context 12, no. 1 (1999), 61–99.

RISKIN, Jessica: Science in the Age of Sensibility: The Sentimental Empiricists of the French Enlightenment. Chicago IL 2002.

ROBERTSDÓTTIR, Hrefna: Wool and Society: Manufacturing Policy, Economic Thought and Local Production in 18th-century Iceland. Lund 2008.

ROBOCK, Alan: Volcanic Eruptions and Climate. In: Reviews of Geophysics 38, no. 2 (2000), 191–219, DOI: 10.1029/1998rg000054.

ROBOCK, Alan; OPPENHEIMER, Clive: Volcanism and the Earth's Atmosphere. Washington DC 2003, DOI: 10.1029/gm139.

RÖGNAVALDARDÓTTIR, Nanna: Icelandic Food and Cookery. New York NY 2002.

ROHR, Christian: Naturkatastrophen. In: JAEGER, Friedrich (ed.): Enzyklopädie der Neuzeit. Vol. 9: Naturhaushalt-Physiokratie. Stuttgart 2009, 10–28.

ROHR, Christian: Von Plinius zu Isidor und Beda Venerabilis. Zur Übernahme antiken Wissens über Witterungsphänomene im Mittelalter. In: DUSIL, Stephan; SCHWEDLER, Gerald; SCHWITTER, Raphael (eds.): Exzerpieren – Kompilieren – Tradieren. Transformationen des Wissens zwischen Spätantike und Frühmittelalter. Berlin, Boston 2017, 49–67.

ROHR, Christian: Ice Jams and their Impact on Urban Communities from a Long-term Perspective (Middle Ages to the Nineteenth Century). In: CHIARENZA, Nicola; HAUG, Annette; MÜLLER, Ulrich (eds.): The Power of Urban Water. Studies in Premodern Urbanism. Berlin, Boston 2020, 197–212.

ROHR, Christian; VLACHOS, Alexandra: Vulkan. In: JAEGER, Friedrich (ed.): Enzyklopädie der Neuzeit. Vol. 14: Renaissance-Signatur. Stuttgart 2010, 467–474.

ROLT, Lionel Thomas Caswell: The Aeronauts. A History of Ballooning, 1783–1903. London 1966.

ROSINSKI, Rosa: Unwetterprophylaxe in Mittelalter und Früher Neuzeit. In: BURHENNE, Verena (ed.): Wetter. Verhext – gedeutet – erforscht. Bönen 2006, 11–23.

RUDWICK, Martin: Geologists' Time: A Brief History. In: LIPPINCOTT, K. (ed.): The Story of Time. Chicago IL 2005, 250–253.

RUDWICK, Martin: Bursting the Limits of Time: The Reconstruction of Geohistory in the Age of Revolution. Chicago IL 2008.

RUDWICK, Martin: Earth's Deep History: How it Was Discovered and Why It Matters. Chicago IL 2014.

RUPPEL, Sophie; STEINBRECHER, Aline: Einleitung. In: RUPPEL, Sophie; STEINBRECHER, Aline (eds.): "Die Natur ist überall bey uns." Mensch und Natur in der Frühen Neuzeit. Zurich 2009, 9–18.

RUSSELL, Andrew J.; DULLER, Robert; MOUNTNEY, Nigel P.: Volcanogenic Jökulhlaups (Glacier Outburst Floods) from Mýrdalsjökull: Impacts on Proglacial Environments. In: SCHOMACKER, Anders; KRÜGER, Johannes; KJÆR, Kurt H. (eds.): Developments in Quaternary Sciences the Mýrdalsjökull Ice Cap, Iceland. Glacial Processes, Sediments and Landforms on an Active Volcano. Amsterdam 2010, 181–207, DOI: 10.1016/s1571-0866(09)01311-6.

RUSSELL, Andrew. J.; FAY, Helen; MARREN, Philip M.; TWEED, Fiona S.; KNUDSEN, Óskar: Icelandic Jökulhlaup Impacts. In: CASELDINE, Christopher; RUSSELL, Andrew; HARDARDÓTTIR, Jórunn; KNUDSEN, Óskar (eds.): Iceland-Modern Processes and Past Environments. Vol. 5. Amsterdam 2005, 153–204.

SANDWELL, David T.: Plate Tectonics: A Martian View. In: ORESKES, Naomi (ed.): Plate Tectonics. An Insider's History of the Modern Theory of the Earth. Boulder CO 2003, 331–346.

SANTEL, Folke; SANTEL, Gregor G.: Chronik für Groß Hesepe Transkription. In: Emsländische Geschichte 2 (1992), 71–94.

SANTEL, Gregor G.: ". . . eine vergieftender thau" 1783 auf S. John Nacht, in Groß Hesepe und ganz Europa. In: Emsländische Geschichte 6 (1997), 108–121.

SARTOR, Joachim: Das Jahrtausendhochwasser der Mosel von 1784. In: Jahrbuch Bernkastel-Wittlich (2010), 73–76.

SAUNDERS, Andy D.; FITTON, J. Godfrey; KERR, Andrew C.; NORRY, M. J.; KENT, Ray W.: The North Atlantic Igneous Province. In: MAHONEY, John J.; COFFIN, Millard F. (eds.): Large Igneous Provinces: Continental, Oceanic and Planetary Flood Volcanism. American Geophysical Union, Geophysical Monograph 100, 1997, 45–93.

SCARTH, Alwyn: Vulcan's Fury: Man Against the Volcano. New Haven CT 1999.

SCHÄFER-WEISS, Dorothea; VERSEMANN, Jens: The Influence of Goethe's Farbenlehre on Early Geological Map Colouring: Goethe's Contribution to Christian Keferstein's General Charte von Teutschland (1821). In: Imago Mundi 57, no. 2 (2005), DOI: 10.1080/03085690500094990.

SCHAFFER, Simon: Natural Philosophy and Public Spectacle in the Eighteenth Century. In: History of Science 21 (1983), 1–43. DOI: 10.1177/007327538302100101.

SCHIFFER, Michael Brian: Draw the Lightning Down: Benjamin Franklin and Electrical Technology in the Age of Enlightenment. Berkeley CA 2003.

SCHMIDT, Andreas: "Wolken krachen, Berge zittern, und die ganze Erde weint . . ." Zur kulturellen Vermittlung von Naturkatastrophen in Deutschland 1755 bis 1855. Münster 1999.

SCHMIDT, Anja: Modelling Tropospheric Volcanic Aerosol. From Aerosol Microphysical Processes to Earth System Impacts. Heidelberg 2013, DOI: 10.1007/978-3-642-34839-6.

SCHMIDT, Anja; CARSLAW, Kenneth S.; MANN, Graham W.; WILSON, Marjorie; BREIDER, T. J.; PICKERING, S. J.; THORDARSON, Thorvaldur: The Impact of the 1783–1784 AD Laki Eruption on Global Aerosol Formation

Processes and Cloud Condensation Nuclei. In: Atmospheric Chemistry and Physics 10, no. 13 (May 2010), 6025–6041, DOI: 10.5194/acp-10-6025-2010.

SCHMIDT, Anja; LEADBETTER, Susan; THEYS, Nicolas; CARBONI, Elisa; WITHAM, Claire S.; STEVENSON, John A.; BIRCH, Cathryn E.; et al.: Satellite Detection, Long-Range Transport, and Air Quality Impacts of Volcanic Sulfur Dioxide from the 2014–2015 Flood Lava Eruption at Bárðarbunga (Iceland). In: Journal of Geophysical Research: Atmospheres 120, no. 18 (2015b), 9739–9757, DOI: 10.1002/2015jd023638.

SCHMIDT, Anja; OSTRO, Bart; CARSLAW, Kenneth S.; WILSON, Marjorie; THORDARSON, Thorvaldur; MANN, Graham W.; SIMMONS, Adrian J.: Excess Mortality in Europe Following a Future Laki-Style Icelandic Eruption. In: Proceedings of the National Academy of Sciences 108, no. 38 (2011), 15710–15715, DOI: 10.1073/pnas.1108569108.

SCHMIDT, Anja; ROBOCK, Alan: Volcanism, the Atmosphere and Climate through Time. In: SCHMIDT, Anja; FRISTAD, Kirsten; ELKINS-TANTON, Linda (eds.): Volcanism and Global Environmental Change. Cambridge UK 2015, 195–207, DOI: 10.1017/cbo9781107415683.017.

SCHMIDT, Anja; SKEFFINGTON, Richard A.; THORDARSON, Thorvaldur; SELF, Stephen; FORSTER, Piers M.; RAP, Alexandru; RIDGWELL, Andy; et al.: Selective Environmental Stress from Sulphur Emitted by Continental Flood Basalt Eruptions. In: Nature Geoscience 9, no. 1 (2015a), 77–82, DOI: 10.1038/ngeo2588.

SCHMIDT, Anja; THORDARSON, Thorvaldur; OMAN, Luke D.; ROBOCK, Alan; SELF, Stephen: Climatic Impact of the Long-Lasting 1783 Laki Eruption: Inapplicability of Mass-Independent Sulfur Isotopic Composition Measurements. In: Journal of Geophysical Research: Atmospheres 117, no. D23 (2012), DOI: 10.1029/2012jd018414.

SCHMIDT, Anja; THORDARSON, Thorvaldur; OMAN, Luke D.; ROBOCK, Alan; SELF, Stephen. Reply to Comment by COLE-DAI et al. on "Climatic Impact of the Long-Lasting Laki Eruption: Inapplicability of Mass-Independent Sulfur Isotope Composition Measurements." In: Journal of Geophysical Research: Atmospheres 119, no. 11 (December 2014b), 6636–6637, DOI: 10.1002/2013jd021440.

SCHMIDT, Anja; WITHAM, Claire S.; THEYS, Nicolas; RICHARDS, Nigel A. D.; THORDARSON, Thorvaldur; SZPEK, Kate; FENG, Wuhu; et al.: Assessing Hazards to Aviation from Sulfur Dioxide Emitted by Explosive Icelandic Eruptions. In: Journal of Geophysical Research: Atmospheres 119, no. 24 (2014a), DOI: 10.1002/2014jd022070.

SCHMIDT, Martin: Hochwasser und Hochwasserschutz in Deutschland vor 1850. Eine Auswertung alter Karten und Quellen. Munich 2000.

SCHNEIDER, David P.; AMMANN, Caspar M.; OTTO-BLIESNER, Bette L.; KAUFMAN, Darrell S.: Climate Response to Large, High-Latitude and Low-Latitude Volcanic Eruptions in the Community Climate System Model. In: Journal of Geophysical Research 114, no. D15101 (January 2009), DOI: 10.1029/2008jd011222.

SCHOTT, Herbert: Das Hochwasser von 1784. In: WAGNER, Ulrich (ed.): Geschichte der Stadt Würzburg. Vom Bauernkrieg 1525 bis zum Übergang an das Königreich Bayern 1815, vol. 2. Stuttgart 2004, 37–39.

SCHRÖDER, Wilfried: The Krakatoa Event and Associated Phenomena. A Historical Review. In: Acta Geodaetica et Geophysica Hungarica 38, no. 4 (2003), 389–395.

SCHWARZBACH, Martin: Island. Geologenfahrten nach Island. Ludwigsburg [5]1983.

SCOTT, Heidi. Apocalypse Narrative, Chaotic System: Gilbert White's *Natural History of Selbourne* and Modern Ecology. In: *Romanticism and Victorianism on the Net* 56 (November 2009), DOI: 10.7202/10011095ar.

SEARLE, Roger: Mid-Ocean Ridges. Cambridge UK 2013.

SELF, Stephen; THORDARSON, Thorvaldur; KESZTHELYI, L.: Emplacement of Continental Flood Basalt Lava Flows. In: MAHONEY, John J.; COFFIN, Millard F. (eds.): Large Igneous Provinces: Continental, Oceanic and Planetary Flood Volcanism. American Geophysical Union, Geophysical Monograph 100, 1997, 381–410.

SELF, Stephen; ZHAO, Jing-Xia; HOLASEK, Rick E.; TORRES, Ronnie C.; KING, Alan J.: The Atmospheric Impact of the 1991 Mount Pinatubo Eruption. In: NEWHALL, Christopher G.; PUNONGBAYAN, Raymundo S. (eds.): Fire and Mud: Eruptions and Lahars of Mount Pinatubo, Philippines. Seattle WA 1996, 1089–1115.

SIGL, Michael; WINSTRUP, M.; McCONNELL, Joseph R.; WELTEN, K. C.; PLUNKETT, Gill; LUDLOW, Francis; BÜNTGEN, Ulf; et al.: Timing and Climate Forcing of Volcanic Eruptions for the Past 2,500 Years. In: Nature 523, no. 7562 (2015), 543–549, DOI: 10.1038/nature14565.

SIGMARSSON, Olgeir; CONDOMINES, Michel; GRÖNVOLD, Karl; THORDARSON, Thorvaldur: Extreme Magma Homogeneity in the 1783–84 Lakagigar Eruption: Origin of a Large Volume of Evolved Basalt in Iceland. In: Geophysical Research Letters 18, no. 12 (1991), 2229–2232, DOI: 10.1029/91gl02328.

SIGMUNDSSON, Freysteinn; SÆMUNDSSON, Kristján: Iceland: A Window on North-Atlantic Divergent Plate Tectonics and Geological Processes. In: Episodes 31, no. 1 (2008), 92–97, DOI: 10.18814/epiiugs/2008/v31i1/013.

SIGURÐSSON, Haraldur; SPARKS, Robert Stephen John: Rifting Episode in North Iceland in 1874–1875 and the Eruptions of Askja and Sveinagja. In: Bulletin Volcanologique 41, no. 3 (1978), 149–167, DOI: 10.1007/bf02597219.

SIGURÐSSON, Haraldur: Melting the Earth: The History of Ideas on Volcanic Eruptions. New York NY 1999.

SIGURÐSSON, Oddur: Variations of Termini of Glaciers in Iceland in Recent Centuries and Their Connection with Climate. In: Developments in Quaternary Sciences Iceland – Modern Processes and Past Environments (2005), 241–255, DOI: 10.1016/s1571-0866(05)80012-0.

SIGURGEIRSSON, Thorbjörn: Lava Cooling. In: WILLIAMS, Richard S. jr (ed.): Lava Cooling Operations During the 1973 Eruption of the Eldfell Volcano, Heimaey, Vestmannaeyjar, Iceland. Denver: U.S. Geological Survey, 1973 (U.S. Geological Survey Open-file Report 97-724).

SIMKIN, Tom; SIEBERT, Lee; McCLELLAND, Lindsay; BRIDGE, D.; NEWHALL, Christopher; LATTER, John H.: Volcanoes of the World. A Regional Directory, Gazetteer, and Chronology of Volcanism During the Last 10,000 Years. Stroudsburg PA 1981.

SMITH, Dwight L.: Josiah Harmar, Diplomatic Courier. In: Pennsylvania Magazine of History and Biography 87, no. 4 (1963), 420–430.

SÖRLIN, Sverker; LANE, Melissa: Historicizing Climate Change – Engaging New Approaches to Climate and History. In: Climatic Change 151, no. 1 (2013), 1–13, DOI: 10.1007/s10584-018-2285-0.

SOUKUPOVA, Jana: Heavy Storms in 1783 in a Historical Documentary Record. In: Meteorologicky Casopis 16 (2013), 13–20.

SPATA, Manfred: Das Jahrtausend-Hochwasser von 1784 in Bonn und Beuel. Beueler Hochwassermarken als Erinnerung an die Eiswasserkatastrophe. Bonn 2017 (Kleine Beiträge zu Denkmal und Geschichte im rechtsrheinischen Bonn 4).

STARK, Peter: Driving to Greenland. Springfield NJ 1994.

STEIN, Peter: Schriftkultur. Eine Geschichte des Schreibens und Lesens. Darmstadt 2006.

STEINLE, Friedrich: Naturwissenschaft. In: JAEGER, Friedrich (ed.): Enzyklopädie der Neuzeit. Vol. 9: Naturhaushalt-Physiokratie. Stuttgart 2009, 54–58.

STEINÞÓRSSON, Sigurður: Annus Mirabilis: 1783 í erlendum heimildum. (Annus Mirabilis: The Year 1783 According to Contemporary Accounts Outside of Iceland). In: Skírnir 166 (1992), 133–159.

STEVENSON, David S.; JOHNSON, C. E.; HIGHWOOD, Ellie J.; GAUCI, Vincent; COLLINS, William J.; DERWENT, R. G.: Atmospheric Impact of the 1783 & 1784 Laki Eruption: Part I Chemistry Modelling. In: Atmospheric Chemistry and Physics 3, no. 1 (2003), 487–507, DOI: 10.5194/acpd-3-551-2003.

STÖBER, Rudolf: Staat und Verleger im 18. Jahrhundert. In: SÖSEMANN, Bernd (ed.): Kommunikation und Medien in Preußen vom 16. bis zum 19. Jahrhundert. Stuttgart 2002, 159–174.

STOCKMAN, Zein: Vor 200 Jahren. Die Große Finsternis und Luftverpestung im Juni 1783. In: Bentheimer Jahrbuch 105 (1984), 220–222.

STOFFEL, Markus; KHODRI, Myriam; CORONA, Christophe; GUILLET, Sébastien; POULAIN, Virginie; BEKKI, Slimane; GUIOT, Joël et al.: Estimates of Volcanic-Induced Cooling in the Northern Hemisphere over the Past 1,500 Years. In: Nature Geoscience 8, no. 10 (2015), 784–788, DOI: 10.1038/ngeo2526.

STOLLBERG-RILINGER, Barbara: Die Aufklärung. Europa im 18. Jahrhundert. Stuttgart ²2011.

STOTHERS, Richard B.: The Great Dry Fog of 1783. In: Climatic Change 32 (1996), 79–89, DOI: 10.1007/BF00141279.

STOTHERS, Richard B.; WOLFF, John A.; SELF, Stephen; RAMPINO, Michael R.: Basaltic Fissure Eruptions, Plume Heights, and Atmospheric Aerosols. In: Geophysical Research Letters 13 (1986), 725–728, DOI: 10.1029/GL013i008p00725.

STREETER, Richard; DUGMORE, Andrew J.; VÉSTEINSSON, Orri: Plague and Landscape Resilience in Premodern Iceland. In: Proceedings of the National Academy of Sciences 109, no. 10 (2012), 3664–3669, DOI: 10.1073/pnas.1113937109.

STRØM, Elin: Naturhistorie-Selskabet in København 1789–1804. Oslo 2006.

STRØM, Elin: Naturelskeren Niels Tønder Lund. Oslo 2017.

STRÖMMER, Elisabeth: Klima-Geschichte. Methoden der Rekonstruktion und historische Perspektive. Ostösterreich 1700 bis 1830. Vienna 2003.

STRUNZ, Sebastian; MARSELLE, Melissa; SCHRÖTER, Matthias: Leaving the 'Sustainability or Collapse' Narrative Behind. In: Sustainability Science 14, no. 6 (August 2019), 1717–1728, DOI: 10.1007/s11625-019-00673-0.

Suðausturland South East 1:200 000. Reykjavík 2013 (Ísland Landshlutakort 7).

SUTHERLAND, Donald: Weather and the Peasantry of Upper Brittany 1780–1790. In: WIGLEY, Tom M. L.; INGRAM, M. J.; FARMER, G. (eds.): Climate and History. Studies in Past Climates and Their Impact on Man. Cambridge UK 1981, 434–449.

SVEINBJÖRNSDÓTTIR, Amy E.; HEINEMEIER, Jan; GUÐMUNDSSON, Garðar: ^{14}C Dating of the Settlement of Iceland. In: Radiocarbon 46, no. 1 (2004), 387–394.

TAYLOR, Kenneth L.: Before Volcanoes Became Ordinary. In: MAYER, Wolf; CLARY, R. M.; AZUELA, Luz Fernanda; MOTA, Teresa Salomé; WOŁKOWICZ, Stanisław (eds.): History of Geoscience: Celebrating 50 Years of INHIGEO. Geological Society of London Special Publications 442, no. 1 (February 2016), 117–126, DOI: 10.1144/sp442.27.

THÉBAUD-SORGER, Marie: Balloons in the Historiography of Aerial Mobility. In: Mobilities in History 4 (2013), 83–88, DOI: 10.3167/mih.2013.040108.

THERRELL, Matthew D.: Tree Rings and "El Año del Hambre." In: Dendrochronologia 22 (2005), 203–207, DOI: 10.1016/j.dendro.2005.04.006.

THOMPSON, Lonnie G.; MOSLEY-THOMPSON, Ellen; DAVIS, Mary E.; BOLZAN, John F.; DAI, Jihong; KLEIN, L.; GUNDESTRUP, N.; YAO, T.; WU, X.; XIE, Z.: Glacial Stage Ice-Core Records from the Subtropical Dunde Ice Cap, China. In: Annals of Glaciology 14 (1990), 288–297.

THOMPSON, Lonnie G.; MOSLEY-THOMPSON, Ellen; DAVIS, Mary E.; BOLZAN, John F.; DAI, Jihong; YAO, T.; GUNDESTRUP, N.; WU, X.; KLEIN, L.; XIE, Z.: Holocene-Late Pleistocene Climatic Ice Core Records from Qinghai-Tibetan Plateau. In: Science 246 (1989), 474–477.

THORDARSON, Thorvaldur: The Eruption Sequence and Eruption Behavior of Laki, 1783–1785, Iceland: Characteristics and Distribution of Eruption Products. MS Thesis, the University of Texas at Arlington, 1990.

THORDARSON, Thorvaldur: Volatile Release and Atmospheric Effects of Basaltic Fissure Eruptions. Department of Geology and Geophysics. Doctoral thesis, Honolulu, University of Hawaii, 1995.

THORDARSON, Thorvaldur: The 1783–1785 AD Laki-Grimsvötn Eruptions I: A Critical Look at the Contemporary Chronicles. In: Jökull 53 (2003), 1–10.

THORDARSON, Thorvaldur: Environmental and Climatic Effects from Atmospheric SO_2 Mass-Loading by Icelandic Flood Lava Eruptions. In: CASELDINE, Christopher; RUSSELL, Andrew; HARDARDÓTTIR, Jórunn; KNUDSEN, Óskar (eds.): Iceland-Modern Processes and Past Environments. Vol. 5. Amsterdam 2005, 205–220.

THORDARSON, Thorvaldur: Perception of Volcanic Eruptions in Iceland. In: Martini, I. Peter; CHESWORTH, Ward (eds.): Landscapes and Societies. New York NY 2010, 285–296.

THORDARSON, Thorvaldur; HÖSKULDSSON, Ármann: Postglacial Volcanism in Iceland. In: Jökull 58 (2008), 197–228.

THORDARSON, Thorvaldur; HÖSKULDSSON, Ármann: Iceland. Edinburgh ²2014.

THORDARSON, Thorvaldur; LARSEN, Guðrún: Volcanism in Iceland in Historical Time: Volcano Types, Eruption Styles and Eruptive History. In: Journal of Geodynamics 43 (2007), 118–152, DOI: 10.1016/j.jog.2006.09.005.

THORDARSON, Thorvaldur; LARSEN, Guðrún; STEINÞÓRSSON, Sigurður; SELF, Stephen: The 1783–1785 AD Laki-Grímsvötn Eruptions II: Appraisal Based on Contemporary Accounts. In: Jökull 51 (2003b), 11–48.

THORDARSON, Thorvaldur; MILLER, D. J.; LARSEN, Guðrún; SELF, Stephen; SIGURDSSON, H.: New Estimates of Sulfur Degassing and Atmospheric Mass-Loading by the 934 AD Eldgjá Eruption, Iceland. In: Journal of Volcanology and Geothermal Research 108 (2001), 33–54, DOI: 10.1016/S0377-0273(00)00277-8.

THORDARSON, Thorvaldur; SELF, Stephen: The Laki (Skaftár Fires) and Grímsvötn Eruptions in 1783–1785. In: Bulletin of Volcanology 55 (1993), 233–263.

THORDARSON, Thorvaldur; SELF, Stephen: Real-Time Observations of the Laki Sulfuric Aerosol Cloud in Europe 1783 as Documented by Professor S. P. van Swinden at Franeker, Holland. In: Jökull 50 (2001), 65–72.

THORDARSON, Thorvaldur; SELF, Stephen: Atmospheric and Environmental Effects of the 1783–1784 Laki Eruption: A Review and Reassessment. In: Journal of Geophysical Research 108 (2003), D14011, 1–29.

THORDARSON, Thorvaldur; SELF, Stephen; MILLER, D. J.; LARSEN, Guðrún; VILMUNDARDÓTTIR, Elsa G.: Sulphur Release from Flood Lava Eruptions in the Veidivötn, Grímsvötn, and Katla Volcanic Systems, Iceland. In: OPPENHEIMER, Clive; PYLE, David M.; BARCLAY, Jennie (eds.): Volcanic Degassing. Geological Society of London Special Publication 213 (2003a), 103–121.

THORDARSON, Thorvaldur; SELF, Stephen; ÓSKARSSON, Niels; HULSEBOSCH, Thomas: Sulfur, Chlorine, and Fluorine Degassing and Atmospheric Loading by the 1783–1784 AD Laki (Skaftár Fires) Eruption in Iceland. In: Bulletin of Volcanology 58, no. 2–3 (1996), 205–225, DOI: 10.1007/s004450050136.

THÜSEN, Joachim von der: Schönheit und Schrecken der Vulkane. Zur Kulturgeschichte des Vulkanismus. Darmstadt 2008.

TILTON, Eleanor M.: Lightning-Rods and the Earthquake of 1755. In: The New England Quarterly 13, no. 1 (March 1940), 85–97.

TINGLEY, Martin P.; HUYBERS, Peter: Recent Temperature Extremes at High Northern Latitudes Unprecedented in the Past 600 Years. In: Nature 496 (2013), 201–205, DOI: 10.1038/nature11969.

TOOHEY, Matthew; SIGL, Michael: Volcanic Stratospheric Sulfur Injections and Aerosol Optical Depth from 500 BCE to 1900 CE. In: Earth System Science Data 9 (2017), 809–831, DOI: 10.5194/essd-9-809-2017.

TRIGO, Ricardo M.; VAQUERO, J. M.; STOTHERS, Richard B.: Witnessing the Impact of the 1783–1784 Laki Eruption in the Southern Hemisphere. In: Climatic Change 99, no. 3–4 (August 2009), 535–546, DOI: 10.1007/s10584-009-9676-1.

TRISCHLER, Helmuth; BUD, Robert: Public Technology: Nuclear Energy in Europe. In: History and Technology 34, no. 3–4 (2018), 187–212, DOI: 10.1080/0734512.2018.1570674.

UEKÖTTER, Frank; LÜBKEN, Uwe (eds.): Managing the Unknown: Essays on Environmental Ignorance. New York, Oxford 2014.

United States Environmental Protection Agency (USEPA): Health and Environmental Effects Profile for Hydrogen Sulfide. In: ECAO-CIN-026A (1980), 118–8.

VALENCIUS, Conevery Bolton: The Lost History of the New Madrid Earthquakes. Chicago, London 2015.

VASEY, Daniel E.: Population, Agriculture, and Famine: Iceland, 1784–1785. In: Human Ecology 19, no. 3 (1991), 323–350.

VASEY, Daniel E.: Population Regulation, Ecology, and Political Economy in Pre-Industrial Iceland. In: American Ethnologist 23, no. 2 (1996), 366–392.

VASEY, Daniel E.: A Quantitative Assessment of Buffers Among Temperature Variations, Livestock, and the Human Population of Iceland, 1784 to 1900. In: Climatic Change 48, no. 1 (2001), 243–263, DOI: 10.1007/978-94-017-3352-6_12.

VASOLD, Manfred: Die Eruptionen des Laki von 1783/84. Ein Beitrag zur deutschen Klimageschichte. In: Naturwissenschaftliche Rundschau 57, no. 11 (2004), 602–608.

Vatnajökull National Park: Lakagígar, Langisjór, and Eldgjá brochure and map. Vatnajökulsþjóðgarður, 2011.

VERBURGT, Lukas M.; BURKE, Peter: Introduction: Histories of Ignorance. In: Journal for the History of Knowledge 2, no. 1 (2021): 1–9. DOI: 10.5334/jhk.45.

VERDENHALVEN, Fritz: Alte Meß- und Währungssysteme aus dem deutschen Sprachgebiet. Was Familien- und Lokalgeschichtsforscher suchen. Insingen [2]2011.

VERMIJ, Rienk: Erschütterung und Bewältigung. Erdbebenkatastrophen in der Frühen Neuzeit. In: JAKUBOWSKI-TIESSEN, Manfred; LEHMANN, Hartmut (eds.): Um Himmels Willen. Religion in Katastrophenzeiten. Göttingen 2003, 235–252.

VÉSTEINSSON, Orri: The Christianization of Iceland: Priests, Power and Social Change, 1000–1300. Oxford 2000.

VÉSTEINSSON, Orri; SVERRISDÓTTIR, Bryndis; YATES, Anna: Reykjavík 871 ± 2: Landnámssyningin; the Settlement Exhibition. Reykjavík 2006.

VIBÉ, Yulia; FRIEDRICH, Anke M.; BUNGE, Hans-Peter; CLARK, S. R.: Correlations of Oceanic Spreading Rates and Hiatus Surface Area in the North Atlantic Realm. In: Lithosphere 10, no. 5 (2018), 677–684, DOI: 10.1130/L736.1.

WAGNER, Sebastian; ZORITA, Eduardo: The Influence of Volcanic, Solar, and CO2 Forcing on the Temperatures in the Dalton Minimum (1790–1830): A Model Study. In: Climate Dynamics 25 (2005), 205–218, DOI: 10.1007/s00382-005-0029-0.

WALLACE, Birgitta Linderoth: L'anse aux Meadows: Gateway to Vinland. The Norse of the North Atlantic. In: Acta Archaeologica 61 (1991), 166–197.

WALTER, Axel E.; BAUDACH, Frank: "Das Tor zur Hölle." Island-Karten aus fünf Jahrhunderten. Eutin 2019.

WALTER, François: Katastrophen: Eine Kulturgeschichte vom 16. bis ins 21. Jahrhundert. Stuttgart 2010.

WAWN, Andrew: John Thomas Stanley and Iceland: The Sense and Sensibility of an 18th Century Explorer. In: Scandinavian Studies 53 (1981), 52–76.

WAWN, Andrew: Gunnlaugs Saga Ormstunga and the Theatre Royal Edinburgh 1812: Melodrama, Mineralogy, and Sir George Mackenzie. In: Scandinavica 21, no. 2 (1982), 139–151.

WAWN, Andrew: The Enlightenment Traveller and the Idea of Iceland: The Stanley Expedition of 1789 Reconsidered. In: Scandinavica 28, no. 1 (1989), 5–16.

WEBER, Christoph Daniel: Vom Gottesgericht zur verhängnisvollen Natur. Darstellung und Bewältigung von Naturkatastrophen im 18. Jahrhundert. Hamburg 2015.

WEBER, Johannes: Straßburg 1605: Die Geburt der Zeitung. In: Jahrbuch für Kommunikationsgeschichte 7 (2005), 3–26.

WEBER, Max: Die protestantische Ethik und der "Geist" des Kapitalismus. Bodenheim 1993.

WEIGL, Engelhard: Entzauberung der Natur durch Wissenschaft – dargestellt am Beispiel der Erfindung des Blitzableiters. In: Jahrbuch der Jean-Paul-Gesellschaft 22 (1987), 7–40.

WEIKINN, Curt: Quellentexte zur Witterungsgeschichte Europas von der Zeitwende bis zum Jahre 1850. Hydrographie 1, part 5 (1751–1800) (eds. BÖRNGEN, Michael; TETZLAFF, Gerd). Berlin, Stuttgart 2000.

WEISBURD, Stefi: Excavating Words: A Geological Tool. In: Science News 127, no. 6 (1985), 91–94.

WELKE, Martin: Die Legende vom "unpolitischen Deutschen." Zeitunglesen im 18. Jahrhundert als Spiegel des politischen Interesses. In: Jahrbuch der Wittheit zu Bremen 25 (1981), 161–188.

WELLBURN, Alan R.: Air Pollution and Climate Change: The Biological Impact. Harlow [2]1994.

WENZLHUEMER, Roland: Connecting the Nineteenth-Century World: The Telegraph and Globalization. Cambridge UK 2013.

WEST, John F. (ed.): The Journals of the Stanley Expedition to the Faroe Islands and Iceland. Tórshavn, Faroe Islands 1970.

WEYER, Monika; KOCH, Christa: Boomende Wissenschaft: Neue Technik, internationale Messnetze und erste Wetterkarten. In: BURHENNE, Verena (ed.): Wetter. Verhext – gedeutet – erforscht. Bönen 2006a, 93–109.

WEYER, Monika; KOCH, Christa: Wenn das Wettermännchen tanzt . . . Messgeräte und Luftexperimente. In: BURHENNE, Verena (ed.): Wetter. Verhext – gedeutet – erforscht. Bönen 2006b, 76–89.

WHITE, Robert S.: A Hot-Spot Model for Early Tertiary Volcanism in the N Atlantic. In: MORTON, A. C.; PARSON, M. (eds.): Early Tertiary Volcanism and the Opening of the NE Atlantic. Geological Society of London Special Publication 39, no. 1 (1988), 3–13, DOI: 10.1144/gsl.sp.1988.039.01.02.

WHITE, Robert; MCKENZIE, Dan: Magmatism at Rift Zones: The Generation of Volcanic Continental Margins and Flood Basalts. In: Journal of Geophysical Research 94, no. B6 (1989), 7685, DOI: 10.1029/jb094ib06p07685.

WHITE, Sam: The Climate of Rebellion in the Early Modern Ottoman Empire. Cambridge UK 2011.

WHITE, Sam: The Real Little Ice Age. In: The Journal of Interdisciplinary History 44, no. 3 (2014), 327–352.

WHITE, Sam; PFISTER, Christian; MAUELSHAGEN, Franz: The Palgrave Handbook of Climate History. London 2018.

WHITE, Sam; PEI, Qing; KLEEMANN, Katrin; DOLÁK, Lukáš; HUHTAMAA, Heli; CAMENISCH, Chantal. New Perspectives on Historical Climatology. In: Wiley Interdisciplinary Reviews: Climate Change 14, no. 1 (2022), e808, DOI: 10.1002/wcc.808.

WHITEHEAD, Thór: The Ally Who Came in from the Cold: A Survey of Icelandic Foreign Policy 1946–1956. Reykjavík 1998.

WIENERS, Claudia E.: Haze, Hunger, Hesitation: Disaster Aid After the 1783 Laki Eruption. In: Journal of Volcanology and Geothermal Research 406 (2020), 107080.

WILKE, Jürgen: Nachrichtenvermittlung und Informationswege im 17. und 18. Jahrhundert in Brandenburg/Preußen. In: SÖSEMANN, Bernd (ed.): Kommunikation und Medien in Preußen vom 16. bis zum 19. Jahrhundert. Stuttgart 2002, 72–84.

WILKE, Jürgen: Korrespondenten und geschriebene Zeitungen. In: ARNDT, Johannes; KÖRBER, Esther-Beate (eds.): Das Mediensystem im Alten Reich der Frühen Neuzeit (1600–1750). Göttingen 2010, 59–74.

WILKENING, Ken: Intercontinental Transport of Dust. Science and Policy, pre-1800s to 1967. In: Environment and History 17, no. 2 (2011), 313–339.

WILLIAMS, Richard S.; MOORE, James G.: Man Against Volcano: The Eruption on Heimaey, Vestmannaeyjar, Iceland. Washington DC 1988.

WILSON, J. Tuzo: A New Class of Faults and Their Bearing on Continental Drift. In: Nature 207 (1965), 343–347.

WINCHESTER, Simon: Krakatoa. The Day the World Exploded. 27th August 1883. London 2005.

WITHAM, Claire S.; OPPENHEIMER, Clive: Mortality in England During the 1783–4 Laki Craters Eruption. In: Bulletin of Volcanology 67, no. 1 (2004), 15–26, DOI: 10.1007/s00445-004-0357-7.

WITZE, Alexandra; KANIPE, Jeff: Island on Fire. The Extraordinary Story of Laki, the Volcano That Turned Eighteenth-Century Europe Dark. Bungay UK 2014.

WOESSNER, Jochen; LAURENTIU, Danciu; GIARDINI, Domenico; et al.: The 2013 European Seismic Hazard Model: Key Components and Results. In: Bulletin of Earthquake Engineering 13 (2015), 3553–3596, DOI: 10.1007/s10518-015-9795-1.

WOOD, Charles A.: The Climatic Effects of the 1783 Laki Eruption. In: HARINGTON, C. Richard (ed.): The Year Without a Summer? World Climate in 1816. Ottawa 1992, 58–77.

WOOD, Gillen D'Arcy: Tambora: The Eruption That Changed the World. Princeton NJ 2014.

WOODS, Andrew W.: A Model of the Plumes Above Basaltic Fissure Eruptions. In: Geophysical Research Letters 20 (1993), 1115–1118, DOI: 10.1029/93GL01215.

World Data Center A for Solid Earth Geophysics: Catalog of Submarine Volcanoes and Hydrological Phenomena Associated With Volcanic Events, 1500 BC to December 21, 1899. Washington DC 1984.

WOZNIAK, Thomas: Zur Wahrnehmung, Darstellung und Instrumentalisierung extremer Naturereignisse und ihrer Folgen in Chroniken und Annalen vom 6. bis 11. Jahrhundert. Habilitation, University of Tübingen, 2017.

Wozniak, Thomas: Naturereignisse im frühen Mittelalter: Das Zeugnis der Geschichtsschreibung vom 6. bis 11. Jahrhundert. Berlin, Boston 2020.

Wrigley, Edward Anthony; Schofield, Roger S.: The Population History of England, 1541–1871: A Reconstruction. Cambridge UK 1989.

Würgler, Andreas: Medien in der Frühen Neuzeit. Munich 2009.

Xoplaki, Elena: European Spring and Autumn Temperature Variability and Change of Extremes over the Last Half Millennium. In: Geophysical Research Letters 32, no. 15 (2005), L15713, DOI: 10.1029/2005gl023424.

Zambri, Brian; Robock, Alan; Mills, Michael J.; Schmidt, Anja: Modeling the 1783–1784 Laki Eruption in Iceland: 1. Aerosol Evolution and Global Stratospheric Circulation Impacts. In: Journal of Geophysical Research: Atmospheres 124, no. 13 (2019a), 6750–6769, DOI: 10.1029/2018jd029553.

Zambri, Brian; Robock, Alan; Mills, Michael J.; Schmidt, Anja: Modeling the 1783–1784 Laki Eruption in Iceland: 2. Climate Impacts. In: Journal of Geophysical Research: Atmospheres 124, no. 13 (2019b), 6770–6790, DOI: 10.1029/2018jd029554.

Zhong, Yafang; Miller, Gifford H.; Otto-Bliesner, Bette L.; Holland, Marika M.; Bailey, David A.; Schneider, David P.; Geirsdóttir, Áslaug: Centennial-Scale Climate Change from Decadally-Paced Explosive Volcanism: a Coupled Sea Ice-Ocean Mechanism. In: Climate Dynamics 37, no. 11–12 (2010), 2373–2387, DOI: 10.1007/s00382-010-0967-z.

Zielinski, Gregory A.: Stratospheric Loading and Optical Depth Estimates of Explosive Volcanism over the Last 2100 Years Derived from the Greenland Ice Sheet Project 2 Ice Core. In: Journal of Geophysical Research 100, no. D10 (1995), 20937, DOI: 10.1029/95jd01751.

Zielinski, Gregory A.; Fiacco, R. J.; Mayewski, Paul A.; Meeker, L. David; Whitlow, Sallie; Twickler, Mark S.; Germani, Mark S.; Endo, K.; Yasui, M.: Climatic Impact of the A. D. 1783 Asama (Japan) Eruption Was Minimal: Evidence from the GISP2 Ice Core. In: Geophysical Research Letters 21, no. 22 (January 1994), 2365–2368, DOI: 10.1029/94gl02481.

Zielinski, Gregory A.; Mayewski, Paul A.; Meeker, L. David; Grönvold, Karl; Germani, Mark S.; Whitlow, Sallie; Twickler, Mark S.; Taylor, Kendrick C.: Volcanic Aerosol Records and Tephrochronology of the Summit, Greenland, Ice Cores. In: Journal of Geophysical Research 102, no. C12 (1997), 26625–26640, DOI: 10.1029/96JC03547.

Zwierlein, Cornel: Introduction: Towards a History of Ignorance. In: Zwierlein, Cornel (ed.): Dark Side of Knowledge, Leiden 2016, 1–47, DOI: 10.1163/9789004325180_002.

Þórarinsson, Sigurður: Tefrokronologiska studier på Island: Þjórsárdalur och dess förödelse. Doctoral thesis 1944, University of Stockholm.

Þórarinsson, Sigurður: Some Aspects of the Grímsvötn Problem. In: Journal of Glaciology 2 (1953), 267–274.

Þórarinsson, Sigurður: Hekla on Fire (translated by Hannesson, Johann). Munich 1956a.

Þórarinsson, Sigurður: The Thousand Years Struggle Against Ice and Fire: 2 Lectures Delivered 21 and 26 February 1952 at Bedford College, London. Reykjavík 1956b.

Þórarinsson, Sigurður: The Öræfajökull Eruption of 1362. In: Acta naturalia Islandica 2, no. 2 (1958).

Þórarinsson, Sigurður: The Lakagigar Eruption 1783. In: Bulletin Volcanologique 33, no. 3 (1969), 910–929. (This paper was read at the IAVCEI International Symposium on Volcanology in 1968.)

Þórarinsson, Sigurður: On the Damage Caused by Volcanic Eruptions with Special Reference to Tephra and Gases. In: Sheets, Payson D.; Grayson, Donald K. (eds.): Volcanic Activity and Human Ecology. New York NY 1979, 125–159.

Þórarinsson, Sigurður: Greetings from Iceland: Ash-Falls and Volcanic Aerosols in Scandinavia. In: Geografiska Annaler 63, no. 3/4 (1981), 109–118.

Þórarinsson, Sigurður: Annáll Skaftárelda. (The Laki Fires). In: Gunnlaugsson, Gísli Ágúst; Guðbergsson, Gylfi Már; Þórarinsson, Sigurður; Rafnsson, Sveinbjörn; Einarsson, Þorleifur (eds.): Skaftáreldar 1783–1784: Ritgerðir og heimildir. Reykjavík 1984, 11–36.

Þórarinsson, Sigurður; Einarsson, Þorleifur; Sigvaldason, Gudmundur E.; Elísson, Gunnlaugur:
The Submarine Eruption off the Vestmann Islands, 1963–64. In: Bulletin of Volcanology 27 (1964),
435–445.

Þórarinsson, Sigurður; Sæmundsson, Kristján: Volcanic Activity in Historical Time. In: Jökull 29 (1979), 29–32.

Þórarinsson, Sigurður; Steinþórsson, Sigurður; Einarsson, Þorleifur; Kristmannsdóttir, Hrefna; Óskarsson,
Niels: The Eruption on Heimaey, Iceland. In: Nature 241 (1973), 372–375.

Online Primary and Secondary Resources

Reindeer Warning in East Iceland. In: Iceland Review, 6 January 2018, https://www.icelandreview.com/
news/reindeer-warning-east-iceland/ (1 March 2020).

Sulfur dioxide (SO2), Air quality fact sheet. In: Australian Government, Department of the Environment
and Heritage, 2005, https://www.environment.gov.au/protection/publications/factsheet-sulfur-diox
ide-so2 (1 March 2020).

America's Historical Newspapers, https://infoweb-newsbank-com.emedien.ub.uni-muenchen.de/
(1 March 2020).

Ámundason, Hallgrímur J.: Af hverju heitir fjallið Laki, sem Lakagígar eru nefndir eftir? (Why is the
Mountain That Gives Lakagígar Its Name Called Laki?), https://www.visindavefur.is/svar.php?id=
77907# (1 March 2020).

ANNO, AustriaN Newspapers Online, http://anno.onb.ac.at/ (1 March 2020).

Bauch, Martin: Die Magdalenenflut 1342 – ein unterschätztes Jahrtausendereignis? In: Mittelalter.
Interdisziplinäre Forschung und Rezeptionsgeschichte (4 February 2014), http://mittelalter.hypotheses.
org/3016 (29 December 2020).

Bauch, Martin: Vulkanisches Zwielicht. Ein Vorschlag zur Datierung des Kuwae-Ausbruchs auf 1464.
In: Mittelalter. Interdisziplinäre Forschung und Rezeptionsgeschichte (2015), https://mittelalter.hypotheses.
org/5697 (29 December 2020).

Bayerische Staatsbibliothek: Allgemeine Deutsche Biographie (ADB) und Neue Deutsche Biographie (NDB),
http://www.deutsche-biographie.de (1 March 2020).

Bible, New International Version. Biblica, the International Bible Society, https://www.biblica.com/bible/
niv/ (1 March 2020).

Bressan, David: How Colors Revolutionized Geological Map Making. In: Scientific American, 13 April 2014,
https://blogs-scientificamerican-com.emedien.ub.uni-muenchen.de/history-of-geology/How-colors-
revolutionized-geological-mapmaking (1 March 2020).

British Geological Survey: Earthquake on 10 August 1783 in Launceston, England. Based on Musson, R.M.
W. Seismicity of Cornwall and Devon. In: British Geological Survey Global Seismology Report, WL/89/
11 (1989), https://web.archive.org/web/20110516173115/http://www.quakes.bgs.ac.uk/earthquakes/
historical/historical_listing.htm (1 March 2020).

British Newspapers, Seventeenth and Eighteenth Century Burney Collection Newspapers, Gale NewsVault,
http://find.gale.com.emedien.ub.uni-muenchen.de/ (1 March 2020).

Broendel, Katie; Hanlon, Shane M.; Lester, Liza; Coleman, Adell; Brauman, Kate: Special Release: Hawaii's
Volcanoes, Water, and . . . Vog? In: Third Pod from the Sun Podcast, produced by the American
Geophysical Union (AGU), podcast, 17:32 minutes, 22 March 2019, https://thirdpodfromthesun.com/
2019/03/22/special-release-hawaiis-volcanoes-water-andvog/ (1 March 2020).

Bryhni, Inge: Amund Helland. In: Norsk biografisk leksikon (13 February 2009), https://nbl.snl.no/Amund_
Helland_-_2 (1 March 2020).

Cabinet Office (UK): National Risk Register of Civil Emergencies. 2017 Edition, https://assets.publishing.ser
    vice.gov.uk/government/uploads/system/uploads/attachment_data/file/644968/UK_National_Risk_
    Register_2017.pdf (1 March 2020).
Catalogue of Icelandic Volcanoes, http://icelandicvolcanos.is/ (1 March 2020). B. Oladóttir, Larsen, G. &
    Guðmundsson, M. T. Catalogue of Icelandic Volcanoes, IMO, UI and CPD-NCIP.
Catalogue of Icelandic Volcanoes: Bárðarbunga, http://icelandicvolcanos.is/?volcano=BAR (1 March 2020).
Catalogue of Icelandic Volcanoes: Grímsvötn, http://icelandicvolcanos.is/?volcano=GRV (1 March 2020).
Catalogue of Icelandic Volcanoes: Hekla, http://icelandicvolcanos.is/?volcano=HEK (1 March 2020).
Catalogue of Icelandic Volcanoes: Reykjanes, http://icelandicvolcanos.is/?volcano=REY (1 March 2020).
Catalogue of Icelandic Volcanoes: Torfajökull, http://icelandicvolcanos.is/?volcano=TOR (1 March 2020).
Catalogue of Icelandic Volcanoes: Vestmannaeyjar, http://icelandicvolcanos.is/?volcano=VES (1 March 2020).
DAVIES, Alex: Why Volcanic Ash is So Terrible for Airplanes. In: WIRED, 22 August 2014, https://www.wired.
    com/2014/08/volcano-ash-planes/ (1 March 2020).
DigiPress, das Zeitungsportal der Bayerischen Staatsbibliothek, https://digipress.digitale-sammlungen.de/
    (1 March 2020).
EBERT, Stephan: Vulkane in der Umweltgeschichte oder das Problem der "Euphorie der Erkenntnis."
    In: Interdisziplinäre Forschung und Rezeptionsgeschichte (2016), https://mittelalter.hypotheses.org/
    7685 (29 December 2020).
ELIAS, Tamar; SUTTON, A. Jeff: Volcanic Air Pollution Hazards in Hawaii. In: Fact Sheet 2017–3017, 2017, DOI:
    10.3133/fs20173017 (1 March 2020).
ELLERTSDÓTTIR, Elin Þóra: Eyjafjallajökull and the 2010 Closure of European Airspace: Crisis Management,
    Economic Impact, and Tackling Future Risk. In: The Student Economic Review vol. 28 (2014), 129–137.
    https://www.tcd.ie/Economics/assets/pdf/SER/2014/elin_thora.pdf (29 December 2020).
Euro-Climhist, https://www.euroclimhist.unibe.ch/en/ (1 March 2020).
FORBES, Keith Archibald: Bermuda's Climate, Weather & Hurricane Conditions. In: Bermuda Online,
    http://www.bermuda-online.org/climateweather.htm (1 March 2020).
GARÐARSDÓTTIR, Ólöf: The Effects of the Volcanic Eruption in Iceland, 1783–1785. In: Conference of the
    European Society for Environmental History in Tallinn, 2019, 293–294. Conference abstracts: https://
    www.tlu.ee/sites/default/files/Abstracts%20ESEH2019%20-aug%2018-1.pdf (1 March 2020).
Gemeindechronik Schnürpflingen, Alb-Donau-Kreis, http://www.schnuerpflingen.de/gemeindeinfo/ge
    schichte/chronik/vom-westfaelischen-frieden-bis-zum-staatsvertr1810.php (1 March 2020).
Geological Map of the Urach-Kirchheimer Region, 2015. This map was created by Ustill and it is licensed
    under a CC BY-SA 3.0 de license, https://de.wikipedia.org/wiki/Schw%C3%A4bischer_Vulkan#/media/
    Datei:Schwaebischer-Vulkan_Geo-Relief_Urach-Kirchheim.jpg (1 March 2020).
GESTSDÓTTIR, Hildur; BAXTER, Peter; GÍSLADÓTTIR, Guðrún Alda: Fluorine Poisoning in Victims of the 1783–1784
    Eruption of the Laki Fissure, Iceland. Eystri Ásar/Búland – Pilot Study Excavation Report. Reykjavík
    2006. https://fornleif.is/wp-content/uploads/2018/01/FS328-04291-Fluorosis.pdf (29 December 2020).
GIBBONS, Brett: Fog, Mist, Smog and Haze: Is there a difference? In: The Weather Channel, 13 November
    2018, https://weather.com/en-GB/unitedkingdom/weather/news/2018-11-13-uk-weather-fog-mist-
    smog-haze-what-is-difference (29 November 2018).
Global Volcanism Program, 2013: Asamayama, (283110) in Volcanoes of the World, v. 4.8.5. VENZKE, E. (ed.).
    Smithsonian Institution, https://volcano.si.edu/volcano.cfm?vn=283110 (15 February 2020),
    DOI: 10.5479/si.GVP.VOTW4-2013.
Global Volcanism Program, 2013: Bárðarbunga, (373030) in Volcanoes of the World, v. 4.8.5. VENZKE, E.
    (ed.). Smithsonian Institution, http://volcano.si.edu/volcano.cfm?vn=373030 (15 February 2020),
    DOI: 10.5479/si.GVP.VOTW4-2013.
Global Volcanism Program, 2013: Etna, (211060) in Volcanoes of the World, v. 4.8.5. VENZKE, E. (ed.).
    Smithsonian Institution, http://volcano.si.edu/volcano.cfm?vn=211060 (15 February 2020),
    DOI: 10.5479/si.GVP.VOTW4-2013.

Global Volcanism Program, 2013: Eyjafjallajökull, (372020) in Volcanoes of the World, v. 4.8.5. Venzke, E. (ed.). Smithsonian Institution, http://volcano.si.edu/volcano.cfm?vn=372020 (15 February 2020), DOI: 10.5479/si.GVP.VOTW4-2013.

Global Volcanism Program, 2013: Grímsvötn, (373010) in Volcanoes of the World, v. 4.8.5. Venzke, E. (ed.). Smithsonian Institution, http://www.volcano.si.edu/volcano.cfm?vn=373010 (15 February 2020), DOI: 10.5479/si.GVP.VOTW4-2013.

Global Volcanism Program, 2013: Hekla, (372070) in Volcanoes of the World, v. 4.8.5. Venzke, E. (ed.). Smithsonian Institution, http://www.volcano.si.edu/volcano.cfm?vn=372070 (15 February 2020), DOI: 10.5479/si.GVP.VOTW4-2013.

Global Volcanism Program, 2013: Katla, (372030) in Volcanoes of the World, v. 4.8.5. Venzke, E. (ed.). Smithsonian Institution, http://www.volcano.si.edu/volcano.cfm?vn=372030 (15 February 2020), DOI: 10.5479/si.GVP.VOTW4-2013.

Global Volcanism Program, 2013: Kīlauea (332010) in Volcanoes of the World, v. 4.8.5. Venzke, E. (ed.). Smithsonian Institution, http://www.volcano.si.edu/volcano.cfm?vn=332010 (15 February 2020), doi:10.5479/si.GVP.VOTW4-2013.

Global Volcanism Program, 2013: Krakatau, (262000) in Volcanoes of the World, v. 4.8.5. Venzke, E. (ed.). Smithsonian Institution, http://volcano.si.edu/volcano.cfm?vn=262000 (15 February 2020), DOI: 10.5479/si.GVP.VOTW4-2013.

Global Volcanism Program, 2013: Krýsuvík-Trölladyngja, (371030) in Volcanoes of the World, v. 4.8.5. Venzke, E. (ed.). Smithsonian Institution, https://volcano.si.edu/volcano.cfm?vn=371030 (21 November 2022), DOI: 10.5479/si.GVP.VOTW4-2013.

Global Volcanism Program, 2013: Öræfajökull, (374010) in Volcanoes of the World, v. 4.8.5. Venzke, E. (ed.). Smithsonian Institution, https://volcano.si.edu/volcano.cfm?vn=374010 (15 February 2020), DOI: 10.5479/si.GVP.VOTW4-2013.

Global Volcanism Program, 2013: Reykjanes (Nýey), (371020) in Volcanoes of the World, v. 4.8.5. Venzke, E. (ed.). Smithsonian Institution, http://www.volcano.si.edu/volcano.cfm?vn=371020 (15 February 2020), DOI: 10.5479/si.GVP.VOTW4-2013.

Global Volcanism Program, 2013: Torfajökull, (372050) in Volcanoes of the World, v. 4.8.5. Venzke, E. (ed.). Smithsonian Institution, http://volcano.si.edu/volcano.cfm?vn=372050 (15 February 2020), DOI: 10.5479/si.GVP.VOTW4-2013.

Global Volcanism Program, 2013: West Eifel Volcanic Field, (210010) in Volcanoes of the World, v. 4.8.5. Venzke, E. (ed.). Smithsonian Institution, https://volcano.si.edu/volcano.cfm?vn=210010 (26 October 2022), DOI: 10.5479/si.GVP.VOTW4–2013.

Global Volcanism Program: Current Eruptions. In: Smithsonian Institution, https://volcano.si.edu/gvp_currenteruptions.cfm (19 January 2023).

Global Volcanism Program: How Many Volcanoes Are There? https://volcano.si.edu/faq/index.cfm?question=activevolcanoes (1 March 2020).

Guffanti, Marianne; Casadevall, T. J.; Budding, Karin: Encounters of Aircraft with Volcanic Ash Clouds; A Compilation of Known Incidents, 1953–2009. In: U. S. Geological Survey Data Series 545, ver. 1.0 (2010). https://pubs.usgs.gov/ds/545/ (30 December 2020).

Halldórsson, Eyþór: The Dry Fog of 1783: Environmental Impact and Human Reaction to the Lakagígar Eruption. Master thesis, University of Vienna, 2013. https://skemman.is/bitstream/1946/17205/1/MA-Thesis.pdf (30 December 2020).

Hellman, Geoffrey: The Laki Volcanic Eruption of 1783–1784: A Reappraisal and Reinterpretation of the Consequences of the Event in Europe. Villain or Fall Guy? Doctoral thesis, Université Rennes 2, 2021. https://tel.archives-ouvertes.fr/tel-03533049/document (13 November 2022).

Historical Demographical Data of the Whole Country of Iceland Between 1703 and 2050 (expected), http://www.populstat.info/Europe/icelandc.htm (23 December 2018).

Icelandic Tourist Board (Ferðamálastofa): Tourism in Iceland in Figures, 2018, https://www.ferdamalas tofa.is/static/files/ferdamalastofa/talnaefni/tourism-in-iceland-2018_2.pdf (21 February 2021).

Íslandskort: Maps of Iceland Sorted by Age and Origin, https://islandskort.is/en/ (1 March 2020).

JAKOBSDÓTTIR, Katrin: The Ice is Leaving. In: New York Times, 17 August 2019, https://www.nytimes.com/2019/08/17/opinion/iceland-glacier-climate-change.html (29 December 2020).

JOHANNSDÓTTIR, Gudrún Eva; THORDARSON, Thorvaldur; GEIRSDÓTTIR, Aslaug: The Widespread 10ka Saksunarvatn Tephra: A Product of Three Separate Eruptions? In: American Geophysical Union, Fall Meeting 2006, abstract id. V33B-0666. https://meetings.copernicus.org/www.cosis.net/abstracts/EGU05/05991/EGU05-J-05991.pdf (30 December 2020).

Katla Geopark Project: Geological Report, http://www.katlageopark.is/media/39154/Geological-report.pdf (1 March 2020).

KLEEMANN, Katrin: Watch Your Step. Moss Conservation in Vatnajökull National Park, Iceland. In: Seeing the Woods, 18 October 2016, https://seeingthewoods.org/2016/10/18/worldview-watch-your-step/ (1 March 2020).

KLEEMANN, Katrin: The Laki Fissure Eruption, 1783–1784. In: Encyclopedia of the Environment | Encyclopédie de l'environnement (2020). https://www.encyclopedie-environnement.org/en/society/laki-fissure-eruption-1783-1784/ (20 April 2021).

KLEMM, Friedrich: Hemmer, Johann Jacob. In: Neue Deutsche Biographie 8 (1969), 510 (Online version). http://www.deutsche-biographie.de/pnd118826190.html (30 December 2020).

KÜBLER, Simon: Active Tectonics of the Lower Rhine Graben (NW Central Europe). Based on New Paleoseismological Constraints and Implications for Coseismic Rupture Processes in Unconsolidated Gravels. Doctoral thesis, LMU Munich, 2012. https://edoc.ub.uni-muenchen.de/15596/1/Kuebler_Simon.pdf (30 December 2020).

MAYR, Helmut: Reck, Hans. In: Neue Deutsche Biographie (NDB). Vol. 21, 232–233. Berlin: Duncker & Humblot, 2003. https://www.deutsche-biographie.de/downloadPDF?url=sfz104643.pdf (30 December 2020).

Meteorological Observations for Arras, Pas-de-Calais, France. Juin 1783. By the Société Royale de Médecine. Histoire de la Société Royale de Médecine, http://meteo.academie-medecine.fr/_app/visualisation.php?id=2414 (1 March 2020).

MICHNOWICZ, Sabina: The Laki Fissure Eruption and UK Mortality Crises of 1783–1784. Master thesis at Aberystwyth University, 2011. https://cadair.aber.ac.uk/dspace/handle/2160/7793 (30 December 2020).

MÖHRING, Christa: Eine Geschichte des Blitzableiters. Die Ableitung des Blitzes und die Neuordnung des Wissens um 1800. Doctoral thesis, University of Weimar, 2005. https://e-pub.uni-weimar.de/opus4/frontdoor/index/index/year/2009/docId/1374 (30 December 2020).

MORGAN, Monique R. The Eruption of Krakatoa (also Known as Krakatau) in 1883. In: Branch Collective, http://www.branchcollective.org/?ps_articles=monique-morgan-the-eruption-of-krakatoa-also-known-as-krakatau-in-1883 (1 March 2020).

MOXHAM, Noah. Crowd-Sourcing Eighteenth-Century Science: The Great Fireball of 1783. In: The Repository, the History of Science Blog of the Royal Society (2013), blogs.royalsociety.org/history-of-science/2013/10/16/crowd-sourcing/ (1 March 2020).

National Aeronautics and Space Administration (NASA): Five Millennium Catalog of Lunar Eclipses, -1999 to +3000 (2000 BCE to 3000 CE), http://astro.ukho.gov.uk/eclipse/1511783/L1783Sep10.pdf (1 March 2020).

National Oceanic and Atmospheric Administration (NOAA): National Geophysical Data Center / World Data Service (NGDC/WDS), Significant Earthquake Database. National Geophysical Data Center, NOAA, DOI: 10.7289/V5TD9V7K (1 March 2020).

National Research Council (US): Committee on Acute Exposure Guideline Levels, 2010, https://www.ncbi.nlm.nih.gov/books/NBK219999/ (1 March 2020).

NEALE, Greg: How an Icelandic Volcano Helped Spark the French Revolution. In: The Guardian, 15 April 2010, https://www.theguardian.com/world/2010/apr/15/iceland-volcano-weather-french-revolution (30 December 2020).

OGILVIE, Astrid; HILL, Brian T.; JÓNSSON, Trausti: Sea Ice as Enemy and Friend: The Case of Iceland and Labrador/Nunatsiavut. In: Proceedings of the Northern Research Forum (2011). https://www.rha.is/static/files/NRF/OpenAssemblies/Hveragerdi2011/proceedings/finalastridogilvienrf2011paper.pdf (30 December 2020).

PITEL, Wilfried; DESARTHE, Jérémy: Les Brouillards d'Islande événements extrêmes et mortalités. In: HISTCLIME: Histoire des sociétés et des territoires face au climat et aux événements extrêmes, http://www.unicaen.fr/histclime/laki.php (1 March 2020).

POWERS, Wendy: The Science of Smell Part 1: Odor Perception and Physiological Response. In: Iowa State University, PM 1963a (2004), http://www.aerisa.com/wp-content/uploads/2014/10/3-Science-of-Smell-Parts-1-4-Iowa-State-PM1963.pdf (1 March 2020).

ROBOCK, Alan: Volcanic Aerosols as an Analog for Geoengineering. In: Presentation, kiss.caltech.edu/workshops/geoengineering/presentations1/robock_sc.pdf (1 March 2020).

Science Direct: Mantle Plume, https://www-sciencedirect-com.emedien.ub.uni-muenchen.de/topics/earth-and-planetary-sciences/mantle-plume (1 March 2020).

SECKEL, Al; EDWARDS, John: Franklin's Unholy Lightning Rod. In: ESD Journal 1984, https://web.archive.org/web/20060526223201/http://www.evolvefish.com/freewrite/franklgt.htm (1 March 2020).

SHEEC: Belledonne Earthquake, 21 June 1783. STUCCHI, M. et al.: The SHARE European Earthquake Catalogue (SHEEC), 1000–1899. In: Journal of Seismology, https://www.emidius.eu/SHEEC/maps/query_eq/external_call.php?eq_id=6127 (1 March 2020), DOI: 10.1007/s10950-012-9335-2.

SHEEC: Luzern [Lucerne] Earthquake, 11 August 1783. STUCCHI, M. et al.: The SHARE European Earthquake Catalogue (SHEEC), 1000–1899. In: Journal of Seismology, https://www.emidius.eu/SHEEC/maps_nomdp/query_eq/external_call.php?eq_id=200976 (1 March 2020), DOI: 10.1007/s10950-012-9335-2.

SHEEC: Vallée de l'Ouche Earthquake, 6 June 1783. STUCCHI, M. et al.: The SHARE European Earthquake Catalogue (SHEEC), 1000–1899. In: Journal of Seismology, https://emidius.eu/SHEEC/maps/query_eq/external_call.php?eq_id=6131 (1 March 2020), DOI: 10.1007/s10950-012-9335-2.

SPENCE, Peter: How a Volcanic Explosion Could Trigger the Next French Revolution. In: The Telegraph, 11 December 2014, https://www.telegraph.co.uk/finance/economics/11285291/How-a-volcanic-explosion-could-trigger-the-next-French-Revolution.html (30 December 2020).

Tambora.org, https://www.tambora.org/index.php/site/index (1 March 2020).

The National Archives of Iceland: Census Database, http://manntal.is/. The1703census, http://manntal.is/leit/_/1703/1/1703. The 1816 census, http://manntal.is/leit/_/1816/1/1816. The 1835 census, http://manntal.is/leit/_/1835/1/1816 (all 1 March 2020).

Time and Date website, used to determine sunrise and sunset times for various locations, https://www.timeanddate.com/ (1 March 2020).

Umweltbundesamt | German Environment Agency: "Das Luftmessungsnetz des Umweltbundesamtes. Langzeitmessungen, Prozessverständnis und Wirkungen ferntransportierter Luftverunreinigungen." 2013, https://www.umweltbundesamt.de/sites/default/files/medien/378/publikationen/das_luftmessnetz_des_umweltbundesamtes_bf_0.pdf (1 March 2020).

United States Geological Survey (USGS): "Does Vog (Volcanic Smog) Impact Plants and Animals?" https://www.usgs.gov/faqs/does-vog-volcanic-smog-impact-plants-and-animals?qt-news_science_products=0#qt-news_science_products (1 March 2020).

United States Geological Survey (USGS): "Historic Earthquakes: Lisbon, Portugal, 1755 November 01," http://earthquake.usgs.gov/earthquakes/world/events/1755_11_01.php (1 March 2020).

United States Geological Survey (USGS): "Preliminary Summary of Kīlauea Volcano's 2018 Lower East Rift Zone Eruption and Summit Collapse." (2018), https://volcanoes.usgs.gov/vsc/file_mngr/file-192/PrelimSum_LERZ-Summit_2018.pdf (1 March 2020).

United States Geological Survey (USGS): "The Modified Mercalli Intensity Scale." (1989), https://www.usgs.gov/natural-hazards/earthquake-hazards/science/modified-mercalli-intensity-scale?qt-science_center_objects=0#qt-science_center_objects (1 March 2020).

United States Geological Survey (USGS): "Volcanic Ashfall Impact Working group: Aviation," https://volcanoes.usgs.gov/volcanic_ash/ash_clouds_air_routes_effects_on_aircraft.html (1 March 2020).

United States Geological Survey (USGS): "Volcanic Gases and Their Effects." http://volcanoes.usgs.gov/hazards/gas/ (1 March 2020).

United States Geological Survey (USGS): "What Health Hazards are Posed by Vog (Volcanic Smog)?" https://www.usgs.gov/faqs/what-health-hazards-are-posed-vog-volcanic-smog?qt-news_science_products=0#qt-news_science_products (1 March 2020).

Veðurstofa Íslands: "Fagradalsfjall Eruption," https://en.vedur.is/volcanoes/fagradalsfjall-eruption (24 May 2021).

VOGRIPA: East Eifel Volcanic Field, Volcano Global Risk Identification and Analysis Project, https://www.bgs.ac.uk/vogripa/searchVOGRIPA.cfc?method=detail&id=156 (1 March 2020).

VOGRIPA: Toba, Volcano Global Risk Identification and Analysis Project, http://www.bgs.ac.uk/vogripa/searchVOGRIPA.cfc?method=detail&id=791 (1 March 2020).

WEICHSELGARTNER, Juergen: Naturgefahren als soziale Konstruktion. Eine geographische Beobachtung der gesellschaftlichen Auseinandersetzung mit Naturrisiken. Doctoral thesis, Rheinische Friedrich-Wilhelms-Universität Bonn, 2001. https://bonndoc.ulb.uni-bonn.de/xmlui/handle/20.500.11811/1728 (30 December 2020).

Wetterkontor, https://www.wetterkontor.de/de/wetter/deutschland/monatswerte-station.asp (1 March 2020).

WILKE, Jürgen: "Das Erdbeben von Lissabon (1755)." Leibniz-Institut für Europäische Geschichte Online (IEG EGO) (ed.), 2014, http://ieg-ego.eu/de/threads/europaeische-medien/europaeische-medienereignisse/juergen-wilke-das-erdbeben-von-lissabon-1755 (1 March 2020).

Wörterbuchnetz, Trier Center for Digital Humanities, www.woerterbuchnetz.de (1 March 2020).

Index

Aachen 149, 292
ABILDGAARD, Peter Christian 249
Académie royale des sciences 170, 328
academy 110, 122, 130, 152, 154, 235
accident (history of geology) 89, 188, 281
ADAIR, William 206
ADAMS, John 95, 216
Adorf 131
Adriatic Sea 117
Aeolian Islands 88, 102
aerosol 10, 14, 121–122
Africa 1, 4–5, 14, 212–213, 266, 285, 289
aftermath 2–3, 7–8, 13, 23, 29, 33–35, 60, 78, 81, 87,
    101–102, 104, 123, 138, 153, 164, 204, 223–224,
    245–246, 250, 257, 269–270, 282, 288, 295
agriculture 49, 52, 55, 83, 106, 203, 205, 211, 283,
    297, 306
ague 151
AHLWARDT, Peter 139
aid 56, 82–84, 87, 106, 117, 221, 224, 284, 293
air traffic 293
Airborne Volcanic Object Imaging Detector
    (AVOID) 294
AL-MUSABI, Muhammad b. Yusef 212
Alaska 13, 15–16, 205, 210–211
albedo effect 10, 12, 21
Albertus Magnus 89
Aleppo 115
ALEXANDER, Bethia 143
Alice Springs 266
almanac 32, 206–207
Alpha Ridge 35
Alps, the 117, 146, 258, 291
Altai Mountains 115, 205, 214
Althing 47, 50
ALVUS, John 158
American Revolutionary War 94–96, 152
Andechs 111, 133
ANDERSON, Johann 241
ANDERSON, Tempest 277–278
animal 18, 49, 52, 55, 58, 59, 64, 69, 73–76, 78,
    82–83, 87, 98, 118, 125, 136, 151, 163, 218, 228,
    236, 283–284
Ankara 134
Annapolis 96
Annonay 1, 97

annus mirabilis 94, 160
anomaly 21, 49, 109, 135, 217
anonymous 30, 121, 130–131, 138–139, 152, 184,
    194, 210, 222, 230
Antarctica 5, 15, 21
Anthropocene 21
anthropogenic climate change 16, 18, 21, 34,
    296–297, 306
anticyclone 134
Antilles 150
Antwerp 168
archives of nature 13, 20
archives of society 20
Arctic 5, 15, 35, 210–211, 238, 283, 285
Arctic Circle 211, 238, 283
Arctic Ocean 35
Ardennes 164
aridification 303
Aristotle 89, 187, 188
Asama 5, 214–215, 283
ash 2, 6, 9, 10, 41, 51, 55, 63, 65, 73–74, 103, 115,
    123, 128, 138, 198, 202, 204, 235, 242, 248, 251,
    260, 265–266, 269, 284, 293–294, 298
Asia 1, 4, 213
Askja 52
asthma 122, 125–126, 128, 153, 297
astrology 23
astronomy 23, 195, 240
Atlantic Ocean 61, 116
atmosphere 1–2, 9, 10–11, 15, 21, 33, 35, 89, 100,
    118, 126–127, 132–136, 147, 162–163, 174,
    186–187, 191, 195–196, 201, 204, 207, 212–213,
    235–236, 260, 265–266, 269, 271–272, 283,
    291, 297
AUBERT, Alexander 192, 194
Augsburg 29–30, 176, 182, 184, 199, 224, 232
Australia 266–267
Austria 131, 144, 147–148, 225, 230, 232, 291
Austrian Netherlands, see Netherlands
authority 30–31, 53, 142, 165, 190, 245
autobiography 28–29, 62, 65, 67, 78, 86
autumn 1, 47, 54, 76–77, 79–80, 82, 108, 125, 140,
    150–151, 159, 164, 187, 234, 284, 295
Avignon 270
Azores 13, 100, 109, 116
Azores High 13

BACHE, Richard 227
Baffin Island 37
Baghdad 115
ballomania 94, 97
Baltic Sea 116–117
Bamberg 109, 199–200, 224
BANKS, Joseph 3, 97, 107, 193, 243, 258–259
Barcelona 200
Bárðarbunga 15, 26, 42, 45, 47, 51, 305, 307
BÁRÐARSON, Guðmundur G. 4
BARDILI, Christoph Gottfried 91, 162, 187, 189, 198
BARKER, Thomas 121, 161, 191, 220
barometer 32, 93, 111, 173, 197, 253
barter 54
Basel 146
Batavia (Jakarta) 266–267, 271
Bavaria 111, 114, 143, 146, 235–236
Bavarian Academy of Sciences and Humanities
    (Bayerische Akademie der
    Wissenschaften) 110, 122, 130, 152
Bayreuth 182
BECK, Johannes 209
Bede the Venerable 89
Bedfordshire 150, 152
BÉGUELIN 114
Belgium 191, 306
Belmont Castle 155
Bentheim 226
Berlin 30, 114, 131, 133, 136, 144, 154, 157, 164, 174,
    176, 184, 186–187, 199
Bermuda 229
Bern 117–118, 199
BEROLDINGEN, Franz von 157, 163, 171, 173–174, 187
Besançon 186
Bible 167, 169, 259–260
biblical Flood 25, 90, 281
Bienne 117
bird 49, 74, 108, 211, 218
bishop 55, 76, 79, 86, 168, 243
BJARNADÓTTIR, Þórunn 257
Black Death 53, 97
BLAGDEN, Charles 195
Boesand 107
Bohemia 138, 159
Boston 189, 205, 208, 228
botany 243, 250
Botnahraun 276
BRAHE, Tycho 195
BRÄKER, Ulrich 140, 218

BRANAGIN, James 209
Bratislava 160
breathing difficulties 16, 73, 123, 126, 184
Breslau 30, 199
Brest 173
Brieg 136, 138
BRIGHT, Richard 259
Britain, see Great Britain
Brittany 152
British and Foreign Bible Society 259
British Isles 37, 46, 191, 240
British Krakatoa Commission 267–269, 288
Brno 159
brouillard sec 113, 165
BRUGMANS, Sebald Justinus 122–124, 174, 197, 204
Brussels 122, 191, 199
BUISSART 121
Burgundy 147
burial records 150, 152–153, 295
Burning Mountain (Brennender Berg) 90, 174,
    178–180

Cairo 213
Caithness 115, 265
Calabria 1, 16, 100, 102–108, 146, 150, 159–160, 164,
    173, 185–189, 200–203, 208, 221, 235, 265,
    282, 289, 296–297
Calais 121, 168, 191
calamity 23, 58, 85, 101, 284
caligine 113
Cambridge 127, 155, 206
Canada 33, 37, 84, 205
Caribbean 115, 229
CAROLUS, Johann 29
cassa sacra 106
catastrophe 23, 26, 85, 224, 248, 262, 271, 284, 300
Catherine the Great 97
Catholic 53, 140, 168
cattle 49, 51, 70, 73, 76, 80, 215, 260, 277, 284
CAVALLO, Tiberius 194
Cenozoic 177
censorship 30, 153, 183
cereal 77
Channel, the 121, 191, 216
Charles Theodore, Prince-elector and Count
    Palatine, Duke of Bavaria 111, 141
CHARLES, Jacques 97, 99
Charleston 228
chemistry 9, 89

cherry blossom 218
Chesapeake Bay 228
China 5, 115, 134, 150, 205, 214–215, 285
chocolate (hraun) 302
CHRIST, Johann Ludwig 164
Christian IV 53
Christian VII 59, 97, 108–109
Christianity 50
chronicle 20, 51, 122, 161, 164, 166, 212, 215, 285, 289
church 24, 31, 53–55, 61–62, 64, 68–69, 78–80, 83, 101, 106, 132, 138–141, 143, 168, 176, 198–199, 208, 223–224, 260, 284, 298
civil protection 295–297
clergy 55, 58, 120–121, 171
climate 5–6, 9–14, 18–22, 25, 27–28, 33–35, 42, 46, 49, 53–54, 86, 214, 220–221, 233–235, 283, 291–293, 296–297, 302, 304, 306–307
climate change 18–20, 22, 34, 86, 296–297, 306–307
climate forcing 10, 12, 14, 235
climate history 19–22, 25, 234, 291–293
climate modeling 10, 12–14, 42
climate variability 12–13, 19, 110, 234
climatology 19, 94
Club of Rome 18
Coevorden 125
collapse 20, 43, 64, 68, 72, 79, 84, 104, 176, 181, 218, 288, 304
Cologne 146, 222–223
colonialism 50–54, 198, 243, 270, 284
comet 88, 161, 195, 204, 236, 269, 296
Common Era 21, 165
Concord 207
cone 2, 66–67, 90, 177, 253, 262, 281
conflict 30, 46, 85, 94, 186, 259, 281, 304
Congress 95–96, 228
connected histories 27
Connecticut 207–208
conservation 18, 299
Constantinople 134
continental crust 36, 41, 291
continental drift 289–290
COOK, James 243
cooling 5, 10, 12–14, 21, 135, 210, 212–215, 232, 293, 296
COOPER, William 191
Copenhagen 53, 58–59, 77, 81–84, 87, 114, 133–134, 155, 198–199, 201–202, 240, 242, 245, 248, 250–253, 256, 264, 284, 288

COPERNICUS, Nicolaus 26
coping 170
correspondence 3, 28–30, 34, 58, 66, 81, 100, 106, 108, 145, 147, 152, 166, 169, 174, 176–178, 181–182, 184, 187, 194, 196–197, 199–200, 203, 206–207, 216–217, 227, 229–230, 243, 245, 248, 251, 256, 263, 277, 285, 297
correspondent 29–30, 100, 121, 123, 131, 136, 138–139, 152–153, 157, 164–166, 173, 176–178, 182–185, 189–190, 195, 198, 230
Cottaberg 1, 175–180, 182, 184, 282
Cottaer Spitzberg, *see* Cottaberg
cough 122, 128, 231
cow, *see* cattle
COWPER, William 151, 166, 289
crater 2, 44, 66–67, 109, 175, 252–254, 264, 272–274, 276–279, 304
Cretaceous 35
Créteil 129
crisis 8, 20, 54, 81–82, 106, 130, 150–151, 159, 170, 211, 234, 283–284, 295
crop failure 159, 211–212
CULLUM, John 121
CUTLER, Manasseh 155
Czech Lands 16, 138–139, 146, 232, 234

DE POEDERLÉ, Eugène-Joseph D'OLMEN, Baron 122, 152, 158, 203–204
Dalkeith 158
Dalton Minimum 22, 234
DALTON, John 22
Danish central administration 58, 63, 81–82, 240, 245, 284
Danish Crown 29, 53–56
Danish Natural History Society 3, 249, 252, 256–257
Danish Royal Society 242
Danube River 221, 225, 232
data 5, 8, 13–14, 18, 20, 24, 26–27, 32, 80, 86, 92, 94, 127, 134, 156, 163, 212, 232–233, 250, 269, 290
death 8, 18, 35, 51, 53, 57, 59, 62, 74, 76–77, 79–80, 84, 86–87, 101, 151–153, 161, 165, 211, 213, 229, 232, 266, 295
Deccan Traps 16
deep geological time 16, 19, 25–26, 93, 283
Delaware 206
Delaware River 227
DELUC, Jean-André 25, 118

dendrochronology 212

Denmark 52–54, 56, 59, 77, 82–83, 87, 96–97, 108–109, 115, 202, 220, 240, 245–246, 250, 267, 271–272

desertification 49, 303

Dessau 30, 182

diet 48–49, 52

Dijon 133, 147, 149–150, 186

dinosaur 35

disaster 1–2, 17, 20, 22–23, 25, 63–64, 81, 85, 176, 211, 284, 289, 297, 307

disaster history 17, 22–23

discourse analysis 17, 25, 286–287

discovery 3, 24, 26–27, 46, 50, 89, 94, 107, 131, 163, 289, 298

disease 2, 9, 53–58, 73–74, 84, 106, 122, 130, 150–152, 172, 202, 215, 284

district governor 86

Djerba (island) 212

Djúpivogur 76

Dominicus de Gravina 162

donation 83

DONAUBAUER, P. 114

Dresden 31, 131, 138, 175

drought 12, 14, 125, 132, 152, 157, 164, 206, 213, 215, 295

duck 98

Dudweiler 180

Duft 113, 162, 167, 181

Dunkirk 191

Dunst 113, 118, 130–131, 138, 157, 173, 185, 190, 208, 230

Düren 102, 146

Düsseldorf 133, 142, 146, 167

dust veil index (DVI) 5

Dutch East India Company 270

Dutch East Indies 3, 270–271, 282

Dutch Krakatoa Commission 269, 288

Dutch Republic 97, 115, 122, 125, 133, 149, 153, 156, 197, 204, 220

dysentery 74, 76, 129, 151–153, 234

Earth 2, 7–10, 12, 21, 25–27, 35–37, 45, 65, 69, 73, 88, 90, 113, 130, 154, 159, 167, 170, 173–174, 177, 185–188, 191, 195–196, 200, 204, 208, 230, 238, 255, 257, 259, 272, 282, 288–292

earth fire 88, 202–203, 245, 247–248

earthquake 1, 16, 23, 34, 43, 64–68, 88, 90, 100–108, 116, 121, 138–139, 144–150, 159–160, 163–164, 172–174, 178, 180, 183, 185–189, 200–204, 208, 235–238, 241, 255, 262, 271, 282, 284–286, 289, 291–292, 296–297, 304–305

East Greenland Current 49, 76

Eastern Volcanic Zone 42–43, 45, 51, 61, 302, 307

eclipse 140, 154–155, 284

economy 51, 59–60, 80–81, 85, 246, 288, 294, 298–299

Edinburgh Royal Society 259

Egypt 14, 205, 212–213, 295

EIRÍKSSON, Jón 64, 245

El Chichón 9

El Niño 213, 235

El Niño-Southern Oscillation 13, 27, 213, 234

Elbe River 226

Elbe Sandstone Mountains 175

Eldfell 295, 299

Eldgjá 47–48, 72, 274, 278, 284, 300–305

eldrit 29, 54, 62–63, 69, 71, 75

electricity 24–25, 89, 92, 141, 188–189, 204, 236

emotion 1, 17, 24, 88, 160, 287

Ems River 123

Emsland 122–123

end time 166–168

England 8, 31, 96, 100, 113, 120–125, 132, 134, 137, 149–160, 169, 189, 191–194, 206, 219, 227, 231–232, 264, 267, 285, 295

engraving 99, 105, 224

Enlightenment 3, 24, 31, 34, 63, 85, 86, 89, 94, 101, 142–143, 157, 161, 166, 169–173, 197, 238, 250, 263, 281, 286–287, 294

ENSO, see El Niño-Southern Oscillation

entangled histories 27

environmental history 16–19, 24–28, 291, 292

environmental impact 4, 7, 9, 73

Ephemerides 32, 111–112, 122, 130, 152, 178, 204, 232, 245

epidemics 53, 85, 213

epizootic 55, 59–60, 283

EPP, Franz Xaver 235–236

erosion 41, 49, 51, 70, 83, 109, 259, 274, 303

eschatological concern (Judgment Day) 161

ESCHELS, Jens Jacob 115–117

Estonia 211

Etna 88–90, 102–104, 107, 241, 258, 276

Eure-et-Loir 152

Europe 1–9, 11–16, 21–24, 28, 30–34, 46, 49, 51, 55, 58, 63, 81, 86–88, 94–97, 100, 102, 106,

108–118, 120, 126–128, 131–137, 141, 144–162, 164, 169–172, 177–178, 183, 187, 189, 191, 194, 198, 201, 203–205, 207–226, 228–236, 238, 240, 243, 257–258, 260, 262, 266, 269–271, 282–289, 293, 295–297, 299–300, 304, 306
European Macroseismic Scale (EMS-98) 100
evacuation 83
EVZ, *see* Eastern Volcanic Zone
expedition 3–4, 34, 82, 109, 241–243, 246, 256–259, 274, 288, 290
Experiment 3, 85, 92, 97, 99, 141, 169, 171, 196–198, 231, 286
explanation 1, 23–24, 31–32, 34, 87, 160, 163–164, 170–174, 180, 183, 187–188, 190, 196, 200, 204, 210, 226, 235–238, 242, 282, 286–287, 291, 296
explosivity, *see* Volcanic Explosivity Index
export 77, 83, 284
eye irritation 129
eyewitness 101, 269, 271
Eyjafjallajökull 6, 35, 47, 293, 297
Eyrarbakki 76

Fagradalsfjall 16, 47, 305
FAHRENHEIT, Daniel Gabriel 93, 129, 218
famine 2, 4, 14, 23, 51, 57–59, 75, 77–82, 84–86, 205, 211, 213–215, 234, 246, 276, 283, 295, 306
Far East 240
farm 50–53, 55, 57–59, 61–71, 80–87, 138–140, 159, 171, 198, 218, 242, 257, 284
farming, *see* agriculture
farmstead, *see* farm
Faroe Islands 46, 115, 202, 210, 235, 263
fear 1, 23–24, 62, 64–65, 69, 85, 102–103, 108, 146, 149, 158, 160–161, 163, 166, 169–170, 183, 218, 222, 234, 238, 286, 294, 296
FENTON, J. 120
Ferdinand IV 106
fertility 157, 159, 170, 189
fever 129–130, 150–153, 189, 234
Fimmvorduhals 47
Finland 101, 211–212, 236
Finnmark 58
FINNSSON, Hannes 76, 79, 86, 243
Fire District 2, 60–61, 64, 245, 248
fire priest (eldprestur) 63, 69, 286
fire sermon (eldmessan) 69
fire treatise (eldrit) 29, 62
fire-spitting mountain 88–91, 174–176, 182, 202, 247–248, 259, 304

fireball, *see* meteor
fires 2, 26, 43, 58, 62, 64–65, 67, 71, 88, 104, 187–188, 199–201, 204, 209–210, 237, 245, 248, 251, 254, 260, 276, 281, 283, 289, 304
firewood 216, 220, 223–224
FISCHER, Johann Nepomuk 171
fish 2, 49, 52, 55, 58, 76–77, 79, 83, 86, 107, 199, 211, 242, 298
fisherman 108, 266, 298
flood basalt 6, 9, 15–17, 26, 45, 47–48, 51, 60, 62–63, 67, 255, 276, 284, 294, 305–306
flood 7–8, 10, 13, 20, 23, 25, 52, 68–69, 90, 130, 153, 167, 173, 189, 213, 216, 219, 221–228, 232, 242, 250, 281, 286, 297–298, 303
floodmark 223–224, 289
Florence 147, 199
fluorosis 73, 76
fluorine 51, 73, 85, 87, 119, 121, 128–129
Föhr 115
food 35, 60, 74–80, 83–86, 151, 214, 216, 218, 223–224, 231, 234, 250, 297
fording 250
forest fire 209, 236
France 1, 7–8, 16, 30–32, 46, 96–97, 100, 110, 113, 120–121, 123, 125, 129, 132, 142–143, 147, 149–153, 156, 159, 161, 165–169, 173, 185–186, 191, 207, 211–212, 221, 226, 233–234, 240, 282, 294–295
Franche-Comté 147
Franeker 122, 133, 197, 204
Frankenstein 262
Frankfurt (am Main) 175–176, 181, 222
FRANKLIN, Benjamin 3, 33, 95–99, 141, 143, 189, 196–197, 203–204, 206, 216, 219–220, 227–231
Frederick IV 54
Frederick the Great 96
French Revolution 7, 96, 110, 142, 233–235, 258
French Revolutionary Wars 111
frequency illusion 188
frost 71, 79–80, 120–121, 123, 126, 137, 207, 211–212, 215, 217, 219–221, 223, 227, 231–233, 248

GALILEI, Galileo 93
GANNETT, Caleb 206
GEIKIE, Archibald 115, 265
Geldingadalir 15, 305
Geneva 114, 117–118, 133, 147, 173, 218, 235
Genkingen 179

geology 5–6, 25–26, 34–39, 46, 88–90, 102, 115,
    188, 243, 250, 262, 264–265, 275, 278–279,
    281, 283, 288, 291–292, 305
George III 96
Georges-Louis Leclerc, Comte de Buffon 101,
    188, 242
geothermal activity 45, 47, 177
German Territories, see Germany
Germany 1, 5, 7–8, 16, 29–32, 46, 92, 102, 106, 115,
    117, 122, 124–125, 130–131, 134, 137–141,
    143–144, 146, 148–149, 153–154, 156, 158, 160,
    166, 172, 174–185, 188, 195, 198, 201, 208, 218,
    221–222, 224, 226, 232, 235–236, 240, 264,
    282, 285–286, 291, 294–295, 297
geyser 238, 243
Gißler, D. 235
Gjálp 305
glacial outburst flood, see jökulhlaup
glacier 21–22, 43–45, 57, 61, 64, 66, 70–72, 199,
    250–256, 260, 264–265, 269, 274–277,
    306–307
glaciology 257
Glatz 145–146
Gleichberg 1, 7, 175–184, 187, 201, 282, 292
global warming, see anthropogenic climate change
God 69, 79, 86, 93, 100, 117, 140, 142, 168, 170,
    189–190, 223, 286
Goethe, Johann Wolfgang von 90
Gorgona (island) 139, 144
Gottesberg 145–146, 178–180, 184, 282
Göttingen 30, 114, 133, 169, 197
governor of Iceland 56, 82
Grand Tour 117, 258–259
grapes 130, 157–159, 162, 168
grass 49, 74–75, 79, 86–87, 125, 199, 202, 220, 283
Graubünden 129, 160
Great Britain 7–8, 16, 23, 32–33, 50, 94–96, 143,
    158, 191, 194–195, 221, 243, 294–295
Great Þjórsá Lava 45
Greenland 2, 5, 11, 14, 21, 35–37, 39, 45, 49–50, 76,
    111, 214, 264, 290
Grímsvötn 4, 42–47, 61, 64–65, 71–72, 115,
    254–255, 276, 305–307
Grindelwald Fluctuation 22
Gronau, Karl Ludwig 164, 217, 230
Groningen 115, 122, 152
Groß Hesepe 122–125
Grove, Jean M. 21
Grumbkow, Ina von 279–281

Grund 120
Guðmundsson, Jón 265
Guerin, H. 202
Gulf of Mexico 229
Gulf Stream 49, 229
gunpowder 122, 138–139
Gunnlaugsson, Björn 274

Haakon VI 52
Hahlrauch 114
hail 144, 174, 189
Halley, Edmond 195
Hallmundarhraun 48
halogen, see volcanic gases
Hamburg 29–30, 123, 133, 141, 199, 222, 226,
    241, 270
Hamilton-Temple-Blackwood, Frederick, Lord
    Dufferin 263
Hamilton, William 107
Hanau 30, 139
Hannesdóttir, Þórunn 62, 79
Hanseatic Trading League 53
hardship 3, 46, 54, 78–80, 85–86, 234,
    257–258, 299
Harmer, Josiah 229
harvest 7, 55–56, 63, 75, 78, 85–86, 93, 151–152,
    156–159, 168, 172, 231, 234
Hawai'i 6, 15, 125
hay 70–71, 75, 78–79, 85, 122, 159
haze 2–4, 7, 16–17, 25, 51, 65, 76, 78, 87–88, 103,
    109, 112–116, 121, 125–128, 132, 140, 155–156,
    164–165, 173, 189–190, 194, 196–199, 204–207,
    210, 212, 214–215, 230–231, 234, 236, 248, 284,
    287, 293, 295
headache 122, 129–130, 284
heat wave 7, 12–13, 125, 132–135, 151, 159, 284, 287
Heerrauch 113
Hehrrauch 158, 162
Heimaey 295–296
Hein, Karel Bernard 137
Heinze 108
Hekla 33, 42, 46, 50–51, 53, 56, 174, 187, 198–199,
    203–204, 238, 241–243, 258–259, 262, 282,
    299, 305
Heldburger Gangschar 177
Heligoland 116
Helland, Amund 3–4, 33, 264, 272–273, 276–278,
    281–282, 288
Hellisá River 251

HEMMER, Johann Jakob 111, 141, 246
Henan 215
HENDERSON, Ebenezer 259–263, 282
Hengill 47
Herbert of Clairvaux 50
Herculaneum 89
HERRMANN, Paul 83
HICKMANN, Dom Robert 164, 173, 200–201
high-latitude eruption 5
high-pressure system 12–13, 126, 134, 151, 221
highlands 2–3, 53, 60–61, 65, 67, 71, 213, 246,
        248, 250–251, 253, 272, 274–275, 283, 288,
        298, 302
Hildburghausen 165, 176–178, 181–182
HILLIGER 174, 184
HILTZHEIMER, Jacob 206
HINDENBURG, Carl Friedrich 165, 203
histoire croisée 27
historical climatology 19, 94
history of science 24, 25, 291
Hnúta Mountain 66
hoax 183–184
HØEGH-GULDBERG, Ove 59
Hoffenthal 156, 208–210
Höhenrauch 113, 265
HOFF, Karl Ernst Adolf von 270
Hólar 55, 62
Holland 122
HOLLAND, Henry 257, 259
HÓLM, Sæmundur Magnússon 29, 112, 201, 236,
        241, 245
Holocene 21, 26, 33, 40–46, 61, 175, 276, 306
Holstein 123–124, 138–139
Holuhraun 15, 47, 305
Holy Roman Empire of the German Nation 1, 96
HOLYOKE, Edward 207
Honshu 214
HOOKER, William Jackson 243, 259
HÖPPEL 164
HORREBOW, Niels 240
horse 51, 64, 76, 78–80, 104, 143, 219–220,
        242–243, 250–253, 279, 284, 303
hot-air balloon 1, 12, 24, 92, 97–100, 195, 236
Hrabanus Maurus 140
hraun (lava field), see lava field
hreppur 57
human health 7, 18, 120, 124, 127, 162, 198, 284
HUMPHREYS, William Jackson 262, 288
Hungary 96, 114, 221

HUTCHINSON, William 191
HUTTON, James 25, 90, 259
Hverfisfljót 60–61, 70, 76, 85, 247, 253–255, 263
hydrogen balloon 1, 24, 100
hygrometer 32, 111–112, 162, 197

ice age 19–22, 45, 306–307
ice core 2, 5, 11, 14, 20, 45, 214, 292–293
ice drift 221, 224, 226, 232
ice fair 220
iceberg 221
Iceland 1–9, 11, 15, 17, 25–26, 29, 33–89, 103,
        108–109, 112–118, 126, 128–129, 151, 159, 164,
        187, 194, 198–204, 215–216, 234–285,
        288–290, 293–307
Iceland mantle plume 16, 26, 35–36, 39–40, 43,
        290, 305
Icelandic highlands 2–3, 53, 60–61, 65–67, 71,
        203, 246–248, 250–253, 272–275, 283, 288,
        298, 302
Icelandic Low 13
igneous rock 37, 85, 263
ignorance 2–3, 238, 281–282, 288
IMBÓ, Guiseppe 107
impact 4–10, 14–17, 27, 34, 46, 51–52, 59, 60, 73,
        124–126, 159, 187, 204, 211–213, 218, 221, 231,
        234, 236, 248, 269, 271, 277, 281–285, 291, 294,
        297, 306
independence, Icelandic 95–96, 143
India 14, 16, 110, 205, 213, 270, 272, 295
Indian Ocean 213, 266–267
Indigenous peoples 205
Indonesia 6, 16
industrial revolution 21
insect 125, 164
intensity (earthquake) 64, 100–104, 114, 149
interdisciplinarity 19, 25, 291–292
Intertropical Convergence Zone 13, 213
interview 157, 161, 165
intraplate region 36, 41, 146
Iraq 110
Ireland 37, 50, 84, 96, 194, 221
Isidore of Seville 89
Isle of Man 8, 71
Íslendingabók 50
Italy 89–90, 96, 102, 106–107, 112–114, 117, 156,
        159, 163, 165, 167, 185–186, 191, 207, 220, 238,
        258, 264
Izmir 134

Jamaica 229
Japan 5, 150, 214, 283
Java 266–271
JAY, John 95, 229
Jersey 8
jet stream 2, 118, 120, 126, 128, 284
jökulhlaup 52, 58, 71–72, 242, 250, 253–254, 256, 292, 306–307
JÓNSSON, Arngrímur 238
JÓNSSON, Steingrímur 62
Joseph II 96, 168
journal, scientific 171
Jura, the 118, 147

Kaldbakur 67
Kalmar Union 53
Kamchatka 15
KANT, Immanuel 3, 101, 188
Kashmir 213
Katla 26, 41–42, 47, 51, 54, 58, 62, 67, 72, 103, 164, 199, 306–307
Kauwerak people 210–211
Kent 151
KERGUELEN TRÉMAREC, Yves-Joseph DE 242
KIESSLING, Johann 268–270
Kīlauea 6, 15
KIRCHER, Athanasius 90–91, 188, 200, 241
Kirkjubæjarklaustur 29, 60–70, 75, 77, 85–86, 246, 251, 252–253, 265, 296, 303
Klaustur, see Kirkjubæjarklaustur
Klofajökull, see Vatnajökull
KNEBEL, Walter VON 278
knowledge production 23, 34, 92
Königsberg 101
Konrad von Megenberg 89
Krafla 72, 262
Krakatau 3, 5, 17, 34, 266–272, 281–282, 288, 294
KRATZENSTEIN, Christian Gottlieb 201, 249
Kristiania (Oslo) 265, 272
KRÜGELSTEIN, Daniel 209
Kuwae 293

L'Anse aux Meadows 50
La Charlière 99, 298
La Niña 13, 213, 235
La Rochelle 115, 129
LA TROBE, Benjamin 154, 208
Laacher See 175
Labrador 33, 154, 205, 208–210, 226

Lakagígar 2, 276, 300
Lake Biel 117
Lake Constance 146
Lake Geneva 117
lake sediments 20
Laki fissure 2, 4, 25, 44, 62–66, 70–72, 86, 115, 128, 199, 204, 246–247, 251–253, 260, 264–266, 272–274, 276–283, 288, 299–303
Laki-style event 8, 13, 297, 305
LALANDE, Joseph Jérôme Lefrançois de 170
LAMANON, Robert Paul de 118, 165
Landnámabók 48, 50
landscape 3, 18, 33, 35, 59, 66, 85, 90, 118, 137, 175, 222, 242, 245, 257–258, 262, 272, 274, 278–285, 288, 291, 299–304
Langensalza 139
language 6, 16, 23, 31–34, 63, 88–89, 92, 106, 166, 190, 204, 240, 243, 246, 263, 272
Lapland 210
large igneous province 16, 26, 37–40, 63, 263, 290, 292
Late Antique Little Ice Age 21
Latin 32, 37, 53, 92, 111, 114, 162, 240
Latvia 211
Lauffell 251–252
LAURENS, Henry 95, 100
Lausanne 167
Lautern 138, 160
lava 2, 6, 9, 15, 17, 37, 43–48, 50, 61, 63, 65, 67–73, 85, 87, 118, 180, 198–199, 236, 242, 246–247, 251–255, 260, 262–263, 272, 274, 277–278, 280, 283–284, 286, 294, 296, 301–306
lava bomb 6
lava field (hraun) 44–45, 48, 61, 67, 70, 85, 248, 251–252, 275, 280, 301–303
lava flow 2, 10, 18, 29, 39, 48, 51, 53, 58, 68–69, 71, 214, 245–247, 253–254, 272–273, 277, 296, 302, 306
layperson 28, 94, 164
Le Globe 97
learned society 1, 4, 17, 24, 31, 92, 94
LEE, Joseph 207
legacy 17, 34, 289
Leicestershire 121, 150, 161, 191
Leiden 191
Leifur Eiríksson 50
LÉMERY, Nicolas 242
letter, see correspondence
LEVETZOW, Hans von 81, 109, 246

lichen 49, 59
LICHTENBERG, Georg Christoph 169, 174, 197
lightning 1, 7, 24, 58, 69, 92, 135, 136–144, 159–160, 172, 174, 178, 188–190, 286, 298
lightning rod 24, 141–143, 189–190
LINDEMAYR, Peter Gottlieb 168
Lisbon 100–103, 189, 199, 226, 271
literacy 31, 285
Little Ice Age 1, 19, 21–22, 24, 52, 54, 86, 109, 234, 283, 293, 307
Liverpool 191
livestock 2, 49, 51–52, 55–56, 58, 60, 73–75, 79–80, 82, 84–85, 130, 143, 159, 228
Livorno 139
locust 213, 215
Loiret 152
London 3, 23, 97, 133, 137, 143, 150, 193–195, 199, 208, 216, 219, 226, 233, 238, 267
Long Island Sound 228
Louis XVI 96, 98
Low Countries 16, 125, 146, 181, 220–221, 226
low-pressure system 125–126
LUBIERES, Charles-Benjamin 173
Lucerne 149
LUDWIG, Christian 184, 203
lunar eclipse 140, 154–155
Lüneburg 174, 222
Luxembourg 199
LYELL, Charles 102, 262
Lyon 147, 159

Maastricht 149
MACKENZIE, George Stuart 257, 259
Madagascar 46
Madrid 188, 199
magma 9, 36, 40, 45, 64, 67–68, 71–72, 107, 175, 305–307
magnitude 16, 46, 59, 64, 100–106, 146–149, 189, 263
MAGNUS, Olaus 238
MAGNÚSSON, Skúli 59, 245
MAGNÚSSON, Þorsteinn 54
Main River 224–225
malaria 129, 151
Maldà anomaly 109
malnutrition 54, 284
Manchester 150, 196, 203, 230

Manchester Literary and Philosophical Society 196, 203, 206
Mannheim 32, 110–111, 130, 132–133, 136–137, 153, 155, 157–158, 162, 166, 217, 221
mantle plume 16, 26, 35–40, 43, 283, 289–290, 292, 305
map 48, 56–57, 110, 146–147, 154–155, 178, 205, 238–239, 243–247, 251–254, 256, 260–261, 267, 272, 274–276, 279, 281, 289, 300
Marie Antoinette 96, 98
Mars 97, 292
Marseille 46, 133
Massachusetts 94, 111, 155, 189, 206–207
MATTHES, François 21
Maunder Minimum 22
MAUNDER, Edward 22
Mauritius 266
maximum latewood density 12, 210
Mecklenburg 165
Meðalland 82, 245
media 16, 23, 29, 32, 187, 285
medicine 159, 203, 242, 248
Medieval Climatic Anomaly 21
Mediterranean, the 88, 117, 135, 221
Meiningen 181
Meißen 175
memory 165–166, 174, 221, 232, 289, 298
Mercalli-Cancani-Sieberg scale (MCS) 104
merchant 53–54, 59, 76, 81–82, 86–87, 108–109, 143, 199, 214, 245, 284
MERCIER, Louis-Sébastien 1, 94
Messina 102–104, 107–108, 116–117, 121, 185–186, 189, 203
metamorphic rock 37
meteor 1, 16, 191–196, 201, 203–204, 283–286, 289, 296
meteora 88–89, 113–114, 188
meteorology 24, 27, 88, 92, 111, 172, 203
Meuse River 225
mice 74
Michigan 206
Mid-Atlantic Ridge 26, 39–43, 60, 283, 289
mid-oceanic ridge 26, 37, 290
Middle Ages 89, 93, 238
Middle East 12, 149, 214
Milanković cycles 21
MINDELBERG, Jörgen 107
Mississippi River 229

mist 1–2, 4, 23, 25, 77–78, 84, 88, 94, 112–118, 120, 130, 155, 194, 206, 208, 210, 212, 269, 276, 282, 288–289, 296
Mittenwald 145–146
Modified Mercalli Intensity Scale (MMI) 100
móðuharðindin 2, 23, 77–80, 83–86, 260, 276, 284
Moldau River 224
monastery 199
money 54, 83, 224, 249, 298
monsoon 10, 14, 213, 295
MONTGOLFIER, Jacques-Étienne 97
MONTGOLFIER, Joseph-Michel 97
Montgolfière 97–98
moon 70, 114, 136, 154, 160, 167, 172, 176–177, 181, 192, 266, 270
Moravia 33, 137, 154, 208, 210, 226
Moravian settlements 33, 226
MORGAN, W. Jason 290
Morocco 101
MORSE, Samuel 271
mortality 8, 23, 78–79, 81, 84–85, 87, 89, 129–130, 150–153, 162, 211, 226, 234, 295
moss 49, 80, 85, 251, 302–305
Mount Erebus 307
Mount Hekla, see Hekla
Mount Laki 2, 66–68, 70, 251–253, 274, 276, 280, 302, 304
Mount St. Helens 9
MOURGUE DE MONTREDON, Jacques Antoine 112, 161, 164–165, 187, 201, 212
Mullingar 194
Munich 30, 49, 117, 131, 152, 154, 158, 169, 198–199, 218, 226
Mýrdalsjökull 48, 52, 60–61, 199, 302
Mýrdalur 64
Mývatn Fires 58, 62

Nagasaki 214
Nain 154, 156, 208–209
NAO, see North Atlantic Oscillation
Naples 106–107, 186, 207
Napoleonic Wars 110, 175, 263
Narbonne 114
Natural History Society (of Denmark) 3, 249–252, 255–257
natural science, see science
naturalist 3, 16–17, 23–25, 31–34, 89, 92–94, 101, 106, 108, 112, 118, 120–121, 131, 141, 152, 157–158, 161–166, 169–173, 175, 177, 185–190,

194–196, 200–204, 235–236, 240, 243, 246, 248–249, 257–258, 276, 281–282, 285–291, 296
nebula 114, 174
Neckar River 221, 224
Neogene 175, 179
Neptunism 90, 259, 281
Netherlands 115, 122, 152, 158, 168, 191, 306
Neuchâtel 197
New Haven 228
New Madrid 188
New Orleans 229
New South Wales 110
New York 47, 208, 229
New Zealand 46
Newfoundland 50, 210
newspaper 1, 7, 17, 23, 28–33, 81, 91–92, 97, 101, 106–108, 117, 130–132, 136–139, 144–147, 149, 153, 156–157, 159–166, 168, 170–186, 190, 198–208, 220, 222, 228, 230, 243, 266, 269, 271, 276, 285, 298
NEWTON, John 166
Niedersgörsdorf 174
Nile River 10, 13–14, 213
noise 65, 67–71, 102, 140, 178, 180, 266–267, 286
Norfolk 120, 132–133, 191
North Africa 1, 4–5, 212, 285
North America 1, 4, 6, 13, 32–33, 39, 42, 46, 50, 84, 94–96, 132, 134, 154, 205–208, 210, 226–228, 262, 267, 285
North Atlantic 16, 34, 36, 39–40, 46, 52, 86, 296
North Atlantic Igneous Province 37, 40
North Atlantic Oscillation 13
North Carolina 206
North Sea 115–116, 138, 191
Northern Hemisphere 4–5, 13–17, 23, 34, 111, 205, 208, 213–214, 231, 284–**285**, 298
Northern Ireland 37
Norway 46, 50–53, 82–83, 96–97, 115, 156, 159, 202, 263–265, 272
Norwegian Sea 115
Norwich 121, 208, 219
Nova Scotia 207
nunatak 72
Nýey 5, 107–109, 198, 200–201, 204, 208, 262, 282, 298

oceanic crust 41, 291
odor 1, 70–71, 74, 122–124, 126, 181
oil of vitriol 112

Okjökull 307
Okkak 155–156, 208–210, 226–227
Okmok 165
ÓLAFSSON, Eggert 242, 257
ÓLAFSSON, Sveinn 299
Oluf II 52
Öræfajökull 52, 58, 255, 306
Öræfi 64
ORTELIUS, Abraham 238–239, 242
Ostend 191
Ottoman Empire 16, 96
outcrop 27, 108, 182

Pacific Islands 240
Pacific Ocean 13
Padua 112, 133, 137, 270
painting 191, 197, 224, 258, 289, 299–301
Paleogene 175
PÁLSSON, Bjarni 242–243, 257
PÁLSSON, Sveinn 3–4, 33, 66, 248–257, 260,
    263–265, 272, 274, 277, 281–282, 288, 298
panic 16, 134, 161, 183–184
paraselena 114, 209
parhelia 114, 209
Paris 95–100, 104, 131, 141–144, 149, 152, 159,
    170–173, 186, 198–200, 202, 207, 216, 220,
    226–229, 270, 298
particulate matter (PM) 9, 73, 130, 153
Pas-de-Calais 121, 168
Passy 96, 98, 196
Patent of Toleration 168
Pater Laurentius 118
Pater Onuphrius 118
PEDERSEN, Peder 108
Peißenberg 111
Pennsylvania 207, 227
PERCIVAL, Thomas 196, 203
Persia 164
Perth 266
PEUCER, Kaspar 242
Philadelphia 206, 227
Philippines, the 6, 9, 150
philosophy 92, 101, 165, 195, 242, 245
photography 277–280, 302
physicotheology 169, 287
physics 89, 92, 165, 169, 197, 201
PILÂTRE DE ROZIER, Jean-François 100
Pilsen 139

Pinatubo 6, 9
plague 53, 65, 73, 97, 178, 186, 234, 283
planet 2, 10, 12, 26, 36, 40, 89–90, 187, 259, 269,
    288–291, 301–302, 305
plate boundary 41
plate tectonics 34, 36, 289–292
Pleistocene 41
Pliny the Elder 89
Plutonism 90, 259, 281
Poland 96, 112, 134, 137, 144–146, 200
pollution, *see* volcanic air pollution
Pomerania 174
Pompeii 89
population 2, 5, 23, 28–29, 31, 46–47, 51–60, 80,
    83–85, 160, 168, 170, 201, 211, 221, 242,
    284–285, 293, 297
Portugal 100–101, 103, 149
poverty 52, 57, 87, 240, 250, 298, 304
powder magazine 138–139, 144
Prague 133, 137, 139, 159, 224, 232
precedent 161, 164–166, 172, 230–231, 270, 285
Prestbakki 62, 70
PREYER, William 263
price 54, 76, 250
priest 50, 53, 55, 57, 62–63, 69, 86, 118, 159,
    168, 286
PRIESTLEY, Joseph 97, 163
Protestant 53, 140, 168, 240
proxy data 5, 20
Prussia 96, 144
punishment 50, 65, 93, 140, 168, 197
Punjab 213
pyroclastic flow 10, 41, 89, 214, 266
Pytheas of Massalia 46

Quebec 208–210

rain 65, 73, 111, 113, 123, 130–133, 136–137, 140,
    143, 159, 168, 173, 189, 201, 209–211, 213, 221,
    230, 252, 281, 286
Rajasthan 213
RÉAUMUR, René-Antoine Ferchault de 132, 162, 218
RECK, Hans 277, 279
Reformation 53, 61
refugee 79
Regensburg 131, 155, 225
Reggio Calabria 103, 107
Regnitz River 224

reindeer 58–59

religion 169, 240

remoteness 3, 10, 34, 59–60, 240, 277–278, 301, 304

Renaissance 172, 238

Rendsburg 138

research 3–9, 16–18, 22–25, 33, 35, 59, 76, 83, 89, 91–93, 110, 123, 153, 175, 180, 231, 234, 246, 249–250, 255, 257, 264, 274, 277, 290, 292, 294–295, 305

respiratory problems 73, 127, 284, 293, 306

revolution 7, 21, 23, 29, 32, 88, 94, 96, 110–111, 142, 152, 184–188, 233–235, 258, 260, 281–282, 286

Reykjanes 4, 16, 50–51, 107–108, 199, 254, 262, 305

Reykjavík 46–47, 55, 64, 75, 80–81, 240–241, 264, 306

Reynistaður 62

Rhine Rift Valley 146

Rhine River 222, 225

Rhode Island 208

Richter scale 64, 101, 105, 146

ríkisdalur (Pl. ríkisdalir) (Icelandic currency) 76, 250

ringing the church bells 139–141, 144, 171

RITTENHOUSE, David 196

ROBERTJOT, l'ábbe 166

ROBESPIERRE, Maximilien 142

ROBINSON, Henry 191–192

rock 37, 41, 65, 67, 85, 89–90, 97, 105, 195, 222, 252, 259, 263, 274, 276, 289

Rodheim 164

Rodrigues 266–267

Romantic period 166

Rome 18, 106, 133, 137, 165, 168, 220

Römhild 176–177

rooster 98

Roßberg 178–180, 184, 282

ROTHE, Christian 256

ROTHWELL, J. 232

Rotterdam 220

ROUSSEAU, Jean-Jacques 101

Royal Academy of the Sciences in Berlin (Preußische Akademie der Wissenschaften) 154

royal land commission, first 59, 83

royal land commission, second 83

Royal Society of London 3, 193, 238, 267

RUDLOFF, Max 278

Russia 95–97

SÆMUNDSSON, Jón 62

Sagan 112, 137, 155, 232

sagas, Icelandic 51, 238, 240, 283

Saint Gotthard 111–112, 118

Saint John's Eve 120–123, 129, 177

Saksunarvatn 46

SALIS-MARSCHLINS, Johann Rudolf von 202–203

Samalas 293

Sämtland 12

SANDBY, Paul 194

sandur plains 60–61, 254, 275

Santorini 109

SAPPER, Karl 277, 281

Saxon Switzerland 175

Saxony 158, 175, 216

Scandinavia 4, 41, 50, 114

SCHELHORN, Johann Georg 140, 329

Schnürpflingen 232

Schuylkill River 227–228

Schweidnitz 144–145

science 1, 3, 7–8, 14, 19, 23–26, 34, 41–42, 85, 88, 91–93, 100, 102, 110–111, 122, 130, 142, 152, 154, 162, 169–171, 193, 197, 201, 235, 237, 242, 250, 256, 287, 291–292, 298

scientific instruments, see barometer, hygrometer, telescope, thermometer

scientific publication 175

Scilla 103, 105

Scotland 37, 96, 101, 115, 121, 155, 158, 191–192, 265

sea ice 21, 49, 52, 58, 76, 208, 215, 283

seal 49, 76

Secularization Decree 168

sedimentary rock 37

seiche 101

Seine River 226

Seine-Maritime 152

seismometer 291

Selbourne 120, 129, 132, 137, 219

SENEBIER, Jean 235

sermon 69, 101, 140

settlement (of Iceland) 33, 35, 46–50, 52, 55, 81–83, 154, 205, 208, 226, 274, 284, 302–303

settlement layer 47

SEWALL, Mary Robie 207

sheep 49, 51, 58–59, 73–74, 76, 80, 98, 138, 215, 231, 277, 284

SHELLEY, Mary 262

sheriff 55, 57, 77

shipping 52, 228

Siberia 4, 13, 212, 231

Sicily 50, 88, 96, 102–107, 159, 173, 186

Síða region 60–61, 64, 67–68, 71–73, 75, 81–82, 108, 248, 251, 274, 276, 281, 284

SILBERSCHLAG, Johann Esaias 154

Silesia 112, 136–138, 144–146, 180, 232

Skaftá River 2, 45, 61, 65, 67–70, 199, 247, 263, 274

Skaftáreldar 2, 6, 276

Skaftárjökull 61, 260, 276

Skálholt 55, 76, 86

smell 16, 71–75, 102–103, 108, 113, 120–126, 129, 174, 178, 180–181, 184, 198, 211, 236, 246, 252, 284, 287, 295

smoke 33, 67–69, 78, 102–103, 106, 108, 112–114, 123, 131, 155, 165, 168, 176–181, 196, 198, 201–204, 209–210, 236–237, 246, 269, 274

snow 21, 66, 211, 215–221, 226–232, 283, 307

Societas Meteorologica Palatina 32, 110–114, 118, 130, 134, 137, 141, 162, 178, 204, 232, 245

Société Royale de Médecine 110, 121, 152

society 6, 18, 20, 22, 46, 56, 81, 84, 110, 246, 288, 292, 304, 306

sol-röken 113

SOLANDO, Daniel Carl 243

Sólheimer eruption 41

sore throat 16, 123, 152, 284, 293

South Africa 266

South America 41, 88, 289

South Asia 213

Southern Hemisphere 2, 15

Spain 95–96, 101, 144

speculation 1, 90, 108, 171–203, 230, 236, 255, 282, 290

Spitsbergen 5, 263

Spörer Minimum 22

spreading axis 43, 60–61, 262

spring 47, 53, 58, 62–63, 66, 71, 79–82, 108–109, 135, 151, 187, 211, 216, 220, 227, 229, 231–234, 242, 246, 248, 284, 295

SPRUENGLI, Johann Jakob 118

St. Mary Magdalene's flood 223–224

St. Petersburg 117, 199

St. Thomas (island) 115

stalagmites 20

stamp 299–301

STANLEY, John Thomas 117, 235, 243, 257–258

stars 88, 111, 114, 167, 191–194, 284

STEINGRÍMSSON, Jón 29, 62–79, 85–86, 246, 248, 253, 286, 295–296

STEPHENSEN, Magnús 29, 82, 109, 246–248, 281

STEPHENSEN, Ólafur 82

Stockholm 199

Stockton 191

stratosphere 10–11, 14–15, 21, 128

Stromboli 88, 90, 102, 107, 276

STUDER, Sigmund Gottlieb 117–118

Stuttgart 163

Stykkishólmur 81

Styria 131

subduction 41, 102

sublime 91

subsistence 52, 85, 159, 233, 283

subterraneous 101, 145, 177, 180, 185–188, 200–201, 248, 274, 284, 286

SUESS, Eduard 291

Suez 212

Suffolk 120–121

sulfur dioxide 2, 5–6, 9–11, 15, 21, 48, 118–120, 123–125, 128, 152, 163, 283, 294

Sumatra 266

summer 1, 4, 6–7, 10, 12–16, 24, 32–34, 53, 66, 70, 79–82, 88, 92, 102, 108, 112–161, 164, 167, 170–171, 174–175, 178, 181, 187, 194, 203–218, 226, 230–236, 243, 250, 258, 262, 264, 277–278, 281, 284–291, 295, 307

sun 1, 23, 32, 64–65, 68, 70, 99, 113–114, 120, 129–130, 133, 136, 157, 159–160, 163–168, 176, 181, 192, 196–203, 209, 212, 250–251, 260, 269–272, 284

SÜNCKENBERG, C. J. 81

Sunda Strait 266

superstition 3, 157, 161–162, 167–169, 181, 190

surface water wave, *see* seiche

Surtsey 299

survey 82, 149, 242

SVENDBORG, Gottfried 108

Swabia 138, 144, 146

Swabian Jura 179

Sweden 12, 53, 96, 113, 156, 159, 220, 236, 240

Switzerland 5, 31, 96, 101, 111–112, 117, 129, 140, 146–149, 160, 168, 175, 197, 218, 235, 259, 262

SYMONS, George James 262, 266–270

Syria 115, 201, 212

sýsla 56–57

Tambora 5–8, 17, 211, 234, 262, 266, 282, 288, 293

taste 74, 78, 122–123, 181, 197, 291

technology 24, 92, 189, 294, 298
Tegernsee 111, 114
teleconnection 13, 27–28, 287
telegraphy 17, 270, 282, 288, 294
telescope 197
temperature 12–15, 21, 49, 109, 111, 129,
    132–137, 151, 210, 212, 215–219, 229,
    230–233, 251
tephra 5, 9, 11, 41, 45–53, 65, 75, 87, 246, 266,
    274, 298
tephrochronology 41–42, 63
THARP, Marie 290
The Hague 199, 216
thermometer 24, 32, 93, 111, 133, 197, 218, 227,
    231, 253
THODAL, Lauritz Andreas 81, 243
Thomas Aquinas 89
Thomas de Cantimpré 89
THORODDSEN, Þorvaldur 3–4, 33, 264–265, 269,
    272–279, 281–282, 288
thunder 1, 65, 68–69, 88, 127, 133, 135–146,
    159–160, 166–169, 171, 178, 180, 189–190, 201,
    207, 230, 284, 287
Thuringia 132
Tibet 21
TITIUS, Johann Daniel 208
TITSINGH, Isaac 214
TOALDO, Guiseppe 159, 165, 186
Toba event 6
Tokyo 214
TØNDER LUND, Niels 256
Tönning 138
TORCIA, Michele 107, 186
TORRICELLI, Evangelista 93
Torsken 64, 108
tourism 35, 257, 263, 299, 302–303, 305
trade monopoly 53, 59, 77, 81, 85, 87
trading post 53, 76–77, 81
travelogue 28, 33–34, 240–241, 246, 260, 263,
    279, 281
Treaty of Paris 95–96, 228–229
tree ring 46, 211–212
Tripoli 149
trockener Nebel 113
TROIL, Uno von 243–244
Trondheim 115
troposphere 10–11, 15, 128
tsunami 101–106, 266, 271
TUFTS, Cotton 206

Tunisia 96, 212
Turin 270
Turks 97

Úlfarsdalur 66, 252–253
Ulmener Maar 175
uniformitarianism 27
United States of America 18, 32–33, 95–96, 100,
    205–208, 215–216, 227–229, 233, 283
university 58, 111, 240, 242, 248–249, 270
unknown eruption (1808/1809) 293
Ural Mountains 111, 212
Utrecht 122
Uttar Pradesh 213

VAHL, Martin 249
VAN BEETHOVEN, Ludwig 228
VAN GEUNS, Mathias 152
VAN SWINDEN, Jan Hendrik 112–115, 122–123, 133,
    164, 197, 204
Vanuatu 293
vapeur 113
vapor 113–114, 130, 155, 162, 174, 202, 208, 230,
    235, 269
Varmárdalur 252–253
VASQUIER, M. du 197
Vatnajökull 40, 43–44, 47, 52, 57, 60–61, 66, 71–72,
    252–256, 260, 272, 274, 280, 300–302, 307
Vatnajökull National Park 300–304
Vechte River 226
Vedde ash 41
vegetation 7, 18, 29, 51, 115, 120–126, 156–159, 260,
    293, 297, 303
VEI, see Volcanic Explosivity Index
Veiðivötn 53
Venus 292
VERBEEK, Rogier Diederik Marius 271
VERDEIL, François 112
Versailles 98–99, 143
Vestmannaeyjar 58, 295–296
Vestur-Skaftafellssýsla 57, 61–62, 64, 80, 82,
    84, 253
Vesuvius 89–90, 107, 241, 258, 276
Vienna 30, 130–131, 157, 199, 220–222, 225
VISSERY de Bois-Valé, Charles Dominique de 142
vog (volcanic fog) 125
Vogtland 146
volcanic air pollution 73–77, 122, 129, 211, 234,
    297, 306

Volcanic Impacts on Climate and Society
    (VICS) 293
Volcanic Explosivity Index (VEI) 5–6, 9, 15, 17, 41,
    45–47, 50–53, 58, 67, 107, 164, 175, 214, 255,
    293, 306
volcanic fog, *see* vog
volcanic gases 9–14, 73, 120–121, 128, 137, 283
volcanic system 43–49, 51, 61, 72, 295, 305
volcanism 16, 26, 33–34, 41–54, 88, 90, 174–175,
    201, 241–242, 283, 289, 293, 304–305
volcanology 5, 9, 88, 265
VOLNEY, Constantin-François 212
Voltaire 101
Vulcano 88, 107
Vulkaneifel (Volcanic Eifel) 175
vulnerability 19–20, 27–28, 56, 101, 211, 294

Wadati-Benioff-Zone 41
Wales 8, 16, 132, 231
Warsaw 199
water 2, 45, 49, 51, 65, 67–68, 74, 76, 86, 89, 101,
    113, 123, 133, 167, 175, 197, 199, 210, 219,
    221–229, 238, 242, 252, 289, 296
WATT, James 21
waves (earthquake) 101–103
weather diary 1, 20, 28, 32, 63, 120, 158, 206–207
weather memory 165–166, 174, 221, 232, 289, 298
weather observer 94, 110, 112, 114, 121, 164,
    209–210, 217, 227
webcam 305
WEGENER, Alfred 289–290
WERNER, Abraham Gottlob 25–26, 90
Western Volcanic Zone 42, 48, 61
Westerwald 136, 138, 178
Wetterläuten, *see* ringing the church bells
Weymouth 206
whale 49, 67–68

WHEELER, Ezra 207
WHITE, Gilbert 32, 120–121, 129, 132, 137, 151, 155,
    158, 160, 219, 231–232
WHITE, Henry 137, 168–169
WIEDEBURG, Johann Ernst Basilius 124, 161, 163,
    185–188, 196–197
WIGGLESWORTH, Edward 206
wilderness 58–59, 300–301
William V 97
WILSON, J. Tuzo 290
wind 2, 11, 13, 41, 65, 71, 113, 115, 117, 121, 130–132,
    138, 153, 167–168, 174, 185–186, 188, 190,
    201–204, 214, 216, 226, 230, 235–236,
    251–252, 269, 271, 294
Windsor 192, 194
wine 80, 158, 232
winter 7–8, 10, 12–13, 28, 32–34, 49, 52, 54–55,
    63–64, 75, 77–79, 86, 96, 124, 151, 155–156,
    167, 215–235, 248, 250, 256, 295, 298
WINTHROP, John 195
WOODFORDE, James 120, 132–133, 191, 219–220
Worcestershire 150–151
Würzburg 133, 178, 224–225
WVZ, *see* Western Volcanic Zone

Yamal Peninsula 212
year without a summer 6, 214, 234
Ykas 209

ZIRKEL, Ferdinand 263
Zuiderzee 220
Zweibrücken 135–136

Þingvellir 47, 243
ÞÓRARINSSON, Sigurður 4, 41–42, 50
ÞÓRARINSSON, Stefán 86
Þórðarhyrn 43

Printed in the USA
CPSIA information can be obtained
at www.ICGtesting.com
JSHW051328141024
71656JS00024B/230

9 783111 620435